Lecture Notes in Mathematics

Editors:
J.-M. Morel, Cachan
F. Takens, Groningen
B. Teissier, Paris

1929

Yuliya S. Mishura

Stochastic Calculus for Fractional Brownian Motion and Related Processes

Springer

Yuliya S. Mishura
Department of Mechanics and Mathematics
Kyiv National Taras Shevchenko University
64 Volodymyrska
01033 Kyiv
Ukraine
myus@mail.univ.kiev.ua

ISBN 978-3-540-75872-3   e-ISBN 978-3-540-75873-0

DOI 10.1007/978-3-540-75873-0

Lecture Notes in Mathematics ISSN print edition: 0075-8434
ISSN electronic edition: 1617-9692

Library of Congress Control Number: 2007939114

Mathematics Subject Classification (2000): 60G15, 60G44, 60G60, 60H05, 60H07, 60H10, 60H40, 91B24, 91B28

© 2008 Springer-Verlag Berlin Heidelberg

This work is subject to copyright. All rights are reserved, whether the whole or part of the material is concerned, specifically the rights of translation, reprinting, reuse of illustrations, recitation, broadcasting, reproduction on microfilm or in any other way, and storage in data banks. Duplication of this publication or parts thereof is permitted only under the provisions of the German Copyright Law of September 9, 1965, in its current version, and permission for use must always be obtained from Springer. Violations are liable to prosecution under the German Copyright Law.

The use of general descriptive names, registered names, trademarks, etc. in this publication does not imply, even in the absence of a specific statement, that such names are exempt from the relevant protective laws and regulations and therefore free for general use.

Cover design: *design & production* GmbH, Heidelberg

Printed on acid-free paper

9 8 7 6 5 4 3 2 1

springer.com

Preface

For several decades the semimartingale processes were the best model in order to implement many ideas. The stochastic calculus for semimartingales and the general theory of stochastic processes, which are closely connected to the theory of stochastic integration and stochastic differential equations, were originated by N. Wiener (Wie23), P. Lévy (Le48), K. Itô (Itô42), (Itô44), (Itô51), A.N. Kolmogorov (Kol31), W. Feller (Fel36), J.L. Doob, M. Loéve, I. Gikhman and A. Skorohod (the list of related papers and books is very long and we do not mention it here in full). Those ideas were developed further by several authors, among them there are K. Bichteler (Bi81), C.S. Chou, P.A. Meyer and C. Stricker (CMS80), K.L. Chung and R.J. Williams (ChW83), C. Dellacherie (Del72), C. Dellacherie and P.A. Meyer (DM82), C. Doléans-Dade and P.A. Meyer (DDM70), H. Föllmer (Fol81a), P.A. Meyer (Me76) and M. Yor (Yor76). These theoretical data were fruitfully discussed and summarized in the monographs of J. Jacod (Jac79), R. Elliott (Ell82), P.E. Kopp (Kop84), M. Métivier and J. Pellaumail (MP80), B. Øksendal (Oks03), P. Protter (Pro90). Limit theorems in the most general semimartingale framework were proved by J. Jacod and A.N. Shiryaev (JS87). A very convenient way to consider financial markets is to insert them into semimartingale models, as perfectly demonstrated by I. Karatzas and S. Shreve (KS98), A.N. Shiryaev (Shi99), F. Delbaen and W. Schachermayer (DS06). The Malliavin calculus for the Wiener process was presented in the books of P. Malliavin (Mal97) and D. Nualart (Nua95). However, in recent years the well-studied theory of semimartingales turns out to be insufficient in order to describe many phenomena. On one hand, telecommunication connections, asset prices and other objects have "long memory". This effect cannot be described with the help of such processes as the Wiener process, which has independent increments and has no memory. On the other hand, the concept of turbulence in hydrodynamics can be described by self-similar fields with stationary (dependent) increments (A.M. Yaglom (Yag57), A. Monin and A.M. Yaglom (MY67) and A.M. Yaglom (Yag87)).

A.N. Kolmogorov (Kol40) was the first to consider continuous Gaussian processes with stationary increments and with the self-similarity property; it means that for any $a > 0$ there exists $b > 0$ such that

$$Law(X(at);\ t \geq 0) = Law(bX(t);\ t \geq 0).$$

It turns out that such processes with zero mean have a special correlation function:

$$EX(t)X(s) = \frac{1}{2}\left(|s|^{2H} + |t|^{2H} - |t-s|^{2H}\right),$$

where $0 < H < 1$. A.N. Kolmogorov called such Gaussian processes "Wiener Spirals" ("Wiener screw-lines"). Later, when the papers of H.E. Hurst (Hur51) and H.E. Hurst, R.P. Black and Y.M. Simaika (HBS65), devoted to long-term storage capacity in reservoirs, were published, the parameter H got the name "Hurst parameter". The stochastic calculus of the processes mentioned above originated with the pioneering work of B.B. Mandelbrot and J.W. van Ness (MvN68) who considered the integral moving average representation of X via the Wiener process on an infinite interval and called this process fractional Brownian motion (fBm). Note that B.B. Mandelbrot worked with fractional processes during a long period and his later results concerned the fractals and scaling were summarized in the book (Man97). Note also that it was proved in the paper (GK05) that the moving average representation of fBm is unique in the class of the right-continuous, nondecreasing concave functions on $\mathbb{R}_+$. The first result where fBm appeared as the limit in the Skorohod topology of stationary sums of random variables was obtained by M. Taqqu (Taq75); another scheme of convergence to fBm in the uniform topology was considered in (Gor77). Spectral properties of fBm were studied by G. Molchan (Mol69), G. Molchan and J. Golosov (MG69), G. Molchan (Mol03), and later by K. Dzhaparidze and H. van Zanten (DvZ05), (RLT95), (SL95).

The next intensive wave of interest in fBm arose in the 1990s. It can be explained by various applications of fBm and other long-memory processes in teletraffic, finances, climate and weather derivatives. The paper (DU95) was one of the first paper devoted to stochastic analysis for fBm. Note that fBm is neither a semimartingale (except the case $H = 1/2$ when it is a Brownian motion) nor a Markov process. However, it is closely connected with fractional calculus and can be represented as a "fractional integral" (with the help of a comparatively complicated hypergeometric kernel) via the Wiener process not only on infinite, but also on finite intervals. This was stated by I. Norros, E. Valkeila and J. Virtamo (NVV99) and C. Bender (Ben03a). Such a representation, together with the Gaussian property of fBm and the Hölder property of its trajectories (fBm with Hurst index H is Hölder up to order H) permits us to create an interesting and specific stochastic calculus for fBm. The development of the theory of long-memory processes moved in several directions: stochastic integration, stochastic differential equations, optimal filtering, financial applications, statistical inference, from one side (these topics create the main points of this book) and a lot of other theoretical problems and applications, from the other side. In our Preface we mention for the most part the papers that are not mentioned and used in the text of the book but play a very important role in the development of the theory of long-memory processes. For example, series, spectral and wavelet analysis for fBm was considered in (AS96), (ALP02a), (Mas93), (Mas96), (RZ91), (DvZ05), (DF02), (DvZ04), (SL95), (Mac81b); local times, the Tanaka formula, the law of the iterated logarithm, maximal properties and the Kallianpur–Robbins law for fBm and related processes were studied in (Ber69), (CNT01), (HO02), (HP04b), (Sin97), (HOS05), (GRV03), (KK97), (KM96), (KO99), (Ros87), (KM96), (Kono96), (Sh96), (ElN93), (Tal96) and (Taq77). Furthermore, stochastic evolution equations driven by fBm were investigated in the papers (AG03), (CD01), (MN03), (TTV03) and some methods of construction of fBm were proposed in (Yor88) and (Sai92).

R.J. Adler and G. Samorodnitsky (AS95) considered super processes connected to fBm. The Clark–Ocone theorem for fBm was established in (BE03) and (AOPU00); forward and symmetric integrals for fBm were constructed in (BO04), (CN02), (Zah02b) (note that the general theory of forward, backward and symmetric integrals was created by F. Russo and P. Vallois in (RV93), (RV95a), (RV95b), (RV98) and (RV00)).

Detection and prediction problems were discussed in the papers (BP88), (GN96), (Dun06); the stochastic maximum principle for a controlled process governed by an SDE involving fBm was proved in (BHOS02); stochastic Fubini theorem for fBm was studied in (KM00); time rescaling for fBm was investigated in (Mac81a); Hausdorff measure and packing dimension connected to fBm were considered in (Tal95), (TX96), (Xiao91), (Xiao96), (Xiao97a), (Xiao97b); estimation of the parameters of long-memory processes, in particular, the estimates of the Hurst parameter are presented in (Ber94), (BGK06), (BG96), (BG98), (GR03a). Markov properties of some functionals connected with an fBm were considered in (CC98).

Rough path analysis for fBm was studied in (CQ02) and some of its applications were considered in the manuscript (HN06); the properties of the Gaussian spaces generated by an fBm were established in (PT01); distribution of functionals connected with fBm was obtained in (CM96), (LN03), (ElN99) (Sin97), (Zha96), (Zha97); the Skorohod–Stratonovich integral for fBm was studied in (Dec01), (ALN01), (AMN01), (AN02); the properties of spectral exponent of fBm were established in (LP95); multi-parameter fractional Brownian fields were studied in (ENO02), (Kam96), (ALP02b), (Lind93), (Gol84), (KK99), (OZ01), (PT02a), (Tal95), (TV03), (TT03), (Tud03), (MisIl03), (MisIl04), (MisIl06), (Mur92); set-parametrized fractional Brownian fields have been studied in the papers (HM06a), (HM06b); asymptotic properties of two-dimensional fractional Brownian fields were considered in (BaNu06). The Malliavin calculus for fBm was developed in (Hu05), (Pri98), (Nua03), (Nua06); fBm in Hilbert space was constructed and investigated in (DPM02). The papers (HN04), (KLeB02), (AHL01), (ALN01), (CKM03) are devoted to stochastic fractional Ornstein–Uhlenbeck, Riesz–Bessel and Lévy type processes. An interesting formula of transformation of fBm with Hurst index H into fBm with index $1 - H$ was obtained in (Jost06). Mention also the papers (DU98), (Daye03) and forthcoming book (BHOZ07).

Note that fBm has a long-memory property only for $H \in (1/2, 1)$. In the case $H \in (0, 1/2)$ it is a process with short memory. The theory of such processes is quite different. FBm with $H \in (0, 1/2)$ was studied in (ALN01), (AMN00), (AI04) and (CN05); simulation of fBm and various applications of fBm were considered in (CM95), (CM96), (Nor95), (Yin96), (Dun00), (Dun01), (DF02), (Seb95) and (Sin94).

Fractional Brownian motion as a model of financial markets was proposed in a large number of papers. (See, for example, (AM06), (BE04), (BSV06), (BO02), (BH05), (Che01b), (Dun04), (EvH01), (EvH03), (Gap04), (HO03), (HOS03), (HOS05), (Rog97), (Sch99), (Shi01), (Sot01), (SV03), (WRL03), (WTT99), (Wyss00) and (Zah02a).) Financial markets with memory were considered in (AI05a), (AI05b), (INA07) and (IN07). Moreover, filtering and prediction problems were considered in (CD99), (INA06), (KKA98b), (LeB98), (KLeBR99), (KLeB99), (KLeBR00), (Dun06) and (GN96). In addition, some related applied problems were studied, e.g., in (MS99), (Nar98), (Nor95), (Nor97), (Nor99). An estimate of ruin probabilities for the models with the long-range dependence was studied in (Mis05), (HP04b). Statistical inferences for the processes related to fBm are a very extended area. The major contributions to this theory were made, among other authors, by M. Taqqu and P.M. Robinson. We mention here also the papers of P. Doukhan, A. Khezour and G. Lang (DKL03), L. Giraitis and P.M. Robinson (GR03b), and the papers (DH03), (HH03), (KS03), (MS03), (BLOPST03), (WTT99). Of course, our list of the papers devoted to the theory of fBm is not exhaustive. The book of P. Doukhan, G. Oppenheim, M. Taqqu (editors): *Theory and Applications of Long-range Dependence* (Birkhäuser, Boston 2003) contains papers devoted to different aspects of stochastic calculus for fractional Brownian motion and related processes. We mention, in particular, the papers of D. Surgailis (Sur03a), (Sur03b) and M. Maejima (Mae03), devoted to central and non-central limit theorems, where the asymptotic distribution is not the classical standard normal and the limit process is not the Wiener process. The processes of moving average type are obtained as the limiting ones for increasing sums of some stationary sequences that do not have finite variance. See also the papers (Ho96), (Dec03), (Do03), (Mol03), (PT03), (Taq03), (SW03) from this edition describing stochastic analysis and other aspects of the processes with long memory; papers concerning statistical problems were mentioned above. It is clear from the aforesaid descriptions and citations that there exists the urgent need to systematize the existing results devoted to fractional Brownian motion, to select the best of them (in the author's opinion) and to present them in appropriate form. Also, some well-known results admit generalizations, and it can be done without great technical difficulties. The present book is devoted to the solution of these two problems. Of course, we cannot claim the complete presentation of all the results concerning fractional

Brownian motion; it is impossible as the reader can see from aforesaid list. So, we choose only the following topics: Wiener and stochastic integration, Itô formula, Fubini and Girsanov theorems, stochastic differential equations, filtering in the mixed Brownian–fractional-Brownian models, financial applications, some statistical inferences for fractional Brownian motion and the stochastic calculus of multi-parameter fractional Brownian processes. These fields coincide with the main directions of our own interest in the long-memory effect.

The book consists of six chapters divided into 41 sections. Chapter 1 is devoted to the Wiener integration (when the integrand is nonrandom) with respect to fractional Brownian motion. Section 1.1 is devoted to the principal definitions from fractional calculus. We recall the notions of fractional integrals and derivatives both for finite and infinite intervals, formulate the Hardy–Littlewood theorem, give the Fourier transformation for fractional integrals and derivatives and calculate the values of some important fractional derivatives. Section 1.2 contains some elementary properties of fractional Brownian motion including the simplest spectral representations. Section 1.3 contains the Mandelbrot–van Ness representation of fractional Brownian motion via the Wiener process and some fractional kernels on real axes. These kernels are the prototypes for the future definition of the Wiener integration w.r.t. fBm. Sections 1.4 and 1.5 describe the construction of fractional Brownian motion and fractional noise on white noise space. Such space is convenient for applications since it is possible to consider mixed Brownian–fractional-Brownian processes and linear combinations of fractional Brownian motions with different Hurst indices on such space and to apply Wick calculus to them. It is proved that any fractional noise with $H \in [1/2, 1]$ belongs to the Hida distribution space S^* (we establish the corresponding estimates for the negative norms). The relations between motion and noise are established as in the usual Wick calculus for the Wiener noise. In Section 1.6 we return to fBm on arbitrary space. The section contains the definition of the Wiener integral with respect to fBm and various relations between different "integrable spaces" related to fBm. Section 1.7 is devoted to (non) completeness of the Gaussian spaces generated by fBm, in connection with their norms. Section 1.8 contains the representation of fBm via the Wiener process on any finite interval $[0,T]$ and some representations for auxiliary processes. Sections 1.9 and 1.10 present moment estimates for Wiener integrals w.r.t. fractional Brownian motion. Using the conditions of continuity of the trajectories of Wiener integrals w.r.t. fBm (Section 1.11) we extend in Section 1.12 the upper moment estimates to solutions of very simple stochastic differential equations containing Wiener integrals. Section 1.13 contains the proof of the stochastic Fubini theorem for the Wiener integrals w.r.t. fractional Brownian motion. Section 1.14 deals with such Gaussian processes that can be transformed into martingales with the help of some kernels (fBm can be transformed into the Wiener process with the help of hypergeometric kernels). Section 1.15 is devoted to different convergence schemes, in which fBm is approximated by the sequence of semimartingales, and even by the continuous processes with bounded variation. In the last case Wiener integrals w.r.t. fractional Brownian motion also can be approximated. Section 1.16 demonstrates the Hölder properties of the Wiener integrals w.r.t. fractional Brownian motion. Section 1.17 contains some auxiliary estimates for fractional derivatives of fBm and for the Wiener integrals w.r.t. Wiener process via the Garsia–Rodemich–Rumsey inequality. Section 1.18 contains one- and two-sided bounds for power variations for fBm and Wiener integrals w.r.t. fBm. Section 1.19 contains the result stating that some conditions of quadratic variation of a stochastic process supply that this process is an fBm; it is kind of generalization of the Lévy theorem for the Wiener process. Section 1.20 concludes; it describes Wiener fields on the plane and related fractional integrals and derivatives.

Chapter 2 is devoted to stochastic integration w.r.t. fractional Brownian motion and other aspects of stochastic calculus of fBm. There exist several approaches to stochastic integration w.r.t. fractional Brownian motion: pathwise integration, Wick integration, Skorohod integration, isometric integration and some others that are not mentioned here. Pathwise stochastic integration in fractional Sobolev-type spaces and in fractional Besov-type spaces is described in Section 2.1 and is generalized to fBm fields in Section 2.2. Wick integration is considered in Section 2.3 and is reduced to the integration w.r.t. white noise. Two approaches to the Skorohod integration and their connections with forward, backward and symmetric integration are discussed in Section 2.4. Isometric integration is the subject of section 2.5. The stochastic Fubini theorem and various versions of the Itô formula and the Girsanov theorem are contained in Sections 2.6–2.8 which conclude Chapter 2.

Chapter 3 is devoted to different properties of stochastic differential equations involving fBm. Section 3.1 contains the conditions of existence and uniqueness of solution of a "pure" stochastic differential equation containing a pathwise integral w.r.t. fBm and the estimates of its solution. Most of the theorems are stated in the spirit of the paper (NR00) but the results of Zähle (Zah99) on existence of local solutions are also presented since they are used later for construction of global solutions in the cases when other results cannot help. Some properties of SDEs with stationary coefficients including differentiability and local differentiability of the solutions are presented in Subsection 3.1.4. Existence and uniqueness of solutions of SDEs with two-parameter fractional Brownian fields is contained in Subsection 3.1.6. Semilinear "pure" and "mixed" SDEs are considered in detail in Subsections 3.1.5 and 3.2.1. The rate of convergence of Euler approximations of solutions of SDEs involving fBm is the subject to Section 3.4. SDEs with fractional white noise are considered in Section 3.3, and a detailed discussion of SDEs with additive Wiener integrals w.r.t. fBm is presented in Section 3.5.

Chapter 4 is devoted to filtering problems in the mixed fractional models. Section 4.1 considers the case when the signal process is modeled by mixed stochastic differential equations involving both fractional Brownian motion and the Wiener process and the observation process is the sum of the fractional

Brownian integral and the term of bounded variation. Optimal filtering in conditionally Gaussian linear systems with mixed signals and fractional Brownian observation is studied in Section 4.2. In these sections we consider only nonrandom integrands in all the stochastic integrals. In Section 4.3 we make an attempt to generalize the model and consider polynomial integrands depending on fBm.

Chapter 5 is devoted to financial models involving fBm. In general, financial markets fairly often have a long memory and it is a natural idea to model them with the help of fBm or with the help of some of its modifications. Nevertheless, it is not so easy to do this because the market model is "good" when it does not admit arbitrage and the models involving fractional Brownian motion are not arbitrage-free. So, this chapter is devoted to some methods of construction of the long-memory arbitrage-free models and to the discussion of different approaches to this problem. In Section 5.1 we introduce the mixed Brownian–fractional-Brownian model and establish conditions that ensure the absence of arbitrage in such a model. In Section 5.2 we consider a fractional version of the Black–Scholes equation for the mixed Brownian-fractional Brownian model which contains pathwise integrals w.r.t. fBm, discuss possible applications of Wick products in fractional financial models and produce Black–Scholes equation for the fractional model involving Wick product w.r.t. fBm.

Chapter 6 is devoted to the solution of some statistical problems involving fBm. The choice of the first problem which is solved in Sections 6.1 and 6.2 was evoked by some financial reasonings considered in Chapter 5. More exactly, we try to determine which of the two geometric Brownian motions from (5.2.6) serves as the better model for the real financial market, i.e. we test the complex hypothesis concerning the shifts in the geometric fBm; one of the shifts corresponds to the pathwise integral, and another to the Wick integral. In Section 6.3 we consider the existence and the properties of estimates of the shift parameter in different "pure" and "mixed" models involving fBm and, possibly, the Wiener process, which can be independent of or, conversely, "linearly dependent" on fractional Brownian motion.

I am grateful to Esko Valkeila who invited me several times to Helsinki University during the period of 1997-2005 and presented a possibility for fruitful work and discussion of the problems connected to fractional Brownian motion and related topics. Also, I am grateful to David Nualart for inviting me to Barcelona University during 2001–2003 when we discussed the problems connected to stochastic differential equations involving fBm. My thanks to all my other coauthors, with whom we have written the series of papers devoted to the stochastic calculus for fractional Brownian motion, especially to Jean Memin, Alexander Kukush, Georgij Shevchenko and Taras Androshchuk. My special thanks to Murad Taqqu and Christian Bender for their useful suggestions concerning contents of the minicourse of the lectures devoted to the stochastic calculus for fBm that I delivered in Helsinki Technology University in May 2005. I wish to thank also Celine Jost who has carefully read a part of the text of this book and made a lot of improvements.

Kiev,
April 24 2007

Yuliya Mishura

Contents

1 Wiener Integration with Respect to Fractional Brownian Motion 1
   1.1 The Elements of Fractional Calculus 1
   1.2 Fractional Brownian Motion: Definition and Elementary Properties 7
   1.3 Mandelbrot–van Ness Representation of fBm 9
   1.4 Fractional Brownian Motion with $H \in (\frac{1}{2}, 1)$ on the White Noise Space 10
   1.5 Fractional Noise on White Noise Space 12
   1.6 Wiener Integration with Respect to fBm 16
   1.7 The Space of Gaussian Variables Generated by fBm 24
   1.8 Representation of fBm via the Wiener Process on a Finite Interval 26
   1.9 The Inequalities for the Moments of the Wiener Integrals with Respect to fBm 35
   1.10 Maximal Inequalities for the Moments of Wiener Integrals with Respect to fBm 41
   1.11 The Conditions of Continuity of Wiener Integrals with Respect to fBm 54
   1.12 The Estimates of Moments of the Solution of Simple Stochastic Differential Equations Involving fBm 55
   1.13 Stochastic Fubini Theorem for the Wiener Integrals w.r.t fBm 57
   1.14 Martingale Transforms and Girsanov Theorem for Long-memory Gaussian Processes 58
   1.15 Nonsemimartingale Properties of fBm; How to Approximate Them by Semimartingales 71
      1.15.1 Approximation of fBm by Continuous Processes of Bounded Variation 71
      1.15.2 Convergence $B^{H,\beta} \to B^H$ in Besov Space $W^\lambda[a,b]$. 73
      1.15.3 Weak Convergence to fBm in the Schemes of Series 78

- 1.16 Hölder Properties of the Trajectories of fBm and of Wiener Integrals w.r.t. fBm ... 87
- 1.17 Estimates for Fractional Derivatives of fBm and of Wiener Integrals w.r.t. Wiener Process via the Garsia–Rodemich–Rumsey Inequality ... 88
- 1.18 Power Variations of fBm and of Wiener Integrals w.r.t. fBm ... 90
- 1.19 Lévy Theorem for fBm ... 94
- 1.20 Multi-parameter Fractional Brownian Motion ... 117
  - 1.20.1 The Main Definition ... 117
  - 1.20.2 Hölder Properties of Two-parameter fBm ... 117
  - 1.20.3 Fractional Integrals and Fractional Derivatives of Two-parameter Functions ... 118

2 Stochastic Integration with Respect to fBm and Related Topics ... 123

- 2.1 Pathwise Stochastic Integration ... 123
  - 2.1.1 Pathwise Stochastic Integration in the Fractional Sobolev-type Spaces ... 123
  - 2.1.2 Pathwise Stochastic Integration in Fractional Besov-type Spaces ... 128
- 2.2 Pathwise Stochastic Integration w.r.t. Multi-parameter fBm ... 131
  - 2.2.1 Some Additional Properties of Two-parameter Fractional Integrals and Derivatives ... 131
  - 2.2.2 Generalized Two-parameter Lebesgue–Stieltjes Integrals ... 132
  - 2.2.3 Generalized Integrals of Two-parameter fBm in the Case of the Integrand Depending on fBm ... 136
  - 2.2.4 Pathwise Integration in Two-parameter Besov Spaces ... 136
  - 2.2.5 The Existence of the Integrals of the Second Kind of a Two-parameter fBm ... 137
- 2.3 Wick Integration with Respect to fBm with $H \in [1/2, 1)$ as S^*-integration ... 141
  - 2.3.1 Wick Products and S^*-integration ... 141
  - 2.3.2 Comparison of Wick and Pathwise Integrals for "Markov" Integrands ... 145
  - 2.3.3 Comparison of Wick and Stratonovich Integrals for "General" Integrands ... 154
  - 2.3.4 Reduction of Wick Integration w.r.t. Fractional Noise to the Integration w.r.t. White Noise ... 157
- 2.4 Skorohod, Forward, Backward and Symmetric Integration w.r.t. fBm. Two Approaches to Skorohod Integration ... 158
- 2.5 Isometric Approach to Stochastic Integration with Respect to fBm ... 162
  - 2.5.1 The Basic Idea ... 162
  - 2.5.2 First- and Higher-order Integrals with Respect to X ... 164
  - 2.5.3 Generalized Integrals with Respect to fBm ... 169

2.6 Stochastic Fubini Theorem for Stochastic Integrals w.r.t.
Fractional Brownian Motion 174
2.7 The Itô Formula for Fractional Brownian Motion 182
  2.7.1 The Simplest Version 182
  2.7.2 Itô Formula for Linear Combination of Fractional
    Brownian Motions with $H_i \in [1/2, 1)$ in Terms of
    Pathwise Integrals and Itô Integral 183
  2.7.3 The Itô Formula in Terms of Wick Integrals 184
  2.7.4 The Itô Formula for $H \in (0, 1/2)$ 185
  2.7.5 Itô Formula for Fractional Brownian Fields 186
  2.7.6 The Itô Formula for $H \in (0, 1)$ in Terms of Isometric
    Integrals, and Its Applications 189
2.8 The Girsanov Theorem for fBm and Its Applications 191
  2.8.1 The Girsanov Theorem for fBm 191
  2.8.2 When the Conditions of the Girsanov Theorem Are
    Fulfilled? Differentiability of the Fractional Integrals ... 193

3 **Stochastic Differential Equations Involving Fractional Brownian Motion** 197
  3.1 Stochastic Differential Equations Driven by Fractional
    Brownian Motion with Pathwise Integrals 197
    3.1.1 Existence and Uniqueness of Solutions: the Results of
      Nualart and Răşcanu 197
    3.1.2 Norm and Moment Estimates of Solution 202
    3.1.3 Some Other Results on Existence and Uniqueness of
      Solution of SDE Involving Processes Related to fBm
      with $(H \in (1/2, 1))$ 204
    3.1.4 Some Properties of the Stochastic Differential
      Equations with Stationary Coefficients 206
    3.1.5 Semilinear Stochastic Differential Equations Involving
      Forward Integral w.r.t. fBm 220
    3.1.6 Existence and Uniqueness of Solutions of SDE with
      Two-Parameter Fractional Brownian Field 223
  3.2 The Mixed SDE Involving Both the Wiener Process
    and fBm 225
    3.2.1 The Existence and Uniqueness of the Solution of the
      Mixed Semilinear SDE 225
    3.2.2 The Existence and Uniqueness of the Solution of the
      Mixed SDE for fBm with $H \in (3/4, 1)$ 227
    3.2.3 The Girsanov Theorem and the Measure
      Transformation for the Mixed Semilinear SDE 238
  3.3 Stochastic Differential Equations with Fractional
    White Noise 240
    3.3.1 The Lipschitz and the Growth Conditions on the
      Negative Norms of Coefficients 240
    3.3.2 Quasilinear SDE with Fractional Noise 241

3.4 The Rate of Convergence of Euler Approximations of
Solutions of SDE Involving fBm 243
  3.4.1 Approximation of Pathwise Equations................ 244
  3.4.2 Approximation of Quasilinear Skorohod-type
         Equations 255
3.5 SDE with the Additive Wiener Integral w.r.t. Fractional
Noise 262
  3.5.1 Existence of a Weak Solution for Regular Coefficients .. 263
  3.5.2 Existence of a Weak Solution for SDE with
         Discontinuous Drift 266
  3.5.3 Uniqueness in Law and Pathwise Uniqueness for
         Regular Coefficients 271
  3.5.4 Existence of a Strong Solution for the Regular Case.... 272
  3.5.5 Existence of a Strong Solution
         for Discontinuous Drift 274
  3.5.6 Estimates of Moments of Solutions for Regular Case
         and $H \in (0, 1/2)$ 278
  3.5.7 The Estimates of the Norms of the Solution in the
         Orlicz Spaces 280
  3.5.8 The Distribution of the Supremum of the Process X
         on $[0, T]$ 284
  3.5.9 Modulus of Continuity of Solution of Equation
         Involving Fractional Brownian Motion 287

4 Filtering in Systems with Fractional Brownian Noise 291
4.1 Optimal Filtering of a Mixed Brownian–Fractional-Brownian
Model with Fractional Brownian Observation Noise 291
4.2 Optimal Filtering in Conditionally Gaussian Linear Systems
with Mixed Signal and Fractional Brownian Observation
Noise 295
4.3 Optimal Filtering in Systems with Polynomial Fractional
Brownian Noise 298

5 Financial Applications of Fractional Brownian Motion 301
5.1 Discussion of the Arbitrage Problem 301
  5.1.1 Long-range Dependence in Economics and Finance 301
  5.1.2 Arbitrage in "Pure" Fractional Brownian Model.
         The Original Rogers Approach 302
  5.1.3 Arbitrage in the "Pure" Fractional Model.
         Results of Shiryaev and Dasgupta 304
  5.1.4 Mixed Brownian–Fractional-Brownian Model:
         Absence of Arbitrage and Related Topics 305
  5.1.5 Equilibrium of Financial Market. The Fractional
         Burgers Equation 321
5.2 The Different Forms of the Black–Scholes Equation 322
  5.2.1 The Black–Scholes Equation for the Mixed
         Brownian–Fractional-Brownian Model................ 322

         5.2.2 Discussion of the Place of Wick Products and Wick–
                Itô–Skorohod Integral in the Problems of Arbitrage
                and Replication in the Fractional Black–Scholes
                Pricing Model................................... 323

6 **Statistical Inference with Fractional Brownian Motion** 327
   6.1 Testing Problems for the Density Process for fBm with
        Different Drifts... 327
        6.1.1 Observations Based on the Whole Trajectory with σ
               and H Known 329
        6.1.2 Discretely Observed Trajectory and σ Unknown 331
   6.2 Goodness-of-fit Test 335
        6.2.1 Introduction 335
        6.2.2 The Whole Trajectory Is Observed and the Parameters
               μ and σ Are Known 335
        6.2.3 Goodness-of-fit Tests with Discrete Observations 337
        6.2.4 On Volatility Estimation 340
        6.2.5 Goodness-of-fit Test with Unknown μ and σ 342
   6.3 Parameter Estimates in the Models Involving fBm 343
        6.3.1 Consistency of the Drift Parameter Estimates in the
               Pure Fractional Brownian Diffusion Model............ 344
        6.3.2 Consistency of the Drift Parameter Estimates in the
               Mixed Brownian–fractional-Brownian Diffusion Model
               with "Linearly" Dependent W_t and B_t^H 349
        6.3.3 The Properties of Maximum Likelihood Estimates
               in Diffusion Brownian–Fractional-Brownian Models
               with Independent Components 354

A **Mandelbrot–van Ness Representation: Some Related
   Calculations** 363

B **Approximation of Beta Integrals and Estimation
   of Kernels** 365

References 369

Index 391

1
Wiener Integration with Respect to Fractional Brownian Motion

1.1 The Elements of Fractional Calculus

Let $\alpha > 0$ (and in most cases below $\alpha < 1$ though this is not obligatory). Define the Riemann–Liouville left- and right-sided fractional integrals on (a,b) of order α by

$$(I_{a+}^\alpha f)(x) := \frac{1}{\Gamma(\alpha)} \int_a^x f(t)(x-t)^{\alpha-1} dt,$$

and

$$(I_{b-}^\alpha f)(x) := \frac{1}{\Gamma(\alpha)} \int_x^b f(t)(t-x)^{\alpha-1} dt,$$

respectively.

We say that the function $f \in \mathcal{D}(I_{a+(b-)}^\alpha)$ (the symbol $\mathcal{D}(\cdot)$ denotes the domain of the corresponding operator), if the respective integrals converge for almost all (a.a.) $x \in (a,b)$ (with respect to (w.r.t.) Lebesgue measure).

The Riemann–Liouville fractional integrals on $\mathbb{R}$ are defined as

$$(I_+^\alpha f)(x) := \frac{1}{\Gamma(\alpha)} \int_{-\infty}^x f(t)(x-t)^{\alpha-1} dt,$$

and

$$(I_-^\alpha f)(x) := \frac{1}{\Gamma(\alpha)} \int_x^\infty f(t)(t-x)^{\alpha-1} dt,$$

respectively.

The function $f \in \mathcal{D}(I_\pm^\alpha)$ if the corresponding integrals converge for a.a. $x \in \mathbb{R}$. According to (SKM93), we have inclusion $L_p(\mathbb{R}) \subset \mathcal{D}(I_\pm^\alpha)$, $1 \leq p < \frac{1}{\alpha}$. Moreover, the following Hardy–Littlewood theorem holds.

Theorem 1.1.1 ((SKM93)). *Let* $1 \leq p, q < \infty$, $0 < \alpha < 1$. *Then the operators* $I_\pm^\alpha$ *are bounded from* $L_p(\mathbb{R})$ *to* $L_q(\mathbb{R})$ *if and only if* $1 < p < \frac{1}{\alpha}$ *and* $q = p(1-\alpha p)^{-1}$. *This means, in particular, that for any* $1 < p < \frac{1}{\alpha}$ *and* $q = \frac{p}{1-\alpha p}$, *there exists a constant* $C_{p,q,\alpha}$ *such that*

$$\left(\int_{\mathbb{R}}\left(\int_{\mathbb{R}}|f(u)||x-u|^{\alpha-1}du\right)^q dx\right)^{\frac{1}{q}} \leq C_{p,q,\alpha}\|f\|_{L_p(\mathbb{R})}. \tag{1.1.1}$$

Fractional integration admits the following composition formulas for fractional integrals:

$$I_{a+}^{\alpha}I_{a+}^{\beta}f = I_{a+}^{\alpha+\beta}f, \quad I_{b-}^{\alpha}I_{b-}^{\beta}f = I_{b-}^{\alpha+\beta}f$$

for $f \in L_1[a,b]$. If $\alpha + \beta \geq 1$ then these equalities hold at any point $x \in (a,b)$, otherwise they hold for a.a. x. Also,

$$I_{\pm}^{\alpha}I_{\pm}^{\beta} = I_{\pm}^{\alpha+\beta}f$$

for $f \in L_p(\mathbb{R})$, $\alpha, \beta > 0$ and $\alpha + \beta < \frac{1}{p}$. Let $f \in L_p[a,b]$, $g \in L_q[a,b]$, $p, q \geq 1$ and $\frac{1}{p} + \frac{1}{q} \leq 1 + \alpha$, or let $p > 1$, $q > 1$ and $\frac{1}{p} + \frac{1}{q} = 1 + \alpha$. Then we have the following integration-by-parts formula for fractional integralsintegration-by-parts formula!for fractional integrals:

$$\int_a^b g(x)(I_{a+}^{\alpha}f)(x)dx = \int_a^b f(x)(I_{b-}^{\alpha}g)(x)dx.$$

Let $f \in L_p(\mathbb{R})$, $g \in L_q(\mathbb{R})$, $p > 1, q > 1$ and $\frac{1}{p} + \frac{1}{q} = 1 + \alpha$. Then

$$\int_{\mathbb{R}} g(x)(I_+^{\alpha}f)(x)dx = \int_{\mathbb{R}} f(x)(I_-^{\alpha}g)(x)dx. \tag{1.1.2}$$

Let $C^{\lambda}(\mathbb{T})$ be the set of Hölder continuous functions $f : \mathbb{T} \to \mathbb{R}$ of order λ, i.e.

$$C^{\lambda}(\mathbb{T}) = \Big\{ f : \mathbb{T} \to \mathbb{R} \,\Big|\, \|f\|_{\lambda} := \sup_{t \in \mathbb{T}} |f(t)|$$
$$+ \sup_{s,t \in \mathbb{T}} |f(s) - f(t)|(t-s)^{-\lambda} < \infty \Big\}.$$

If $\alpha > 0$ and $\alpha p > 1$, then $I_{\pm}^{\alpha}(L_p(\mathbb{R})) \subset C^{\lambda}[a,b]$ for any $-\infty < a < b < \infty$ and $0 < \lambda \leq \alpha - \frac{1}{p}$.

The next result is evident.

Lemma 1.1.2. *Let $0 < \alpha < 1$, $f \in L_p(\mathbb{R})$, $1 \leq p < \frac{1}{\alpha}$ and $I_{\pm}^{\alpha}f = 0$. Then $f(x) = 0$ for a.a. $x \in \mathbb{R}$.*

For $p \geq 1$, denote by $I_{\pm}^{\alpha}(L_p(\mathbb{R}))$ the class of functions f, that can be presented as Riemann–Liouville integrals, more exactly, $f = I_{\pm}^{\alpha}\varphi$ for some $\varphi \in L_p(\mathbb{R})$, $p \geq 1$. Lemma 1.1.2 ensures the uniqueness of such function φ. For $0 < \alpha < 1$ it coincides for a.a. $x \in \mathbb{R}$ with the left- (right-) sided Riemann–Liouville fractional derivative of f of order α. These derivatives are denoted by

$$(I_+^{-\alpha}f)(x) = (D_+^{\alpha}f)(x) := \frac{1}{\Gamma(1-\alpha)} \frac{d}{dx} \int_{-\infty}^{x} f(t)(x-t)^{-\alpha} dt,$$

and

$$(I_-^{-\alpha}f)(x) = (D_-^{\alpha}f)(x) := \frac{-1}{\Gamma(1-\alpha)} \frac{d}{dx} \int_{x}^{\infty} f(t)(t-x)^{-\alpha} dt,$$

respectively.

For $p > 1$, the class $I_{\pm}^{\alpha}(L_p(\mathbb{R}))$ coincides with the class of those functions $f \in L_r(\mathbb{R})$, $r = \frac{p}{1-\alpha p}$, for which the integrals

$$\int_{-\infty}^{x-\varepsilon} (f(x) - f(t))(x-t)^{-\alpha-1} dt$$

and

$$\int_{x+\varepsilon}^{\infty} (f(x) - f(t))(t-x)^{-\alpha-1} dt,$$

respectively, converge in $L_p(\mathbb{R})$ as $\varepsilon \to 0$. Thus, for $f \in I_{\pm}^{\alpha}(L_p(\mathbb{R}))$ with $p > 1$ the Riemann–Liouville derivatives coincide with the Marchaud fractional derivatives

$$(\tilde{D}_+^{\alpha}f)(x) := \frac{1}{\Gamma(1-\alpha)} \int_{\mathbb{R}_+} (f(x) - f(x-y)) y^{-\alpha-1} dy,$$

and

$$(\tilde{D}_-^{\alpha}f)(x) := \frac{1}{\Gamma(1-\alpha)} \int_{\mathbb{R}_+} (f(x) - f(x+y)) y^{-\alpha-1} dy,$$

respectively. If $\alpha > 0$ and $\alpha p < 1$, then $I_{\pm}^{\alpha}(L_p(\mathbb{R})) \subset L_q(\mathbb{R})$ for $\frac{1}{q} = \frac{1}{p} - \alpha$.

The Riemann–Liouville fractional derivatives can be considered on any interval $[a, b] \subset \mathbb{R}$ in the following way: we introduce the class $I_{\pm}^{\alpha}(L_p[a, b])$ of functions f that can be presented as $f = I_{a+}^{\alpha}\varphi$ ($f = I_{b-}^{\alpha}\varphi$) for $\varphi \in L_p[a, b]$, $p \geq 1$, where we denote

$$(I_{a+}^{-\alpha}f)(x) = (D_{a+}^{\alpha}f)(x) = \frac{1}{\Gamma(1-\alpha)} \frac{d}{dx} \int_a^x f(t)(x-t)^{-\alpha} dt,$$

and

$$(I_{b-}^{-\alpha}f)(x) = (D_{b-}^{\alpha}f)(x) = -\frac{1}{\Gamma(1-\alpha)} \frac{d}{dx} \int_x^b f(t)(t-x)^{-\alpha} dt,$$

respectively. In this case the Riemann–Liouville fractional derivatives $D_{a+}^{\alpha}f$ and $D_{b-}^{\alpha}f$ admit the Weyl representation of fractional derivatives (we suppose that $f = 0$ outside (a, b)):

$$(D_{a+}^{\alpha}f)(x) = \frac{1}{\Gamma(1-\alpha)} \Big(f(x)(x-a)^{-\alpha}$$
$$+ \alpha \int_a^x (f(x) - f(t))(x-t)^{-\alpha-1} dt \Big) \cdot \mathbf{1}_{(a,b)}(x),$$

and
$$(D^\alpha_{b-}f)(x) = \frac{1}{\Gamma(1-\alpha)} \left(f(x)(b-x)^{-\alpha} \right.$$
$$\left. + \alpha \int_x^b (f(x)-f(t))(t-x)^{-\alpha-1} dt \right) \cdot \mathbf{1}_{(a,b)}(x),$$

respectively, where the convergence of the integrals holds pointwise for a.a. $x \in (a,b)$ for $p=1$ and in $L_p[a,b]$ for $p>1$.

According to (SKM93, Theorem 13.4), we have that $f = I^\alpha_{a+}\varphi$ for some $\varphi \in L_p[a,b]$, where $1 < p < \infty$, if and only if $f(x)(x-a)^{-\alpha} \in L_p[a,b]$ and
$$\sup_{\varepsilon>0} \int_{a+\varepsilon}^b |\psi_\varepsilon(x)|^p dx < \infty,$$
where $\psi_\varepsilon(x) = \int_a^{x-\varepsilon} \frac{f(x)-f(t)}{(x-t)^{1+\alpha}} dt$, $a+\varepsilon \leq x \leq b$. Let $f \in I^\alpha_\pm(L_p(\mathbb{R}))$, $0 < \alpha < 1$ and $p \geq 1$. Then
$$I^\alpha_\pm I^{-\alpha}_\pm f = f; \tag{1.1.3}$$
moreover, for $f \in L_1(\mathbb{R})$ we have that
$$I^{-\alpha}_\pm I^\alpha_\pm f = f. \tag{1.1.4}$$
We set $I^0_\pm f := f$.

The composition formula for fractional derivatives has the form
$$D^\alpha_{a+} D^\beta_{a+} f = D^{\alpha+\beta}_{a+} f, \tag{1.1.5}$$
where $\alpha \geq 0$, $\beta \geq 0$ and $f \in I^{\alpha+\beta}_{a+}(L_1(\mathbb{R}))$.

Also, under the assumptions $0 < \alpha < 1$, $f \in I^\alpha_{a+}(L_p[a,b])$ and $g \in I^\alpha_{b-}(L_q[a,b])$, $1/p + 1/q \leq 1 + \alpha$ we have the integration-by-parts formula for fractional derivatives
$$\int_a^b (D^\alpha_{a+}f)(x) g(x) dx = \int_a^b f(x)(D^\alpha_{b-}g)(x) dx. \tag{1.1.6}$$

For $0 < \alpha < 1$ and $f \in C^1[a,b]$, the derivatives $D^\alpha_{a+}f$ and $D^\alpha_{b-}f$ exist, belong to $L_r[a,b]$ for $1 \leq r < 1/\alpha$, and have the form
$$D^\alpha_{a+}f = \frac{1}{\Gamma(1-\alpha)} \left(f(a)(x-a)^{-\alpha} + \int_a^x f'(t)(x-t)^{-\alpha} dt \right),$$
and
$$D^\alpha_{b-}f = \frac{1}{\Gamma(1-\alpha)} \left(f(b)(b-x)^{-\alpha} - \int_x^b f'(t)(t-x)^{-\alpha} dt \right),$$
respectively.

Let the general indicator function be given by

$$\mathbf{1}_{(a,b)}(t) = \begin{cases} 1, & a \le t < b, \\ -1, & b \le t < a, \\ 0, & \text{otherwise.} \end{cases}$$

Lemma 1.1.3. *Let $H \in (0, \tfrac{1}{2}) \cup (\tfrac{1}{2}, 1)$ and $\alpha = H - \tfrac{1}{2}$. Then, for all $t \in \mathbb{R}$, we have the equality*

$$(I_-^\alpha \mathbf{1}_{(0,t)})(x) = \frac{1}{\Gamma(1+\alpha)}((t-x)_+^\alpha - (-x)_+^\alpha).$$

Proof. Let $H \in (\tfrac{1}{2}, 1)$ and, for example, $x < 0 < t$ (the other cases can be considered similarly). Then,

$$(I_-^\alpha \mathbf{1}_{(0,t)})(x) = \frac{1}{\Gamma(\alpha)} \int_x^\infty \mathbf{1}_{(0,t)}(u)(u-x)^{\alpha-1} du$$

$$= \frac{1}{\Gamma(\alpha)} \int_0^t (u-x)^{\alpha-1} du = \frac{1}{\Gamma(\alpha+1)} \left((t-x)^\alpha - (-x)^\alpha \right). \quad (1.1.7)$$

Let $H \in (0, \tfrac{1}{2})$. According to the definition of the fractional derivative and (1.1.3), we must prove that

$$\int_x^\infty ((t-u)_+^\alpha - (-u)_+^\alpha)(u-x)^{-\alpha-1} du = \Gamma(-\alpha)\Gamma(\alpha+1)\mathbf{1}_{(0,t)}(x). \quad (1.1.8)$$

Let, for example, $0 < x < t$. Then the left-hand side of (1.1.8) equals

$$\int_x^t (t-u)^\alpha (u-x)^{-\alpha-1} du \, \mathbf{1}_{(0,t)}(x)$$

$$= B(\alpha+1, -\alpha)\mathbf{1}_{(0,t)}(x) = \Gamma(-\alpha)\Gamma(\alpha+1)\mathbf{1}_{(0,t)}(x).$$

The other cases can be considered similarly. □

Remark 1.1.4. Obviously, $(I_+^\alpha \mathbf{1}_{(a,b)}(x)) = \frac{1}{\Gamma(1+\alpha)}((b-x)_+^\alpha - (a-x)_+^\alpha)$, $-\infty < a < b < \infty$.

Let $f \in L_1(\mathbb{R})$. The Fourier transform of f is defined as

$$(\mathcal{F}f)(x) = \widehat{f}(x) = \int_\mathbb{R} e^{ixt} f(t) dt.$$

Denote by $S(\mathbb{R})$ the class of smooth, i.e. infinitely differentiable, and rapidly decreasing functions.

Theorem 1.1.5 ((SKM93)). (i) *For any $0 < \alpha < 1$ and $f \in L_1(\mathbb{R})$ it holds that*

$$\mathcal{F}(I_\pm^\alpha f) = \widehat{f}(x) \cdot (\mp ix)^{-\alpha},$$

where $(\mp ix)^\alpha = |x|^\alpha \exp\left\{ \mp \frac{\alpha \pi i}{2} \operatorname{sign} x \right\}$.

(ii) *For any $0 < \alpha < 1$ and $f \in S(\mathbb{R})$ it holds that*

$$\mathcal{F}(I_\pm^{-\alpha} f) = \widehat{f}(x) \cdot (\mp ix)^\alpha.$$

For $H \in (0,1)$ we introduce the set

$$\mathcal{F}_H := \left\{ f \in L_2(\mathbb{R}), f : \mathbb{R} \to \mathbb{R} \,\Big|\, \int_\mathbb{R} |\widehat{f}(x)|^2 |x|^{-2\alpha} dx < \infty \right\}$$

with the norm

$$\|f\|_{\mathcal{F}_H}^2 = \int_\mathbb{R} |\widehat{f}(x)|^2 \cdot |x|^{-2\alpha} dx.$$

Here and throughout the whole text $\alpha := H - 1/2$. The set $\mathcal{F}_H$ will be considered in detail in Sections 1.6 and 1.7.

We say that f is step function, or elementary function, if there exist a finite number of points $t_k \in \mathbb{R}$, $0 \le k \le n-1$, and $a_k \in \mathbb{R}$, $1 \le k \le n$, such that

$$f(t) = \sum_{k=1}^{n} a_k \mathbf{1}_{[t_{k-1}, t_k)}(t).$$

Lemma 1.1.6. *Let $f \in \mathcal{F}_H$. Then there exists a sequence of step functions f_n, such that*

$$\|f - f_n\|_{\mathcal{F}_H} \to 0, \; n \to \infty.$$

Theorem 1.1.7 ((PT00b)). *For $H \in (0,1)$, the set $\mathcal{F}_H$ is a linear space with inner product*

$$(f,g)_{\mathcal{F}_H} = \int_\mathbb{R} \widehat{f}(x) \overline{\widehat{g}(x)} |x|^{-2\alpha} dx, \quad \alpha = H - 1/2.$$

Moreover, the set of elementary functions belongs to $\mathcal{F}_H$, and it is dense in $\mathcal{F}_H$.

Proof. The first statement is evident. Furthermore, for any $-\infty < a < b < \infty$, it holds that $\mathbf{1}_{(a,b)} \in \mathcal{F}_H$, because $\int_\mathbb{R} |\widehat{\mathbf{1}}_{(a,b)}(x)|^2 |x|^{-2\alpha} dx = \int_\mathbb{R} |e^{ixb} - e^{ixa}|^2 |x|^{-2-2\alpha} dx$, and the latter integral is equivalent to the convergent integral $\int |x|^{-2-2\alpha} dx$, in the neighborhood of $\pm\infty$, and equivalent to the convergent integral $\int |x|^{-2\alpha} dx$ in the neighborhood of 0. Therefore, any step function belongs to $\mathcal{F}_H$. The second statement then follows from Lemma 1.1.6.

Lemma 1.1.8 ((PT00b)). *Let $f \in L_2(\mathbb{R})$. Then, for any $H \in (0,1)$, there exists a sequence of step functions f_n such that*

$$\int_\mathbb{R} |\widehat{f}(x) - \widehat{f}_n(x)| x|^{-2\alpha}|^2 dx \to 0, \quad n \to \infty. \tag{1.1.9}$$

Proof. Indeed, for $\varepsilon > 0$, put $\widehat{f}_\varepsilon(x) := \widehat{f}(x)\mathbf{1}_{\{|x|>\varepsilon\}}$. Then $\int_\mathbb{R} |\widehat{f}(x) - \widehat{f}_\varepsilon(x)|^2 dx \to 0$, $\varepsilon \to 0$. Let $H \in (0, \frac{1}{2})$. Then $\widehat{f}_\varepsilon(x) = (\widehat{f}(x)|x|^\alpha \mathbf{1}_{\{|x|>\varepsilon\}})|x|^{-\alpha} = \widehat{g}_\varepsilon(x)|x|^{-\alpha}$, where $g_\varepsilon \in L_2(\mathbb{R})$, $\alpha = H - 1/2$. Now (1.1.9) follows from Lemma 1.1.6. In the case $H \in [\frac{1}{2}, 1)$ the proof is similar. □

1.2 Fractional Brownian Motion: Definition and Elementary Properties

Let $(\Omega, \mathcal{F}, P)$ be a complete probability space.

Definition 1.2.1. The (two-sided, normalized) fractional Brownian motion (fBm) with Hurst index $H \in (0, 1)$ is a Gaussian process $B^H = \{B_t^H, t \in \mathbb{R}\}$ on $(\Omega, \mathcal{F}, P)$, having the properties

(i) $B_0^H = 0$,
(ii) $EB_t^H = 0, t \in \mathbb{R}$,
(iii) $EB_t^H B_s^H = \frac{1}{2}(|t|^{2H} + |s|^{2H} - |t-s|^{2H}), s, t \in \mathbb{R}$.

Remark 1.2.2. Since $E(B_t^H - B_s^H)^2 = |t-s|^{2H}$ and B^H is a Gaussian process, it has a continuous modification, according to the Kolmogorov theorem. Indeed, for all $n \geq 1$ it holds that $E|B_t^H - B_s^H|^n = \frac{2^{\frac{n}{2}}}{\pi^{\frac{1}{2}}}\Gamma(\frac{n+1}{2})|t-s|^{nH}$.

Remark 1.2.3. For $H = 1$, we set $B_t^H = B_t^1 = t\xi$, where ξ is a standard normal random variable.

Remark 1.2.4. It is possible to consider the fBm only on $\mathbb{R}_+$ (one-sided fBm) with evident changes in Definition 1.2.1.

The characteristic function has the form

$$\varphi_\lambda(t) := E\exp\left\{i\sum_{k=1}^n \lambda_k B_{t_k}^H\right\} = \exp\left\{-\frac{1}{2}(C_t\lambda, \lambda)\right\},$$

where $C_t = (EB_{t_k}^H B_{t_i}^H)_{1 \leq i,k \leq n}$ and $(\cdot, \cdot)$ is the inner product on $\mathbb{R}^n$.

Therefore, it follows from item (iii) of Definition 1.2.1, that for any $\beta > 0$

$$\varphi_\lambda(\beta t) = \exp\left\{-\frac{1}{2}\beta^{2H}(C_t\lambda, \lambda)\right\}. \tag{1.2.1}$$

Definition 1.2.5. A stochastic process $X = \{X_t, t \in \mathbb{R}\}$ is called b-self-similar if

$$\{X_{at}, t \in \mathbb{R}\} \stackrel{d}{=} \{a^b X_t, t \in \mathbb{R}\}$$

in the sense of finite-dimensional distributions.

From Definition 1.2.5 and (1.2.1) it follows that B^H is H-self-similar. Note that

$$E(B_t^H - B_s^H)(B_u^H - B_v^H) = \frac{1}{2}(|s-u|^{2H} + |t-v|^{2H} - |t-u|^{2H} - |s-v|^{2H}). \quad (1.2.2)$$

It follows from (1.2.2) that the process B^H has stationary increments (evidently, it is not stationary itself). Let $H = \frac{1}{2}$. Then the increments of B^H are non-correlated, and consequently independent. So B^H is a Wiener process which we denote further by B or W. For $H \in (0, \frac{1}{2}) \cup (\frac{1}{2}, 1)$ and $t_1 < t_2 < t_3 < t_4$, it follows from (1.2.2) for $\alpha = H - 1/2$ that

$$E(B_{t_4}^H - B_{t_3}^H)(B_{t_2}^H - B_{t_1}^H) = 2\alpha H \int_{t_1}^{t_2} \int_{t_3}^{t_4} (u-v)^{2\alpha-1} du\, dv.$$

Therefore, the increments are positively correlated for $H \in (\frac{1}{2}, 1)$ and negatively correlated for $H \in (0, \frac{1}{2})$. Furthermore, for any $n \in \mathbb{Z} \setminus \{0\}$, the autocovariance function is given by

$$r(n) := EB_1^H(B_{n+1}^H - B_n^H) = 2\alpha H \int_0^1 \int_n^{n+1} (u-v)^{2\alpha-1} du\, dv$$

$$\sim 2\alpha H |n|^{2\alpha-1}, \quad |n| \to \infty.$$

If $H \in (0, \frac{1}{2})$, then $\sum_{n \in \mathbb{Z}} |r(n)| \sim \sum_{n \in \mathbb{Z} \setminus \{0\}} |n|^{2\alpha-1} < \infty$.

If $H \in (\frac{1}{2}, 1)$, then $\sum_{n=1}^{\infty} |r(n)| \sim \sum_{n \in \mathbb{Z} \setminus \{0\}} |n|^{2\alpha-1} = \infty$. In this case we say that fBm B^H has the property of long-range dependence. For the spectral density function of $\{X_n^H := B_{n+1}^H - B_n^H, n \in \mathbb{Z}\}$, which is denoted by $f_H(\lambda)$, it holds that (BG96; DvZ05),

$$f_H(\lambda) = C_H^{(0)} |e^{i\lambda} - 1|^2 \sum_{k \in Z} |\lambda + 2\pi k|^{-2-2\alpha}, \quad \lambda \in [-\pi, \pi],$$

where $C_H^{(0)}$ is some constant depending on H. It is easy to see that

$$f_H(\lambda) \sim C_H^{(0)} |\lambda|^2 |\lambda|^{-2-2\alpha} = C_H |\lambda|^{-2\alpha}$$

as $\lambda \to 0$. Therefore, for $H \in (\frac{1}{2}, 1)$ it holds that $f_H(\lambda) \to \infty$ as $\lambda \to 0$, and, for $H \in (0, \frac{1}{2})$, it holds that $f(\lambda) \to 0$ as $\lambda \to 0$.

According to (PT00b) and (ST94), B^H admits the spectral representation $\{B_t^H, t \in \mathbb{R}\} \stackrel{d}{=} \{C_H^{(1)} \int_{\mathbb{R}} (e^{itx} - 1)(ix)^{-1}|x|^{-\alpha} d\widetilde{B}(x), t \in \mathbb{R}\}$, where $\widetilde{B} = B_1 + iB_2$ is a complex Gaussian measure with $B_1(A) = B_1(-A)$, $B_2(A) = -B_2(-A)$ and $E(B_1(A))^2 = E(B_2(A))^2 = \frac{\text{mesh}(A)}{2}$ for any Borel set A of finite Lebesgue measure $\text{mesh}(A)$ and $C_H^{(1)} = \left(\frac{\Gamma(2H+1)\sin(1/2 \cdot \pi(H+1/2))}{2\pi}\right)^{\frac{1}{2}}$.

1.3 Mandelbrot–van Ness Representation of fBm

Let $W = \{W_t, t \in \mathbb{R}\}$ be the two-sided Wiener process, i.e. the Gaussian process with independent increments satisfying $EW_t = 0$ and $EW_t W_s = s \wedge t$, $s, t \in \mathbb{R}$. Evidently, $W = B^{\frac{1}{2}}$. Denote $k_H(t, u) := (t-u)_+^\alpha - (-u)_+^\alpha$, where $\alpha = H - \frac{1}{2}$. The following representation is due to Mandelbrot and van Ness (MvN68).

Theorem 1.3.1. *The process* $\overline{B}^H = \{\overline{B}^H_t, t \in \mathbb{R}\}$ *defined by*

$$\overline{B}^H_t := C^{(2)}_H \int_{\mathbb{R}} k_H(t, u) dW_u, \quad H \in \left(0, \frac{1}{2}\right) \cup \left(\frac{1}{2}, 1\right),$$

where $C^{(2)}_H = \left(\int_{\mathbb{R}_+} \left((1+s)^\alpha - s^\alpha\right)^2 ds + \frac{1}{2H} \right)^{-\frac{1}{2}} = \dfrac{(2H \sin \pi H \Gamma(2H))^{1/2}}{\Gamma(H+1/2)},$

has a continuous modification which is a normalized two-sided fBM.

Remark 1.3.2. The constant $C^{(2)}_H$ is calculated in Appendix A.

Proof. Evidently, $\overline{B}^H$ is a Gaussian process with $\overline{B}^H_0 = 0$ and $E\overline{B}^H_t = 0$. Furthermore, it holds that for $t > 0$,

$$E(\overline{B}^H_t)^2 = (C^{(2)}_H)^2 \left(\int_{-\infty}^0 k_H^2(t, u) du + \int_0^t (t-u)^{2\alpha} du \right) = t^{2H}.$$

For $t < 0$ we have that

$$E(\overline{B}^H_t)^2 = (C^{(2)}_H)^2 \left(\int_{-\infty}^t k_H^2(t, u) du + \int_t^0 (-u)^{2\alpha} du \right) = (-t)^{2H}.$$

Furthermore, for $h > 0$, it holds that

$$\overline{B}^H_{s+h} - \overline{B}^H_s = C^{(2)}_H \int_{-\infty}^s \left(k_H(s+h, u) - k_H(s, u) \right) dW_u$$

$$+ \int_s^{s+h} k_H(s+h, u) \, dW_u =: I_1 + I_2. \quad (1.3.1)$$

Note that the terms I_1 and I_2 on the right-hand side of (1.3.1) are independent, and the Wiener process W has stationary increments. Therefore,

$$I_1 \stackrel{d}{=} \int_{-\infty}^0 \left(k_H(s, u) - k_H(0, u) \right) dW_u, \quad I_2 \stackrel{d}{=} \int_0^h k_H(h, u) dW_u,$$

and $E(\overline{B}^H_{s+h} - \overline{B}^H_s)^2 = E(\overline{B}^H_h)^2 = h^{2H}$. By combining these results, we obtain that

$$E\overline{B}_s^H \overline{B}_t^H = \frac{1}{2}\Big(E(\overline{B}_s^H)^2 + E(\overline{B}_t^H)^2 - E(\overline{B}_t^H - \overline{B}_s^H)^2\Big)$$

$$= \frac{1}{2}\big(|t|^{2H} + |s|^{2H} - |t-s|^{2H}\big). \quad (1.3.2)$$

The proof follows immediately from Definition 1.2.1 and Remark 1.2.2. □

Define the operator

$$M_\pm^H f := \begin{cases} C_H^{(3)} I_\pm^\alpha f, & H \in (0, \tfrac{1}{2}) \cup (\tfrac{1}{2}, 1), \\ f, & H = \tfrac{1}{2}, \end{cases} \quad (1.3.3)$$

where $C_H^{(3)} = C_H^{(2)} \Gamma(H + \tfrac{1}{2})$.

Corollary 1.3.3. *It follows from Lemma 1.1.3 and Theorem 1.3.1, that for any $H \in (0,1)$ the process*

$$B_t^H = \int_\mathbb{R} (M_-^H \mathbf{1}_{(0,t)})(s) dW_s \quad (1.3.4)$$

is a normalized fractional Brownian motion.

A little later we shall establish (see Corollary 1.6.11) that any fBm B^H can be presented in the form (1.3.4) with a suitable Brownian motion W.

Remark 1.3.4. It is easy to see that the domain $\mathcal{D}(M_-^H)$ of the operator M_-^H has a form

$$\mathcal{D}(M_-^H) = \begin{cases} \cup_{1 \le p < \frac{1}{\alpha}} L_p(\mathbb{R}), & H \in (\tfrac{1}{2}, 1),\ \alpha = H - \tfrac{1}{2}, \\ \cup_{p \ge 1} I_\pm^{-\alpha}(L_p(\mathbb{R})), & H \in (0, \tfrac{1}{2}), \\ \text{all measurable functions}, & H = \tfrac{1}{2}. \end{cases}$$

1.4 Fractional Brownian Motion with $H \in (\tfrac{1}{2}, 1)$ on the White Noise Space

Consider the probability space of the white noise. Namely, recall that $S(\mathbb{R})$ denotes the Schwartz space of rapidly decreasing infinitely differentiable real-valued functions, and let $S'(\mathbb{R})$ be the dual space of $S(\mathbb{R})$, i.e., the space of tempered distributions with weak* topology. We consider $S'(\mathbb{R})$ as a probability space Ω with the σ-algebra $\mathcal{F}$ of Borel sets. According to the Bochner–Minlos theorem, there exists the probability measure P on $(\Omega, \mathcal{F})$, such that for any function $f \in S(\mathbb{R})$ with the norm $\|f\|_{L_2(\mathbb{R})}$, it holds that

$$E \exp(i\langle f, \omega \rangle) = \exp\left\{-\frac{1}{2}\|f\|_{L_2(\mathbb{R})}^2\right\}, \quad (1.4.1)$$

where $\langle \cdot, \cdot \rangle$ denotes the dual operation.

Note that from (1.4.1), we obtain that

1.4 Fractional Brownian Motion with $H \in (\frac{1}{2}, 1)$ on the White Noise Space

$$E\langle f, \omega\rangle = 0, \quad E\langle f, \omega\rangle^2 = \|f\|^2_{L_2(\mathbb{R})}, \qquad (1.4.2)$$

where $f \in S(\mathbb{R})$, and the duality $\langle f, \omega\rangle$ can be extended by isometry to $f \in L_2(\mathbb{R})$. Note that from (1.4.1)–(1.4.2), it follows that the process $W_t := \langle \mathbf{1}_{[0,t]}, \omega\rangle$ is a standard Brownian motion.

Now, let $H \in [\frac{1}{2}, 1)$, $f_1 \in L_2(\mathbb{R})$ and $f_2 \in L_{\frac{1}{H}}(\mathbb{R})$. Then $M_+^H f_1 \in L_{\frac{1}{1-H}}(\mathbb{R})$, $M_-^H f_2 \in L_2(\mathbb{R})$, therefore, we can consider on $L_2(\mathbb{R})$ the inner product of the form

$$(f_1, M_-^H f_2)_{L_2(\mathbb{R})} = \int_{\mathbb{R}} f_1(x)(M_-^H f_2)(x)dx.$$

By (1.1.1) and (1.3.3), it holds that

$$(f, M_-^H f_2)_{L_2(\mathbb{R})} = (M_+^H f_1, f_2)_{L_2(\mathbb{R})}.$$

According to (SKM93), denote the spaces

$$\Phi(\mathbb{R}) = \{\phi \mid \phi \in S(\mathbb{R}), \widehat{\phi}^k(0) = 0, k \geq 0\}$$
$$= \{\phi \mid \phi \in S(\mathbb{R}), (\phi, t^k)_{L_2(\mathbb{R})} = 0, k \geq 0\}.$$

It was proved in (SKM93) that $M_\pm^H(\Phi(\mathbb{R})) \subset \Phi(\mathbb{R})$ and that the space $\Phi(\mathbb{R})$ is closed in $S(\mathbb{R})$.

Now, define two stochastic processes

$$B_\pm^H(t)(\omega) := \langle M_\pm^H \mathbf{1}_{(0,t)}, \omega\rangle, \quad t \in \mathbb{R}.$$

Then the processes $B_\pm^H(t)$ are Gaussian, $EB_+^H(t) = EB_-^H(t) = 0$. For the covariance function, it holds that

$$EB_\pm^H(t)B_\pm^H(s) = \int_{\mathbb{R}} (M_\pm^H \mathbf{1}_{(0,t)})(x)(M_\pm^H \mathbf{1}_{(0,s)})(x)dx. \qquad (1.4.3)$$

By considering the sign "$-$", we obtain from (1.3.4) that the right-hand side of (1.4.3) coincides with

$$EB_t^H B_s^H = \int_{\mathbb{R}} (M_-^H \mathbf{1}_{(0,t)})(x)(M_-^H \mathbf{1}_{(0,s)})(x)dx$$
$$= \frac{1}{2}(|t|^{2H} + |s|^{2H} - |t-s|^{2H}).$$

One obtains the same result if one considers the sign "$+$". Therefore, each of the processes $B_\pm^H(t)$ has a modification that is a normalized fBm. The process $B_-^H(t)$ is called a *"backward" fBm*. It coincides with usual Mandelbrot–van Ness representation and depends only on the past, i.e. on $\{W_s, s \in (-\infty, t)\}$. Indeed, $B_-^H(t) = \int_{\mathbb{R}} (M_-^H \mathbf{1}_{(0,t)})(s)dW_s$, where $W_t(\omega) = \langle \mathbf{1}_{(0,t)}, \omega\rangle$. The process $B_+^H(t)$ is called a *"forward" fBm*; it admits the representation

$$B_+^H(t) = C_H^{(3)} \int_t^\infty (u_+^\alpha - (u-t)_+^\alpha)dW_u = \int_{\mathbb{R}} (M_+^H \mathbf{1}_{(0,t)})(s)dW_s,$$

and depends on future values of W, i.e. on $\{W_s, s \in (t, +\infty)\}$.

The case $H \in (0, 1/2)$ can be considered similarly. Also, it is possible to consider the linear combinations of the operators $M_\pm^{H_k}$ and of fractional Brownian motions with different Hurst indices (in what follows we consider only the case $H_k \in [1/2, 1)$):

$$M_\pm f(x) := \sum_{k=1}^m \sigma_k M_\pm^{H_k} f(x), \quad \sigma_k > 0$$

and

$$B_\pm^M(t) = \sum_{k=1}^m \sigma_k B_\pm^{H_k}(t) = \langle M_\pm \mathbf{1}_{(0,t)}, \omega \rangle. \tag{1.4.4}$$

Clearly, the operators $M_\pm$ are mutually adjoint in the same way as $M_\pm^H$. Indeed,

$$(f_1, M_- f_2)_{L_2(\mathbb{R})} = (M_+ f_1, f_2)_{L_2(\mathbb{R})}$$

for appropriate functions f_1, f_2.

1.5 Fractional Noise on White Noise Space

Let $\mathbb{N}_0 = \mathbb{N} \cup \{0\}$ and $\mathcal{I}$ be the set of all finite multiindices $\alpha = (\alpha_1, \ldots, \alpha_n)$ with $\alpha_i \in \mathbb{N}_0$. Denote $|\alpha| = \alpha_1 + \cdots + \alpha_n$, $\alpha! := \alpha_1! \cdots \alpha_n!$. (Of course, in this and similar situations α as a multiindex differs from our $\alpha = H - 1/2$ but it will not lead to misunderstanding.) Define the Hermite polynomials by

$$h_n(x) := (-1)^n e^{x^2} \frac{d^n}{dx^n} (e^{-x^2})$$

and Hermite functions

$$\widetilde{h}_n(x) := \pi^{-1/4} (n!)^{-1/2} 2^{-n/2} h_n(x) e^{-x^2/2}, \quad n \geq 0.$$

It is well-known that the functions $\{\widetilde{h}_n, n \geq 1\}$ form an orthonormal basis in $L_2(\mathbb{R})$ with Fourier transform

$$\int_{\mathbb{R}} e^{i\lambda x} \widetilde{h}_n(x) dx = (2\pi)^{1/2} i^n \widetilde{h}_n(x), \quad n \geq 1.$$

Define

$$\mathcal{H}_\alpha(\omega) := \prod_{i=1}^n h_{\alpha_i}(\langle \widetilde{h}_i, \omega \rangle),$$

the product of Hermite polynomials and consider a random variable

$$F = F(\omega) \in L_2(\Omega) := L_2(S'(\mathbb{R}), \mathcal{F}, P).$$

Then, according to (HOUZ96, Theorem 2.2.4), $F(\omega)$ admits the representation

$$F(\omega) = \sum_{\alpha \in \mathcal{I}} c_\alpha \mathcal{H}_\alpha(\omega), \tag{1.5.1}$$

and

$$\|F\|_{L_2(\Omega)}^2 = \sum_{\alpha \in \mathcal{I}} \alpha! \, c_\alpha^2 < \infty.$$

Next, we introduce the following dual spaces.
(i) $F \in S$ if the coefficients from expansion (1.5.1) satisfy

$$\|F\|_k^2 = \sum_{\alpha \in \mathcal{I}} \alpha! c_\alpha^2 (2\mathbb{N})^{k\alpha} < \infty$$

for any $k \geq 1$, where $(2\mathbb{N})^\gamma = \prod_{j=1}^m (2j)^{\gamma_j}, \gamma = (\gamma_1, \ldots, \gamma_m \in \mathcal{I})$.
(ii) $F \in S^*$ if F admits the formal expansion (1.5.1) with finite negative norm

$$\|F\|_{-q}^2 = \sum_{\alpha \in \mathcal{I}} \alpha! \, c_\alpha^2 (2\mathbb{N})^{-q\alpha} < \infty$$

for at least one $q \in \mathbb{N}$ (in this case we say that $F \in S_{-q}$).
For $F = \sum_\alpha c_\alpha H_\alpha \in S$, $G = \sum_\alpha d_\alpha H_\alpha \in S^*$, we define

$$\langle\!\langle F, G \rangle\!\rangle := \sum_{\alpha \in \mathcal{I}} \alpha! \, c_\alpha d_\alpha.$$

Taking into account the Parseval identity, we can also define

$$L_{M_\pm}^2(\mathbb{R}) = \{f : M_\pm f \in L_2(\mathbb{R})\} = \{f : \widehat{M_\pm f} \in L_2(\mathbb{R})\},$$

where, according to our notations, $\widehat{g}(\lambda) = \int_{\mathbb{R}} e^{i\lambda y} g(y) dy$ is the Fourier transform of the function g.

The inner product in $L_{M_\pm}^2(\mathbb{R})$ is defined by

$$(f, g)_{M_\pm} := \int_{\mathbb{R}} M_\pm f(x) M_\pm g(x) dx = (M_\pm f, M_\pm g)_{L_2(\mathbb{R})}.$$

Also, define an inverse operator $M_\pm^{-1}$ in terms of the Fourier transform. For $g(x) = M_\pm^{-1} f(x) \in L_2(\mathbb{R})$, it holds that $f(x) = M_\pm g(x)$, and, according to Theorem 1.1.5, we have the equalities

$$\widehat{f}(\lambda) = \widehat{g}(\lambda) \sum_{k=1}^{\infty} \sigma_k C_{H_k}^{(\lambda)} |\lambda|^{-\alpha_k},$$

where $C_{H_k}^{(\lambda)} = \exp\left\{\frac{\alpha_k \pi i}{2} \operatorname{sign} \lambda\right\} C_{H_k}^{(3)}$ and $\alpha_k = H_k - 1/2$. Hence,

$$(\widehat{M_\pm^{-1} f})(\lambda) = \Big(\sum_{k=1}^m \sigma_k C_{H_k}^{(\lambda)} |\lambda|^{-\alpha_k}\Big)^{-1} \widehat{f}(\lambda).$$

Lemma 1.5.1. *The functions $e_k^\pm := M_\pm^{-1}\widetilde{h}_k, k \geq 1$, exist and form an orthonormal basis in $L^2_{M_\pm}(\mathbb{R})$.*

Proof. Let, for simplicity, $m = 1$, so that $M_\pm = M_\pm^H$ and $\sigma_1 = \sigma$. Consider, for example, the sign "$-$". Then it holds that

$$\widehat{e_k^-}(\lambda) = (\sigma C_H(\lambda))^{-1}|\lambda|^\alpha \widehat{\widetilde{h}_k}(\lambda) = (\sigma C_H(\lambda))^{-1} i^k \sqrt{2\pi} |\lambda|^\alpha \widetilde{h}_k(\lambda), \quad \alpha = H - 1/2.$$

Therefore, e_k^- exists and belongs to $S(\mathbb{R})$. The second assertion is evident. $\square$

Now we want to present the linear combination $B_\pm^M(t)$ of fBms in terms of $\widetilde{h}_k, k \geq 1$.

Lemma 1.5.2. *It holds that*

$$B_\pm^M(t) = \sum_{k=1}^\infty \int_0^t M_\mp \widetilde{h}_k(x) dx \langle \widetilde{h}_k, \omega \rangle, \quad t \in \mathbb{R}, \quad \omega \in S'(\mathbb{R}), \tag{1.5.2}$$

and the series converges in $L_2(\Omega)$.

Proof. Let $\omega \in S(\mathbb{R})$. Then, from equality (1.4.4) it follows that

$$B_\pm^M(t) = \langle M_\pm \mathbf{1}_{(0,t)}, \omega \rangle = \langle \mathbf{1}_{(0,t)}, M_\mp \omega \rangle,$$

and $M_\mp \omega \in S(\mathbb{R})$. Since $\mathbf{1}_{(0,t)} \in L^2_{M_\pm}(\mathbb{R})$, it admits the expansion

$$\mathbf{1}_{(0,t)} = \sum_{k=1}^\infty \langle \mathbf{1}_{(0,t)}, e_k^\pm \rangle_{M_\pm} e_k^\pm,$$

where the series converges in $L^2_{M_\pm}(\mathbb{R})$. Then,

$$\langle \mathbf{1}_{(0,t)}, M_\mp \omega \rangle = \sum_{k=1}^\infty \langle \mathbf{1}_{(0,t)}, e_k^\pm \rangle_{M_\pm} \langle e_k^\pm, M_\mp \omega \rangle,$$

and the series converges in $L_2(\Omega)$. Furthermore,

$$\sum_{k=1}^\infty \langle \mathbf{1}_{(0,t)}, e_k^\pm \rangle_{M_\pm} \langle e_k^\pm, M_\mp \omega \rangle = \sum_{k=1}^\infty \int_\mathbb{R} M_\pm \mathbf{1}_{(0,t)}(x) M_\pm e_k^\pm(x) dx \langle M_\pm e_k^\pm, \omega \rangle$$

$$= \sum_{k=1}^\infty \int_\mathbb{R} \mathbf{1}_{(0,t)}(x) M_\mp \widetilde{h}_k(x) dx \langle \widetilde{h}_k, \omega \rangle = \sum_{k=1}^\infty \int_0^t M_\mp \widetilde{h}_k(x) dx \langle \widetilde{h}_k, \omega \rangle,$$

i.e. we obtain (1.5.2) for $\omega \in S(\mathbb{R})$. Moreover, we can extend (1.5.2) on $S'(\mathbb{R})$ since $S(\mathbb{R})$ is dense in $S'(\mathbb{R})$ in weak* topology, and this topology generates the weak convergence. Since

$$\langle \widetilde{h}_k, \omega \rangle = \mathcal{H}_{\varepsilon_k}(\omega),$$

where $\varepsilon_k = (0, \ldots, 1, \ldots, 0)$, where 1 is in kth place, we have that

$$\sum_{k=1}^{\infty} \left| \int_0^t M_{\mp} \widetilde{h}_k(x) dx \right|^2 (\varepsilon_k!)^2 = \sum_{k=1}^{\infty} \left| \int_0^t M_{\mp} \widetilde{h}_k(x) dx \right|^2 = \|\mathbf{1}_{(0,t)}\|_{L^2_{M_{\mp}}}$$
$$\leq 2^{m-1} \sum_{k=1}^m \sigma_k^2 t^{2H_k} < \infty.$$

□

Now, we introduce the fractional noise $\dot{B}^H$ as the formal expansion

$$\dot{B}^H_x(\omega) = \sum_{k=1}^{\infty} M^H_+ \widetilde{h}_k(x) \langle \widetilde{h}_k, \omega \rangle,$$

and the linear combination of fractional noises as

$$\dot{B}^M_x(\omega) = \sum_{k=1}^{\infty} M_+ \widetilde{h}_k(x) \langle \widetilde{h}_k, \omega \rangle.$$

Recall, that here we consider only $H \in [1/2, 1)$ and that $\dot{B}_x(\omega) = \sum_{k=1}^{\infty} \widetilde{h}_k(x) \langle \widetilde{h}_k, \omega \rangle$ is white noise.

Lemma 1.5.3. *The fractional noise $\dot{B}^H_x$ and the linear combination $\dot{B}^M_x$ of such noises belong to S^* for any $x \in \mathbb{R}$.*

Proof. It is sufficient to consider $\dot{B}^H$. By using the Fourier transform and Theorem 1.1.5, we obtain that

$$\left| M^H_+ \widetilde{h}_k(x) \right| = C_{H,k} \left| \int_{\mathbb{R}} e^{-ixt} \widehat{\widetilde{h}_k}(t)(it)^{-\alpha} dt \right| \leq C_{H,k} \left| \int_{|t| \leq 1} \right| + C_{H,k} \left| \int_{|t| > 1} \right|,$$

where $C_{H,k}$ denotes suitable constants. We have that $\widehat{\widetilde{h}_k}(\lambda) = C_k \widetilde{h}_k(\lambda)$, $C_k = i^k \sqrt{2\pi}$, and

$$|\widetilde{h}_k(\lambda)| \leq \begin{cases} Ck^{-1/12} & \text{for } |\lambda| \leq 2\sqrt{k} \\ Ce^{-\gamma \lambda^2} & \text{for } |\lambda| > 2\sqrt{k} \end{cases}$$

where $C > 0$ and $\gamma > 0$ do not depend on λ and k. Therefore,

$$\left| M^H_+ \widetilde{h}_k(x) \right| \leq C \left(\int_{|t| \leq 1} k^{-1/12} |t|^{-\alpha} dt \right.$$
$$\left. + \int_{1 < |t| \leq 2\sqrt{k}} k^{-1/12} |t|^{-\alpha} dt + \int_{|t| > 2\sqrt{k}} |t|^{-\alpha} e^{-\gamma t^2} dt \right) \quad (1.5.3)$$
$$\leq C \left(k^{-1/12} + k^{-1/12} k^{3/4 - H/2} + e^{-2\gamma \sqrt{k}} \right) \leq C k^{2/3 - H/2}.$$

From (1.5.3) it follows that

$$\|\dot{B}^H_x\|^2_{-q} = \sum_{k=1}^{\infty} |M^H_+ \widetilde{h}_k(x)|^2 (2k)^{-q} \leq C \sum_{k=1}^{\infty} k^{4/3 - H - q} < \infty$$

for any $q > 7/3 - H$. So, for $q > 7/3$, it holds that $\|\dot{B}^H_x\|^2_{-q} < \infty$ for any $x \in \mathbb{R}$. This completes the proof. □

1.6 Wiener Integration with Respect to fBm

Now we return to an arbitrary complete probability space $(\Omega, \mathcal{F}, P)$, and continue the considerations of Sections 1.1–1.3.

Consider the space $L_2^H(\mathbb{R}) := \{f : M_-^H f \in L_2(\mathbb{R})\}$ equipped with the norm $\|f\|_{L_2^H(\mathbb{R})} = \|M_-^H f\|_{L_2(\mathbb{R})}$.

Definition 1.6.1. Let $f \in L_2^H(\mathbb{R})$. Then *the Wiener integral w.r.t. fBm is defined as*

$$I_H(f) := \int_{\mathbb{R}} f(s) dB_s^H := \int_{\mathbb{R}} (M_-^H f)(s) dW_s. \qquad (1.6.1)$$

Here, B_s^H and W_s are connected as in (1.3.4). As a particular case, consider the step function $f : \mathbb{R} \to \mathbb{R}$ given by

$$f(t) = \sum_{k=1}^{n} a_k \mathbf{1}_{[t_{k-1}, t_k)}(t),$$

where $t_0 < t_1 < \cdots < t_n \in \mathbb{R}$ and $a_k \in \mathbb{R}, 1 \le k \le n$. Then, from the linearity of the operator M_-^H, we have that

$$I_H(f) = \sum_{k=1}^{n} a_k \int_{\mathbb{R}} M_-^H \mathbf{1}_{[t_{k-1}, t_k)}(s) dW_s = \sum_{k=1}^{n} a_k (B_{t_k}^H - B_{t_{k-1}}^H), \qquad (1.6.2)$$

and the latter sum coincides with the usual Riemann–Stieltjes sum. A question arises: in which sense can we consider formula (1.6.1) as the extension of the sum (1.6.2)? Note, that for a step function, it holds that

$$\|I_H(f)\|_{L_2(\Omega)}^2 = \sum_{i,k=1}^{n} a_i a_k \int_{\mathbb{R}} M_-^H \mathbf{1}_{[t_{k-1}, t_k)}(x) M_-^H \mathbf{1}_{[t_{i-1}, t_i)}(x) dx \qquad (1.6.3)$$
$$= \|M_-^H f\|_{L_2(\mathbb{R})}^2 = 2\alpha H \int_{\mathbb{R}^2} f(u) f(v) |u-v|^{2\alpha - 1} du\, dv,$$

where the last equality holds for $H \in (1/2, 1)$ but not for $H \in (0, 1/2)$. Nevertheless, for any $0 < H < 1$ we have the following:

Lemma 1.6.2 ((Ben03a)). *For $0 < H < 1$, it holds that the linear span of the set $\{M_-^H \mathbf{1}_{(u,v)}, u, v \in \mathbb{R}\}$ is dense in $L_2(\mathbb{R})$.*

Proof. (i) Let $H \in (1/2, 1)$ (for $H = 1/2$ the assertion is evident). Since $(b+x)^{-\alpha} - x^{-\alpha} \sim Cx^{-1/2-H}$ as $x \to \infty$, we have that the function $(b-x)_+^{-\alpha} - (-x)_+^{-\alpha} \in L_{1/H}(\mathbb{R})$. Therefore, for any $a < b$ it holds that $g(x) := M_-^{1-H} \mathbf{1}_{(a,b)}(x) \in L_{1/H}(\mathbb{R})$. Therefore, $\mathbf{1}_{(a,b)} = M_-^\alpha g \in I_-^\alpha(L_{1/H}(\mathbb{R}))$, and this is true also for step functions. Since the class of step functions is dense in $L_2(\mathbb{R})$, it follows that $I_-^\alpha(L_{1/H}(\mathbb{R}))$ is dense in $L_2(\mathbb{R})$. Let $h \in I_-^\alpha(L_{1/H}(\mathbb{R}))$, $h = M_-^H g$, $g \in L_{1/H}(\mathbb{R})$. Then there exists the sequence of step functions

$g_n \to g$ in $L_{1/H}(\mathbb{R})$. From the Hardy–Littlewood theorem (Theorem 1.1.1) it follows that

$$\|M_-^H g_n - h\|_{L_2(\mathbb{R})} \leq C\|g_n - g\|_{L_{1/H}(\mathbb{R})} \to 0, n \to \infty.$$

So, the linear span of $\{M_-^H \mathbf{1}_{(u,v)}, u, v \in \mathbb{R}\}$ is dense in $I_-^\alpha(L_{1/H}(\mathbb{R}))$, and therefore it is dense in $L_2(\mathbb{R})$.

(ii) Let $H \in (0, 1/2)$. Due to the Parceval identity, it is sufficient to prove that the linear span of the functions $\widehat{M_-^H \mathbf{1}_{(a,b)}}$ is dense in $L_2(\mathbb{R})$. According to Theorem 1.1.5, we have that

$$\widehat{M_-^H \mathbf{1}_{(a,b)}}(x) = C_H^{(3)} C_H(x) \widehat{\mathbf{1}_{(a,b)}}(x)|x|^{-\alpha},$$

where $C_H(x) = \exp\{i\pi \operatorname{sign} x \alpha/2\}$. According to Lemma 1.1.8, for any $\varphi \in L_2(\mathbb{R})$ there exists a sequence of step functions φ_n such that

$$\int_\mathbb{R} (C_H^{(3)})^{-1} |C_H(-x)\widehat{\varphi}(x) - \widehat{\varphi_n}(x)|x|^{-\alpha}|^2 dx \to 0, \quad n \to \infty,$$

because $(C_H^{(3)})^{-1} C_H(-x)\widehat{\varphi}(x) = \widehat{g}(x)$ for some $g \in L_2(\mathbb{R})$. Then, we obtain that

$$\int_\mathbb{R} |\widehat{\varphi}(x) - C_H^{(3)} C_H(x) \widehat{\varphi_n}(x)|x|^{-\alpha}|^2 dx$$
$$= \int_\mathbb{R} |(C_H^{(3)})^{-1} C_H(-x)\widehat{\varphi}(x) - \widehat{\varphi_n}(x)|x|^{-\alpha}|^2 dx \to 0, \quad n \to \infty.$$

□

Remark 1.6.3. Let $H \in (0, 1/2)$. Then the operator M_-^H defines an isometric isomorphism from $L_2^H(\mathbb{R})$ to $L_2(\mathbb{R})$. Indeed, the operator $I_-^{-\alpha}$ is bounded from $L_2(\mathbb{R})$ to $L_{1/H}^2(\mathbb{R})$, according to Theorem 1.1.1. Let f_n be a Cauchy sequence in $L_2^H(\mathbb{R})$ and $\varphi_n = M_-^H f_n$. Then

$$\|f_n - f_m\|_{L_2^H(\mathbb{R})} = \|\varphi_n - \varphi_m\|_{L_2(\mathbb{R})} \to 0, m, n \to \infty,$$

whence $\varphi_n \to \varphi \in L_2(\mathbb{R})$, and $f_n = (M_-^H)^{-1}\varphi_n \to (M_-^H)^{-1}\varphi =: f$ in $L_{1/H}(\mathbb{R})$. We have that

$$\|f\|_{L_2^H(\mathbb{R})} = \|\varphi\|_{L_2(\mathbb{R})} < \infty,$$

and

$$\|f_n - f\|_{L_2^H(\mathbb{R})} = \|\varphi_n - \varphi\|_{L_2(\mathbb{R})} \to 0.$$

It means that $L_2^H(\mathbb{R})$ is complete, i.e., it is a Hilbert space, and equals the closure of the step functions under L_2^H-norm. By (1.6.3), there exists a unique continuous extension of fractional Wiener integrals for the step functions to the space $L_2^H(\mathbb{R})$. For any $f \in L_2^H(\mathbb{R})$ and the approximating sequence of step functions f_n

$$\int_{\mathbb{R}} f(s) dB_s^H = \lim_{n \to \infty} \int_{\mathbb{R}} f_n(s) dB_s^H \quad \text{in} \quad L_2(\mathbb{R}). \qquad (1.6.4)$$

Remark 1.6.4. Now, let $H \in (1/2, 1)$. Then, the domain of the operator M_-^H coincides with

$$\mathcal{D}(I_-^{-\alpha}) = \mathcal{D}(D_-^\alpha) = \cup_{p \geq 1} I_-^\alpha(L_p(\mathbb{R})),$$

and, according to Theorem 1.1.1 we can take here only $1 \leq p < \alpha^{-1}$ since

$$L_2^H(\mathbb{R}) = \{f \in \mathcal{D}(I_-^{-\alpha}) : M_-^H f \in L_2(\mathbb{R})\}.$$

Note, that

$$L_2(\mathbb{R}) \neq \cup_{1 \leq p < \alpha^{-1}} I_-^\alpha(L_p(\mathbb{R})). \qquad (1.6.5)$$

Indeed, it was proved in (SKM93) that all spaces $I_-^\alpha(L_p(\mathbb{R}))$ coincide for $1 < p < \alpha^{-1}$ and $I_-^\alpha(L_p(\mathbb{R}))$ does not coincide with any space $L_r(\mathbb{R}), 1 \leq r \leq \infty$. The description of $I_-^\alpha(L_p(\mathbb{R}))$ for $1 < p < 1/\alpha$ and for $p = 1$ is contained in (SKM93, Theorems 6.2 and 6.3) and (1.6.5) follows from these theorems.

Theorem 1.6.5. *The space L_2^H is incomplete for $H \in (1/2, 1)$.*

Proof. The operator $M_-^H : L_2^H(\mathbb{R}) \to L_2(\mathbb{R})$ is isometric. So, $L_2^H(\mathbb{R})$ can be identified with its image in $L_2(\mathbb{R})$. According to Lemma 1.6.2, $L_2^H(\mathbb{R})$ is dense in $L_2(\mathbb{R})$, but Remark 1.6.4 demonstrates that $L_2^H(\mathbb{R}) \neq L_2(\mathbb{R})$. Therefore, the image $M_-^H(L_2^H(\mathbb{R}))$, and hence $L_2^H(\mathbb{R})$ itself, is incomplete. $\square$

In spite of the incompleteness of $L_2^H(\mathbb{R})$ for $H \in (1/2, 1)$, due to Lemma 1.6.2, we can approximate any $f \in L_2^H(\mathbb{R})$ by step functions f_n in $L_2^H(\mathbb{R})$. Then $M_-^H f_n \to M_-^H f$ in $L_2(\mathbb{R})$, and we have that

$$I_H(f) := \int_{\mathbb{R}} f(x) dB_s^H = \int_{\mathbb{R}} (M_-^H f)(s) dW_s$$

$$= \lim_{n \to \infty} \int_{\mathbb{R}} (M_-^H f_n)(s) dW_s = \lim_{n \to \infty} \int_{\mathbb{R}} f_n(s) dB_s^H,$$

where the convergence is in $L_2(\Omega)$. Furthermore, for $H \in (1/2, 1)$, we have that

$$E |I(f)|^2 = \int_{\mathbb{R}} |(M_-^H f)(x)|^2 dx$$

for $f \in L_2^H(\mathbb{R})$; however, in general, it does not hold (compare with (1.6.3)) that

$$E |I_H(f)|^2 = 2\alpha H \int_{\mathbb{R}^2} f(u) f(v) |u - v|^{2\alpha - 1} du\, dv,$$

1.6 Wiener Integration with Respect to fBm

even if the last integral is finite. This equality can be obtained only if we can apply the Fubini theorem or if we can prove that the integral $\int_{\mathbb{R}^2} f_n(u)f_n(v)|u-v|^{2\alpha-1}du\,dv$ with step functions f_n converges to $\int_{\mathbb{R}^2} f(u)f(v)|u-v|^{2\alpha-1}du\,dv$. Both things need some additional assumptions.

For $H \in (\frac{1}{2}, 1)$, define the space of measurable functions by

$$|R_H| := \left\{ f : \mathbb{R} \to \mathbb{R} \,\Big|\, \int_{\mathbb{R}_+^2} |f(u)||f(v)||u-v|^{2\alpha-1}du\,dv < \infty \right\},$$

with the norms

$$\|f\|_{|R_H|,1}^2 = 2\alpha H \int_{\mathbb{R}_+^2} f(u)f(v)|u-v|^{2\alpha-1}du\,dv \qquad (1.6.6)$$

and

$$\|f\|_{|R_H|,2}^2 = 2\alpha H \int_{\mathbb{R}_+^2} |f(u)||f(v)||u-v|^{2\alpha-1}du\,dv. \qquad (1.6.7)$$

For $H \in (0, 1)$, we introduce one more space,

$$\mathcal{F}_H := \left\{ f : \mathbb{R} \to \mathbb{R} \,\Big|\, f \in L_2(\mathbb{R}), \int_{\mathbb{R}} |\hat{f}(x)|^2 |x|^{-2\alpha} dx < \infty \right\},$$

with the norm

$$\|f\|_{\mathcal{F}_H}^2 = \int_{\mathbb{R}} |\hat{f}(x)|^2 |x|^{-2\alpha} dx. \qquad (1.6.8)$$

Moreover, consider $L_2^H(\mathbb{R})$ with the norm

$$\|f\|_{L_2^H(\mathbb{R})}^2 = \int_{\mathbb{R}} |(M_-^H f)(x)|^2 dx. \qquad (1.6.9)$$

Below we study the most important features of these spaces. (The space $\mathcal{F}_H$ was partially considered in Theorem 1.1.7.) Note, at first, that the norms defined in (1.6.6)–(1.6.9) are all generated by corresponding inner products. Namely,

$$(f, g)_{|R_H|,1} = 2\alpha H \int_{\mathbb{R}_+^2} f(u)g(v)|u-v|^{2\alpha-1}du\,dv, \qquad (1.6.10)$$

$$(f, g)_{|R_H|,2} = 2\alpha H \int_{\mathbb{R}_+^2} |f(u)||g(v)||u-v|^{2\alpha-1}du\,dv, \qquad (1.6.11)$$

$$(f, g)_{\mathcal{F}_H} = \int_{\mathbb{R}} \hat{f}(x)\overline{\hat{g}(x)}|x|^{1-2H}dx \qquad (1.6.12)$$

and

$$(f, g)_{L_2^H(\mathbb{R})} = \int_{\mathbb{R}} (M_-^H f)(x)(M_-^H g)(x)dx. \qquad (1.6.13)$$

Thus, all these spaces are spaces with inner products. Furthermore, (1.6.6) is indeed a norm on $|R_H|$. Indeed, we can apply the Fubini theorem, use the following relation from (GN96):

$$\int_{-\infty}^{s \wedge t} (s-u)^{\alpha-1}(t-u)^{\alpha-1} du = C_H^{(4)} |t-s|^{2\alpha-1},$$

where $C_H^{(4)} = \frac{\Gamma(H-\frac{1}{2})\Gamma(1-2\alpha)}{\Gamma(1-\alpha)}$, and rewrite (1.6.6) as

$$2\alpha H \int_R f(u)f(v)|u-v|^{2\alpha-1} du\, dv$$
$$= (C_H^{(4)})^{-1} 2\alpha H \int_{\mathbb{R}^2_+} f(u)f(v) \int_{-\infty}^{u \wedge v} (u-z)^{\alpha-1}(v-z)^{\alpha-1} dz\, du\, dv$$
$$= (C_H^{(4)})^{-1} 2\alpha H \int_\mathbb{R} \int_z^\infty f(u)(u-z)^{\alpha-1} du \int_z^\infty f(v)(v-z)^{\alpha-1} dv\, dz$$
$$= (C_H^{(4)})^{-1} 2H\alpha (C_H^{(3)})^{-2} \|M_-^H f\|^2_{L_2(\mathbb{R})} = 2\alpha H (C_H^{(4)})^{-1}(C_H^{(3)})^{-2} \|f\|^2_{L_2^H(\mathbb{R})}.$$
$$\tag{1.6.14}$$

Note that the relation $f \in L_2^H(\mathbb{R})$ means, in particular, that the interior integral $\int_x^\infty |f(u)|(u-x)^{\alpha-1} du$ is finite for a.a. $x \in \mathbb{R}$.

Lemma 1.6.6. *We have that the space* $L_1(\mathbb{R}) \cap L_2(\mathbb{R}) \subset L_{\frac{1}{H}}(\mathbb{R}) \subset |R_H|$ *for any* $H \in (\frac{1}{2}, 1)$.

Proof. It is enough to prove that for any $f \in L_1(\mathbb{R}) \cap L_2(\mathbb{R})$ the iterated integral is finite,

$$I := \int_\mathbb{R} |f(u)| \left(\int_\mathbb{R} |f(v)||u-v|^{2\alpha-1} dv \right) du < \infty.$$

From Theorem 1.1.1 with $\alpha = 2H - 1$, $p = \frac{1}{H}$ and $q = \frac{p}{1-2\alpha p} = \frac{1}{1-H}$ we obtain that

$$I \leq \left(\int_\mathbb{R} |f(u)|^{\frac{1}{H}} du \right)^H \left(\int_\mathbb{R} \left(\int_\mathbb{R} |f(v)||u-v|^{2H-1} dv \right)^{\frac{1}{1-H}} du \right)^{1-H}$$
$$\leq \|f\|_{L_{\frac{1}{H}}(\mathbb{R})} C_{1/H, 1/1-H, 2H-1} \|f\|_{L_{\frac{1}{H}}(\mathbb{R})} = C_H \|f\|^2_{L_{\frac{1}{H}}(\mathbb{R})}.$$

Obviously, $L_1(\mathbb{R}) \cap L_2(\mathbb{R}) \subset L_{\frac{1}{H}}(\mathbb{R})$ for $H \in (\frac{1}{2}, 1)$, whence the claim follows. □

Lemma 1.6.7. *The inclusion* $L_1(\mathbb{R}) \cap L_2(\mathbb{R}) \subset \mathcal{F}_H$ *is valid if and only if* $H \in (\frac{1}{2}, 1)$.

Proof. Assume that $H \in (\frac{1}{2}, 1)$. Since $|\widehat{f}(x)| \leq \|f\|_{L_1(\mathbb{R})}$ for any $x \in \mathbb{R}$, we have that

$$\int_{\mathbb{R}}|\widehat{f}(x)|^2|x|^{-2\alpha}dx = \int_{|x|\geq 1}|\widehat{f}(x)|^2|x|^{-2\alpha}dx + \int_{|x|<1}|\widehat{f}(x)|^2|x|^{-2\alpha}dx$$

$$\leq \int_{\mathbb{R}}|\widehat{f}(x)|^2 dx + \|f\|_{L_1(\mathbb{R})}^2 \int_{|x|<1}|x|^{-2\alpha}dx \leq \|f\|_{L_2(\mathbb{R})}^2 + (1-H)^{-1}\|f\|_{L_1(\mathbb{R})}^2.$$

Let $H \in (0, \frac{1}{2})$. According to (PT00b), take the function $f(u) = \operatorname{sign} u \frac{e^{-|u|}}{|u|^p}$ with $p \in (H, \frac{1}{2})$. Evidently, $f \in L_1(\mathbb{R}) \cap L_2(\mathbb{R})$. Nevertheless, due to (GR80, p. 491),

$$\widehat{f}(\lambda) = 2\Gamma(1-p)(\lambda^2+1)^{\frac{p-1}{2}}\sin\left((1-p)\arctan\lambda\right) \sim |\lambda|^{p-1}$$

as $|\lambda| \to \infty$, and $2p - 2 > 2\alpha - 1 > -1$, which means that $\|f\|_{\mathcal{F}_H} = +\infty$. □

Lemma 1.6.8. *For any $H \in (0,1)$, we have that $\mathcal{F}_H \subset L_2^H(\mathbb{R})$.*

Proof. For $H = \frac{1}{2}$, the statement is evident and $\mathcal{F}_{\frac{1}{2}} = L_2^{\frac{1}{2}}(\mathbb{R}) = L_2(\mathbb{R})$. Let $H \in (\frac{1}{2}, 1)$ and $f \in \mathcal{F}_H$. Then, in particular, $f \in L_2(\mathbb{R})$, and, therefore, according to Theorem 1.1.1, the operator $I_-^{\alpha} f$ is well-defined and bounded from $L_2(\mathbb{R})$ to $L_{\frac{1}{1-H}}(\mathbb{R})$. Moreover, according to Theorem 1.1.5 and since $\int_{\mathbb{R}}|\widehat{f}(x)|^2|x|^{-2\alpha}dx < \infty$, it follows that $I_-^{\alpha} f \in L_2(\mathbb{R})$. Therefore, $f \in L_2^H(\mathbb{R})$. Let $H \in (0, \frac{1}{2})$. We must prove, that for any $f \in L_2(\mathbb{R})$ with $\int_{\mathbb{R}}|\widehat{f}(x)|^2|x|^{-2\alpha}dx < \infty$, there exists $\widetilde{\varphi} \in L_2(\mathbb{R})$, such that

$$\widetilde{\varphi} = M_-^H f = C_H^{(3)} D_-^{-\alpha} f. \tag{1.6.15}$$

Consider the function $\psi(x) = \widehat{f}(x)|x|^{-\alpha}C_H(x)$. Since $|C_H(x)| = 1$, $\psi \in L_2(\mathbb{R})$ and $\overline{\psi(x)} = \psi(-x)$, we conclude that $\psi(x) = \widehat{\varphi}(x)$ for some function $\varphi \in L_2(\mathbb{R})$. Now we prove that $C_H^{(3)}\varphi$ satisfies (1.6.15). Indeed,

$$\widehat{f}(x) = \widehat{\varphi}(x)|x|^{\alpha}C_H(-x), \tag{1.6.16}$$

whence $|\widehat{f}(x)|^2 = |\widehat{\varphi}(x)|^2|x|^{2\alpha}$. Since $\widehat{f} \in L_2(\mathbb{R})$, we have that $\varphi \in \mathcal{F}_{1-H}$, and from Theorem 1.1.5 and (1.6.16), it follows that

$$f = I_-^{-\alpha}\varphi.$$

Therefore, $\widetilde{\varphi}(x) = C_H^{(3)}\varphi(x)$ satisfies (1.6.15), whence the claim follows. □

Next, by using Lemma 1.6.8 and an example from (PT00b) with a slightly modified proof, we establish that $|R_H| \subset L_2^H(\mathbb{R})$.

Lemma 1.6.9. *Let $H \in (\frac{1}{2}, 1)$. Then $(|R_H|, \|\cdot\|_{|R_H|,1}) \subset L_2^H(\mathbb{R})$ and this inclusion is proper.*

Proof. The inclusion itself follows from (1.6.14). We prove that the inclusion is strict if we find a function $f \in \mathcal{F}_H \setminus |R_H|$. Let $f(u) = \text{sign}\, u \cdot |u|^{-p} \cdot \sin u$, $\frac{1}{2} < p < 1$. Then $f \in L_2(\mathbb{R})$, $\widehat{f} \in L_2(\mathbb{R})$. For calculation of $\widehat{f}$ we consider the approximations $f_n(u) = f(u)\mathbf{1}_{\{|u|<n\}} \to f$ in $L_2(\mathbb{R})$. The function $\widehat{f}_n$ satisfies the relations

$$\widehat{f}_n(\lambda) = 2\int_0^n \cos \lambda u |u|^{-p} \sin u\, du$$

$$= \int_0^n u^{-p} \sin((\lambda+1)u)\, du - \int_0^n u^{-p} \sin((\lambda-1)u)\, du$$

$$= \text{sign}(\lambda+1)|\lambda+1|^{p-1} \int_0^{n|\lambda+1|} v^{-p} \sin v\, dv$$

$$- \text{sign}(\lambda-1)|\lambda-1|^{p-1} \int_0^{n|\lambda-1|} v^{-p} \sin v\, dv$$

$$\to \left(\text{sign}(\lambda+1)|\lambda+1|^{p-1} - \text{sign}(\lambda-1)|\lambda-1|^{p-1}\right) \int_0^\infty v^{-p} \sin v\, dv$$

$$= \widehat{f}(\lambda).$$

Since $\frac{1}{2} < p < 1$, we have that

$$\int_\mathbb{R} |\widehat{f}(\lambda)|^2 |\lambda|^{1-2H}\, d\lambda$$

$$\leq C\left(\int_\mathbb{R} |\lambda|^{1-2H}|\lambda+1|^{2p-2}\, d\lambda + \int_\mathbb{R} |\lambda|^{1-2H}|\lambda-1|^{2p-2}\, d\lambda\right) < \infty$$

and it means that $f \in \mathcal{F}_H$. Now, let $\frac{1}{2} < p < H$. We shall use the inequalities $|\sin u| > \frac{1}{2}$ for $u \in (\pi k + \frac{\pi}{4}, \pi k + \frac{3\pi}{4})$, $(u + \frac{\pi}{2})^{-p} > (2u)^{-p}$ for $u > \frac{3\pi}{4}$, $(u + \frac{\pi}{2} - x)_+^{\alpha-1} > (2u-x)_+^{\alpha-1}$ for $x > \pi$, $u > \frac{3\pi}{4}$ and $(u-x)_+^{\alpha-1} > (2u-x)_+^{\alpha-1}$ for $x > 0$. Consider

$$\int_\mathbb{R} \left(\int_x^\infty |f(u)|(u-x)^{\alpha-1} du\right)^2 dx = \int_\mathbb{R} \left(\int_x^\infty |u|^{-p}|\sin u|(u-x)^{\alpha-1} du\right)^2 dx$$

$$\geq \int_\mathbb{R} \left(\int_{x \vee \frac{\pi}{4}}^\infty |u|^{-p}|\sin u|(u-x)^{\alpha-1} du\right)^2 dx$$

$$\geq \frac{1}{2}\int_\mathbb{R} \left(\sum_{k=0}^\infty \int_{\pi k + \frac{\pi}{4}}^{\pi k + \frac{3\pi}{4}} u^{-p}(u-x)_+^{\alpha-1} du\right)^2 dx$$

$$\geq \frac{1}{4}\int_0^\infty \left(\sum_{k=0}^\infty \int_{\pi k + \frac{\pi}{4}}^{\pi k + \frac{3\pi}{4}} u^{-p}(u-x)_+^{\alpha-1} du\right)^2 dx$$

$$+ \frac{1}{4}\int_0^\infty \left(\sum_{k=1}^\infty \int_{\pi k - \frac{\pi}{4}}^{\pi k + \frac{\pi}{4}} \left(u + \frac{\pi}{2}\right)^{-p} \left(u + \frac{\pi}{2} - x\right)_+^{\alpha-1} du\right)^2 dx$$

$$\geq \frac{1}{4}\int_\pi^\infty \Big(\sum_{k=0}^\infty \int_{\pi k+\frac{\pi}{4}}^{\pi k+\frac{3\pi}{4}} u^{-p}(2u-x)_+^{\alpha-1}du\Big)^2 dx$$

$$+\frac{2^{-2p}}{4}\int_\pi^\infty \Big(\sum_{k=1}^\infty \int_{\pi k-\frac{\pi}{4}}^{\pi k+\frac{\pi}{4}} u^{-p}(2u-x)_+^{\alpha-1}du\Big)^2 dx$$

$$\geq \frac{2^{-2p}}{8}\int_\pi^\infty \Big(\sum_{k=0}^\infty \int_{\pi k+\frac{\pi}{4}}^{\pi k+\frac{3\pi}{4}} u^{-p}(2u-x)_+^{\alpha-1}du$$

$$+\sum_{k=1}^\infty \int_{\pi k-\frac{\pi}{4}}^{\pi k+\frac{\pi}{4}} u^{-p}(2u-x)_+^{\alpha-1}du\Big)^2 dx$$

$$= \frac{2^{-2p}}{8}\int_\pi^\infty \Big(\int_{\frac{\pi}{4}}^\infty u^{-p}(2u-x)_+^{\alpha-1}du\Big)^2 dx$$

$$= \frac{2^{-2p}}{8}\int_\pi^\infty \Big(\int_{\frac{\pi}{4x}}^\infty v^{-p}(2v-1)_+^{\alpha-1}du\Big)^2 x^{2\alpha-2p}dx$$

$$\geq \frac{2^{-2p}}{8}\int_\pi^\infty x^{2\alpha-2p}dx \Big(\int_{\frac{1}{2}}^\infty v^{-p}(2v-1)^{\alpha-1}du\Big)^2 = \infty$$

for $H > p$. □

Now we consider the representation of the Wiener process via fBm, i.e., the relation which is inverse to the relation (1.6.1).

Lemma 1.6.10. *Let $0 < H < 1$. Then $M_-^{1-H}\mathbf{1}_{(0,t)} \in L_2^H(\mathbb{R})$ for all $t \in \mathbb{R}$, and the underlying Wiener process W admits the representation*

$$W_t = \widetilde{C_H}\int_\mathbb{R} M_-^{1-H}\mathbf{1}_{(0,t)}(s)dB_s^H,$$

where $\widetilde{C_H} = (C_H^{(3)}C_{1-H}^{(3)})^{-1}$.

Proof. We must check that $M_-^{1-H}\mathbf{1}_{(0,t)} \in L_2^H(\mathbb{R})$. Indeed,

$$M_-^H \cdot M_-^{1-H}\mathbf{1}_{(0,t)} = C_H^{(3)}C_{1-H}^{(3)}I_-^{H-\frac{1}{2}}(I_-^{\frac{1}{2}-H}\mathbf{1}_{(0,t)}) = (\widetilde{C_H})^{-1}\mathbf{1}_{(0,t)} \in L_2(\mathbb{R}).$$

Furthermore, according to Definition 1.6.1, it holds that

$$\widetilde{C_H}\int_\mathbb{R}(M_-^{1-H}\mathbf{1}_{(0,t)})(s)dB_s^H = \widetilde{C_H}\int_\mathbb{R}(M_-^H M_-^{1-H}\mathbf{1}_{(0,t)})(s)dW_s$$

$$= \int_\mathbb{R}\mathbf{1}_{(0,t)}(s)dW_s = W_t. \qquad (1.6.17)$$

□

Corollary 1.6.11. *Any fBm B^H admits a Mandelbrot–van Ness representation with respect to the Wiener process W from representation (1.6.17).*

1.7 The Space of Gaussian Variables Generated by fBm.

Denote
$$\mathcal{B}_H = \overline{\mathrm{span}}\{B_t^H, t \in \mathbb{R}\},$$
where the closure is taken in $L_2(\Omega)$. We are interested in the following question: which classes of integrands in the definition of the Wiener integral w.r.t. fBm are isometric to $\mathcal{B}_H$ or to some of its subspaces? The following theorem from (PT00b) gives the general answer to this question.

Theorem 1.7.1. *Let $\mathcal{I}$ be some class of integrands and let $\mathcal{I}_s \subset \mathcal{I}$ be the class of step functions. Under the assumptions*

(i) *$\mathcal{I}$ is a space with inner product $(f,g)_\mathcal{I}$, $f,g \in \mathcal{I}$,*
(ii) *for $f, g \in \mathcal{I}_s$ $(f,g)_\mathcal{I} = EI(f)I(g)$,*
(iii) *the set $\mathcal{I}_s$ is dense in $\mathcal{I}$,*

we have the following:

(a) *there is an isometry between the space $\mathcal{I}$ and a linear subspace of $\mathcal{B}_H$ which is an extension of the map $f \to I(f)$ for $f \in \mathcal{I}_s$;*
(b) *$\mathcal{I}$ is isometric to $\mathcal{B}_H$ if and only if $\mathcal{I}$ is complete.*

Proof. (a) Let $f \in \mathcal{I}$. By (iii), there exists $f_n \in \mathcal{I}_s$, such that $\{f_n, n \geq 1\}$ is a Cauchy sequence in $\mathcal{I}$ with norm $\|\cdot\|_\mathcal{I} = (\cdot,\cdot)_\mathcal{I}$. According to (ii), $I(f_n)$ is a Cauchy sequence in $L_2(\Omega)$, hence it converges to some r.v. $\xi \in L_2(\Omega)$. We set $I(f) := \xi$. Since $I(f_n) \in \mathcal{B}_H$ and $\mathcal{B}_H$ is a closed subspace of $L_2(\Omega)$, we obtain that $I(f) \in \mathcal{B}_H$. So, we can define the map $I: \mathcal{I} \to \mathcal{B}_H$. For any $f, g \in \mathcal{I}$ it holds that

$$(f,g)_\mathcal{I} = \lim_{n\to\infty}(f_n,g_n)_\mathcal{I} = \lim_{n\to\infty} EI(f_n)I(g_n) = EI(f)I(g).$$

Moreover, ξ does not depend on the choice of the sequence $f_n \to f$ in $\mathcal{I}$. Since the map I is linear, we get an isometry between $\mathcal{I}$ and some subspace of $\mathcal{B}_H$.
(b) Since $\mathcal{B}_H$ is complete as a closed subspace of the complete space $L_2(\Omega)$, it follows that $\mathcal{I}$ is complete if I is an isometry between $\mathcal{I}$ and $\mathcal{B}_H$. Conversely, let $\mathcal{I}$ be complete. Then, for any $\eta \in \mathcal{B}_H$, it holds that $\eta = \lim \eta_n$, $\eta_n = I(f_n) \in \mathrm{span}\{B_t^H, t \in \mathbb{R}\}$, $f_n \in \mathcal{I}_s$. So, $I(f_n) \to \eta$ in $L_2(\Omega)$. Therefore, from (ii) it follows that f_n is a Cauchy sequence in $\mathcal{I}$, and from completeness, $f_n \to f$ in $\mathcal{I}$, $\eta = I(f)$. □

Corollary 1.7.2. *From Lemma 1.6.2, Remark 1.6.3 and Theorem 1.6.5, we obtain the following: the space $\mathcal{I} = L_2^H(\mathbb{R})$ is complete for $H \in (0, \frac{1}{2})$ and incomplete for $H \in (\frac{1}{2}, 1)$. Step functions are dense in $L_2^H(\mathbb{R})$ for any $H \in (0,1)$. Therefore, $L_2^H(\mathbb{R})$ is isometric to $\mathcal{B}_H$ for $H \in (0, \frac{1}{2})$ and isometric to a subspace of $\mathcal{B}_H$ for $H \in (\frac{1}{2}, 1)$.*

Theorem 1.7.3. *The space* $(|R_H|, \|\cdot\|_{L_2^H(\mathbb{R})})$ *is incomplete for* $H \in (\frac{1}{2}, 1)$, *the space* $(\mathcal{F}_H, \|\cdot\|_{\mathcal{F}_H})$ *is incomplete unless* $H = \frac{1}{2}$, *and the space* $(|R_H|, \|\cdot\|_{|R_H|,2})$, $H \in (\frac{1}{2}, 1)$, *is complete.*

Proof. (i) Consider the space $(|R_H|, \|\cdot\|_{|R_H|,1})$, $H \in (\frac{1}{2}, 1)$. Evidently, if some space is dense in an incomplete space, then it is also incomplete. From Lemma 1.6.9, it follows that $|R_H| \subset L_2^H(\mathbb{R})$, and from Theorem 1.6.5, we have that $L_2^H(\mathbb{R})$ is incomplete. So, it is enough to establish that $|R_H|$ is dense in $L_2^H(\mathbb{R})$. If the function $f \in L_2^H(\mathbb{R})$, then $g := M_-^H f \in L_2(\mathbb{R})$. Therefore, there exists a sequence of step functions $\{g_n, n \geq 1\} \subset L_2(\mathbb{R})$ such that $\|g_n - g\|_{L_2(\mathbb{R})} \to 0$. Evidently, any step function g_n can be expressed as $g_n = M_-^H \varphi_n$, where φ_n is a linear combination of functions $M_-^{1-H} \mathbf{1}_{(a,b)}$, $-\infty < a < b < \infty$, and φ_n can be determined via Lemma 1.1.3. Note that

$$\|f - \varphi_n\|_{L_2^H(\mathbb{R})} = \|M_-^H f - M_-^H \varphi_n\|_{L_2(\mathbb{R})} \to 0,$$

$n \to \infty$, so it is enough to prove that $\varphi_n \in |R_H|$. As will be established in Corollary 1.9.3, there exists some constant C such that $\|\varphi_n\|_{|R_H|,2} \leq C\|\varphi_n\|_{L_{\frac{1}{H}}(\mathbb{R})}$, and as mentioned in the proof of Lemma 1.6.2, we have that $M_-^{1-H} \mathbf{1}_{(a,b)} \in L_{\frac{1}{H}}(\mathbb{R})$ for all $-\infty < a < b < \infty$. Therefore, $(|R_H|, \|\cdot\|_{|R_H|,1})$ is dense in $L_2^H(\mathbb{R})$, and hence incomplete.

(ii) Consider the space $\mathcal{F}_H$, $H \neq \frac{1}{2}$. Let $0 < H < \frac{1}{2}$, and let $\{f_n, n \geq 1\}$ be the sequence of functions

$$\widehat{f_n}(x) = |x|^{-p} \mathbf{1}_{\{\frac{1}{n} < |x| < 1\}}(x), \quad \frac{1}{2} < p < 1 - H.$$

Evidently, $\widehat{f_n} \in L_2(\mathbb{R})$ and $\overline{\widehat{f_n}(x)} = \widehat{f_n}(-x)$. Therefore, $\widehat{f_n}$ is the Fourier transform of some $f_n \in L_2(\mathbb{R})$. Moreover, $f_n \in \mathcal{F}_H$ and since $-1 < -2p - 2\alpha$, we have for $n > m$ that

$$\|f_n - f_m\|_{\mathcal{F}_H}^2 = \int_{\mathbb{R}} (\widehat{f_n}(x) - \widehat{f_m}(x))^2 |x|^{-2\alpha} dx$$

$$= \int_{\mathbb{R}} |x|^{-2p-2\alpha} \mathbf{1}_{\{1/n < x < 1/m\}} dx \to 0.$$

Suppose that there exist $f_n \in \mathcal{F}_H$ such that $\|f - f_n\|_{\mathcal{F}_H} \to 0, n \to \infty$. Then, there exists a subsequence $\widehat{f_{n_k}}(x)$ such that $\widehat{f_{n_k}}(x) \to \widehat{f}(x)$ for a.a. $x \in \mathbb{R}$, whence $\widehat{f}(x) = |x|^{-p} \mathbf{1}_{\{|x|<1\}}$. Since $-2p < -1$ we have that $\widehat{f} \notin L_2(\mathbb{R})$, therefore $f \notin L_2(\mathbb{R})$. For $H \in (1/2, 1)$, we can take the sequence $\widehat{f_n}(x) = |x|^{-p} \mathbf{1}_{\{1<|x|<n\}}$, with $p > 1 - H$.

(iii) Lastly, consider the space $(|R_H|, \|\cdot\|_{|R_H|,2})$, $H \in (1/2, 1)$. Let $\{f_n, n \geq 1\} \subset (|R_H|, \|\cdot\|_{|R_H|,2})$ be a Cauchy sequence. Then there exists a subsequence $f_{n_k}(x) \to f(x)$ for a.a. $x \in \mathbb{R}$, where f is some function. Indeed,

$$0 \leftarrow \|f_n - f_m\|_{|R_H|,2} \geq (2r)^{2\alpha-1} \|f_n - f_m\|_{L_2[-r,r]}^2 \quad \text{as} \quad n, m \to \infty$$

whence the above statement easily follows. Moreover, by the Fatou lemma, we have that
$$\|f\|_{|R_H|,2} \leq \varliminf_{n\to\infty} \|f_{n_k}\|_{|R_H|,2} < \infty$$
and
$$\|f - f_n\|_{|R_H|,2} \leq \varliminf_{k\to\infty} \|f_n - f_{n_k}\|_{|R_H|,2} \to 0, n \to \infty. \quad \square$$

1.8 Representation of fBm via the Wiener Process on a Finite Interval

Sometimes it is convenient to consider a "one-sided" fBm $B^H = \{B^H_t, t \geq 0\}$ and to represent it as a functional of the form $B^H_t = \varphi(B_s, 0 \leq s \leq t)$, of some Wiener process $B = \{B_t, t \geq 0\}$, instead of (1.3.4). For this purpose consider the kernels
$$l_H(t,s) = C_H^{(5)} s^{-\alpha}(t-s)^{-\alpha} I_{\{0<s<t\}},$$
and
$$m_H(t,s) = C_H^{(6)}\left(\left(\frac{t}{s}\right)^\alpha (t-s)^\alpha - \alpha s^{-\alpha}\int_s^t u^{\alpha-1}(u-s)^\alpha du\right),$$
where
$$C_H^{(5)} = \left(\frac{\Gamma(2-2\alpha)}{2H\Gamma(1-\alpha)^3\Gamma(1+\alpha)}\right)^{\frac{1}{2}}, \; C_H^{(6)} = \left(\frac{2H\Gamma(1-\alpha)}{\Gamma(1-2\alpha)\Gamma(\alpha+1)}\right)^{\frac{1}{2}},$$
and $\alpha = H - \frac{1}{2}$, $H \in (0,1)$. Throughout the book we shall use the notations $\tilde{\alpha} = (1-\alpha)^{1/2}$, $\hat{\alpha} = (1-\alpha)^{-1/2}$.

(i) Let $H \in (\frac{1}{2}, 1)$. Then, by using the equality
$$\int_0^1 t^{-\mu}(1-t)^{-\mu}|x-t|^{2\mu-1}dt = B(\mu, 1-\mu), \quad (1.8.1)$$
that was established in (NVV99, Lemma 2.2) for any $\mu \in (0,1), x \in (0,1)$, we obtain that for any $t > 0$

$$\|l_H(t,\cdot)\|_{|R_H|,2}$$
$$= \left(C_H^{(5)}\right)^2 2H\alpha \int_0^t\int_0^t (t-u)^{-\alpha}(t-s)^{-\alpha}u^{-\alpha}s^{-\alpha}|u-s|^{2\alpha-1}du\,ds$$
$$= t^{1-2\alpha}\left(C_H^{(5)}\right)^2 2H\alpha \int_0^1 u^{-\alpha}(1-u)^{-\alpha}\left(\int_0^1 (1-s)^{-\alpha}s^{-\alpha}|u-s|^{2\alpha-1}ds\right)du$$
$$= t^{1-2\alpha}\left(C_H^{(5)}\right)^2 2H\alpha B(\alpha, 1-\alpha)B(1-\alpha, 1-\alpha)$$
$$= t^{1-2\alpha}\frac{\Gamma(2-2\alpha)\Gamma(\alpha)\Gamma(1-\alpha)^3}{\Gamma(1-\alpha)^3\Gamma(\alpha)\Gamma(2-2\alpha)} = t^{1-2\alpha} < \infty.$$
$$(1.8.2)$$

1.8 Representation of fBm via the Wiener Process on a Finite Interval

Therefore, we can consider the integral

$$
\begin{aligned}
I_t^H(l_H) &= \int_0^t l_H(t,s)dB_s^H := \int_{\mathbb{R}} l_H(t,s)dB_s^H \\
&= \int_{\mathbb{R}} (M_-^H l_H)(t,\cdot)(x)dW_x,
\end{aligned}
\tag{1.8.3}
$$

where $W = \{W_x, x \in \mathbb{R}\}$ is the underlying Wiener process. Similarly to (1.8.2), for any $0 < t < t'$, we obtain that

$$
\begin{aligned}
\mathbf{E} I_t^H(l_H) I_{t'}^H(l_H) &= (l_H(t,\cdot), l_H(t',\cdot))_{|R_H|,2} \\
&= (C_H^{(5)})^2 2H\alpha \int_0^t (t-u)^{-\alpha} u^{-\alpha} \left(\int_0^{t'} (t'-s)^{-\alpha} s^{-\alpha} |u-s|^{2\alpha-1} ds \right) du \\
&= (C_H^{(5)})^2 2H\alpha t^{1-2\alpha} B(\alpha, 1-\alpha) B(1-\alpha, 1-\alpha) = t^{1-2\alpha}.
\end{aligned}
\tag{1.8.4}
$$

From (1.8.3), it follows that $\{I_t^H, t \geq 0\}$ is a centered Gaussian process. Moreover, from (1.8.4), we obtain for any $0 < s < t \leq s' < t'$ that

$$\mathbf{E} \left(I_{t'}^H(l_H) - I_{s'}^H(l_H) \right) \left(I_t^H(l_H) - I_s^H(l_H) \right) = 0.$$

Thus, the increments of $I_t^H(l_H)$ are uncorrelated, and hence independent. It follows that $I_t^H(l_H)$ is a martingale w.r.t. its natural filtration

$$\mathcal{F}_t^H := \sigma \left\{ I_s^H(l_H), 0 \leq s \leq t \right\},$$

having angle bracket $\langle I_t^H(l_H) \rangle = t^{1-2\alpha}$ and $I_0^H(l_H) = 0$. By the Lévy theorem, there exists some Wiener process $B = \{B_t, t \geq 0\}$, such that

$$M_t^H := I_t^H(l_H) = \tilde{\alpha} \int_0^t s^{-\alpha} dB_s. \tag{1.8.5}$$

The process M^H is called the Molchan martingale, or the fundamental martingale, since it was considered originally in the papers (Mol69; MG69). See also (NVV99).

(ii) Now, let $H \in (0, \frac{1}{2})$. In this case we need some preliminaries.

1) Let $f \in BV[0,T]$, where $BV[0,T]$ is the class of functions of bounded variation on $[0,T]$, and $f = 0$ outside $[0,T]$.
Let us calculate $M_-^H f$. For $\alpha = H - \frac{1}{2}$ it holds that

$$
(M_-^H f)(x) = \begin{cases} 0, & x > T \\ C_H^{(2)} \int_x^T (u-x)^\alpha df(u) - (T-x)^\alpha f(T-), & 0 < x < T \\ C_H^{(2)} \alpha \int_0^T f(u)(u-x)^{\alpha-1} du, & x < 0. \end{cases}
$$

Let $I_T^H := \int_0^T f(s) dB_s^H$. Then

$$\mathbf{E}\left|I_T^H(f)\right|^2 = \left(C_H^{(2)}\right)^2 \left(\alpha^2 \int_{-\infty}^0 \left|\int_0^T f(u)(u-x)^{\alpha-1}du\right|^2 dx\right.$$

$$+ \int_0^T \left|\int_x^T (u-x)^\alpha df(u) - (T-x)^\alpha f(T-)\right|^2 dx\bigg)$$

$$= \left(C_H^{(2)}\right)^2 \left(\alpha^2 \int_0^T \int_0^T f(u)f(s) \left(\int_{-\infty}^0 (u-x)^{\alpha-1}(s-x)^{\alpha-1} dx\right) du\, ds\right.$$

$$+ \int_0^T \int_0^T \left(\int_0^{s \wedge u} (u-x)^\alpha (s-x)^\alpha dx\right) df(u) df(s)$$

$$+ f^2(T-)\frac{T^{2\alpha+1}}{2\alpha+1} - 2\int_0^T \left[\int_0^u (u-x)^\alpha (T-x)^\alpha dx\right] df(u) \cdot f(T-)\bigg).$$
(1.8.6)

Evidently, the function f and its variation $\mathrm{var}\, f$ are bounded on $[0,T]$: there exists $C > 0$ such that $|f(u)| \leq C$ and $\psi_u := \mathrm{var}_{[0,u]} f \leq C$, $0 \leq u \leq T$. Therefore, on the one hand, it holds that

$$\int_0^T \int_0^T |f(u)||f(s)| \left(\int_{-\infty}^0 (u-x)^{\alpha-1}(s-x)^{\alpha-1} dx\right) du\, ds$$

$$\leq C^2 \int_0^T \int_0^T \left(\int_{-\infty}^0 (u-x)^{\alpha-1}(s-x)^{\alpha-1} dx\right) du\, ds \qquad (1.8.7)$$

$$= C^2 \alpha^{-2} \int_{-\infty}^0 ((T-x)^\alpha - (-x)^\alpha)^2 dx < \infty.$$

On the other hand, we obtain that

$$\int_0^T \int_0^T \left(\int_0^{s \wedge u} (u-x)^\alpha (s-x)^\alpha dx\right) d\psi_u d\psi_s$$

$$\leq \int_0^T \int_0^s \left(\int_0^u (u-x)^\alpha (s-x)^\alpha dx\right) d\psi_u d\psi_s$$

$$+ \int_0^T \int_s^T \left(\int_0^s (u-x)^\alpha (s-x)^\alpha dx\right) d\psi_u d\psi_s \qquad (1.8.8)$$

$$\leq (2\alpha+1)^{-1} \left(\int_0^T \int_0^s u^{2\alpha+1} d\psi_u d\psi_s + \int_0^T \int_s^T s^{2\alpha+1} d\psi_u d\psi_s\right)$$

$$\leq \frac{T^{2\alpha+1}}{2\alpha+1} C^2 < \infty.$$

Clearly, the last integral in (1.8.6) is finite. It follows from (1.8.7) and (1.8.8) that the integrals in (1.8.6) are well defined, $\mathbf{E}\left|I_T^H(f)\right|^2 < \infty$ and the limits of integration in (1.8.6) are changed correctly. Moreover, the integral

1.8 Representation of fBm via the Wiener Process on a Finite Interval

$\int_0^T f(s)dB_s^H$ exists for any $f \in BV[0,T]$. Now, let $\{f_n, n \geq 1\}$ be the sequence of functions satisfying the assumptions

(a) $f_n \in BV[0,T]$ and there exists $C > 0$ such that $\sup_n \mathrm{var}_{[0,T]} f_n \leq C$;
(b) $f_n \to 0$ pointwise on $[0,T]$.

Then, we can repeat estimates (1.8.6)–(1.8.8) with f_n instead of f and obtain from the Helly theorem and the Lebesgue dominated convergence theorem that $E|I_T^H(f_n)|^2 \to 0, n \to \infty$. Finally, let $f \in BV[0,T] \cap C[0,T]$ and $\widetilde{f}_n(t) = \sum_{k=1}^n f\left(\frac{kT}{n}\right) \mathbf{1}_{\{\frac{(k-1)T}{n} \leq t < \frac{kT}{n}\}}$. Then the functions $f_n := f - \widetilde{f}_n$ satisfy the assumptions (a) and (b), whence

$$I_T^H(f) = \lim_{n \to \infty} \sum_{k=1}^n f\left(\frac{kT}{n}\right) \Delta B_k^H \quad \text{in} \quad L_2(\Omega), \tag{1.8.9}$$

where

$$\Delta B_k^H = B_{\frac{kT}{n}}^H - B_{\frac{(k-1)T}{n}}^H.$$

But

$$\sum_{k=1}^n f\left(\frac{kT}{n}\right) \Delta B_k^H = f(T)B_T^H - \sum_{k=1}^n B_{\frac{kT}{n}}^H \Delta f_k \to f(T)B_T^H - \int_0^T B_t^H df(t). \tag{1.8.10}$$

We obtain from (1.8.9) and (1.8.10) that $I_T^H(f) = f(T)B_T^H - \int_0^T B_t^H df(t)$ for any $f \in BV[0,T] \cap C[0,T]$.

2) Evidently, for any fixed $t > 0$ the kernel $l_H(t, \cdot) \in BV[0,t] \cap C[0,t]$, if $H \in (0, \frac{1}{2})$. Therefore,

$$I_t^H(l_H) = \int_0^t l_H(t,s)dB_s^H = \int_0^t B_s^H dl_H(t,s) = \int_0^t B_s^H (l_H)'_s(t,s)ds$$
$$= -\alpha C_H^{(5)} \int_0^t B_s^H s^{-\alpha-1}(t-s)^{-\alpha-1}(t-2s)ds,$$

and this integral is obviously a Gaussian random variable. By using the fact that l_H vanishes at the endpoints, we can easily show that
$EI_t^H(l_H)I_{t'}^H(l_H) = t^{1-2\alpha}$ for any $0 < t < t'$:

$$\mathbf{E}I_t^H(l_H)I_{t'}^H(l_H)$$

$$= \frac{1}{2}\int_0^t \int_0^{t'} (u^{2H} + s^{2H} - |u-s|^{2H})(l_H)'_s(t,s)(l_H)'_u(t',u)du\,ds$$

$$= -\frac{1}{2}\int_0^t \int_0^{t'} |u-s|^{2H}(l_H)'_s(t,s)(l_H)'_u(t',u)du\,ds$$

$$= -\frac{1}{2}\int_0^t (l_H)'_s(t,s)\left(\int_0^s (s-u)^{2H}(l_H)'_u(t',u)du\right)ds$$

$$- \frac{1}{2}\int_0^t (l_H)'_s(t,s)\left(\int_s^{t'} (u-s)^{2H}(l_H)'_u(t',u)du\right)ds \qquad (1.8.11)$$

$$= -H\int_0^t l_H(t,s)\left(\int_0^s (s-u)^{2\alpha}(l_H)'_u(t',u)du\right)ds$$

$$+ H\int_0^t l_H(t,s)\left(\int_s^{t'} (u-s)^{2\alpha}(l_H)'_u(t',u)du\right)ds$$

$$= -\alpha H C_H^{(5)} \int_0^t l_H(t,s)$$

$$\times \int_0^{t'} |u-s|^{2\alpha}\operatorname{sign}(u-s)u^{-\alpha-1}(t'-u)^{-\alpha-1}(t'-2u)du\,ds.$$

From (NVV99, Proposition 2.1), we can obtain that

$$\int_0^{t'} |u-s|^{2\alpha}\operatorname{sign}(u-s)u^{-\alpha-1}(t'-u)^{-\alpha-1}(t'-2u)du$$
$$= -2/\alpha \cdot \Gamma(1-\alpha)\Gamma(1+\alpha).$$

Therefore, $\mathbf{E}I_t^H(l_H)I_{t'}^H(l_H) = t^{1-2\alpha}$. We can conclude, similarly to part (i), that $I_t^H(l_H)$ is a martingale w.r.t. its natural filtration, and

$$I_t^H(l_H) = \tilde{\alpha}\int_0^t s^{-\alpha}dB_s \qquad (1.8.12)$$

for some Wiener process B. Thus, we have proved the following result.

Theorem 1.8.1. *Let B^H be an fBm with $H \in (0,1)$, and let*

$$M_t^H := I_t^H(l_H) = \int_0^t l_H(t,s)dB_s^H. \qquad (1.8.13)$$

Then there exists a Wiener process B such that (1.8.12) holds. Moreover, $\sigma\{B_s^H, 0 \le s \le t\} = \sigma\{B_s, 0 \le s \le t\}$.

The inverse relation can be obtained for any $H \in (0,1)$ in the following way: evidently, for any $t > 0$ the random variable $Y_t := \int_0^t s^{-\alpha}dB_s^H$ is well

1.8 Representation of fBm via the Wiener Process on a Finite Interval

defined. It can be proved similarly (but more easily) as the existence of $I_t^H(l_H)$. Furthermore, $\widehat{f}(s) := s^{-\alpha} \in BV[0,t] \cap C[0,t]$ for any $t > 0$ and $H \in (0, \frac{1}{2})$. Therefore, it holds that

$$Y_t = t^{-\alpha} B_t^H + \alpha \int_0^t B_s^H s^{-\alpha-1} ds. \quad (1.8.14)$$

Now, let $H \in (\frac{1}{2}, 1)$, $f \in BV[0,t] \cap C[0,t]$ and

$$I_t(f) := \int_0^t f(s) dB_s^H.$$

Then

$$E|I_t(f)|^2 = 2H\alpha \int_0^t \int_0^t f(u)f(s)|u-s|^{2\alpha-1} du\, ds < \infty,$$

and it is easy to see, similarly to (1.8.10) that

$$I_t(f) = B_t^H f(t) - \int_0^t B_s^H df(s).$$

Let $\widehat{f}_\varepsilon(s) = \widehat{f}(s)\mathbf{1}_{\{\varepsilon < s < \infty\}}$ for some $\varepsilon > 0$. Then

$$\int_0^t f_\varepsilon(s) dB_s^H = \int_\varepsilon^t s^{-\alpha} dB_s^H$$

$$= B_t^H t^{-\alpha} - B_\varepsilon^H \varepsilon^{-\alpha} - \alpha \int_0^t B_s^H s^{-\alpha-1} ds.$$

Note that the trajectories of B^H belong to $C^{H-\rho}[0,T]$ for any $0 < \rho < H$, (see Section 1.16). Therefore $B_\varepsilon^H \varepsilon^{-\alpha} \to 0, \varepsilon \to 0$ a.s. By similar reasoning, $\int_\varepsilon^t B_s^H s^{-\alpha} ds \to \int_0^t B_s^H s^{-\alpha} ds, \varepsilon \to 0$ a.s.

Evidently, $E\left|\int_0^\varepsilon f(s) dB_s^H\right|^2 \to 0, \varepsilon \to 0$, and we obtain (1.8.14) for $H \in (\frac{1}{2}, 1)$. But (1.8.14) is an integral equation with respect to $\{B_s^H, 0 \leq s \leq t\}$ and its solution has the form

$$B_t^H = t^\alpha Y_t - \alpha \int_0^t s^{\alpha-1} Y_s ds = \int_0^t s^\alpha dY_s. \quad (1.8.15)$$

Let $M_t^H := I_t^H(l_H)$ be the Molchan martingale. Then, for $H \in (0, \frac{1}{2})$, integration by parts leads to the equality

$$M_t^H = C_H^{(5)} \int_0^t (t-s)^{-\alpha} s^{-\alpha} dB_s^H = -\alpha C_H^{(5)} \int_0^t (t-s)^{-\alpha-1} Y_s ds,$$

whence

$$\int_0^t (t-u)^\alpha M_u^H du = -\alpha C_H^{(5)} \int_0^t Y_s \left(\int_s^t (t-u)^\alpha (u-s)^{-1-\alpha} du \right) ds$$

$$= -\alpha C_H^{(5)} B(\alpha+1, -\alpha) \int_0^t Y_s ds,$$

and

$$Y_t = C_H^{(6)} \widehat{\alpha} \int_0^t (t-u)^\alpha dM_u^H. \qquad (1.8.16)$$

Therefore,

$$B_t^H = \widehat{\alpha} C_H^{(6)} \left(t^\alpha \int_0^t (t-u)^\alpha dM_u^H \right.$$
$$\left. - \alpha \int_0^t s^{\alpha-1} \left(\int_0^s (s-u)^\alpha dM_u^H \right) ds \right) = \int_0^t m_H(t,s) dB_s. \qquad (1.8.17)$$

Let $H \in (\frac{1}{2}, 1)$. Then, by using Theorem 1.8.1, we obtain that

$$\int_0^t (t-u)^\alpha dM_u^H = \alpha \int_0^t (t-u)^{\alpha-1} M_u^H du$$
$$= C_H^{(5)} \alpha \int_0^t (t-u)^{\alpha-1} \int_0^u (u-s)^{-\alpha} s^{-\alpha} dB_s^H du \qquad (1.8.18)$$
$$= C_H^{(5)} \alpha \int_0^t \left(\int_s^t (t-u)^{\alpha-1} (u-s)^{-\alpha} du \right) s^{-\alpha} dB_s^H$$
$$= C_H^{(5)} \alpha B(\alpha, 1-\alpha) Y_t = (C_H^{(6)})^{-1} \widehat{\alpha} Y_t,$$

i.e. we have (1.8.16) and obtain (1.8.17). In this case the kernel $m_H(t,s)$ can be simplified to $m_H(t,s) = \alpha C_H^{(6)} s^{-\alpha} \int_s^t u^\alpha (u-s)^{\alpha-1} du$.

Remark 1.8.2. It easily follows from (1.8.17) and (1.8.18) that the process B^H satisfying (1.8.17) is an fBm. Indeed, it is a Gaussian process with zero mean and covariance

$$E B_t^H B_s^H = \int_0^{t \wedge s} m_H(t,u) m_H(s,u) du = \frac{1}{2}(t^{2H} + s^{2H} - |t-s|^{2H}).$$

Now we state a result of Le Breton (LeB98), see also (KLeBR00), demonstrating how the Wiener integral $\int_0^t f(s) dB_s^H$ can be presented as an integral with respect to fundamental martingale M^H:

Theorem 1.8.3. *Let $f \in L_2^H(\mathbb{R})$ vanish outside $[0,T]$, where $H \in (\frac{1}{2}, 1)$. Furthermore, let*

$$K_H^f(t,s) := C_H^{(7)} \int_s^t f(u) u^\alpha (u-s)^{\alpha-1} du,$$

1.8 Representation of fBm via the Wiener Process on a Finite Interval

where $C_H^{(7)} = \left(\frac{2H\alpha}{(1-2\alpha)B(1-2\alpha,\alpha)} \right)^{1/2}$.

Then,
$$\int_0^t f(s) dB_s^H = \int_0^t K_H^f(t,s) dM_s^H. \tag{1.8.19}$$

Proof. Note that

$$\int_0^t (K_H^f(t,s))^2 d\langle M^H \rangle_s = (C_H^{(7)})^2 (1-2\alpha) \int_0^t (K_H^f(t,s))^2 s^{-2\alpha} ds$$

$$= (C_H^{(7)})^2 (1-2\alpha) \int_0^t \int_0^t f(u) f(v) u^\alpha v^\alpha \tag{1.8.20}$$

$$\times \int_0^{u \wedge v} (u-s)^{\alpha-1} (v-s)^{\alpha-1} s^{-2\alpha} ds\, du\, dv.$$

Further, Lemma 2.2 from (NVV99) states that

$$\int_0^1 t^{\mu-1}(1-t)^{\nu-1}(c-t)^{-\mu-\nu} dt = c^{-\nu}(c-1)^{-\mu} B(\mu,\nu)$$

for $\mu, \nu > 0$, $c > 1$. Hence for $u < v$, $\mu = 1 - 2\alpha$, $\nu = \alpha$ we have that

$$\int_0^u (u-s)^{\alpha-1}(v-s)^{\alpha-1} s^{-2\alpha} ds$$

$$= u^{-1} \left(\frac{v}{u}\right)^{-\alpha} \left(\frac{v}{u} - 1\right)^{2\alpha-1} B\left(1 - 2\alpha, \alpha\right)$$

$$= B\left(1 - 2\alpha, \alpha\right)(uv)^{-\alpha}(v-u)^{2\alpha-1}.$$

Moreover, for $v < u$ it holds that

$$\int_0^v (v-s)^{\alpha-1}(u-s)^{\alpha-1} s^{-2\alpha} ds = B\left(1 - 2\alpha, \alpha\right)(uv)^{-\alpha}(u-v)^{2\alpha-1}.$$

By substituting these equalities into (1.8.20), we obtain for the integral on the right-hand side that

$$(C_H^{(7)})^2 (1-2\alpha) B\left(1 - 2\alpha, \alpha\right) \int_0^t \int_0^t f(u) f(v) |u-v|^{2\alpha-1} du\, dv$$

$$= 2H\alpha \int_0^t \int_0^t f(u) f(v) |u-v|^{2\alpha-1} du\, dv = E|I_H(f)|^2 < \infty.$$

Moreover, the system $(I_s(f), M_s^H, 0 \leq s \leq T)$ is Gaussian and M^H is Gaussian martingale. Therefore it follows from Theorem 7.16 (LS01) that

$$I_t(f) = \int_0^t \left(\frac{d}{d\langle M^H \rangle_u} E(M_u^H I_t(f)) \right) dM_u^H, \quad t \in [0,T].$$

For $u \leq t$, we have that

$$E(M_u^H I_t(f))$$
$$= C_H^{(5)} 2H\alpha \int_0^t \int_0^t f(v) s^{-\alpha}(u-s)^{-\alpha} \mathbf{1}_{\{s<u\}} |v-s|^{2\alpha-1} dv\, ds \qquad (1.8.21)$$
$$= C_H^{(5)} 2H\alpha \int_0^t f(v) \int_0^u s^{-\alpha}(u-s)^{-\alpha} |v-s|^{2\alpha-1} ds$$
$$\times (\mathbf{1}_{\{v<u\}} + \mathbf{1}_{\{v\geq u\}}) dv.$$

The first integral on the right-hand side of (1.8.21) equals, according to (1.8.1), $2C_H^{(5)} H\alpha B(\alpha, 1-\alpha) \int_0^u f(v) dv$. Moreover, according to the equality ((NVV99)):

$$\int_0^1 t^{\mu-1}(1-t)^{\nu-1}(c-t)^{-\mu-\nu+1} dt$$
$$= (\mu+\nu-1) B(\mu,\nu) c^{-\nu+1} \int_0^1 s^{\mu+\nu-2} \cdot (c-s)^{-\mu} ds,\ c>1,\ \mu>0,\ \nu>0,$$

the second integral equals, for $\mu = 1-\alpha$ and $\nu = 1-\alpha$, to

$$C_H^{(5)} 2H\alpha(1-\alpha) B\big(1-\alpha,\ 1-\alpha\big) \int_u^t f(v) v^\alpha \int_0^u z^{1-2H}(v-z)^{\alpha-1} dz\, dv.$$

Therefore, the derivative in u of the right-hand side of (1.8.21) equals

$$C(H) B\big(\alpha,\ 1-\alpha\big) f(u) - C(H)(1-2\alpha) B\big(1-\alpha,\ 1-\alpha\big) f(u) B\big(1-2\alpha,\ \alpha\big)$$
$$+ C(H)(1-\alpha) B\big(1-\alpha,\ 1-\alpha\big) u^{-2\alpha} \int_u^t f(v) v^\alpha (v-u)^{\alpha-1} dv,$$

where $C(H) = 2H\alpha C_H^{(5)}$. It is easy to check that

$$(1-2\alpha) B\big(1-\alpha,\ 1-\alpha\big) B\big(1-2\alpha,\ \alpha\big) = B\big(\alpha,\ 1-\alpha\big).$$

Therefore,

$$\frac{dE(M_u^H I_t(f))}{du} = C(H) B\big(1-\alpha,\ 1-\alpha\big) \cdot (1-2\alpha) u^{-2\alpha}$$
$$\times \int_u^t f(v) v^\alpha \cdot (v-u)^{\alpha-1} dv$$
$$= C_H^{(7)} (1-2\alpha) u^{-2\alpha} \int_u^t f(v) v^\alpha \cdot (v-u)^{\alpha-1} dv.$$

Hence $\frac{dE(M_u^H I_t(f))}{d\langle M^H\rangle_u} = K_H^f(t,u)$, and the theorem is proved. $\square$

1.9 The Inequalities for the Moments of the Wiener Integrals with Respect to fBm

These inequalities were originated with paper (MMV01). Indeed, the Hardy–Littlewood theorem has an immediate consequence, namely, the estimates for the moments of the Wiener integrals with respect to fBm.

Theorem 1.9.1. (i) Let $H \in (0, \frac{1}{2})$. Then $L_2^H(\mathbb{R}) \subset L_{\frac{1}{H}}(\mathbb{R})$ and there exists a constant $C_H > 0$ such that for any $f \in L_2^H(\mathbb{R})$, it holds that

$$\|f\|_{L_{\frac{1}{H}}(\mathbb{R})} \leq C_H \|f\|_{L_2^H(\mathbb{R})}. \qquad (1.9.1)$$

(ii) Let $H \in (\frac{1}{2}, 1)$. Then $L_{\frac{1}{H}}(\mathbb{R}) \subset L_2^H(\mathbb{R})$ and there exists a constant $C_H > 0$ such that for any $f \in L_{\frac{1}{H}}(\mathbb{R})$ it holds that

$$\|f\|_{L_2^H(\mathbb{R})} \leq C_H \|f\|_{L_{\frac{1}{H}}(\mathbb{R})}. \qquad (1.9.2)$$

Proof. (i) Let $f \in L_2^H(\mathbb{R})$. This means that $M_-^H f = C_H^{(3)} D_-^{-\alpha} f \in L_2(\mathbb{R})$. Evidently, $f = I_-^{-\alpha} D_-^{-\alpha} f$ and from the Hardy–Littlewood theorem (Theorem 1.1.1 with $q = \frac{1}{H}$, $p = 2$ and $\alpha = \frac{1}{2} - H$), it follows that

$$\|f\|_{L_{\frac{1}{H}}(\mathbb{R})} = \|I_-^{-\alpha} D_-^{-\alpha} f\|_{L_{\frac{1}{H}}(\mathbb{R})} \leq C_{2,\frac{1}{H},-\alpha} \|D_-^{-\alpha} f\|_{L_2(\mathbb{R})} = C_H \|f\|_{L_2^H(\mathbb{R})}.$$

(ii) We directly apply the Hardy–Littlewood theorem with $p = \frac{1}{H}$, $\alpha = H - \frac{1}{2}$ and $q = 2$:

$$\|f\|_{L_2^H(\mathbb{R})} = \|M_-^H f\|_{L_2(\mathbb{R})} \leq C_H \|f\|_{L_{\frac{1}{H}}(\mathbb{R})}.$$

□

Corollary 1.9.2. Let $f \in L_2^H(\mathbb{R})$. Then there exists $I(f) = \int_{\mathbb{R}} f(s) dB_s^H$ and $\mathbf{E}|I(f)|^2 = \|f\|_{L_2^H(\mathbb{R})}^2$. Therefore, we have for $H \in (0, \frac{1}{2})$ that $\mathbf{E}|I(f)|^2 \geq C_H^{-2} \|f\|_{L_{\frac{1}{H}}(\mathbb{R})}^2$ and, for $H \in (\frac{1}{2}, 1)$, it holds that $\mathbf{E}|I(f)|^2 \leq C_H^2 \|f\|_{L_{\frac{1}{H}}(\mathbb{R})}^2$. Since $I(f)$ is a Gaussian random variable, we obtain the following inequalities for the moments of the Wiener integrals with respect to fBm: for any $r > 0$, there exists a constant $C(H, r)$, such that for $H \in (\frac{1}{2}, 1)$

$$\mathbf{E}|I(f)|^r \leq C(H, r) \|f\|_{L_{\frac{1}{H}}(\mathbb{R})}^r$$

and such that for $H \in (0, \frac{1}{2})$, we have that

$$\|f\|_{L_{\frac{1}{H}}(\mathbb{R})}^r \leq C(H, r) \mathbf{E}|I(f)|^r.$$

Corollary 1.9.3. *Let $H \in (\frac{1}{2}, 1)$ and $f \in L_{\frac{1}{H}}(\mathbb{R})$. Then it follows from Theorem 1.9.1, (ii), (1.6.7) and (1.6.14), that*

$$\|f\|_{|R_H|,2} \leq C\|f\|_{L_{\frac{1}{H}}(\mathbb{R})}.$$

Corollary 1.9.4. *Let $f \in L_{\frac{1}{H}}[a,b]$ and $f = 0$ outside (a,b). Then we obtain the following estimates: for any $r > 0$, there exists a constant $C(H,r)$, such that for $H \in (\frac{1}{2}, 1)$, it holds that*

$$\mathbf{E}\left|\int_a^b f(s)dB_s^H\right|^r \leq C(H,r)\|f\|_{L_{\frac{1}{H}}[a,b]}^r$$

and

$$\mathbf{E}\left|\int_a^b f(s)dB_s^H \int_a^b g(s)dB_s^H\right|^r \leq C(H,r)\|f\|_{L_{\frac{1}{H}}[a,b]}^r \|g\|_{L_{\frac{1}{H}}[a,b]}^r.$$

Furthermore, for $H \in (0, \frac{1}{2})$ the opposite inequality holds:

$$\|f\|_{L_{\frac{1}{H}}[a,b]}^r \leq C(H,r)\mathbf{E}\left|\int_a^b f(s)dB_s^H\right|^r$$

Remark 1.9.5. Let $H \in (\frac{1}{2}, 1)$ and $f \in |R_H|$. Then, from Hölder inequality, we obtain the estimate

$$\|f\|_{|R_H|,2}^2 = \int_{\mathbb{R}} |f(s)| \left(\int_{\mathbb{R}} |f(u)||s-u|^{2\alpha-1}du\right) ds$$

$$\leq \left(\int_{\mathbb{R}} |f(s)|^{\frac{1}{H}}ds\right)^H \left(\int_{\mathbb{R}} ds \left(\int_{\mathbb{R}} |f(u)||s-u|^{2\alpha-1}du\right)^{\frac{1}{1-H}}\right)^{1-H}.$$

Further, from the Hardy–Littlewood theorem with $\alpha = 2H - 1$, $q = \frac{1}{1-H}$ and $p = \frac{1}{H}$, we obtain that

$$\left(\int_{\mathbb{R}} ds \left(\int_{\mathbb{R}} |f(u)||s-u|^{2\alpha-1}du\right)^{\frac{1}{1-H}}\right)^{1-H} \leq C_H \|f\|_{L_{\frac{1}{H}}(\mathbb{R})}.$$

Therefore,

$$\|f\|_{|R_H|,2} \leq C_H \|f\|_{L_{\frac{1}{H}}(\mathbb{R})}.$$

Remark 1.9.6. Next, we show that the lower inequality in the case $H \in (\frac{1}{2}, 1)$ fails. Indeed, let $f(u) = \operatorname{sign} u \cdot |u|^{-p} \sin u$ with $\frac{1}{2} < p < H$. Then according to the proof of Lemma 1.6.9, it holds that $f \in L_2^H(\mathbb{R})$. Nevertheless,

$$\int_0^\infty |f(u)|^{\frac{1}{H}} du = \int_0^\infty \frac{|\sin u|^{\frac{1}{H}}}{|u|^{\frac{p}{H}}} du = \infty, \quad \text{since} \quad \frac{p}{H} < 1.$$

Therefore, the inclusion $L_{\frac{1}{H}}(\mathbb{R}) \subset L_2^H(\mathbb{R})$ is proper. Moreover, consider the function $f_\varepsilon(u) = u^{\varepsilon - H}$, $0 \le u \le 1$, $0 < \varepsilon < H$. Then

$$\|f_\varepsilon\|_{\frac{1}{H}}^2 = \left(\frac{H}{\varepsilon}\right)^{2H}, \quad \|f_\varepsilon\|_{L_2^H(\mathbb{R})}^2 = \frac{1}{\varepsilon} \frac{\Gamma(1 - H + \varepsilon)\Gamma(2\alpha)}{\Gamma(H - \varepsilon)} \sim \frac{C_0}{\varepsilon}, \quad \varepsilon \to 0,$$

where $C_0 = B(1 - H, 2\alpha)$. Since $\dfrac{\frac{1}{\varepsilon}}{\frac{1}{\varepsilon^{2H}}} = \varepsilon^{2\alpha}$ and we can let ε tend to 0, it follows that the inequality

$$\|f\|_{L_2^H(\mathbb{R})} \ge C_H \|f\|_{L_{\frac{1}{H}}(\mathbb{R})}$$

is impossible for $H \in (\frac{1}{2}, 1)$.

Remark 1.9.7. It is very easy to check that the function $f(u) = u^{-H} \notin |R_H|$ for any $H \in (\frac{1}{2}, 1)$. Indeed,

$$\int_0^T \int_0^T u^{-H} s^{-H} |u - s|^{2\alpha - 1} du\, ds = \int_0^1 \int_0^1 u^{-H} s^{-H} |u - s|^{2\alpha - 1} du\, ds,$$

for any $T > 0$, and this is possible only in the case when these integrals are infinite.

Now, let $H \in (0, \frac{1}{2})$. As mentioned in (SKM93), the domain of the operator $D_-^{-\alpha}$ does not coincide with any space $L_r(\mathbb{R}), 1 \le r \le +\infty$. Therefore, the inclusion $L_{\frac{1}{H}}(\mathbb{R}) \subset L_2^H(\mathbb{R})$ is strict. Moreover, let $f(u) = u^{\varepsilon - H}$ with $\varepsilon > \alpha$ (note that ε can be negative). By direct computations, we get

$$\|f\|_{L_2(\mathbb{R})} = (2\varepsilon - 2\alpha)^{-\frac{1}{2}}$$

and

$$\|I_-^{-\alpha} f\|_{L_{\frac{1}{H}}(\mathbb{R})} = K_{\varepsilon, H}(2\varepsilon - 2\alpha)^{-H},$$

where

$$K_{\varepsilon, H} = \frac{\Gamma(\varepsilon - \alpha)}{\Gamma(\varepsilon - 2\alpha + \frac{1}{2})}(2H)^H, \quad \alpha = H - \frac{1}{2}.$$

Therefore,

$$\frac{\|f\|_{L_2(\mathbb{R})}}{\|I_-^{-\alpha} f\|_{L_{\frac{1}{H}}(\mathbb{R})}} \uparrow +\infty, \quad \varepsilon \downarrow (\alpha).$$

Set $g = I_-^{-\alpha} f$, $f = D_-^\alpha g$, then $\|f\|_{L_2(\mathbb{R})} = \|g\|_{L_2^H(\mathbb{R})}$ and

$$\frac{\|g\|_{L_2^H(\mathbb{R})}}{\|g\|_{L_{\frac{1}{H}}(\mathbb{R})}} \uparrow +\infty, \quad \varepsilon \downarrow \alpha.$$

38   1 Wiener Integration with Respect to Fractional Brownian Motion

So, we cannot obtain the inverse inequality to (1.9.1).

Consider now the upper bound for the moments of $I(f)$ with $H \in (0, \frac{1}{2})$. As always, $\alpha = H - \frac{1}{2}$.

Let $W_2^2(\mathbb{R})$ be the standard Sobolev space

$$W_2^2(\mathbb{R}) = \{f : \mathbb{R} \to \mathbb{R} \mid \|f\|_{L_2(\mathbb{R})} + \|f'\|_{L_2(\mathbb{R})} < \infty\}.$$

Theorem 1.9.8. *Let $f \in C^1(\mathbb{R}) \cap W_2^2(\mathbb{R})$ and $|f(x)| + |f'(x)| \leq C_0 |x|^{\alpha-1-\varepsilon}$ for some $\varepsilon > 0$, as $|x| \to \infty$. Then $f \in L_2^H(\mathbb{R})$, and there exists a constant $C(H)$ depending only on H, such that*

$$\|f\|_{L_2^H(\mathbb{R})} \leq C(H) \|f\|_{W_2^2(\mathbb{R})}.$$

Proof. Now, we have that

$$\|f\|_{L_2^H(\mathbb{R})} = \left(\int_{\mathbb{R}} |(M_-^H f)(t)|^2 dt\right)^{1/2} = C_H^{(3)} \left(\int_{\mathbb{R}} |(I_-^\alpha f)(t)|^2 dt\right)^{1/2}$$

$$= C_H^{(3)} \left(\int_{\mathbb{R}} |(D_-^{\frac{1}{2}-H} f)(t)|^2 dt\right)^{1/2}$$

$$= C_H^{(2)} \left(\int_{\mathbb{R}} \left|\frac{d}{dx} \int_{-\infty}^0 f(x-u)(-u)^\alpha du\right|^2 dx\right)^{1/2}$$

$$= C_H^{(2)} \left(\int_{\mathbb{R}} \left|\int_0^\infty f'(x+u) u^\alpha du\right|^2 dx\right)^{1/2}$$

$$\leq \sqrt{2} C_H^{(2)} \left(\left(\int_{\mathbb{R}} \left|\int_x^{x+1} f'(u)(u-x)^\alpha du\right|^2 dx\right)^{1/2}\right. \quad (1.9.3)$$

$$\left. + \left(\int_{\mathbb{R}} \left|\int_{x+1}^\infty f'(u)(u-x)^\alpha du\right|^2 dx\right)^{1/2}\right).$$

Further, it holds that

$$\left(\int_{\mathbb{R}} \left|\int_x^{x+1} f'(u)(u-x)^\alpha du\right|^2 dx\right)^{1/2}$$

$$\leq (2H)^{-1/2} \left(\int_{\mathbb{R}} \int_x^{x+1} |f'(u)|^2 du\, dx\right)^{1/2} = (2H)^{-1/2} \|f'\|_{L_2(\mathbb{R})}, \quad (1.9.4)$$

and

$$\left(\int_{\mathbb{R}} \left|\int_{x+1}^\infty f'(u)(u-x)^\alpha du\right|^2 dx\right)^{1/2}$$

$$= \left(\int_{\mathbb{R}} \left|f(x+1) - \alpha \int_{x+1}^\infty f(u)(u-x)^{\alpha-1} du\right|^2 dx\right)^{1/2} \quad (1.9.5)$$

$$\leq \sqrt{2} \|f\|_{L_2(\mathbb{R})} + \sqrt{2} |\alpha| \left(\int_{\mathbb{R}} \left|\int_{x+1}^\infty f(u)(u-x)^{\alpha-1} du\right|^2 dx\right)^{1/2}.$$

1.9 The Inequalities for the Moments of Wiener Integrals 39

From the generalized Minkowsky inequality, we obtain that

$$\left(\int_{\mathbb{R}}\left|\int_{x+1}^{\infty}f(u)(u-x)^{\alpha-1}du\right|^2 dx\right)^{1/2} = \left(\int_{\mathbb{R}}\left|\int_1^{\infty}f(u+x)u^{\alpha-1}du\right|^2 dx\right)^{1/2}$$

$$\leq \int_1^{\infty} u^{\alpha-1}du\left(\int_{\mathbb{R}}|f(u+x)|^2 dx\right)^{1/2} \leq -1/\alpha \|f\|^2_{L_2(\mathbb{R})}. \tag{1.9.6}$$

The claim follows now immediately from (1.9.3)–(1.9.6). □

Now we turn to the case when $f = 0$ outside some interval $[0,T]$. In this case the conditions on f can be much less restrictive. Indeed, then

$$(I^{\alpha}_- f)(x) = \begin{cases} 0, & x \geq T, \\ -\frac{1}{\Gamma(H+\frac{1}{2})}\frac{d}{dx}\int_x^T f(t)(t-x)^{\alpha}dt, & x \in (0,T), \\ -\frac{\alpha}{\Gamma(H+\frac{1}{2})}\int_0^T f(t)(t-x)^{\alpha-1}dt, & x \leq 0. \end{cases} \tag{1.9.7}$$

Consider some partial cases. Let $f \in I^{-\alpha}_-(L_p[0,T])$, for some $p > 1$, i.e. we can present f as a fractional integral $f(x) = \frac{1}{\Gamma(-\alpha)}\int_x^T \varphi(t)(t-x)^{-1-\alpha}dt$, $\varphi \in L_p[0,T]$. Then, according to (SKM93), for any $x \in (0,T)$ it holds that

$$-\frac{d}{dx}\int_x^T f(t)(t-x)^{\alpha}dt = f(x)(T-x)^{\alpha} + \alpha \int_x^T (f(x) - f(t))(t-x)^{\alpha-1}dt. \tag{1.9.8}$$

The same equality holds for $f \in C^{\beta}[0,T]$ for $\alpha + \beta > 0$.

From (1.9.7) and (1.9.8) it follows immediately that for $f \in I^{-\alpha}_-(L_p[0,T])$, in particular, for $f \in C^{\beta}[0,T]$ with $\alpha + \beta > 0$ we have that

$$E\left|\int_0^T f(t)dB_t^H\right|^2 = \alpha^2(C_H^{(2)})^2 \int_{-\infty}^0 \left|\int_0^T f(t)(t-x)^{\alpha-1}dt\right|^2 dx$$

$$+ (C_H^{(2)})^2 \int_0^T \left|f(x)(T-x)^{\alpha} + \alpha\int_x^T (f(t)-f(x))(t-x)^{\alpha-1}dt\right|^2 dx. \tag{1.9.9}$$

Introduce now some classes of functions vanishing outside $[0,T]$:

$$L_2^H[0,T] = \left\{f : [0,T] \to \mathbb{R} | \int_{\mathbb{R}} |(M^H_- f)(x)|^2 dx < \infty\right\},$$

and

$$D_H[0,T] := \left\{f : [0,T] \to \mathbb{R} \Big| \right.$$

$$\|f\|^2_{D_H[0,T]} := \int_0^T \left(\int_x^T |f(x)-f(t)|(t-x)^{\alpha-1}dt\right)^2 dx < \infty\right\}.$$

Theorem 1.9.9. (i) *The following inclusion holds: for any $p > \frac{1}{H}$ it holds that*
$$E_p^H[0,T] := I_-^{-\alpha}(L_p[0,T]) \cap D_H[0,T] \subset L_2^H[0,T].$$

Moreover, there exists a constant $C(H,p)$, such that for any $f \in E_p^H[0,T]$ we have that
$$\left(E \left| \int_0^T f(t) dB_t^H \right|^2 \right)^{1/2} \leq C(H,p) \left(\|f\|_{L_p[0,T]} T^{H-\frac{1}{p}} + \|f\|_{D_H[0,T]} \right). \tag{1.9.10}$$

(ii) $C^\beta[0,T] \subset L_2^H[0,T]$ *and there exists a constant $C(H,\beta)$ such that for any $f \in C^\beta[0,T]$*
$$\left(E \left| \int_0^T f(t) dB_t^H \right|^2 \right)^{1/2} \leq C(H,\beta) \|f\|_{C^\beta[0,T]} \left(T^H + T^{H+\beta} \right). \tag{1.9.11}$$

Proof. (i) Let $f \in E_p^H[0,T]$. Then $E \left| \int_0^T f(t) dB_t^H \right|^2$ equals the right-hand side of (1.9.9) and also $f \in L_p[0,T]$. We have the following estimate:
$$\mathcal{I}_f := \int_0^T \int_0^T |f(t)| |f(s)| \int_{-\infty}^0 ((t-x)(s-x))^{\alpha-1}) dx \, ds \, dt$$
$$\leq \int_0^T \int_0^T |f(t)| |f(s)| \int_{-\infty}^0 (\sqrt{st} - x)^{2\alpha-2} dx \, ds \, dt$$
$$\leq \frac{1}{2(1-H)} \left(\int_0^T |f(t)| t^{H-1} dt \right)^2$$
$$\leq \frac{1}{2(1-H)} \left(\frac{p-1}{Hp-1} \right)^{\frac{2(p-1)}{p}} \|f\|_{L_p[0,T]}^2 T^{2H-\frac{2}{p}}. \tag{1.9.12}$$

Therefore, for $f \in L_p[0,T]$ it holds that $\mathcal{I}_f < \infty$. Then the Fubini theorem implies that the first term on the right-hand side of (1.9.9) equals, up to a constant
$$\int_0^T \int_0^T f(t) f(s) \int_{-\infty}^0 (t-x)(s-x)^{\alpha-1} dx \, ds \, dt,$$
and can be estimated by the right-hand side of (1.9.12). Moreover, the Hölder inequality implies that
$$\int_0^T |f(x)|^2 (T-x)^{2\alpha} dx \leq \|f\|_{L_p[0,T]}^2 T^{2H-\frac{2}{p}} \frac{(p-2)^{\frac{p-2}{p}}}{(2\alpha p + p - 2)^{\frac{p-2}{p}}}. \tag{1.9.13}$$

From (1.9.12) and (1.9.13) we obtain (1.9.10) with

$$C(H,p) = C_H^{(2)}\left(\left((2(1-H))^{-1/2}\alpha\left(\frac{p-1}{Hp-1}\right)^{\frac{(p-1)}{p}}\right.\right.$$
$$\left.\left.+ \sqrt{2}\left(\frac{p-2}{2\alpha p + p - 2}\right)^{\frac{p-2}{2p}}\right) \vee \sqrt{2}\alpha\right).$$

(ii) In this case,

$$\mathcal{I}_f \le \frac{1}{2(1-H)}\left(\int_0^T f(t)t^{H-1}dt\right)^2 \le \frac{1}{2(1-H)H^2}\|f\|_{C^\beta[0,T]}^2 T^{2H}, \quad (1.9.14)$$

$$\int_0^T |f(x)|^2 (T-x)^{2\alpha} dx \le \frac{T^{2H}}{2H}\|f\|_{C^\beta[0,T]}^2 T^{2H}, \quad (1.9.15)$$

and

$$\int_0^T \left(\int_x^T |f(x)-f(t)|(t-x)^{\alpha-1}dt\right)^2 dx \le \frac{T^{2H+2\beta}}{2(\alpha+\beta)^2(H+\beta)}\|f\|_{C^\beta[0,T]}^2.$$

Thus, we obtain (1.9.11) with

$$C_H = \left(\frac{1}{H^2(1-H)} + \frac{1}{2H}\right) \vee \left(\frac{1}{2(\alpha+\beta)^2(H+\beta)}\right).$$

□

1.10 Maximal Inequalities for the Moments of Wiener Integrals with Respect to fBm

For any fixed $T > 0$, denote $\zeta_T^* = \sup_{0 \le t \le T} |\zeta_t|$, where ζ_t is any function on $[0,T]$. If $B^H = \{B_t^H, t \ge 0\}$ is a fractional Brownian motion, then from its self-similar properties we obtain that $\mathbf{E}((B^H)_T^*)^p = \widehat{C}(H,p)T^{pH}$, where $\widehat{C}(H,p) = \mathbf{E}((B^H)_1^*)^p$. (It is an interesting and open problem how to compute this maximal moment.) Now, let $f \in L_2^H(\mathbb{R})$. We try to find possible bounds for the process $I_t = I_t(f) := \int_0^t f(s)dB_s^H$ both on random and nonrandom intervals. Denote $\|I_T^*\|_p := (E(I_T^*)^p)^{1/p}$.

(i) *Upper bound on nonrandom interval,* $H \in (\frac{1}{2}, 1)$. Note that the process $I_t(f)$ is Gaussian, therefore it admits entropy maximal estimates. In this context, suppose that $f \in |R_H|$ and consider on $[0,T]$ the semi-metric ρ_I generated by the process I, i.e.

$$\rho_I^2(s,t) := \mathbf{E}(I_t - I_s)^2 = E\left|\int_s^t f(u)dB_u^H\right|^2.$$

For any $\varepsilon > 0$ denote by $\mathcal{N}([0,T],\varepsilon)$ the metric ε-capacity of $([0,T],\rho)$, or the minimal number of points in the ε-net of the interval $[0,T]$ in the semi-metric ρ_I, i.e. the minimal number of centers of closed ε-balls covering $[0,T]$.

Also, let $\mathcal{H}([0,T],\varepsilon) := \log \mathcal{N}([0,T],\varepsilon)$ be the metric ε-entropy of this interval in the semi-metric ρ_I, and let $D(T,\varepsilon) = \int_0^\varepsilon \mathcal{H}([0,T],u)^{\frac{1}{2}} du$ be the Dudley integral.

Lemma 1.10.1. *Let $\rho(s,t)$ be some semi-metric on $[0,T]$ and let $\varphi(x)$, $x > 0$, be a continuous increasing function, such that $\varphi(0) = 0$. Also, let g be a function with $g(v) \geq 0$, $g \in L_1[0,T]$, such that for any $0 \leq s < t \leq T$, it holds that $\varphi(\rho(s,t)) \leq \int_s^t g(v)dv$. Then*

$$\mathcal{N}([0,T],u) \leq 1 + \frac{\int_0^T g(v)dv}{\varphi(2u)}.$$

Proof. Consider $0 = s_0 < s_1 < \ldots < s_M < T$, where $|s_{k+1} - s_k| = 2u$, $0 \leq k \leq M-1$, $|T - S_M| \leq 2u$. Such a partition exists, because our condition ensures the continuity of $\rho(s,t)$. Evidently, $\varphi(2u) \leq \int_{s_k}^{s_{k+1}} g(v)dv$, $0 \leq k \leq M-1$, and $\mathcal{N}([0,T],u) \leq M+1$. So,

$$M\varphi(2u) \leq \sum_{k=0}^{M-1} \int_{s_k}^{s_{k+1}} g(v)dv = \int_a^{s_M} g(u)du \leq \int_0^T g(v)dv,$$

i.e. $M \leq \int_0^T g(v)dv \cdot (\varphi(2u))^{-1}$. □

Lemma 1.10.2. *The Dudley integral admits the estimate*

$$D(T,\varepsilon) \leq \int_0^\varepsilon \left[\log\left(1 + u^{-\frac{1}{H}} \widetilde{C}_H \int_0^T |f(v)|^{\frac{1}{H}} dv \right) \right]^{\frac{1}{2}} du,$$

where $\widetilde{C}_H$ is some constant.

Proof. According to (1.9.2) and Corollary 1.9.2, it holds that

$$\mathbf{E}\left| \int_s^t f(u)dB_u^H \right|^2 \leq C(H,2) \|f\|^2_{L_{\frac{1}{H}}[s,t]}.$$

If we choose $\varphi(u) = u^{\frac{1}{H}}$ and $g(v) = |f(v)|^{\frac{1}{H}}$, then $\varphi(\rho_I(s,t)) \leq \int_s^t g(v)dv$. We obtain from Lemma 1.10.1, that for any $u > 0$ the metric u-entropy of the interval $[0,T]$ does not exceed $\log\left(1 + u^{-\frac{1}{H}}(C(H,2))^{\frac{1}{2H}} \cdot 2^{-\frac{1}{H}} \int_0^T |f(v)|^{\frac{1}{H}} dv \right)$. From here the claim follows with $\widetilde{C}_H = 2^{-\frac{1}{H}}(C(H,2))^{\frac{1}{2H}}$. □

Theorem 1.10.3. *For any $p > 0$, there exists a constant $C_p(H)$ such that*

$$\|I_T^*\|_p \leq C_p(H) \|f\|_{L_{\frac{1}{H}}[0,T]}.$$

1.10 Maximal Inequalities for Wiener Integrals w.r.t. fBm

Proof. Denote
$$\sigma^2 := \sup_{0 \leq t \leq T} \mathbf{E} I_t^2.$$

Then according to (Lif95, Theorem 1, p. 141) and its corollary, for any $r > 4\sqrt{2}D(T, \frac{\sigma}{2})$, we have the inequality

$$P\{I_T^* > r\} \leq 2\left(1 - \Phi\left(\frac{r - 4\sqrt{2}D(T, \frac{\sigma}{2})}{\sigma}\right)\right), \quad (1.10.1)$$

where $\Phi(x) = \frac{1}{\sqrt{2\pi}} \int_{-\infty}^{x} e^{\frac{-y^2}{2}} dy$. Since

$$\mathbf{E}(I_T^*)^p \leq p \int_0^\infty x^{p-1}(1 - F(x))dx,$$

where $F(x) = P\{I_T^* < x\}$, we obtain from (1.10.1) that for $D = D(T, \frac{\sigma}{2})$ it holds that

$$\begin{aligned}
\mathbf{E}(I_T^*)^p &\leq p \int_0^{4\sqrt{2}D} x^{p-1}(1 - F(x))dx \\
&+ p \int_{4\sqrt{2}D}^\infty x^{p-1}(1 - F(x))dx \leq (4\sqrt{2}D)^p \\
&+ 2p \int_0^\infty (x + 4\sqrt{2}D)^{p-1} \left(1 - \Phi\left(\frac{x}{\sigma}\right)\right) dx \\
&\leq (4\sqrt{2}D)^p + p2^p \int_0^\infty x^{p-1}\left(1 - \Phi\left(\frac{x}{\sigma}\right)\right) dx \\
&+ p2^p(4\sqrt{2}D)^{p-1} \int_0^\infty \left(1 - \Phi\left(\frac{x}{\sigma}\right)\right) dx \\
&\leq (4\sqrt{2}D)^p + p2^p \sigma^p C_1(p) + 2^p p(4\sqrt{2}D)^{p-1}\sigma C_1(1),
\end{aligned} \quad (1.10.2)$$

where $C_1(p) = \int_0^\infty x^{p-1}(1 - \Phi(x)) dx$. Now we estimate $D = D(T, \frac{\sigma}{2})$. From Lemma 1.10.2 and Corollary 1.9.4,

$$\begin{aligned}
D &\leq \int_0^{\frac{\sigma}{2}} \left[\log\left(1 + u^{-\frac{1}{H}}\widetilde{C}_H \int_0^T |f(v)|^{\frac{1}{H}} dv\right)\right]^{\frac{1}{2}} du \\
&\leq H(\widehat{C}_H)^H \int_{\log 2}^\infty z^{\frac{1}{2}} \frac{\exp z \, dz}{(\exp z - 1)^{H+1}},
\end{aligned} \quad (1.10.3)$$

where $\widehat{C}_H = \widetilde{C}_H \int_0^T |f(v)|^{\frac{1}{H}} dv$. Therefore, $D \leq \overline{C}_H \|f\|_{L_{\frac{1}{H}}[0,T]}$, where $\overline{C}_H = (\widetilde{C}_H)^H H \int_{\log 2}^\infty z^{\frac{1}{2}} \frac{\exp z \, dz}{(\exp z - 1)^{H+1}}$. Evidently, $\sigma \leq (C(H,2))^{\frac{1}{2}} \|f\|_{L_{\frac{1}{H}}[0,T]}$. By substituting these two estimates into (1.10.2), we obtain the proof. □

(ii) *Lower bound on nonrandom interval,* $H \in (\frac{1}{2}, 1)$. According to Remark 1.9.6, the reverse inequality of (1.9.2) fails. Therefore, we obtain the lower bound under stronger assumptions. We suppose here that $f = f(s) > 0$ on $[0,T]$. Denote $g(t) = \frac{1}{f(t)}$, $g_T^* = \operatorname{ess\,sup}_{0 \le s \le T} g(s)$ and assume that $g_T^* < \infty$.

Theorem 1.10.4. *For any $p > 0$, we have an estimate*

$$\|I_T^*\|_p \ge c_p(H) T^H (g_T^*)^{-1}.$$

Proof. According to the lower bound obtained by Sudakov (Lif95, Theorem 5, p. 152), for any $\varepsilon > 0$ it holds that

$$\mathbf{E}(I_T^*)^p \ge (\mathbf{E}I_T^*)^p \ge C_p H([0,T],\varepsilon)^{\frac{p}{2}} \varepsilon^p,$$

where $H([0,T],\varepsilon) = \log N([0,T],\varepsilon)$. Evidently, $N([0,T],\varepsilon) \ge 1 \vee T(2g_T^*\varepsilon)^{-\frac{1}{H}}$. Therefore,

$$H([0,T],\varepsilon) \ge \log\left(1 \vee T(2g_T^*\varepsilon)^{-\frac{1}{H}}\right).$$

Indeed, take an arbitrary partition $\pi = \{0 = s_0 < s_1 < \cdots < s_n = T\}$ such that $\left(E\left|\int_{s_{k-1}}^{s_k} f(s) dB_s^H\right|^2\right)^{\frac{1}{2}} \le 2\varepsilon$. Then

$$E\left|\int_{s_{k-1}}^{s_k} f(s) dB_s^H\right|^2 = \left\|M_-^H(\mathbf{1}_{(s_{k-1},s_k)}f)\right\|_{L_2(\mathbb{R})}^2 \ge (g_T^*)^{-2} E|B_{s_k} - B_{s_{k-1}}|^2$$
$$= (g_T^*)^{-2}(s_k - s_{k-1})^{2H},$$

so, $(g_T^*)^{-\frac{1}{H}}(s_k - s_{k-1}) \le (2\varepsilon)^{\frac{1}{H}}$. Hence $N([0,T],\varepsilon) \ge 1 \vee T(2g_T^*\varepsilon)^{-1/H}$.

For the function $\varphi(\varepsilon) = \left(\log\left(1 \vee T(2g_T^*\varepsilon)\right)^{-\frac{1}{H}}\right)^{\frac{1}{2}} \cdot \varepsilon$, with $\varepsilon > 0$, it holds that

$$\max_{\varepsilon < T^H(2g_T^*)^{-1}} \varphi(\varepsilon) = \frac{1}{2} e^{-\frac{1}{2}} T^H (2g_T^*)^{-1},$$

whence the claim follows. □

(iii) *Lower bound on nonrandom interval,* $H \in (0, \frac{1}{2})$. This case is very simple, due to inequality (1.9.1). As an immediate consequence, we obtain the following statement (see also Corollary 1.9.3). Let $f : [0,T] \to \mathbb{R}$ be a measurable function.

Theorem 1.10.5. *For any $p > 0$, there exists a constant $C(H,p)$ such that*

$$\|I_T^*\|_p \ge C(H,p) \|f\|_{L_{\frac{1}{H}}[0,T]}.$$

(iv) *Upper bound on nonrandom interval,* $H \in (0, \frac{1}{2})$.

1.10 Maximal Inequalities for Wiener Integrals w.r.t. fBm

Theorem 1.10.6. *Let $f : [0,T] \to \mathbb{R}$, $f \in L_p[0,T] \cap D_p^H[0,T]$ for some $p > \frac{1}{H}$, where $D_p^H[0,T] = \{f : [0,T] \to \mathbb{R} \mid \int_0^T (\int_x^T \varphi(x,t)dt)^p dx < \infty\}$ and $\varphi(x,t) = \frac{|f(t)-f(x)|}{(t-x)^{1-\alpha}} \cdot \mathbf{1}_{\{0<x<t\leq T\}}$. Then there exists a constant $C_1(H,p)$, such that*

$$\|I_T^*\|_p \leq C_1(H,p) G_p^1(0,T,f), \qquad (1.10.4)$$

where

$$G_p^1(0,T,f) := \left(\|f\|_{L_p[0,T]} \cdot T^{H-\frac{1}{p}} + T^{\frac{1}{2}-\frac{1}{p}} \left(\int_0^T \left(\int_x^T \varphi(x,t)dt \right)^p dx \right)^{\frac{1}{p}} \right).$$

Proof. According to (1.10.2), it holds that

$$\mathbf{E}(I_T^*)^p \leq (4\sqrt{2}D)^p + p2^p \sigma^p C_1(p) + p2^p (4\sqrt{2}D)^{p-1} \sigma C_1(1), \qquad (1.10.5)$$

where $\sigma^2 = \sup_{0 \leq t \leq T} \mathbf{E} I_t^2$, $D = D(T, \frac{\sigma}{2}) = \int_0^{\frac{\sigma}{2}} \mathcal{H}([0,T], u)^{1/2} du$ and $C_1(p) = \int_0^\infty x^{p-1}(1-\Phi(x))dx$. Further, from (1.9.10), we have that

$$\sigma \leq C(H,p) \left(\|f\|_{L_p[0,T]} \cdot T^{H-1/p} + \left(\int_0^T \left(\int_x^T \varphi(x,t)dt \right)^2 dx \right)^{1/2} \right)$$

$$\leq C(H,p) \left(\|f\|_{L_p[0,T]} T^{H-1/p} + T^{1/2-1/p} \left(\int_0^T \left(\int_x^T \varphi(x,t)dt \right)^p dx \right)^{1/p} \right). \qquad (1.10.6)$$

From Lemma 1.10.1 it follows that

$$(\rho_I(s,t))^p \leq 2^{p-1} C^p(H,p) \left(\int_s^t |f(u)|^p du \cdot T^{pH-1} \right.$$
$$\left. + \int_s^t \left(\int_x^T \varphi(x,t)dt \right)^p dx \cdot T^{p/2-1} \right).$$

So, we can put $\varphi(x) = x^p$,

$$g(u) = 2^p C^p(H,p) \left(|f(u)|^p \cdot T^{pH-1} + \left(\int_u^T \varphi(u,t)dt \right)^p \cdot T^{p/2-1} \right),$$

and obtain the estimate

$$\mathcal{N}([0,T], u) \leq 1 + \frac{\int_0^T g(v)dv}{\varphi(2u)} + C^p(H,p)u^{-p}\left(\int_0^T |f(v)|^p dv \cdot T^{pH-1} \right.$$
$$\left. + \int_0^T \left(\int_v^T \varphi(v,t)dt \right)^p dv \cdot T^{p/2-1} \right) =: 1 + u^{-p}(G_p^1(0,T,f))^p.$$

Therefore,

$$D \leq \int_0^{\frac{\sigma}{2}} \left(\log(1 + u^{-p}(G_p^1(0,T,f))^p)\right)^{1/2} du = p^{-1} G_p^1(0,T,f) \cdot C_p, \quad (1.10.7)$$

where $C_p = \int_{\log 2}^{\infty} z^{1/2} \frac{e^z}{(e^z-1)^{1/p+1}} dz$. By substituting (1.10.6) and (1.10.7) into (1.10.5), we obtain the proof. □

Remark 1.10.7. 1. Let $f \in C^\beta[0,T]$ with $\beta > -\alpha$. Then

$$\|f\|_{L_p[0,T]} \leq \|f\|_{C^\beta[0,T]} T^{1/p},$$

$$\left(\int_0^T \left(\int_x^T \varphi(x,t)dt\right)^p dx\right)^{\frac{1}{p}} \leq C \|f\|_{C^\beta[0,T]} T^{\alpha+\beta+\frac{1}{p}},$$

and

$$\|I_T^*\|_p \leq C_2(H,p) \|f\|_{C^\beta[0,T]} (T^H + TH + \beta),$$

where $I_t^* := \sup_{0 \leq s \leq t} |I_s|$.

2. Similarly to Theorem 1.10.6, we can suppose that $f \in L_p[0,T] \cap D_p^H[0,T]$ and obtain the estimate for $\|I_T^*\|_r$, $r > 0$. Indeed, the estimate (1.10.5) holds for any $r > 0$, and we obtain from (1.10.6) and (1.10.7) that

$$E(I_T^*)^r \leq (4\sqrt{2} C_p p^{-1} G_p^1(0,T,f))^r + r2^r (C(H,p) G_p^1(0,T,f))^r C_1(r)$$
$$+ r2^r (C_p p^{-1} G_p^1(0,T,f))^{r-1} \cdot C_1(1) \cdot C(H,p) G_p^1(0,T,f)$$
$$\leq (C(H,p,r))^r (G_p^1(0,T,f))^r.$$

From here $\|I_T^*\|_r \leq C(H,p,r) G_p^1(0,T,f)$, where $C(H,p,r) \leq 4\sqrt{2} C_p \cdot p^{-1} + 2C(H,p) \cdot \frac{2^{1/2-1/r}}{\pi^{1/2r}} \cdot (\Gamma(\frac{r+1}{2}))^{1/r} + r^{1/r} 2 \cdot C_p^{\frac{r-1}{r}} \cdot p^{-\frac{r-1}{r}} (C_1(1) C(H,p))^{1/r}$. Evidently, $C(H,p,r)$ can be estimated as $C(H,p,r) \leq C(H,p)(\Gamma(\frac{r+1}{2}))^{1/r}$ for some constant $C(H,p)$ depending only on H and p.

We continue now with random intervals. Let $\mathcal{F} = \{\mathcal{F}_t, t \geq 0\}$ be the natural filtration generated by the fBm B^H and let τ be any stopping time with respect to this filtration, i.e., the event $\{\tau \leq t\} \in \mathcal{F}_t$ for any $t \geq 0$.

(v) *Upper bound on random interval*, $H \in (\frac{1}{2}, 1)$. Let f be a measurable positive function on $\mathbb{R}$, $\alpha = H - \frac{1}{2}$.

Theorem 1.10.8. *Let the function $s^\alpha f(s)$ be nondecreasing on $\mathbb{R}$. Then, for any $p > 0$, there exists a constant $C(H,p)$ such that for any stopping time τ we have that*

$$\|I_\tau^*\|_p \leq C(H,p)(E((f(\tau))^{\frac{pH}{2\alpha}} \tau^{pH}))^{\frac{2\alpha}{pH}} (E(\tau^{pH}))^{\frac{1-H}{pH}}.$$

Remark 1.10.9. For a bounded positive function f with $f(x) \leq f^* < \infty$, $x \in \mathbb{R}$, we obtain that

$$\|I_\tau^*\|_p \leq C(p,H) f^* (E\tau^{pH})^{1/p}.$$

In particular, for $f(s) \equiv 1$, we obtain the upper bound from (NV98, Theorem 1.2).

Proof. Denote $Y_t = \int_0^t s^{-\alpha} dB_s^H$. Then $B_t^H = \int_0^t s^\alpha dY_s$ and $I_t = \int_0^t s^\alpha f(s) dY_s$. Integration by parts gives the following upper bound for I_t^*:

$$I_t^* = \sup_{0 \le s \le t} |I_s| = \sup_{0 \le s \le t} \left| t^\alpha f(t) Y_t - \int_0^t Y_s d(s^\alpha f(s)) \right| \le 2f(t) t^\alpha Y_t^*.$$

Now we use the representation (1.8.16) for Y_t,

$$Y_t = \widehat{C}_H \int_0^t (t-s)^\alpha dM_s^H = \alpha \widehat{C}_H \int_0^t (t-s)^{\alpha-1} M_s^H ds, \qquad (1.10.8)$$

whence $Y_t^* \le \widehat{C}_H t^\alpha (M_t^H)^*$. Here $\widehat{C}_H = C_H^{(6)} \widehat{\alpha}$, $\widehat{\alpha} = (1-\alpha)^{-1/2}$.
From these two estimates, we obtain for any $t > 0$ that
$I_t^* \le 2\widehat{C}_H t^{2\alpha} f(t) (M_t^H)^*$, and for the random stopping time τ it holds that

$$I_\tau^* \le 2\widehat{C}_H \tau^{2\alpha} f(\tau) (M_\tau^H)^*.$$

Therefore, for any $p > 0$

$$E(I_\tau^*)^p \le (2\widehat{C}_H)^p E(\tau^{2\alpha p} (f(\tau))^p ((M_\tau^H)^*)^p). \qquad (1.10.9)$$

From the Hölder inequality it follows that

$$E(\tau^{2\alpha p}(f(\tau))^p((M_\tau^H)^*)^p) \le (E(\tau^{2\alpha pq}(f(\tau))^{pq}))^{\frac{1}{q}} (E((M_\tau^H)^*)^{pr})^{\frac{1}{r}}, \qquad (1.10.10)$$

where $q = \frac{H}{2\alpha} > 1$ and $r = \frac{H}{1-H}$.
From the Burkholder–Davis–Gundy inequalities for martingales, it follows that for any $p > 0$ there exist constants $c_p, C_p > 0$, such that

$$c_p E \langle M^H \rangle_\tau^{\frac{p}{2}} \le E((M_\tau^H)^*)^p \le C_p E \langle M^H \rangle_\tau^{\frac{p}{2}}.$$

But $\langle M^H \rangle_t = t^{1-2\alpha}$, and

$$E((M_\tau^H)^*)^p \le C_p E \tau^{p(1-H)}.$$

Therefore,
$$E((M_\tau^H)^*)^{pr} \le C_{pr} E \tau^{pH}, \qquad (1.10.11)$$

and the proof follows from (1.10.8)–(1.10.11) with

$$C(H,p) = (2\widehat{C}_H)^p C_{pr}^{\frac{1}{p}}, \quad r = \frac{H}{1-H}.$$

□

(vi) *Lower bound on random interval*, $H \in (\frac{1}{2}, 1)$. Let f be, as before, a positive measurable function, $T > 0$ be fixed, $g(t) = \frac{1}{f(t)}$ and $g_T^* = \sup_{0 \le s \le T} g(s)$. In order to proceed, we need the following auxiliary result from (NV98). Denote $\xi_t := t^{2\alpha} |M_t^H|$.

Lemma 1.10.10. *For any $p > 0$ there exists a constant $c_p > 0$, such that for any stopping time τ, it holds that*

$$E(\xi_\tau^*)^p \geq c_p E \tau^{pH}. \tag{1.10.12}$$

Proof. Let $p = 2$. From the Itô formula we obtain that $\xi_t^2 = \int_0^t (s^{2\alpha} + 4\alpha s^{4\alpha-1}(M_s^H)^2) ds + 2 \int_0^t s^{4\alpha} M_s^H dM_s^H$.

Therefore, for any bounded stopping time τ, it holds that

$$E\xi_\tau^2 \geq E \int_0^\tau s^{2\alpha} ds = (2H)^{-1} E\tau^{2H}. \tag{1.10.13}$$

For arbitrary stopping time τ, we obtain by applying (1.10.13) to bounded stopping time $\tau \wedge n$, that

$$E\xi_{\tau \wedge n}^2 \geq (2H)^{-1} E(\tau \wedge n)^{2H},$$

and the Fatou lemma gives (1.10.12) with $p = 2$. Let $p < 2$. Inequality (1.10.12) with $p = 2$ means that continuous and hence predictable process $(\xi_t^*)^2$ dominates the (nonrandom) process $\varphi(t) = t^{2H}$. Then, from the Lenglart inequality, for $p < 2$, we obtain that

$$E(\xi_\tau^*)^p \geq c_p E\tau^{pH}$$

with $c_p = \frac{(2H)^{-p}(4-p)}{2-p}$ (DM82, VI, p. 113).

Finally, let $p > 2$. Set $k > 0$, $\delta > 0$ and define a process with positive values by

$$\eta_t = \delta + kt^{2H} + \xi_t^2.$$

Then, from the Itô formula, for $p > 2$, we obtain that

$$\eta_t^{\frac{p}{2}} = \delta^{\frac{p}{2}} + \int_0^t \left(\frac{p}{2} \eta_s^{\frac{p}{2}-1}((1+2kH)s^{2\alpha} + 4\alpha s^{2H-3}(M_s^H)^2) \right.$$

$$\left. + \frac{1}{2} p(p-2) \eta_s^{\frac{p}{2}-2} s^{6\alpha} (M_s^H)^2 \right) ds + \int_0^t p\eta_s^{\frac{p}{2}-1} s^{4\alpha} M_s^H dM_s^H.$$

Therefore, for any bounded stopping time τ

$$E\eta_\tau^{\frac{p}{2}} \geq \frac{p}{2} E \int_0^\tau \eta_s^{\frac{p}{2}-1}(1+2kH)s^{2\alpha} ds$$

$$\geq \frac{p}{2} E \int_0^\tau k^{\frac{p}{2}-1} s^{2H(\frac{p}{2}-1)} s^{2\alpha} ds \cdot (1+2kH) \tag{1.10.14}$$

$$\geq \frac{k^{\frac{p}{2}-1}}{2H}(1+2kH) E\tau^{pH}.$$

From the Fatou lemma, applied, for any stopping time τ, to $\tau \wedge n$, we obtain (1.10.14) for $\tau \wedge n$ and for $\delta = 0$.

So,
$$E(k\tau^{2H} + \xi_\tau^2)^{\frac{p}{2}} \geq \frac{k^{\frac{p}{2}-1}(1+2kH)}{2H} E\tau^{pH}.$$

From the inequality
$$(k\tau^{2H} + \xi_\tau^2)^{\frac{p}{2}} \leq 2^{\frac{p}{2}-1}(k^{\frac{p}{2}}\tau^{pH} + \xi_\tau^p),$$

we obtain that
$$E\xi_\tau^p \geq \left(2^{1-\frac{p}{2}} \frac{k^{\frac{p}{2}-1}(1+2kH)}{2H} - k^{\frac{p}{2}}\right) E\tau^{pH}.$$

This means that (1.10.12) holds with
$$c_p = k^{\frac{p}{2}}\left(2^{1-\frac{p}{2}} \frac{(\frac{1}{k}+2H)}{2H} - 1\right) > 0$$

for $k < \frac{1}{H(2^{\frac{p}{2}}-2)}$. □

Now we are in a position to establish the lower bound on a random interval for $H \in (\frac{1}{2}, 1)$.

Theorem 1.10.11. *Let, for any $t \in [0,T]$, the function $\varphi(s) := s^{-\alpha}(t-s)^{-\alpha} g(s)$ be nondecreasing on $[0,t]$. Then, for any $p > 0$, there exists a constant $c(H,p) > 0$, such that for any stopping time $\tau \leq T$ it holds that*
$$\|I_\tau^*\|_p \geq c(H,p)(g_T^*)^{-1}(E\tau^{pH})^{1/p}.$$

Remark 1.10.12. Either of the following conditions (a) and (b) is sufficient for Theorem 1.10.11:

(a) $g \in C^1[0,T]$ and for any $s \in (0,T)$, it holds that $g'(s) \geq g(s)(\frac{\alpha}{s} - \frac{\alpha}{T-s})$.
(b) The function $g(s)s^{-\alpha}$ is nondecreasing on $[0,T]$ (or the function $f(s)s^\alpha$ is nonincreasing on [0, T]; compare with the condition of Theorem 1.10.8).

Remark 1.10.13. The class of functions satisfying the condition of Theorem 1.10.11 is nonempty. For example, $f(s) = s^{-\gamma} e^{-\beta s}$ with $\gamma \geq \alpha$ and $\beta \geq 0$ belongs to this class. (In this case assumption (b) is satisfied.)

Proof. Let $0 < a < b < 1$. Then the martingale M_t^H can be represented as
$$M_t^H = \int_0^{at} l_H(t,s) dB_s^H + \int_{at}^{bt} l_H(t,s) dB_s^H + \int_{bt}^{t} l_H(t,s) dB_s^H$$
$$:= M_t^H(a) + \int_{at}^{bt} l_H(t,s) g(s) dI_s + M_t^H(1-b). \quad (1.10.15)$$

The middle term can be integrated by parts, and we obtain from the condition of the theorem that

$$\left| \int_{at}^{bt} l_H(t,s)g(s)dI_s \right|$$

$$= \left| l_H(t,bt)g(bt)I_{bt} - l_H(t,at)g(at)I_{at} - \int_{at}^{bt} I_s d(l_H(t,s)g(s)) \right| \qquad (1.10.16)$$

$$\leq C_H^{(5)} I_t^* g_t^* t^{-2\alpha}((1-b)^{-\alpha}b^{-\alpha} + (1-a)^{-\alpha}a^{-\alpha}).$$

Therefore, the process $\xi_t = t^{2\alpha}|M_t^H|$ can be estimated as $\xi_t \leq t^{2\alpha}|M_t^H(a) + M_t^H(1-b)| + C_H I_t^* g_t^*$, where $C_H = 2C_H^{(5)}(((1-b)^{-\alpha})b^{-\alpha} + (1-a)^{-\alpha}a^{-\alpha})$. Now we use Lemma 1.10.10 and obtain

$$C_p E \tau^{pH} \leq E(\xi_\tau)^p$$
$$\leq 2^{p-1} E \tau^{2p\alpha}|M_\tau^H(a) + M_\tau^H(1-b)|^p + 2^{p-1} C_H^{p-1} E(I_\tau^*)^p (g_\tau^*)^p. \qquad (1.10.17)$$

Further, from (1.10.8) we have that

$$|M_t^H(a)| \leq C_H^{(5)} \left| (t(1-a))^{-\alpha} Y_{at} - \int_0^{at} Y_s d(t-s)^{-\alpha} \right|$$

$$\leq C_H^{(5)} Y_{at}^* \cdot 2(t(1-a))^{-\alpha} \leq 2C_H^{(5)} \widehat{C}_H \frac{a^\alpha}{(1-a)^\alpha} (M_{at}^H)^*.$$

Hence

$$|M_\tau^H(a)| \leq C_H \frac{a^\alpha}{(1-a)^\alpha} (M_\tau^H)^*, \qquad (1.10.18)$$

where $C_H = 2C_H^{(5)} \widehat{C}_H$.

In order to estimate $M_t^H(1-b)$, note at first that for fixed t, the process $\widetilde{B}_s^H := B_t^H - B_{t-s}^H$, $0 \leq s \leq t$, is a fractional Brownian motion with Hurst index H. Therefore,

$$M_t^H(1-b) = C_H^{(5)} \int_{tb}^t (t-s)^{-\alpha} s^{-\alpha} dB_s^H = C_H^{(5)} \int_0^{t(1-b)} u^{-\alpha}(t-u)^{-\alpha} d\widetilde{B}_u^H,$$

and similarly as in the above estimates (1.10.15), we obtain that

$$|M_\tau^H(1-b)| \leq C_H \left(\frac{1-b}{b} \right)^\alpha (\widetilde{M}_\tau^H)^*, \qquad (1.10.19)$$

where $\widetilde{M}^H$ is the Molchan martingale for $\widetilde{B}_H$. But the symmetry of the kernel $l_H(t,s)$ leads to the equality $\widetilde{M}_t^H = \int_0^t l_H(t,s) dB_{t-s}^H = \int_0^t l_H(t,s) dB_s^H = M_t^H$. Hence,

$$|M_\tau^H(1-b)| \leq C_H (\frac{1-b}{b})^\alpha (M_\tau^H)^*. \qquad (1.10.20)$$

From (1.10.7), (1.10.18), (1.10.20), (1.10.10) and (1.10.11) with $f \equiv 1$ we obtain that

1.10 Maximal Inequalities for Wiener Integrals w.r.t. fBm

$$2^{p-1}C_H^{p-1}E(I_\tau^*)^p \cdot (g_\tau^*)^p$$
$$\geq C_p E\tau^{pH} - 2^{p-1}C_H E\tau^{2p\alpha} \cdot (M_\tau^*)^p \left(\frac{a^\alpha}{(1-a)^\alpha} + \frac{(1-b)^\alpha}{b^\alpha}\right) E\tau^{pH} \cdot c_p.$$

By choosing a sufficiently small and b close to 1, we obtain that

$$E(I_\tau^*)^p \geq (g_T^*)^{-p} E\tau^{pH} \cdot C_{p,H},$$

where

$$C_{p,H} = 2^{1-p}C_H^{1-p}\left[C_p - 2^{p-1}C_H c_p \left(\frac{a^\alpha}{(1-a)^\alpha} + \frac{(1-b)^\alpha}{b}\right)\right] > 0. \quad \square$$

(vii) *Upper and lower bounds for power functions and* $H \in (\frac{1}{2}, 1)$.
The function $f(s) \equiv 1$ does not satisfy the condition of Theorem 1.10.11. To cover this case, we consider the power functions $f(s) = s^\gamma$, $\gamma > -2\alpha$, and obtain a better result than in Theorems 1.10.11 and 1.10.6:

Theorem 1.10.14. *Let* $f(s) = s^\gamma$ *with* $\gamma > -2\alpha$. *Then, for any* $p > 0$, *there exist constants* $c_{p,H}$ *and* $C_{p,H}$, *such that for any stopping time* τ *it holds that*

$$c_{p,H}(E\tau^{p(H+\gamma)})^{1/p} \leq \|I_\tau^*\|_p \leq C_{p,H}(E\tau^{p(H+\gamma)})^{1/p}.$$

Proof. Consider the upper bound. Now inequality (1.10.4) has the form

$$E(I_\tau^*)^p \leq (2C_H^{(6)})^p E(\tau^{(2\alpha+\gamma)p} M_\tau^*)^p.$$

By applying Hölder's inequality with $q = \frac{1+2\alpha+2\gamma}{4\alpha+2\gamma} > 1$ and $r = \frac{1+2\alpha+2\gamma}{1-2\alpha} = \frac{H+\gamma}{1-H}$, and the Burkholder–Davis–Gundy inequalities, we obtain that

$$E(I_\tau^*)^p \leq (2C_H^{(6)})^p (E\tau^{\frac{1+2\alpha+2\gamma}{2}p})^{\frac{1}{q}} (E(M_\tau^*)^{pr})^{\frac{1}{r}}$$
$$\leq (2C_H^{(6)})^p C_{p,H}(E\tau^{(H+\gamma)p})^{\frac{1}{q}}(E\tau^{(H+\gamma)p})^{\frac{1}{r}} = C_{p,H} E\tau^{(H+\gamma)p}.$$

Consider the lower bound. We use expansion (1.10.15) and estimate its middle term similarly to the first part of (1.10.16) with $g(s) = s^{-\gamma}$:

$$\left|\int_{at}^{bt} l_H(t,s)g(s)dI_s\right|$$
$$= |l_H(t,bt)g(bt)I_{bt}| + |l_H(t,at)g(at)I_{at}| + \left|\int_{at}^{bt} I_s d(P_H(t,s)g(s))\right|$$
$$\leq C_H^{(5)} b^{-\gamma-\alpha}(1-b)^{-\alpha} t^{-2\alpha-\gamma} I_t^* + C_H^{(5)} a^{-\gamma-\alpha}(1-a)^{-\alpha} I_t^* \cdot t^{-2\alpha-\gamma}$$
$$+ C_H^{(5)} I_t^* \int_{at}^{bt} |d((t-s)^{-\alpha} s^{-\alpha-\gamma})|.$$

The function $\varphi(s) := (t-s)^{-\alpha} s^{-\alpha-\gamma}$ has the following derivative on (at, bt):

$$\varphi'(s) = s^{-\alpha-\gamma-1}(t-s)^{-\alpha-1}((\gamma+2\alpha)s - (\gamma+\alpha)t).$$

For $\gamma > -\alpha$, on the interval $[0, t]$, the function $\varphi(s)$ has an extremal point $s_{\max} = \rho t$, where $\rho = \frac{\gamma+\alpha}{\gamma+2\alpha}$, and for $-2\alpha < \gamma < -\alpha$, no extremal point exists. Therefore, the variation of $\varphi(s)$ on the interval $[at, bt]$ can be estimated as

$$\int_{at}^{bt} |d((t-s)^{-\alpha} s^{-\alpha-\gamma})|$$
$$\leq t^{-2\alpha-\gamma}\left(b^{-\gamma-\alpha}(1-b)^{-\alpha} + 2|\rho|^{-\gamma-\alpha}(1-\rho)^{-\alpha} + a^{-\gamma-\alpha}(1-a)^{-\alpha}\right).$$

From here,

$$\left|\int_{at}^{bt} l_H(t,s) s^{-\gamma} dI_s\right| \leq C(a,b,H,\gamma) t^{-2\alpha-\gamma} I_t^*,$$

where

$$C(a,b,H,\gamma) = 2C_H^{(5)}\left(b^{-\gamma-\alpha}(1-b)^{-\alpha} + a^{-\gamma-\alpha}(1-a)^{-\alpha} + |\rho|^{-\gamma-\alpha}(1-\rho)^{-\alpha}\right).$$

Therefore, for the process $\widetilde{\xi}_t := t^{2\alpha+\gamma}|M_t^H|$, we have that

$$\widetilde{\xi}_t \leq t^{2\alpha+\gamma}|M_t^H(a) + M_t^H(1-b)| + C(a,b,H,\gamma) I_t^*,$$

whence for any stopping time τ and $p > 0$, it holds that

$$E(\widetilde{\xi}_\tau)^p \leq (C(a,b,H,\gamma))^p E(I_\tau^*)^p + E(\tau^{2\alpha+\gamma}|M_\tau^H(a) + M_\tau^H(1-b)|)^p. \quad (1.10.21)$$

Similarly to Lemma 1.10.10, we can establish the following bound for $\widetilde{\xi}_\tau$:

$$E(\widetilde{\xi}_\tau)^p \geq c_p E \tau^{p(H+\gamma)}.$$

Further, we apply (1.10.11), the bounds (1.10.15) and (1.10.17), and Hölder's inequality with $q = \frac{1+2\alpha+2\gamma}{4\alpha+2\gamma} > 1$ and $r = \frac{1+2\alpha+2\gamma}{1-2\alpha} > 1$, where $\frac{1}{q} + \frac{1}{r} = 1$, and obtain the bounds of the pth moment of $\tau^{2\alpha+\gamma} M_\tau^H(a)$ and $\tau^{2\alpha+\gamma} M_\tau^H(1-b)$:

$$E(\tau^{2\alpha+\gamma} M_\tau^H(a))^p \leq (C_H)^p \frac{a^{\alpha p}}{(1-a)^{\alpha p}} E((\tau)^{2\alpha+\gamma}(M_\tau^H)^*)^p$$
$$\leq \widetilde{C}_H (E\tau^{(2\alpha+\gamma)pq})^{\frac{1}{q}} (E((M_\tau^H)^*)^{pr})^{\frac{1}{r}} \leq \widetilde{C}_H E\tau^{p(H+\gamma)}, \quad (1.10.22)$$

where $\widetilde{C}_H = (C_H)^p \left(\frac{\alpha}{1-\alpha}\right)^{\alpha p}$. Similarly,

$$E(\tau^{2\alpha+\gamma} M_\tau^H(1-b))^p \leq \widehat{C}_H E\tau^{p(H+\gamma)}, \quad (1.10.23)$$

1.10 Maximal Inequalities for Wiener Integrals w.r.t. fBm

$$\widehat{C}_H = (C_H)^p \left(\frac{1-b}{b}\right)^{\alpha p}.$$

From (1.10.21)–(1.10.23)

$$E(I_\tau^*)^p \geq c_{p,H} E\tau^{p(H+\gamma)},$$

where

$$c_{p,H} = \left(\frac{c_p - (C_H)^p \left(\left(\frac{a}{1-a}\right)^{\alpha p} + \left(\frac{1-b}{b}\right)^{\alpha p}\right)}{(C(a,b,H,\gamma))^p}\right)^{\frac{1}{p}} > 0$$

for sufficiently small a and $1-b$. □

(viii) *Lower bound on random interval, $H \in (0, \frac{1}{2})$.*

Here, we consider only power functions $f(s) = s^\gamma$, $s > 0$. According to (1.8.6), the integral $\int_0^t s^\gamma dB_s^H$ exists, if

$$\int_0^t \int_0^t u^\gamma s^\gamma \left(\int_{-\infty}^0 (u-x)^{\alpha-1}(s-x)^{\alpha-1} dx\right) du\, ds < \infty$$

and

$$\int_0^t \int_0^t u^{\gamma-1} s^{\gamma-1} \left(\int_0^{s \wedge u} (u-x)^\alpha (s-x)^\alpha dx\right) du\, ds < \infty.$$

If we choose $\gamma > -H$, then both of these inequalities hold.

Theorem 1.10.15. *Let $H \in (0, \frac{1}{2})$ and $f(s) = s^\gamma$ with $\gamma \in (-H, -\alpha)$. Then, for any $p > 0$, there exists a constant $c(H, p)$ such that*

$$\|I_t^*\|_p \geq c(H,p)(E\tau^{p(H+\gamma)})^{1/p}.$$

Proof. We estimate the Molchan martingale from above:

$$(M_t^H)^* = C_H^{(5)} \left(\int_0^t s^{-\alpha-\gamma}(t-s)^{-\alpha} dI_s\right)^* \leq C_H^{(5)} I_t^* \int_0^t |d(s^{-\alpha-\gamma}(t-s)^{-\alpha})|.$$

The last integral exists when $-\alpha - \gamma > -1$ or $\gamma < 1 - \alpha$. As before, the derivative of the function $\varphi(s) = s^{-\alpha-\gamma}(t-s)^{-\alpha}$, $s \in (0,t)$, equals

$$\varphi'(s) = s^{-\alpha-\gamma-1}(t-s)^{-\alpha-1}((\gamma + 2\alpha)s - (\gamma + \alpha)t).$$

So, for $\gamma \in (-H, -\alpha)$, the function $\varphi(s)$ has the unique extremal point $s = \frac{\gamma+\alpha}{\gamma+2\alpha} t$, and $\int_0^t |d(s^{-\alpha-\gamma}(t-s)^{-\alpha})| \leq C_\alpha t^{-2\alpha-\gamma}$, where

$$C_\alpha := \left(\frac{\alpha}{\gamma+2\alpha}\right)^{-\alpha} \left(\frac{\gamma+\alpha}{\gamma+2\alpha}\right)^{-\alpha-\gamma}.$$

Hence, for any stopping time τ and any $\widetilde{p} > 0$ it holds that

$$((M_\tau^H)^*)^{\widetilde{p}} \leq (C_H^{(5)}C_\alpha)^{\widetilde{p}}(I_\tau^*)^{\widetilde{p}}\tau^{(-2\alpha-\gamma)\widetilde{p}}.$$

Further, from the Burkholder–Davis–Gundy inequalities we obtain that

$$E(M_\tau^*)^{\widetilde{p}} \geq \widetilde{C}_{\widetilde{p}} E\tau^{\widetilde{p}(1-H)}.$$

Hence,

$$\widetilde{C}_{\widetilde{p}} E\tau^{\widetilde{p}(1-H)} \leq \widetilde{C}(H,\widetilde{p})(E(I_\tau^*)^{\widetilde{p}q})^{\frac{1}{q}} \cdot (E\tau^{(-2\alpha-\gamma)\widetilde{p}r})^{\frac{1}{r}},$$

where $\widetilde{C}(H,\widetilde{p}) = (C_H^{(5)}C_\alpha)^{\widetilde{p}}$.

Now, we choose $r = \frac{1-H}{1-2H-\gamma} > 1$, $q = \frac{1-H}{H+\gamma} > 1$ and $\widetilde{p} = \frac{p(H+\gamma)}{1-H}$, and obtain for

$$c(H,p) = \left(\frac{\widetilde{C}_{\widetilde{p}}}{\widetilde{C}(H,\widetilde{p})}\right)^q$$

that $c(H,p)E\tau^{p(H+\gamma)} \leq E(I_\tau^*)^p$. □

1.11 The Conditions of Continuity of Wiener Integrals with Respect to fBm

Consider the case $H \in (\frac{1}{2},1)$. Let $f \in L_{\frac{1}{H}}[0,t]$, $t \in [0,T]$. Then in particular, the integral $I_t(f) = \int_0^t f(s)dB_s^H$ exists on $[0,T]$ and $E(I_t(f))^2 = \|f\|^2_{L_2^H[0,t]} \leq C_H\|f\|^2_{L_{\frac{1}{H}}[0,t]}$. According to (Lif95), a sufficient condition for the continuity of separable modification of $I_t(f)$ on $[0,T]$ is the finiteness of the Dudley integral $\int_0^\varepsilon H([0,T],u)^{\frac{1}{2}}du$. But in our case, from (1.10.3) with ε instead of $\frac{\sigma}{2}$, it follows that

$$\int_0^\varepsilon \mathcal{H}([0,T],u)^{\frac{1}{2}}du \leq \int_0^\varepsilon \left(\log\left(1+u^{-\frac{1}{H}}\widetilde{C}_H \int_0^T |f(u)|^{\frac{1}{H}}du\right)\right)^{\frac{1}{2}}du$$

$$\leq \int_0^\varepsilon u^{-\frac{1}{2H}}du \cdot \left(\widetilde{C}_H \int_0^t |f(u)|^{\frac{1}{H}}du\right)^{\frac{1}{2}} < \infty$$

for $H \in (\frac{1}{2},1)$.

This means that the separable modification of the Wiener integral w.r.t. fBm with $H \in (\frac{1}{2},1)$ is continuous if $f \in L_{\frac{1}{H}}[0,T]$.

Now, let $H \in (0,1/2)$. Then, according to (1.10.7) with ε instead of $\frac{\sigma}{2}$, we have that $\int_0^\varepsilon \mathcal{H}([0,T],u)^{1/2}du$ is finite for any $f \in L_p[0,T] \cap D_p^H[0,T], p > \frac{1}{H}$. So, for such f, a separable modification of $I_t(f)$ is continuous on $[0,T]$.

1.12 The Estimates of Moments of the Solution of Simple Stochastic Differential Equations Involving fBm

(i) Let $H \in (\frac{1}{2}, 1)$ and $\mathcal{F}_t = \sigma\{B_s^H, 0 \leq s \leq t\}$.
Consider a stochastic differential equation of the form

$$dX_t = b(t, X_t)dt + f(t)dB_t^H, \quad t \geq 0. \tag{1.12.1}$$

Here, X_0 is $\mathcal{F}_0$-measurable random variable, $E|X_0|^{p_0} < \infty$ for some $p_0 > 1$ and $b(t,x) : \mathbb{R}_+ \times \mathbb{R} \longrightarrow \mathbb{R}$ is a measurable Lipschitz function, i.e.

$$|a(t,x) - a(t,y)| \leq C|x-y| \tag{1.12.2}$$

with some constant C. Furthermore, b is of linear growth, meaning that

$$|b(t,x)| \leq C(1+|x|) \tag{1.12.3}$$

and

$$f \in L_{\frac{1}{H}}[0,T]. \tag{1.12.4}$$

Theorem 1.12.1. *Let b satisfy (1.12.2), (1.12.3) and f satisfy (1.12.4). Then equation (1.12.1) has a unique solution.*

Proof. We establish now that for any $p \leq p_0$ the map

$$(AX)_t := X_0 + \int_0^t b(s, X_s)ds + I_t(f)$$

is a contraction in the space

$$S_p := \left\{ \xi(t,\omega), \; t \in [0, T_p] \,\middle|\, \xi(t,\cdot) \text{ is } \mathcal{F}_t\text{-measurable}, \; \sup_{t \in [0, T_p]} E|\xi_t|^p < \infty \right\},$$

with the norm

$$\|\xi\|_{S_p} := \sup_{t \in [0, T_p]} (E|\xi_t|^p)^{\frac{1}{p}},$$

where T_p is a number such that $T_p < C^{-1}$.
Indeed, from (1.12.2)–(1.12.4) it follows that

$$E|(AX)_t|^p \leq 3^p \left(E|X_0|^p + E|I_t(f)|^p + 2^p t^{p-1} C \left(1 + \int_0^t E|X_s|^p ds\right) \right).$$

This means that $AX \in S_p$ if $X \in S_p$. Further, for $t \leq T_p$

$$E|(AX)_t - (AY)_t|^p \leq E \left| \int_0^t (b(s, X_s) - b(s, Y_s))ds \right|^p$$

$$\leq C^p E \left(\int_0^t |X_s - Y_s| ds \right)^p \leq C^p T_p^{p-1} E \int_0^t |X_s - Y_s|^p ds,$$

56 1 Wiener Integration with Respect to Fractional Brownian Motion

i.e., $\|AX - AY\|_{S_p} \leq L \|X - Y\|_{S_p}$, where $L = C^p T_p^p < 1$. Therefore, on the interval $[0, T_p]$ equation (1.12.1) has unique solution. If we obtain this solution X_t by the method of successive approximations, and the initial process is some continuous process $X_s^{(0)} \in S_p$, then by the continuity of the process $I(b)$ and the equicontinuity of the integral $\int_0^t b(s,\cdot)ds$, the solution X_t is continuous on $[0, T_p]$. The proof of the theorem is obtained by extension of the solution from $[0, kT_p]$ to $[0, (k+1)T_p]$ via the relation

$$X_t = X_{kT_p} + \int_{kT_p}^t b(s, X_s)ds + (I_t - I_{kT_p}), \qquad (1.12.5)$$

where $k \in N$ and X_{kT_p} is the solution of the "previous" equation taken at the point $t = kT_p$. Existence, uniqueness and continuity of the solution of (1.12.5) is established similarly to previous estimates. □

Now we establish the upper bound for the solution of equation (1.12.1) on a random interval.

Theorem 1.12.2. *Let the functions b and f satisfy the conditions of Theorem 1.12.1, $E|X_0|^p < \infty$ for any $p > 0$ and the function $s^\alpha f(s)$ be nondecreasing on $\mathbb{R}$. Then,*

(a) for any $T > 0$, $p > 0$ and stopping time $\tau \in [0, T]$, we have the estimate

$$E(X_\tau^*)^p \leq 4^p e^{4^p C^p T^{p-1}} (E|X_0|^p + C^p E\tau^p)$$
$$+ (C(H,p))^p (E((f(\tau))^{\frac{pH}{2\alpha}} \tau^{pH}))^{\frac{2\alpha}{H}} (E\tau^{pH})^{\frac{1-H}{H}}),$$

where a constant $C(H, p)$ appeared in Theorem 1.10.8.
(b) If, in addition, the function f is bounded, i.e. $|f(x)| \leq f^* < \infty$, then

$$E(X_\tau^*)^p \leq 4^p e^{4^p C^p T^{p-1}} \left(E|X_0|^p + C^p E\tau^p + (C(H,p))^p (b^*)^p E\tau^{pH} \right).$$

Proof. Let $\tau \in [0, T]$ and $\tau_n = \tau \wedge \inf\{t > 0 : |X_t| \geq n\}$. Then

$$(X_{\tau_n}^*)^p \leq (|X_0| + C\tau_n + C \int_0^{\tau_n} X_s^* ds + I_{\tau_n}^*(f))^p$$

$$\leq 4^p (|X_0|^p + C^p \tau_n^p + C^p \int_0^{\tau_n} (X_s^*)^p ds \cdot \tau_n^{p-1} + (I_{\tau_n}^*(f))^p).$$

Therefore, by Gronwall's inequality, we obtain that

$$(X_{\tau_n}^*)^p \leq 4^p e^{4^p C^p \tau_n^{p-1}} (|X_0|^p + C^p \tau_n^p + (I_{\tau_n}^*(f))^p).$$

Hence,

$$E(X_{\tau_n}^*)^p \leq 4^p e^{4^p C^p T^{p-1}} (E|X_0|^p + C^p E\tau_n^p + E(I_{\tau_n}^*(f))^p).$$

By applying Theorem 1.10.6, we obtain (a) and (b) for $\tau = \tau_n$, $n \geq 1$. By taking $n \to \infty$, we obtain the proof. □

Remark 1.12.3. Exponential estimates for the solution of the more simple version of equation (1.12.1), were obtained in (TV03). We shall return to this problem in Section 3.5.

1.13 Stochastic Fubini Theorem for the Wiener Integrals w.r.t fBm

We consider now only the case $H \in (1/2, 1)$. Let $\mathcal{P}_T = [0,T]^2$.

Theorem 1.13.1. *Let the measurable function $f = f(t,s) : \mathcal{P}_T \to \mathbb{R}$ satisfy the conditions*

$$\int_{[0,T]^3} |f(t,u)|\,|f(t,s)|\,|s-u|^{2\alpha-1}\,ds\,du\,dt < \infty \tag{1.13.1}$$

and

$$\int_{[0,T]^4} |f(t_1,u)|\,|f(t_2,s)|\,|s-u|^{2\alpha-1}\,ds\,du\,dt_1\,dt_2 < \infty. \tag{1.13.2}$$

Then both the repeated integrals $I_1 := \int_0^T (\int_0^T f(t,s)dt)dB_s^H$ and $I_2 := \int_0^T (\int_0^T f(t,s)dB_s^H)dt$ exist and $I_1 = I_2$ with probability 1.

Proof. The existence of the integral I_1 is evident, due to (1.13.2). As to I_2, $\int_0^T f(t,s)dB_s^H$ exists a.e. (mod λ), where λ is the Lebesgue measure, and according to (1.13.1), it holds that

$$E\int_0^T \left|\int_0^T f(t,s)dB_s^H\right|dt \leq T^{1/2}\left(E\int_0^T \left|\int_0^T f(t,s)dB_s^H\right|^2 dt\right)^{1/2}$$

$$\leq \left(T2\alpha H \int_{[0,T]^3} |f(t,s)||f(t,u)||s-u|^{2\alpha-1}du\,ds\,dt\right)^{1/2} < \infty.$$

We consider at first only the measurable and bounded functions. Let $f^* := \sup_{(t,s)\in[0,T]^2} |f(t,s)| < \infty$. Then there exists the sequence of simple and totally bounded functions $f_n = f_n(t,s)$, such that $f_n \to f$ uniformly on $\mathcal{P}_T$. The statement of the theorem is evident for f_n. Further, denote $g_n(t,s) := f(t,s) - f_n(t,s)$ and obtain the estimate

$$|I_1 - I_2| \leq \left|\int_0^T \left(\int_0^T g_n(t,s)dt\right)dB_s^H\right| + \left|\int_0^T \left(\int_0^T g_n(t,s)dB_s^H\right)dt\right|$$

$$=: I_{1n} + I_{2n}.$$

Furthermore,

$$E|I_{1n}|^2 = 2\alpha H \int_{\mathcal{P}_T} \left(\int_0^T g_n(t_1,s)dt_1\right)\left(\int_0^T g_n(t_2,u)dt_2\right)|s-u|^{2\alpha-1}ds\,du$$

$$\leq 2\alpha H T^2 \sup_{(t,s)\in[0,T]^2} |g_n(t,s)|^2 \int_{\mathcal{P}_T} |s-u|^{2\alpha-1}ds\,du$$

$$= T^{2H+2} \sup_{(t,s)\in\mathcal{P}_T} |g_n(t,s)|^2 \to 0,$$

and

$$E|I_{2n}|^2 \leq T\int_0^T E\left|\int_0^T g_n(t,s)dB_s^H\right|^2 dt \leq \sup_{(t,s)\in\mathcal{P}_T} |g_n(t,s)|^2 T^{2H+2} \to 0,$$

as $n \to \infty$, and we obtain the proof for bounded f. Now, let f satisfy (1.13.1) and (1.13.2). For $f_n(t,s) := f(t,s)\mathbf{1}_{\{|f(t,s)|\leq n\}}$, $n \geq 1$ the theorem is already proved. Define

$$C_n := \{(t,s,u) \in [0,T]^3 \mid |f(t,s)| \geq n, |f(t,u)| \geq n\}, \quad \overline{f}_n = f - f_n.$$

Then for any $n \geq 1$ we have that

$$|I_1 - I_2| \leq \left|\int_0^T \left(\int_0^T f(t,s)\mathbf{1}_{\{|f(t,s)|>n\}}dt\right)dB_s^H\right|$$
$$+ \left|\int_0^T \left(\int_0^T f(t,s)\mathbf{1}_{\{|f(t,s)|>n\}}dB_s^H\right)dt\right| =: I'_{1n} + I'_{2n}.$$

Furthermore, we have that

$$E|I'_{1n}|^2 = 2\alpha H \int_{[0,T]^2} \left(\int_0^T \overline{f}_n(t_1,s)dt_1\right)\left(\int_0^T \overline{f}_n(t_2,s)dt_2\right)|s-u|^{2\alpha-1}ds\,du$$

$$\leq 2\alpha H \int_{[0,T]^4} |\overline{f}_n(t_1,s)||\overline{f}_n(t_2,s)||s-u|^{2\alpha-1}ds\,du\,dt_1\,dt_2 \to 0,$$

as $n \to \infty$, according to (1.13.2), and

$$E|I'_{2n}|^2 \leq T2\alpha H \int_{[0,T]^3} |\overline{f}_n(t,s)||\overline{f}_n(t,u)||s-u|^{2\alpha-1}ds\,du\,dt \to 0,$$

as $n \to \infty$, according to (1.13.1). $\square$

1.14 Martingale Transforms and Girsanov Theorem for Long-memory Gaussian Processes

According to Section 1.8, the process

1.14 Martingale Transforms and Girsanov Theorem

$$M_t^H := C_H^{(5)} \int_0^t s^{-\alpha}(t-s)^{-\alpha} dB_s^H$$

is a square integrable martingale, and $B_t := \widehat{\alpha} \int_0^t s^\alpha dM_s^H$ is a Wiener process. In turn, $B_t^H = C_H^{(6)} \int_0^t m_H(t,s) dB_s$. Moreover, the process

$$Y_t = C_H^{(6)} \int_0^t (t-s)^\alpha s^{-\alpha} dB_s \qquad (1.14.1)$$

has the property that $M_t^H = C_H^{(5)} \int_0^t (t-s)^{-\alpha} dY_s$ is square-integrable martingale. All these processes are Gaussian. Therefore, in some sense, it is more convenient to consider the processes of a form similar to Y_t and M_t, and to avoid fractional Brownian motion itself. In this section we consider long-memory Gaussian processes that can be presented as integrals $V_t = \int_0^t h(t-s)\varphi(s) dW_s$ with some Wiener process W_t and establish the conditions allowing us to transform these processes, similarly to Y_t, into square-integrable martingales.

Let $\{W_t, \mathcal{F}_t^W, t \geq 0\}$ be the standard Wiener process on a complete probability space $(\Omega, \mathcal{F}, P)$ with $\mathcal{F} = \mathcal{F}_\infty := \bigvee_{t \geq 0} \mathcal{F}_t^W$. Define the convolution of two measurable integrable functions φ_1 and $\varphi_2 : \mathbb{R}_+ \to \mathbb{R}$ by $(\varphi_1 * \varphi_2)(t) = \int_0^t \varphi_1(t-s)\varphi_2(s) ds, t \in \mathbb{R}_+$. Let h and φ satisfy the assumption

$$\varphi \in L_2(0,t), \quad (h^2 * \varphi^2)_t < \infty, \quad t > 0. \qquad (1.14.2)$$

Define the Gaussian process $V_t = \int_0^t h(t-s)\varphi(s) dW_s$. Evidently, $EV_t = 0$. In the case when $h(s) = s^\alpha, \varphi(s) = s^{-\alpha}$ and $H \in (1/2, 1)$, the covariance function between distant increments of the process V_t vanishes at a power rate. More precisely,

$$EV_t(V_{t+k} - V_k) = \int_0^t (t-s)^\alpha ((t+k-s)^\alpha - (k-s)^\alpha) s^{-2\alpha} ds$$
$$\geq \alpha t \int_0^t (t-s)^\alpha (t+k-s)^{\alpha-1} s^{-2\alpha} ds$$
$$\geq \alpha t^{2-\alpha} B(\alpha+1, \alpha) k^{\alpha-1},$$

and the series $\sum_{k=1}^\infty k^{\alpha-1}$ diverges for $H \in (1/2, 1)$. Due to this reason, according to the generally accepted terminology (CCM03; Ber94; WTT99), such processes are said to have a long memory. Compare this to the notion of long-range dependence from Section 1.2.

Denote by $\mathbb{R}_{uv} = EV_u V_v$ the correlation function. Then we have that

$$\mathbb{R}_{uv} = \int_0^{u \wedge v} h(u-s) h(v-s) \varphi^2(s) ds.$$

Let $\mathcal{F}_t^X = \sigma\{X_s, 0 \leq s \leq t\}$ and $\mathcal{H}_t^X = \mathcal{H}\{X_s, 0 \leq s \leq t\}$ be, correspondingly, σ-fields and Gaussian subspaces, generated by the process X

on the interval $(0,t]$, $X = W, V$. It follows from (CCM03, Proposition 15) that $\mathcal{F}_t^V = \mathcal{F}_t^W$, $t \in \mathbb{R}_+$ if and only if $\mathcal{H}_t^V = \mathcal{H}_t^W$. A necessary and sufficient condition for this coincidence can be formulated as

$$\text{the only function } f \text{ such that } \forall t \in \mathbb{R}_+ \\ f \in L_2(0,t) \text{ and } ((f \cdot \varphi) * h)_t = 0 \text{ is the zero function.} \tag{1.14.3}$$

Evidently, in this case $\mathcal{F}_\infty^V = \mathcal{F}_\infty^W$. We give one sufficient condition for the latter relation. Denote by

$$F_f(\lambda) := \int_0^\infty e^{-\lambda s} f(s) \, ds, \quad \lambda > 0$$

the Laplace transform of f. The following result is a direct consequence of (CCM03, Proposition 17).

Lemma 1.14.1. *Let the following condition hold*

$$0 < |F_h(\lambda)| < \infty, \quad F_{|\varphi|}(\lambda) < \infty, \quad F_\varphi(\lambda) \neq 0 \tag{1.14.4}$$

on some interval $\lambda \in (a,b) \subset (0,\infty)$. *Then* $\mathcal{F}_\infty^V = \mathcal{F}_\infty^W$.

Now, let (1.14.3) hold. Denote by $L_2(V) = L_2(W) = L_2(\Omega, \mathcal{F}_\infty, P)$ the space of $\mathcal{F}_\infty$-measurable ξ with $E\xi^2 < \infty$. Let $\mathcal{H}(V)$ be the closed subspace of $L_2(V)$ consisting of linear functionals of V. Suppose that the function $R : \mathbb{R}_+^2 \to \mathbb{R}$ has a bounded variation $|R|_t := \text{var}_{\mathcal{P}_t} R$ on any rectangle $\mathcal{P}_t$, $t \in \mathbb{R}_+^2$, and consider the measurable function $g : \mathbb{R}_+ \to \mathbb{R}$ such that

$$\int_{\mathcal{P}_{(s,t)}} |g(s-u)| \, |g(t-v)| \, d\,|R|_{uv} < \infty, \quad s, t \in \mathbb{R}_+. \tag{1.14.5}$$

As stated by (HC78), we have an isomorphism I between $\Lambda_2(R)$ and $\mathcal{H}(V)$. Here $\Lambda_2(R)$ is the completion of the space Λ of step functions $f(t) = \sum_{k=1}^N \alpha_k \mathbf{1}_{[t_{k+1}, t_k)}(t)$ in the norm generated by a scalar product

$$\langle f, g \rangle = \int_\mathbb{R} f(u) \, g(v) \, dR_{uv}, \quad I(f) = \sum_{k=1}^N \alpha_k (V_{t_{k+1}} - V_{t_k}).$$

Denote by $I(f) = \int_\mathbb{R} f \, dV \in \mathcal{H}(V)$ the image of $f \in \Lambda_2(\mathbb{R})$ and let

$$M_t := \int_0^t g(t-u) \, dV_u := I(\tilde{g}),$$

where $\tilde{g}(s) = g(t-s) \mathbf{1}_{\{s \leq t\}}$, $t \geq 0$. Then $\{M_t, \mathcal{F}_t^W, t \geq 0\}$ is a Gaussian process and

$$E M_s M_t = \int_{\mathcal{P}_{(s,t)}} g(s-u) \, g(t-v) \, dR_{uv}.$$

Moreover, under the condition:

the double Riemann integral $\int_{\mathcal{P}_{(s,t)}} g(s-u)\, g(t-v)\, dR_{uv}$ exists, (1.14.6)

the process M_t can be considered for any $t \geq 0$ as a limit of Riemann sums in the mean-square sense. Note that the following condition is sufficient for (1.14.6): the derivative $h'(s)$, $s > 0$, exists, $h(0) = 0$, and R_{uv} admits a representation

$$R_{uv} = \int_{\mathcal{P}_{(u,v)}} \left[\int_0^{u_1 \wedge v_1} h'(u_1 - z)\, h'(v_1 - z)\, \varphi^2(z)\, dz \right] du_1\, dv_1 \quad (1.14.7)$$

and

$$\int_{\mathcal{P}_{(s,t)}} |g(s-u)|\, |g(t-v)| \left[\int_0^{u \wedge v} \left| h'(u-z)\, h'(v-z) \right| \varphi^2(z)\, dz \right] du\, dv < \infty.$$

Now we are in a position to study conditions on φ, h and g supplying martingale properties of M_t.

Definition 1.14.2. Gaussian process V is called (g)-transformable if the process

$$M_t := \int_0^t g(t-s)\, dV_s$$

is a martingale.

Remark 1.14.3. Since M_t is a Gaussian process, it is a square-integrable martingale if V is (g)-transformable.

Denote $U = \{f : \mathbb{R}_+ \to \mathbb{R} \mid (f * q)_t = 0,\ t \in \mathbb{R}_+,\ \text{for such } q : \mathbb{R}_+ \to \mathbb{R}$ that $(|f| * |q|)_t < \infty,\ t \geq 0,\ \text{if and only if } q = 0\}$,
$AC[0,t] = \{f : \mathbb{R}_+ \to \mathbb{R} \mid f(s) = \int_0^s f'(u)\, du;\ 0 \leq s \leq t\ \text{with}\ \int_0^t |f'(u)|\, du < \infty\}$. Theorems 1.14.4 and 1.14.5 contain two groups of sufficient conditions on the functions φ, h, g ensuring (g)-transformability of V_t (statements 1) and 3)). Statements 2) and 4) demonstrate that these conditions are, in some sense, necessary.

Theorem 1.14.4. 1) *Let φ, h, g satisfy conditions (1.14.2), (1.14.3), (1.14.7) and*

$$(|g| * |h'|)_t < \infty, \quad t > 0, \quad (1.14.8)$$
$$(g * h')_t = C_0, \quad t > 0 \quad \text{for some} \quad C_0 \in \mathbb{R}. \quad (1.14.9)$$

Then V_t is (g)-transformable and $\langle M \rangle_t = C_0^2 \int_0^t \varphi^2(s)\, ds$.

2) Let φ, h, g satisfy conditions (1.14.2), (1.14.3), (1.14.7) and (1.14.8), $h \in U$, $\varphi \neq 0$ (mod λ) (λ is the Lebesgue measure), $(g * h')_t \in C(0, \infty)$, V_t be (g)-transformable.
Then $(g * h')_t = C_0$, $t > 0$, for some $C_0 \in \mathbb{R}$.

Theorem 1.14.5. 3) Let φ and h satisfy (1.14.2) and (1.14.3), $\varphi \neq 0$ (mod λ), g satisfies (1.14.6) and

$$g \in AC[0, t], \quad t \geq 0, \quad g(0) = 0, \tag{1.14.10}$$

$$(|g'| * (h^2 * \varphi^2)^{1/2})_t < \infty, \quad t > 0, \tag{1.14.11}$$

$$(g' * h)_t = C_0, \quad t > 0 \quad \text{for some} \quad C_0 \in \mathbb{R}. \tag{1.14.12}$$

Then V_t is (g)-transformable and $\langle M \rangle_t = C_0^2 \int_0^t \varphi^2(s)\, ds$.

4) Let φ and h satisfy (1.14.2), (1.14.3), $\varphi \neq 0$ a.e. (mod λ), the process V_t is (g)-transformable with g satisfying (1.14.10), (1.14.11), $(g' * h)_t \in C(0, \infty)$.
Then $(g' * h)_t = C_0$, $t > 0$, for some $C_0 \in \mathbb{R}$.

Remark 1.14.6. Conditions (1.14.9) and (1.14.12) mean, in particular, that corresponding convolutions have jumps at zero, so at least one of the functions involved is singular at 0.

Remark 1.14.7. Let $h(s) = s^\alpha, \varphi(s) = s^{-\alpha}, g(s) = s^{-\alpha}$. Then statement 1) holds for $H \in (1/2, 1)$ and statement 3) holds for $H \in (0, 1/2)$.

Proof of Theorem 1.14.4. 1) It follows from (1.14.7) that

$$f_t(z) := \int_0^t g(t-v) \left[\int_0^{z \wedge v} h'(v-r)\, h'(z-r)\, \varphi^2(r)\, dr \right] dv, \quad 0 \leq z \leq t$$

is defined for a.a. $z \leq t$ for any $t \in \mathbb{R}_+$ fixed. Condition (1.14.7) ensures the Fubini theorem for f_t, and from (1.14.8)–(1.14.9) we obtain that

$$f_t(z) = \int_0^z g(t-v) \left(\int_0^v h'(v-r)\, h'(z-r)\, \varphi^2(r)\, dr \right) dv$$

$$+ \int_z^t g(t-v) \left(\int_0^z h'(v-r)\, h'(z-r)\, \varphi^2(r)\, dr \right) dv$$

$$= \int_0^z h'(z-r)\, \varphi^2(r) \left(\int_r^t h'(v-r)\, g(t-v)\, dv \right) dr$$

$$= C_0 \int_0^z h'(z-r)\, \varphi^2(r)\, dr,$$

i.e. f_t does not depend on $t \geq z$. Further, for any $0 \leq s \leq t$ we have that

$$E(M_t - M_s)M_s = \int_0^s g(s-u)\left(f_t(u) - f_s(u)\right) du = 0.$$

1.14 Martingale Transforms and Girsanov Theorem

It means that the Gaussian process M_t with $EM_t = 0$ has uncorrelated, thus independent, increments. Hence, M_t is a Gaussian martingale, and it holds that

$$\langle M \rangle_t = \int_0^t g(t-u) \left(\int_0^t g(t-v) \int_0^{u \wedge v} h'(u-r) h'(v-r) \varphi^2(r) \, dr \right) du$$

$$= C_0 \int_0^t g(t-u) \left(\int_0^v h'(v-r) \varphi^2(r) \, dr \right) dv = C_0^2 \int_0^t \varphi^2(r) \, dr.$$

2) Let $M_t = \int_0^t g(t-s) \, dV_s$ be a square integrable martingale with g satisfying (1.14.7) and (1.14.8). Then

$$E(M_t - M_s) V_s = 0, \quad 0 \le s < t,$$

or

$$0 = \int_0^s \left(\int_0^v h'(v-r) \varphi^2(r) \left(\int_r^t h'(u-r) g(t-u) \, du \right. \right.$$
$$\left. \left. - \int_r^s h'(u-r) g(s-u) \, du \right) dr \right) dv = (h * (\varphi^2 \cdot \zeta))_s,$$

where

$$\zeta(r) = \int_0^{t-r} h'(u) g(t-r-u) \, du - \int_0^{s-r} h'(u) g(s-r-u) \, du.$$

Since $h \in U$, we obtain $\varphi^2 \cdot \zeta = 0$, and, taking into account that $\varphi \neq 0$, we derive that $\zeta(r) = 0 \pmod \lambda$, $r \le s \le t$. Together with continuity of $h' * g \in C(0, \infty)$ it means that $(h' * g)_t = C_0$, $t > 0$, for some $C_0 \in \mathbb{R}$. □

Proof of Theorem 1.14.5. 3) Under condition (1.14.6) the integral M_t is a mean-square limit of Riemann sums, and condition (1.14.10) permits us to transform the sum:

$$M_t = \underset{|\lambda_N| \to 0}{\text{l.i.m.}} \sum_{i=0}^{N-1} g(t - s_i) (V_{s_{i+1}} - V_{s_i})$$

$$= \underset{|\lambda_N| \to 0}{\text{l.i.m.}} \sum_{i=0}^{N-1} V(s_{i+1}) (g(s_{i+1}) - g(s_i))$$

$$= \int_0^t g'(t-s) V_s \, ds = \int_0^t g'(t-s) \left(\int_0^s h(s-z) \varphi(z) dW_z \right) ds,$$

where $|\lambda_N| = \max_{0 \le i \le N-1} |g(s_{i+1}) - g(s_i)|$, and the last integral is the limit of Riemann sums in the mean-square sense. Further, condition (1.14.11), according to (Pro90, p. 160) or (Leb95), permits to apply to M_t the stochastic Fubini theorem, and we obtain from (1.14.12) that

$$M_t = \int_0^t \varphi(z) \left(\int_z^t g'(t-u)\, h(u-z)\, du \right) dW_s = C_0 \int_0^t \varphi(z)\, dW_z. \quad (1.14.13)$$

4) If the process M_t is a square-integrable martingale, then from (1.14.13) it follows that for any $0 \le s \le t$

$$0 = E(M_t - M_s / \mathcal{F}_s^W) = \int_0^s \varphi(z)\, \eta(z)\, dW_z,$$

where

$$\eta(z) = (g' * h)_{t-z} - (g' * h)_{s-z}.$$

Hence $\int_0^s \varphi^2(z)\, \eta^2(z)\, dz = 0$, and, arguing similarly to the completion of the proof of Theorem 1.14.4, part 2), we obtain that $(g' * h)_t = C_0$ for some $C_0 \in \mathbb{R}$. □

Consider some examples of the functions φ, h satisfying conditions 1) or 3). (One example is contained in Remark 1.14.7.)

Example 1.14.8. Let

$$g(x) = x^{-1/2} \cosh(ax^{1/2}),$$

$$h'(x) = \int_0^x s^{\nu/2} I_\nu(as^{1/2}) (x-s)^\gamma ds,$$

where $-1 < \nu < -\tfrac{1}{2}$,

$$I_\nu(y) = \frac{y^\nu}{2^\nu} \sum_{k=0}^\infty \frac{(-1)^k y^{2k} 2^{-2k}}{k!\, \Gamma(\nu+k+1)}$$

is the Bessel function of the first kind, $\gamma + \nu = -\tfrac{3}{2}$.

The Laplace transforms of these functions equal

$$F_g(\lambda) = (\pi/\lambda)^{1/2} \exp(a^2/4\lambda),\ F_{h'}(\lambda) = \Gamma(\gamma+1) 2^{-\nu-1} a^\nu \lambda^{-\nu-1}$$

$$\times \exp(-a^2/4\lambda) \lambda^{-\gamma-1} = \Gamma(\gamma+1) 2^{-\nu-1} a^\nu \lambda^{-1/2} \exp(-a^2/4\lambda),$$

$$F_g(\lambda) F_{h'}(\lambda) = \Gamma(\gamma+1) 2^{-\nu-1} \pi a^\nu \lambda^{-1},\ \lambda > 0,$$

whence $(g * h')_t = \Gamma(\gamma+1) 2^{-\nu-1} \pi a^\nu$, $t > 0$, and condition (1.14.9) holds. (For the details of the theory of Bessel functions of the first kind and their Laplace transforms see (Wat95) and (GR80).)

Condition (1.14.8) is fulfilled since $|h'(x)| \le C x^{\nu+\gamma+1}$ on any interval $(0, t)$, where C depends on t.

Conditions (1.14.2) and (1.14.7) hold for any $\varphi \in L_2(0,t)$, $t > 0$; condition (1.14.3), according to Lemma 1.14.1, holds for any φ such that $F_{|\varphi|}(\lambda) < \infty$, $F_\varphi(\lambda) \ne 0$ for $\lambda \in (a,b) \subset (0,\infty)$. In this case V_t is (g)-transformable, according to part 1) of Theorem 1.14.4.

Example 1.14.9. Let $g(x) = x^{-1/2} \cosh(ax^{1/2})$, $h(x) = \int_0^x t^{-1/2} \cos(at^{1/2}) \, dt$. Then $F_g(\lambda) = (\pi/\lambda)^{1/2} \exp(a^2/4\lambda)$, $F_{h'}(\lambda) = (\pi/\lambda)^{1/2} \exp(-a^2/4\lambda)$, $F_g(\lambda) F_{h'}(\lambda) = \pi/\lambda$, $\lambda > 0$, so $(g * h')_t = \pi$, $t > 0$. Since $|h(x)| \leq Cx^{1/2}$, we can conclude as in Example 1.14.8.

Example 1.14.10. Let $g'(x) = \int_0^x t^{-1/2} \cosh(at^{1/2})(x-t)^\gamma dt$, $h(x) = x^{\nu/2} I_\nu(ax^{1/2})$ with $\gamma \in (-1, -\frac{1}{2})$, $\nu \in (-1, 0)$, $\gamma + \nu = -\frac{3}{2}$. Then $F_{g'}(\lambda) = \pi^{1/2} \lambda^{-\gamma - 3/2} \exp(a^2/4\lambda)$, $F_h(\lambda) = \lambda^{-\nu-1} \exp(-a^2/4\lambda)$, $F_{g'}(\lambda) F_h(\lambda) = \pi^{1/2} \lambda^{-1}$.

Conditions (1.14.2), (1.14.3) and (1.14.11) hold for $\varphi \in L_2(0, t)$, $t > 0$, $F_{|\varphi|}(\lambda) < \infty$, $F_\varphi(\lambda) \neq 0$ for some interval $(a, b) \subset (0, \infty)$, (1.14.10) is evident, (1.14.6) is fulfilled at least for $\varphi \in C(\mathbb{R}_+)$. So, if $\varphi > 0$, $\varphi \in C(\mathbb{R}_+)$ and $F_{|\varphi|}(\lambda) < \infty$ we have part 3) of Theorem 1.14.5.

Remark 1.14.11. According to Proposition 7 from (HC78), under the condition $h' \in L_2(0, t), t > 0, \varphi \equiv 1$, V_t is a semimartingale. In this case we transform semimartingale into martingale by (g)-transformation. For example, let $h(x) = x^\varepsilon$, $1/2 < \varepsilon < 1$, $\varphi(x) = 1$. Then

$$V_t = \int_0^t h(t-s) dW_s = \varepsilon \int_0^t \left(\int_0^s (s-u)^{\varepsilon - 1} dW_u \right) ds$$

is a semimartingale, more precisely, a process of bounded variation. Put $g(x) = x^{-\varepsilon}$. Then $M_t = \varepsilon \int_0^t (t-s)^{-\varepsilon} (\int_0^s (s-u)^{\varepsilon-1} dW_u) ds = \varepsilon B(\varepsilon, 1-\varepsilon) W_t$, where $B(\cdot, \cdot)$ is the beta-function.

Now, let V_t be equal to Y_t from (1.14.1). Recall that $B_t^H = \int_0^t s^\alpha dV_s$ is an fBm with Hurst index H, and in this case B_t^H can be presented as $B_t^H = \int_0^t m_H(t, s) dB_s$, where B is a Wiener process and the kernel $m_H(t, s)$ is defined in Section 1.8. Consider general conditions on function $\psi : \mathbb{R}_+ \to \mathbb{R}$ for the process $N_t := \int_0^t \psi_s dV_s$ to be presented in a similar way.

Theorem 1.14.12. *Let conditions (1.14.2), (1.14.3) hold and also*

$$\lim_{\varepsilon \downarrow 0} \psi^2(\varepsilon) \int_0^\varepsilon h^2(\varepsilon - u) \varphi^2(u) \, du = 0; \quad (1.14.14)$$

the Riemann integral $\int_{[0,(s,t)]} \psi(u) \psi(v) \, dR_{uv}$ *exists, $s, t > 0$;* (1.14.15)

there exists a derivative $\psi'(s)$, $s > 0$ and

$$(h^2 * \varphi^2)^{1/2} \psi' \in L_1(0, t), \quad (|h| * |\psi'|)_t < \infty, \quad t > 0. \quad (1.14.16)$$

Then

$$\int_0^t \psi(s) \, dV_s = \int_0^t m(t, s) \varphi(s) \, dW_s, \ t > 0, \ a.s.,$$

where

$$m(t,s) = \psi(t)h(t-s) - \int_s^t h(u-s)\psi'(u)\,du,$$

W is a Wiener process.

If (1.14.16) is strengthened to

$$(h^2 * \varphi^2)^{1/2}\psi' \in L_2(0,t),\ t > 0, \tag{1.14.17}$$

then $E(\int_0^t \psi(s)\,dV_s)^2 < \infty$.

Proof. Under (1.14.14)–(1.14.16), we can consider the integral $\int_0^t \psi(u)\,dV_u$ as a mean-square limit of Riemann sums, and integrating by parts, we obtain the following limits in the mean-square sense

$$\int_0^t \psi(u)\,dV_u = \lim_{\varepsilon \downarrow 0} \int_\varepsilon^t \psi(u)\,dV_u$$

$$= \psi(t)V(t) - \lim_{\varepsilon \downarrow 0}\psi(\varepsilon)V(\varepsilon) - \int_0^t \psi'(u)V(u)\,du$$

$$= \psi(t)V(t) - \int_0^t \psi'(u)\left(\int_0^u h(u-s)\varphi(s)dW_s\right)du.$$

Due to (1.14.16), the stochastic Fubini theorem can be applied to the last integral, and we obtain

$$\int_0^t \psi(u)\,dV_u = \int_0^t \psi(t)h(t-s)\varphi(s)ds - \int_0^t \varphi(s)\left(\int_s^t h(u-s)\psi'(u)\,du\right)dW_s$$

$$= \int_0^t m(t,s)\varphi(s)dW_s.$$

The second statement is evident. □

Now let P and $\widehat{P}$ be two probability measures on $(\Omega, \mathcal{F})$. Denote by P_t ($\widehat{P}_t$) the restriction of P ($\widehat{P}$) on $\mathcal{F}_t$ and suppose that $\widehat{P} \overset{\text{loc}}{\ll} P$ (it means that $\widehat{P}_t \ll P_t$, $t \in \mathbb{R}_+$). Consider the density process $Z_t = \mathcal{E}(X_t) := \exp\left\{X_t - \frac{1}{2}\langle X^c\rangle_t\right\}\prod_{0 \le s \le t}(1+\Delta X_s)e^{-\Delta X_s}$, X is a local martingale.

As before, we consider the Gaussian process $V_t = \int_0^t h(t-s)\varphi(s)\,dW_s$ and suppose that V_t is (g)-transformable by the function g; moreover, the conditions (1.14.8)–(1.14.9) or (1.14.10)–(1.14.12) hold. Let $M_t = C_0\int_0^t \varphi(s)\,dW_s$ with C_0 depending on g. Since M_t has continuous modification, the process $[M,X]$ has P-locally bounded variation (see (JS87, Lemma 3.14)).

Denote by $A_t := \langle M,X\rangle_t$ the P-compensator of $[M,X]$. Suppose further that the function ψ satisfies conditions (1.14.14)–(1.14.16) of Theorem 1.14.12.

Lemma 1.14.13. *The integral $\int_0^t m(t,s)\,dA_s$ exists for any $t > 0$ P- and $\widehat{P}$-a.s.*

Proof. Since $m(t,s) = \psi(t)h(t-s) - \int_s^t h(u-s)\psi'(u)\,du$, we consider $\int_0^t h(t-s)\,dA_s$ and $\int_0^t \left(\int_s^t h(u-s)\psi'(u)\,du\right) dA_s$ individually. From Kunita's inequality and (1.14.2),

$$\int_0^t |h(t-s)|\,d|A|_s \le \left(\int_0^t |h(t-s)|^2\,d\langle M\rangle_s \cdot \langle X\rangle_t\right)^{\frac{1}{2}}$$

$$= C_0 \left(\int_0^t |h(t-s)|^2 \varphi^2(s)\,ds \langle X\rangle_t\right)^{\frac{1}{2}} < \infty$$

P- and $\widehat{P}$-a.s.

Similarly,

$$\int_0^t \left|\int_s^t \psi'(u)h(u-s)\,du\right| d|A|_s$$

$$\le C_0 \left(\int_0^t \left|\int_s^t \psi'(u)h(u-s)\,du\right|^2 \varphi^2(s)\,ds \cdot \langle X\rangle_t\right)^{\frac{1}{2}}$$

$$\le C_0 \left(\int_0^t (h^2 * \varphi^2)_u |\psi'(u)|^2\,du \cdot \langle X\rangle_t\right)^{\frac{1}{2}} < \infty,$$

P and $\widehat{P}$-a.s. □

Theorem 1.14.14. *Let V_t be (g)-transformable with g satisfying (1.14.8)–(1.14.9) or (1.14.10)–(1.14.12), ψ satisfying (1.14.14)–(1.14.16), $\varphi \ne 0$ a.e. (mod λ). Then $\widehat{N}_t := N_t - C_0^{-1}\int_0^t m(t,s)\,dA_s$ is a Gaussian process w.r.t. $\widehat{P}$ and admits the representation $\widehat{N}_t = \int_0^t m(t,s)\varphi(s)\,d\widehat{W}_s$, where $\widehat{W}_t$ is a Wiener process w.r.t. $\widehat{P}$.*

Remark 1.14.15. Consider the case where $\varphi(s) = s^{-\alpha}$, $h(s) = C_1 s^\alpha$, $g(s) = C_2 s^{-\alpha}$, V_t is defined by $V_t = C_1 \int_0^t (t-s)^\alpha s^{-\alpha}\,dW_s$, $\psi(s) = s^\alpha$, and $B_t^H = \int_0^t s^\alpha\,dV_s$ is an fBm with Hurst index H. Then we obtain that $\widehat{B}_t^H := B_t^H - C_0^{-1}\int_0^t m_H(t,s)\,d\langle X,M\rangle_s$ is an fBm w.r.t. $\widehat{P}$, $M_t = C_4 \int_0^t s^{-\alpha}\,dW_s = C_1 C_2 \int_0^t (t-s)^{-\alpha}\,dV_s$, $C_0 = C_4 = \pi|\alpha|\,|\cos\pi H|^{-1} C_1 \cdot C_2$.

Proof. According to the classical Girsanov theorem, $\widehat{M}_t := M_t - \langle M, X\rangle_t$ is a $\widehat{P}$-local martingale with the angle bracket $\langle \widehat{M}\rangle_t = \langle M\rangle_t = C_0^2 \int_0^t \varphi^2(s)\,ds$. Therefore, $\widehat{M}_t$ is a continuous square-integrable $\widehat{P}$-martingale. Since $\varphi \ne 0$ a.e. (mod λ), we obtain from the Lévy theorem that $\widehat{M}_t = C_0 \int_0^t \varphi_s\,d\widehat{W}_s$, $\widehat{W}$ is $\widehat{P}$-Wiener process. According to Theorem 1.14.12, $\widehat{B}_t = C_0^{-1}\int_0^t z(t,s)\,d(M_s - \langle M,X\rangle_s) = C_0^{-1}\int_0^t m(t,s)\,d\widehat{M}_s = \int_0^t m(t,s)\varphi(s)\,d\widehat{W}_s$. □

According to the Theorem 1.14.14, we obtain that the drift has the form $D_t := C_0^{-1}\int_0^t m(t,s)\,dA_s$ in the case when the density process Z_t is known. Consider also the question: what "drifts" are admissible?

Theorem 1.14.16. *Let (1.14.14)–(1.14.16) and one of the following sets of conditions hold:*

1) *conditions (1.14.2), (1.14.3), (1.14.7)–(1.14.9) and $\varphi \neq 0$ a.e. (mod λ);*
2) $\int_s^t |h'(v-s)| \, |\psi'(v)| \, dv < \infty, \quad 0 \leq s \leq t$ *a.s;*
3) *a process $\{D_t, \mathcal{F}_t^W, \, t \geq 0\}$ has a.s. bounded variation $|D|_t = \mathrm{var}_{[0,t]} D$, $t > 0$, $D_0 = 0$;*
4) $\psi \neq 0$, *the integral $\int_0^t |g(t-s)| \, |\psi^{-1}(s)| \, d|D|_s < \infty$ a.s., $t > 0$, and we have a representation*

$$\int_0^t g(t-s) \, \psi^{-1}(s) \, dD_s = \int_0^t \delta_s ds, \quad \text{where} \quad \int_0^t |\delta_s| \, ds < \infty \text{ a.s.}$$

$$E \int_0^t \varphi_s^{-2} \delta_s^2 ds < \infty, \quad t > 0;$$

5) $E\mathcal{E}(X_t) = 1$, *where*

$$X_t = C_0^{-1} \int_0^t \varphi_s^{-1} \delta_s dW_s, \quad \mathcal{E}(X_t) = \exp\left\{X_t - \frac{1}{2}\langle X \rangle_t\right\};$$

or:

6) *conditions (1.14.2), (1.14.3), (1.14.6), (1.14.10)–(1.14.12);*
7) *conditions 3)–5);*
8) *a process $E_t = \int_0^t m(t,s) \delta_s ds$ has bounded variation and*

$$\int_0^t |g(t-s)| \, |\psi^{-1}(s)| \, d|E|_s < \infty, \quad \text{a.s.}, \quad t > 0;$$

9) $g' \in U$.

Then the process $\widehat{B}_t = B_t - D_t$ is Gaussian and admits the representation $\widehat{B}_t = \int_0^t m(t,s)\varphi(s) d\widehat{W}_s$ under the measure $\widehat{P} \overset{\mathrm{loc}}{\ll} P$ such that $\left.\frac{d\widehat{P}}{dP}\right|_{\mathcal{F}_t^W} = \mathcal{E}(X_t)$.

Proof. In our case $A_t = \langle M, X \rangle_t = \int_0^t \delta_s ds$, therefore from Theorem 1.14.14 the "drift" equals $C_0^{-1} E_t$.

It is enough to establish that $D_t = C_0^{-1} E_t$. If conditions 1)–5) hold, then

$$\int_0^t m(t,s) \, \delta_s ds = \int_0^t \left(\psi(t) h(t-s)\right.$$

$$\left. - \int_s^t h(u-s) \psi'(u) du\right) d\left(\int_0^t g(t-s) \psi^{-1}(s) dD_s\right)$$

$$= \int_0^t \psi(t) h'(t-s) \left(\int_0^s g(s-u) \psi^{-1} dD_u\right) ds - \int_0^t \left(\int_s^t h'(v-s) \psi'(v) dv\right)$$

$$\times \left(\int_0^s g(s-u)\,\psi^{-1}(u)\,dD_u \right) ds$$

$$= \psi(t) \int_0^t \left(\int_u^t h'(t-s)\,g(s-u)\,ds \right) \psi^{-1}(u)\,dD_u$$

$$- \int_0^t \int_0^t \int_0^t h'(v-s)\,\psi'(v)\,g(s-u)\,\psi^{-1}(u)\,I\{u \le s \le v \le t\}\,dv\,ds\,dD_u$$

$$= C_0\psi(t) \int_0^t \psi^{-1}(u)\,dD_u - C_0 \int_0^t \int_0^t \psi'(v)\,\psi^{-1}(u)\,I\{u \le v \le t\}\,dv\,dD_u$$

$$= C_0\psi(t) \int_0^t \psi^{-1}(u)\,dD_u - C_0 \int_0^t (\psi(t) - \psi(u))\,\psi^{-1}(u)\,dD_u = C_0 D_t.$$

If conditions 6)–9) hold, then for any $t > 0$

$$\int_0^t g(t-s)\,\psi^{-1}(s)\,dE_s = \int_0^t \left(g'(t-s)\,\psi^{-1}(s) + g(t-s)\,\psi'(s)\,\psi^{-2}(s) \right)$$

$$\times \left(\psi(s) \int_0^s h(s-u)\,\delta_u du - \int_0^s \left(\int_u^s h(v-u)\,\psi'(v)\,dv \right) \delta_u du \right) ds. \tag{1.14.18}$$

The right-hand side of (1.14.18) contains four integrals. Consider them separately. From (1.14.12),

$$\int_0^t g'(t-s) \int_0^s h(s-u)\,\delta_u du\,ds = C_0 \int_0^t \delta_u du.$$

Further,

$$\int_0^t g(t-s)\frac{\psi'(s)}{\psi^2(s)}\left(\psi(s) \int_0^s h(s-u)\delta_u du \right.$$

$$\left. - \int_0^s \left(\int_u^s h(v-u)\psi'(v)dv \right) \delta_u du \right) ds$$

$$= \int_0^t g'(t-s) \left(\int_0^s \frac{\psi'(z)}{\psi^2(z)} \left(\psi(z) \int_0^z h(z-u)\,\delta_u\,du \right. \right.$$

$$\left. \left. - \int_0^z \left(\int_u^z h(v-u)\,\psi'(v)\,dv \right) \delta_u du \right) dz \right) ds.$$

It is sufficient to prove that

$$\sigma_s := \int_0^s \frac{\psi'(z)}{\psi^2(z)}\left(\psi(z) \int_0^z h(z-u)\delta_u du \right.$$

$$\left. - \int_0^z \left(\int_u^z h(v-u)\psi'(v)dv \right) \delta_u du \right) dz$$

$$= \psi^{-1}(s) \int_0^s \left(\int_u^s h(v-u)\,\psi'(v)\,dv \right) \delta_u du =: \bar\sigma_s, \tag{1.14.19}$$

and then it follows that the right-hand side of (1.14.18) equals $C_0 \int_0^t \delta_u du$.
But $\sigma_0 = \bar{\sigma}_0$, and the derivative

$$\bar{\sigma}'_s = -\frac{\psi'(s)}{\psi^2(s)} \int_0^s \left(\int_u^s h(v-u)\,\psi'(v)\,dv \right) \delta_u du$$
$$+ \psi^{-1}(s) \int_0^s h(s-u)\,\delta_u du \cdot \psi'(s) = \sigma'_s.$$

We obtain that

$$\int_0^t g(t-s)\,\psi^{-1}(s) dD_s = \int_0^t g(t-s)\,\psi^{-1}(s)\,d\left(C_0^{-1} \int_0^s z(s,u)\,\delta_u\,du \right),$$

or

$$\int_0^t g'(t-s) \int_0^s \psi^{-1}(u)\,dD_u\,ds = C_0^{-1} \int_0^t g'(t-s)\left(\int_0^s \psi^{-1}(u)\,dE_u \right) ds.$$

If $g' \in U$ then $\int_0^s \psi^{-1}(u)\,d(D-E)_u = 0$, whence $D_t = \int_0^t \psi_s \cdot \psi_s^{-1} dD_s - \psi_t \cdot \int_0^t \psi_s^{-1} dE_s - \int_0^t \psi'_s \cdot \int_0^s \psi_u^{-1} dE_u ds = E_t$. □

Theorem 1.14.16 permits us to calculate the Hellinger process for P and $\widehat{P}$.

Let $\widehat{P} \ll P$ and $Y_t = \mathcal{E}(X_t)$, X_t be a continuous square-integrable martingale. According to (JS87, Corollary 1.37) the Hellinger process in a narrow sense of order β equals $h_t(\beta) = \frac{1}{2}\beta(1-\beta)\langle X \rangle_t$.

Theorem 1.14.17. *Let one of conditions 1)–5) or 6)–9) hold, then*

$$h_t(\beta) = \frac{\beta(1-\beta)}{2C_0^2} \int_0^t \varphi_s^{-2} \delta_s^2 ds$$
$$= \frac{\beta(1-\beta)}{2C_0^2} \int_0^t \varphi_s^{-2} \left(\frac{d}{ds} \int_0^s g(t-u)\,\psi^{-1}(u)\,dD_u \right)^2 ds.$$

The proof follows immediately from Theorem 1.14.16.

Remark 1.14.18. It is possible to study if the process $V_t = \int_0^t h(t-s)\varphi(s)dW_s$ is itself a semimartingale. In the case when $\varphi \equiv 1$ this question is investigated in (CCM98).

Theorem 1.14.19. *Let the function h be differentiable on $\mathbb{R}_+$, $\int_0^t |h'(u)|du < \infty, t \geq 0$, and $\int_0^t (h'(t-u)\varphi(u))^2 du < \infty, t \geq 0$. Then the process $\{V_t, \mathcal{F}_t^W, t \geq 0\}$ is a semimartingale.*

Proof. We have the representation $h(t) = h(0) + \int_0^t h'(u)du$, which together with the Fubini theorem supplies the following transformations:

$$V_t = \int_0^t h(t-s)c(s)dW_s = h(0) \int_0^t c(s)dW_s + \int_0^t \left(\int_0^{t-s} h'(u)du\varphi(s) \right) dW_s$$

$$= h(0) \int_0^t \varphi(s)dW_s + \int_0^t \int_s^t h'(v-s)\varphi(s)dv\, dW_s = h(0) \int_0^t \varphi(s)dW_s$$
$$+ \int_0^t \int_0^v h'(v-s)\varphi(s)dW_s\, dv.$$

□

1.15 Nonsemimartingale Properties of fBm; How to Approximate Them by Semimartingales

A process $\{X_t, \mathcal{F}_t, t \geq 0\}$ is called semimartingale, if it admits the representation
$$X_t = X_0 + M_t + A_t,$$
where M is an $\mathcal{F}_t$-local martingale with $M_0 = 0$, A is a process of locally bounded variation, X_0 is $\mathcal{F}_0$-measurable. Evidently, any semimartingale has locally bounded quadratic variation; if X is continuous, then M and A are continuous. Let $X_t = B_t^H$ with $H \in (0, 1/2)$. Then its quadratic variation is infinite, therefore, it is not a semimartingale. If $H \in (1/2, 1)$, then the quadratic variation of X is zero, and if we suppose that X is semimartingale, then the quadratic variation of $M_t = X_t - X_0 - A_t$ is zero, and M is zero. But $X_t \neq A_t$ since X has unbounded variation. Therefore, $X_t = B_t^H$ is not a semimartingale for any $H \neq 1/2$. (There are many another elegant proofs of this fact.) Nevertheless, there are many approaches to how to approximate fBm by a sequence of semimartingales.

1.15.1 Approximation of fBm by Continuous Processes of Bounded Variation

We follow here the approach of and (And05) and (AM06). According to (1.8.5) and (1.8.18), we can represent $\{B_t^H, t \geq 0\}$ with Hurst index $H \in (1/2, 1)$ as
$$B_t^H = \int_0^t s^\alpha dY_s,$$
where
$$Y_t = C_H^{(8)} \int_0^t (t-s)^\alpha s^{-\alpha} dB_s,$$
$\{B_t, t \geq 0\}$ is a Wiener process, $C_H^{(8)} = C_H^{(6)} \widetilde{\alpha}$.

We can rewrite Y_t as
$$Y_t = C_H^{(8)} \alpha \int_0^t \left(\int_s^t (u-s)^{\alpha-1} du \right) s^{-\alpha} dB_s. \qquad (1.15.1)$$

If we formally apply the stochastic Fubini theorem to the right-hand side of (1.15.1), we obtain that

$$Y_t = C_H^{(8)} \alpha \int_0^t \left(\int_0^u (u-s)^{\alpha-1} s^{-\alpha} dB_s \right) du. \tag{1.15.2}$$

But the right-hand side of (1.15.2) does not exist, since the variance of interior integral is infinite,

$$\int_0^u (u-s)^{2\alpha-2} s^{-2\alpha} ds = \infty.$$

Thereupon, we introduce the "truncated" process for $\beta \in (0,1)$,

$$Y_t^\beta = C_H^{(8)} \alpha \int_0^t \left(\int_0^{\beta s} (s-u)^{\alpha-1} u^{-\alpha} dB_u \right) ds,$$

and

$$B_t^{H,\beta} = \int_0^t s^\alpha dY_s^\beta = C_H^{(8)} \alpha \int_0^t s^\alpha \left(\int_0^{\beta s} (s-u)^{\alpha-1} u^{-2\alpha} dB_u \right) ds \tag{1.15.3}$$

is a process of bounded variation which will serve as an approximation of B_t^H.

Theorem 1.15.1. *We have that*

$$E(B_t^H - B_t^{H,\beta})^2 \le c_1 t^{2H} (1-\beta)^{2\alpha},$$

where $c_1 = c_1(H)$ is some constant, independent of t and β.

Proof. First, we want to change the limits of the integration in (1.15.3) and consider the process

$$Z_t^\beta := \alpha C_H^{(8)} \int_0^{\beta t} \left(\int_{u/\beta}^t (s-u)^{\alpha-1} ds \right) u^{-\alpha} dB_u$$

$$= C_H^{(8)} \left(\int_0^{\beta t} (t-u)^\alpha u^{-\alpha} dB_u - \left(\frac{1-\beta}{\beta} \right)^\alpha B_{\beta t} \right). \tag{1.15.4}$$

We cannot apply here the stochastic Fubini theorem (Pro90, Theorem IV.4.5), because it is valid if the integral $\int_0^{\beta t} \int_{u/\beta}^t (s-u)^{2\alpha-2} u^{-2\alpha} ds\, du$ is finite but it is infinite. Therefore, we must go an indirect way. We consider the integral $Y_t^{\beta,\varepsilon} = D \int_\varepsilon^t \left(\int_{\beta\varepsilon}^{\beta s} (s-u)^{\alpha-1} u^{-\alpha} dB_u \right) ds$, where $D = \alpha C_H^{(8)}$, and the Fubini theorem ensures the equality

$$Y_t^{\beta,\varepsilon} = Z_t^{\beta,\varepsilon} := D \int_{\beta\varepsilon}^{\beta t} \left(\int_{u/\beta}^t (s-u)^{\alpha-1} ds \right) u^{-\alpha} dB_u.$$

Furthermore,

$$E|Y_t^{\beta,\varepsilon} - Y_t^{\beta}| \le D\Big(\int_0^{\varepsilon}\Big(\int_0^{\beta s}(s-u)^{2\alpha-2}u^{-2\alpha}du\Big)^{1/2}ds$$

$$+\int_{\varepsilon}^{t}\Big(\int_0^{\beta\varepsilon}(s-u)^{2\alpha-2}u^{-2\alpha}du\Big)^{1/2}ds\Big) \le D\Big(\int_0^{\varepsilon}u^{-1/2}du\Big(\int_0^{\beta}(1-u)^{2\alpha-2}$$

$$\times u^{-2\alpha}du\Big) + \widehat{\alpha}(\beta\varepsilon)^{1/2-\alpha}\int_{\varepsilon}^{t}(s-\beta\varepsilon)^{\alpha-1}ds\Big) \to 0$$

and

$$E|Z_t^{\beta,\varepsilon} - Z_t^{\beta}|^2 \le D^2\int_0^{\beta\varepsilon}\Big(\int_{u/\beta}^{t}(s-u)^{\alpha-1}ds\Big)^2 u^{-2\alpha}du \le CD^2\beta\varepsilon^{1-2\alpha} \to 0$$

as $\varepsilon \to 0$, where $C > 0$ is some constant. This means that $Y_t^{\beta} = Z_t^{\beta}$ a.s. for any $t \in [0, T]$. Therefore, for $1/2 < \beta < 1$

$$E(Y_t - Y_t^{\beta})^2 = (C_H^{(8)})^2 E\Big(\int_{\beta t}^{t}(t-u)^{\alpha}u^{-\alpha}dB_u + \Big(\frac{1-\beta}{\beta}\Big)^{\alpha}B_{\beta t}\Big)^2$$

$$\le 2(C_H^{(8)})^2\int_{\beta t}^{t}(t-u)^{2\alpha}u^{-2\alpha}du + 2(C_H^{(8)})^2\Big(\frac{1-\beta}{\beta}\Big)^{2\alpha}\beta t$$

$$\le H^{-1}(C_H^{(8)})^2(\beta t)^{-2\alpha}t^{2H}(1-\beta)^{2H} + 2(C_H^{(8)})^2\Big(\frac{1-\beta}{\beta}\Big)^{2\alpha}\beta t$$

$$\le c_2 t(1-\beta)^{2\alpha} \quad \text{with } c_2 = (C_H^{(8)})^2 \cdot 2^{2\alpha-1}(H^{-1}+2). \quad (1.15.5)$$

Integration by parts gives us

$$B_t^H - B_t^{H,\beta} = t^{\alpha}(Y_t - Y_t^{\beta}) - \alpha\int_0^{t}(Y_s - Y_s^{\beta})s^{\alpha-1}ds$$

whence we obtain from (1.15.5) that

$$E(B_t^H - B_t^{H,\beta})^2 \le 2t^{2\alpha}E(Y_t - Y_t^{\beta})^2 + 2\alpha^2 t\int_0^{t}E(Y_s - Y_s^{\beta})^2 s^{2\alpha-2}ds$$

$$\le 2c_2 t^{2H}(1-\beta)^{2\alpha} + 2\alpha^2 t\int_0^{t}s^{2\alpha-1}ds \cdot c_2(1-\beta)^{2\alpha},$$

and we can put $c_1 = 2c_2(\alpha + 1)$. □

1.15.2 Convergence $B^{H,\beta} \to B^H$ in Besov Space $W^{\lambda}[a, b]$.

For $\lambda \in (0, 1/2)$ define the Besov space $W^{\lambda}[a, b]$ as the space of measurable functions $f : [a, b] \to \mathbb{R}$ such that

$$\|f\|_{a,b,\lambda} := \int_a^{b}\frac{|f(s)|}{(s-a)^{\lambda}}ds + \int_a^{b}\int_a^{s}\frac{|f(s)-f(y)|}{(s-y)^{\lambda+1}}dy\,ds < \infty.$$

Theorem 1.15.2. *For any* $\lambda \in (0, 1/2)$, $H \in (1/2, 1)$ *and any* $[a, b] \subset [0, T]$
$$E\|B^H - B^{H,\beta}\|_{a,b,\lambda} \leq c_1(H, \lambda, T)(1-\beta)^\alpha.$$

Proof. Denote $\overline{B}_t^{H,\beta} := B_t^H - B_t^{H,\beta}$. We have
$$E\|\overline{B}^{H,\beta}\|_\lambda = E \int_a^b \frac{|\overline{B}_s^{H,\beta}|}{(s-a)^\lambda} ds + E \int_a^b \int_a^s \frac{|\overline{B}_s^{H,\beta} - \overline{B}_y^{H,\beta}|}{(s-y)^{\lambda+1}} dy\, ds. \quad (1.15.6)$$

From Theorem 1.15.1,
$$E \int_a^b \frac{|\overline{B}_s^{H,\beta}|}{(s-a)^\lambda} ds \leq \int_a^b \frac{(E(\overline{B}_s^{H,\beta})^2)^{1/2}}{(s-a)^\lambda} ds \leq c_1^{1/2}(1-\beta)^\alpha \int_a^b \frac{s^H}{(s-a)^\lambda} ds$$
$$\leq c_1(H, \lambda, T)(1-\beta)^\alpha, \quad (1.15.7)$$

with $c_1(H, \lambda, T) = c_1^{1/2} \cdot T^{H-\lambda+1} \cdot (H - \lambda + 1)^{-1}$. Consider the second term in the right-hand side of (1.15.6). Rewrite the difference in the numerator as
$$\overline{B}_s^{H,\beta} - \overline{B}_y^{H,\beta} = (B_s^H - B_s^{H,\beta}) - (B_y^H - B_y^{H,\beta})$$
$$= \int_y^s u^\alpha d(Y_u - Y_u^\beta) = \int_y^s u^\alpha d\overline{Y}_u^\beta, \quad (1.15.8)$$

where $\overline{Y}_u^\beta = Y_u - Y_u^\beta$. Equality (1.15.8) and integration by parts give us the estimates
$$\int_a^b \int_a^s \frac{|\overline{B}_s^{H,\beta} - \overline{B}_y^{H,\beta}|}{(s-y)^{\lambda+1}} dy\, ds$$
$$= \int_a^b \int_a^s (s-y)^{-\lambda-1} \left| s^\alpha \overline{Y}_s^\beta - y^\alpha \overline{Y}_y^\beta + \alpha \int_y^s \overline{Y}_u^\beta u^\alpha du \right| dy\, ds$$
$$\leq \int_a^b \int_a^s (s-y)^{-\lambda-1} s^\alpha \left| \overline{Y}_s^\beta - \overline{Y}_y^\beta \right| dy\, ds$$
$$+ \int_a^b \int_a^s (s-y)^{-\lambda-1} (s^\alpha - y^\alpha) |\overline{Y}_y^\beta| dy\, ds$$
$$+ \alpha \int_a^b \int_a^s (s-y)^{-\lambda-1} \left(\int_y^s |\overline{Y}_u^\beta| u^{\alpha-1} du \right) dy\, ds$$
$$=: I_1(\beta) + I_2(\beta) + \alpha I_3(\beta).$$

Now we estimate $I_2(\beta)$:
$$EI_2(\beta) \leq \alpha \int_a^b \int_a^s y^{\alpha-1}(s-y)^{-\lambda}(E(\overline{Y}_y^\beta)^2)^{1/2} dy\, ds$$
$$\leq c_2^{1/2} \alpha \int_a^b \int_a^s y^{\alpha-1}(s-y)^{-\lambda} y^{1/2} dy\, ds \cdot (1-\beta)^\alpha$$
$$\leq c_2(H, \lambda, T)(1-\beta)^\alpha, \quad (1.15.9)$$

where $c_2(H, \lambda, T) = c_2^{1/2} \alpha T^{1-\lambda}$. Similarly,

$$\begin{aligned}
E\, I_3(\beta) &\leq \int_a^b \int_a^s (s-y)^{-\lambda-1} \left(\int_y^s (E(\overline{Y}_u^\beta)^2)^{1/2} u^{\alpha-1} du \right) dy\, ds \\
&\leq c_2^{1/2} \int_a^b \int_a^s (s-y)^{-\lambda-1} \left(\int_y^s u^{\alpha-1/2} du \right) dy\, ds \cdot (1-\beta)^\alpha \quad (1.15.10) \\
&\leq c_3(H, \lambda, T)(1-\beta)^\alpha,
\end{aligned}$$

where $c_3(H, \lambda, T) = c_2^{1/2} \dfrac{T^{H-\lambda+1}}{H(H-\lambda)(H-\lambda+1)}$. Now we use the representation (1.15.4) to estimate $I_1(\beta)$:

$$|\overline{Y}_s^\beta - \overline{Y}_y^\beta| \leq C_H^{(8)} \left| \int_{\beta s}^s (s-u)^\alpha u^{-\alpha} dB_u - \int_{\beta y}^y (s-u)^\alpha u^{-\alpha} dB_u \right|$$
$$+ C_H^{(8)} \left(\frac{1-\beta}{\beta} \right)^\alpha |B_{\beta s} - B_{\beta y}|,$$

therefore

$$\begin{aligned}
I_1(\beta) &\leq C_H^{(8)} \int_a^b \int_a^s (s-y)^{-\lambda-1} s^\alpha \\
&\quad \times \left| \int_{\beta s}^s (s-u)^\alpha u^{-\alpha} dB_u - \int_{\beta y}^y (y-u)^\alpha u^{-\alpha} dB_u \right| dy\, ds \quad (1.15.11) \\
&\quad + C_H^{(8)} \left(\frac{1-\beta}{\beta} \right)^\alpha \int_a^b \int_a^s s^\alpha (s-y)^{-\lambda-1} |B_{\beta s} - B_{\beta y}| dy\, ds \\
&=: \mathcal{I}_1(\beta) + \mathcal{I}_2(\beta).
\end{aligned}$$

Further,

$$\begin{aligned}
E\, \mathcal{I}_2(\beta) &\leq C_H^{(8)} \left(\frac{1-\beta}{\beta} \right)^\alpha \int_a^b \int_a^s s^\alpha (s-y)^{-\lambda-1/2} dy\, ds\, \beta^{1/2} \quad (1.15.12) \\
&= c_4(H, \lambda, T)(1-\beta)^\alpha,
\end{aligned}$$

where $c_4(H, \lambda, T) = C_H^{(8)} 2^\alpha \cdot \dfrac{T^{H-\lambda+1}}{1/2-\lambda}$. (Here we see that indeed λ must be less than $1/2$.) Next, we decompose $\mathcal{I}_1(\beta)$ into two integrals

$$\mathcal{I}_1(\beta) = C_H^{(8)} \int_a^b \int_a^{(\beta s)\vee a} + C_H^{(8)} \int_a^b \int_{(\beta s)\vee a}^s =: \mathcal{I}_3(\beta) + \mathcal{I}_4(\beta).$$

$$E\mathcal{I}_3(\beta) \leq C_H^{(8)} \int_a^b \int_a^{(\beta s)\vee a} (s-y)^{-\lambda-1} s^\alpha$$

$$\times \left(E\left(\int_{\beta s}^s (s-u)^\alpha u^{-\alpha} dB_u - \int_{\beta y}^y (y-u)^\alpha u^{-\alpha} dB_u \right)^2 \right)^{1/2} dy\, ds$$

$$\leq \sqrt{2} C_H^{(8)} \int_a^b \int_a^{(\beta s)\vee a} (s-y)^{-\lambda-1} s^\alpha$$

$$\times \left(\int_{\beta s}^s (s-u)^{2\alpha} u^{-2\alpha} du + \int_{\beta y}^y (y-u)^{2\alpha} u^{-2\alpha} du \right)^{1/2} dy\, ds$$

$$\leq 2^\alpha H^{-1/2} C_H^{(8)} \int_a^b \int_a^{(\beta s)\vee a} (s-y)^{-\lambda-1}(s+y)^{1/2} s^\alpha dy\, ds \cdot (1-\beta)^H$$

$$\leq c(H,\lambda,T)(1-\beta)^{H-\lambda} \tag{1.15.13}$$

with $c(H,\lambda,T) = \frac{2^H T^{1+H-\lambda}}{\lambda(1-\lambda)H^{1/2}}$. Finally,

$$E\mathcal{I}_4(\beta) \leq C_H^{(8)} \int_a^b \int_{(\beta s)\vee a}^s (s-y)^{-\lambda-1} s^\alpha \left(E\left| \int_0^s ((s-u)^\alpha u^{-\alpha} \mathbf{1}_{(\beta s,s)}(u) \right. \right.$$

$$\left. \left. - (y-u)^\alpha u^{-\alpha} \mathbf{1}_{(\beta y,y)}(u) \right) dB_u \right|^2 \right)^{1/2} dy\, ds$$

$$= C_H^{(8)} \int_a^b \int_{(\beta s)\vee a}^s (s-y)^{-\lambda-1} s^\alpha \left(\int_0^s ((s-u)^\alpha \mathbf{1}_{(\beta s,s)}(u) \right.$$

$$\left. - (y-u)^\alpha \mathbf{1}_{(\beta y,y)}(u) \right)^2 u^{-2\alpha} du \right)^{1/2} dy\, ds.$$

The interior integral equals

$$\int_{\beta s}^y ((s-u)^\alpha - (y-u)^\alpha)^2 u^{-2\alpha} du + \int_y^s (s-u)^{2\alpha} u^{-2\alpha} du$$

$$+ \int_{\beta y}^{\beta s} (y-u)^{2\alpha} u^{-2\alpha} du =: \mathcal{I}_5(\beta),$$

and via some routine calculations can be estimated as

$$\mathcal{I}_5(\beta) \leq C_H (1-\beta)^{2\alpha}(s-y),$$

where $C_H = 1 + 2^{2\alpha} + \frac{\alpha}{1-2\alpha}$.

Therefore

$$E\mathcal{I}_4(\beta) \leq C_H^{(8)} (C_H)^{1/2}(1-\beta)^\alpha \int_a^b s^\alpha \int_{(\beta s)\vee a}^s (s-y)^{-\lambda-1/2} dy\, ds$$

$$\leq C_H^{(8)} (C_H)^{1/2}(1-\beta)^\alpha \int_a^b s^{H-\lambda} ds \int_\beta^1 (1-y)^{-\lambda-1/2} dy$$

$$\leq C(H,\lambda,T)(1-\beta)^{H-\lambda} \tag{1.15.14}$$

with $C(H,\lambda,T) = C_H^{(8)}(C_H)^{1/2}\frac{T^{H-\lambda+1}}{(H-\lambda+1)(1/2-\lambda)}$. Summarizing (1.15.9), (1.15.10), (1.15.12)–(1.15.14), we obtain the proof. □

We obtain another approximation, considering the "truncated" process of the form

$$Y_t^\beta := C_H^{(8)}\alpha \int_0^t \left(\int_0^{(s-\beta)_+} (s-u)^{\alpha-1}u^{-\alpha}dB_u\right)ds$$

and

$$B_t^{H,\beta} = \int_0^t s^\alpha dY_s^\beta, \quad t \geq 0, \quad H \in (1/2, 1). \tag{1.15.15}$$

Evidently, we intend to obtain the approximation while $\beta \to 0$.

Theorem 1.15.3. *The process $B^{H,\beta}$ satisfies the relations*

$$E(B_t^H - B_t^{H,\beta})^2 \leq c(H)\begin{cases} t^{2H}, & t < \beta \\ \beta^{2\alpha}t(1+\ln\frac{t}{\beta}), & t \geq \beta, \end{cases}$$

and for $2 < m < \frac{1}{1-H}$

$$E|B_t^H - B_t^{H,\beta}|^m \leq c(H,m)\begin{cases} t^{mH}, & t < \beta \\ \beta^{m\alpha}t^{m/2} + \beta^{m(H-1)+1}t^{m-1}, & t \geq \beta. \end{cases}$$

Proof. Using the stochastic Fubini theorem we obtain

$$Y_t^\beta = C_H^{(8)}\alpha \int_0^{(t-\beta)_+} \left(\int_{u+\beta}^t (s-u)^{\alpha-1}ds\right) u^{-\alpha}dW_u$$

$$= C_H^{(8)}\left(\int_0^{(t-\beta)_+}(t-u)^\alpha u^{-\alpha}dW_u - \beta^\alpha \int_0^{(t-\beta)_+} u^{-\alpha}dW_u\right),$$

whence

$$E(Y_t - Y_t^\beta)^2 = (C_H^{(8)})^2 E\left(\int_{(t-\beta)_+}^t (t-u)^\alpha u^{-\alpha}dW_u\right.$$

$$\left. + \beta^\alpha \int_0^{(t-\beta)_+} u^{-\alpha}dW_u\right)^2$$

$$= (C_H^{(8)})^2\left(\int_{(t-\beta)_+}^t (t-u)^{2\alpha}u^{-2\alpha}du + \beta^{2\alpha}\int_0^{(t-\beta)_+} u^{-2\alpha}du\right)$$

$$\leq (C_H^{(8)})^2\begin{cases} \int_0^t (t-u)^{2\alpha}u^{-2\alpha}du, & t < \beta \\ \beta^{2\alpha}\int_0^t u^{1-2\alpha}du, & t \geq \beta \end{cases}$$

$$= c(H)\begin{cases} t, & t < \beta \\ \beta^{2\alpha}t^{-2\alpha}, & t \geq \beta, \end{cases} \tag{1.15.16}$$

where $c(H) = (C_H^{(8)})^2 \max(B(2H, 1-2\alpha), \frac{1}{1-2\alpha})$. Since $Y_t - Y_t^\beta$ is a Gaussian random variable with zero mean, for $m \geq 0$

$$E|Y_t - Y_t^\beta|^m = \pi^{-1/2}\Gamma\left(\frac{m+1}{2}\right)(2\sigma^2)^{m/2},$$

where $\sigma^2 = E(Y_t - Y_t^\beta)^2$, therefore, from (1.15.16)

$$E|Y_t - Y_t^\beta|^m \leq c(m, H) \begin{cases} t^{m/2}, & t < \beta \\ \beta^{m\alpha}t^{m(1-H)}, & t \geq \beta. \end{cases} \quad (1.15.17)$$

As before, integration by parts gives us

$$B_t^H - B_t^{H,\beta} = t^\alpha(Y_t - Y_t^\beta) - \alpha \int_0^t (Y_s - Y_s^\beta)s^{\alpha-1}ds. \quad (1.15.18)$$

From (1.15.17) and (1.15.18) we obtain for $m \geq 1$:

$$E\,|B_t^H - B_t^{H,\beta}|^m \leq 2^{m-1}\left(t^{m\alpha}E|Y_t - Y_t^\beta|^m \right.$$

$$\left. + \alpha^m t^{m-1}\int_0^t E|Y_s - Y_s^\beta|^m s^{m(\alpha-1)}ds\right)$$

$$\leq c(m, H) \begin{cases} t^{mH}, & t < \beta \\ \beta^{m\alpha}t^{m/2} + t^{m-1}\int_0^\beta s^{m(H-1)}ds + t^{m-1}\beta^{m\alpha}\int_\beta^t s^{-m/2}ds, & t \geq \beta. \end{cases}$$

The integrals in the last expression converge for $m < \frac{1}{1-H}$. For $m = 2$ we get

$$E(B_t^H - B_t^{H,\beta})^2 \leq c(2, H) \begin{cases} t^{2H}, & t < \beta \\ \beta^{2\alpha}t + \beta^{2\alpha}t \ln \frac{t}{\beta}, & t \geq \beta, \end{cases}$$

and for $2 < m < \frac{1}{1-H}$ we obtain

$$E|B_t^H - B_t^{H,\beta}|^m \leq c(m, H) \begin{cases} t^{mH}, & t < \beta \\ \beta^{m\alpha}t^{m/2} + \beta^{mH-m+1}t^{m-1} + \beta^{m\alpha}t^{m/2}, & t \geq \beta, \end{cases}$$

whence the proof follows. □

Remark 1.15.4. Note that the approximation of fBm with the sequence of semimartingales was considered in (Thao03).

1.15.3 Weak Convergence to fBm in the Schemes of Series

We formulate in this section some results concerning weak convergence to fBm in different schemes of series.

(i) *Convergence of the piecewise linear processes to fBm.* Let $\{\xi_k, k \in \mathbb{Z}\}$ be a sequence of i.i.d. random variables, and $\{a_{kn}\}_{k \in \mathbb{Z}, n \geq 0}$ be a matrix with real elements satisfying the following assumptions:

$$E\xi_0 = 0, \quad E\xi_0^2 = 1, \quad E|\xi_0|^p < \infty \quad \text{for some } p > 2, \qquad (1.15.19)$$

$$\gamma_n = \max_{k \in \mathbb{Z}} |a_{kn}| \to 0 \ (n \to \infty), \quad \sum_{k \in \mathbb{Z}} a_{kn}^2 = 1. \qquad (1.15.20)$$

Also, let $\{\varphi_n(t), n \geq 1, t \in [0,1]\}$ be a sequence of real functions on the unit interval. Denote $\theta(x) = \frac{1}{2}(x^H + x^{-H} - |x^{1/2} - x^{-1/2}|^{2H})$, $x > 0$, $H \in (0,1)$. We construct the sequence of continuous random piecewise linear processes $\xi_n(t)$, $t \in [0,1]$, such that

$$\xi_n\left(\frac{m}{n}\right) = \varphi_n\left(\frac{m}{n}\right) \sum_{k \in \mathbb{Z}} a_{km} \xi_k, \quad 0 \leq m \leq n.$$

Theorem 1.15.5 ((Gor77)). *Let conditions* (1.15.19)–(1.15.20) *hold and*

$$\sum_k a_{kl} a_{km} \to \theta(x) \quad \text{as} \quad l \to \infty, \ l/m \to x,$$

$$\sup_n \sup_{|(l-m)/n| \leq h} \left(\sum_k (\varphi_n(l/n) a_{kl} - \varphi_n(m/n) a_{km})^2 \right)^{p/2} \leq ch^{1+\varepsilon}, \ \text{for some } \varepsilon > 0.$$

Then the sequence of processes $\{\xi_n(t), n \geq 1, t \in [0,1]\}$ weakly converges in $C[0,1]$ endowed with uniform topology to the fBm with Hurst index H.

(ii) *Convergence of the Weierstrass–Mandelbrot process to complex fBm.* Consider the complex-valued Gaussian process

$$\widetilde{B}_t^H := c_H \int_{\mathbb{R}_+} (e^{itx} - 1) x^{-H-1/2} (dW_1(x) + i dW_2(x)), \ t \in \mathbb{R},$$

where W_1 and W_2 are two independent standard Brownian motions. Evidently, $\widetilde{B}_0^H = 0$, $E\widetilde{B}_t^H = 0$, $E|\widetilde{B}_{t+s}^H - \widetilde{B}_s^H|^2 = c_H^2 \cdot |t|^{2H} \int_{\mathbb{R}} \sin^2 x \cdot x^{-2\alpha-2} dx \cdot 2^{1-2\alpha} = t^{2H}$, if we choose $c_H = 2^{H-1}(\int_{\mathbb{R}_+} \sin^2 x \cdot x^{-2\alpha-2} dx)^{-1/2}$. Therefore, with this choice of c_H $\widetilde{B}_t^H$ is a normalized complex-valued fBm.

Now, suppose that (ξ_n, η_n), $n \in \mathbb{Z}$, is a sequence of independent random variables with $E\xi_n^2 = E\eta_n^2 = 1$, $E\xi_n = E\eta_n = 0$, and either
1) $\zeta_n := \xi_n + i\eta_n$, $n \in \mathbb{Z}$ are identically distributed random vectors, or
2) $\sup_n (E|\xi_n|^{2+\delta} + E|\eta_n|^{2+\delta}) < \infty$ for some $\delta > 0$.
Also, let $f(t, u) : \mathbb{R}^2 \to \mathbb{C}$, $t \in \mathbb{R}$, be such a function that for all $t \in \mathbb{R}$
3) $f(t, \cdot) \in C^1(\mathbb{R})$;
4) $|f(t, u)| = 0(|u|^{-l})$ as $u \to \infty$ for some $l > 1/2$.

Theorem 1.15.6 ((PT00b)).
1. Under conditions 1)–4) *the following convergence (in the sense of convergence of finite-dimensional distributions) takes place:*

$$\xi_a^f(t) := a^{-1/2} \sum_{n \in \mathbb{Z}} f\left(t, \frac{n}{a}\right)(\xi_n + i\eta_n) \xrightarrow{d} \xi^f(t) := \int_{\mathbb{R}} f(t,u)(dW_1(u) + i dW_2(u)),$$

as $a \to \infty$, where W_1 and W_2 are two independent standard Brownian motions.

2. If, in addition, $f(0, u) = 0$,

$$|f(t,u) - f(s,u)| \leq c|f(t-s,u)| \quad \text{for all} \quad s, t, u \in \mathbb{R},$$

$$|f(t,u)| \leq ct^H |f(1, u + \ln t)| \quad \text{for some} \quad 0 < H < 1$$

and

$$\sup_{n \in \mathbb{Z}} E(|\xi_n|^{2k} + |\eta_n|^{2k}) < \infty \quad \text{for some} \quad k > \frac{1}{2H},$$

then for any $T > 0$ $\xi_a(t)$ converges weakly to $\xi(t)$ in the space $C[0,T]$ endowed with the uniform topology.

Corollary 1.15.7. Let $\widetilde{f}(t,u) = (e^{ie^u t} - 1)c^{-Hu}$. Then the corresponding process $\xi_a^{\widetilde{f}}(t)$ is called the normalized Weierstrass–Mandelbrot process and, according to Theorem 1.15.6, it converges weakly to the process

$$\xi^{\widetilde{f}}(t) := \int_{\mathbb{R}} (e^{ie^u t} - 1)e^{-Hu}(dW_1(u) + dW_2(u)).$$

Moreover, the processes $\xi^{\widetilde{f}}(t)$ and $\widetilde{B}_t^H$ have identical finite-dimensional distributions because they are both Gaussian, have zero mean and the same covariance functions.

Remark 1.15.8. The proof of Theorem 1.15.6 is based on the Functional Central Limit Theorem.

(iii) *Weak convergence of random walks to fBm in Besov spaces (in the scheme of series).* Consider a random walk $\{X_n\}_{n \geq 1}$ consisting of stationary Gaussian random variables with zero mean and correlations $r(i-j) := EX_i X_j$. Recall that a positive function $\varphi(x), x \geq a$ for some $a > 0$ is said to be slowly varying at ∞ if for all $t > 0$ $\lim_{x \to \infty} \varphi(tx)/\varphi(x) = 1$. Denote by $\mathcal{D} = \mathcal{D}[0,1]$ the Skorohod space of right-continuous functions on the interval $[0,1]$ that have left-hand limits, and equip D with the metric

$$d(x,y) := \inf\{\varepsilon > 0 : \exists \lambda \in \Lambda \quad \text{such that} \quad \|\lambda\| < \varepsilon$$
$$\text{and} \sup_t |x(t) - y(\lambda(t))| \leq \varepsilon\}.$$

Here $\|\lambda\| := \sup_{s \neq t} |\log(\lambda(t) - \lambda(s))/(t-s)|$ and

$$\Lambda := \{\lambda : [0,1] \to [0,1], \lambda \text{ is strictly increasing and continuous mapping of } [0,1] \text{ into itself}\}.$$

Under this metric $\mathcal{D}$ is a separable and complete metric space, and we denote by $\xrightarrow{D}$ the convergence in the Skorohod topology, which is the weak topology induced by this metric. That is, $X^n \xrightarrow{D} X$ if $E\psi(X^n) \to E\psi(X)$ as $n \to \infty$ for any bounded and continuous $\psi : \mathcal{D} \to \mathbb{R}$. We start with the following result of Taqqu:

Lemma 1.15.9 ((Taq75)). *Let $\{X_n\}_{n\geq 1}$ be a stationary Gaussian sequence with mean 0 and correlations $r(i-j) = EX_iX_j$. Assume that*

$$\sum_{i,j=1}^{n} r(i-j) \sim n^{2H}\varphi(n) \quad \text{as } n \to \infty, \qquad (1.15.21)$$

with $0 < H < 1$, φ slowly varying.

Then $Z_n \xrightarrow{D} \widetilde{B}^H$, where $Z_n(t) = d_n^{-1} \sum_{i=1}^{[nt]} X_i$ with $d_n \sim n^{2H}\varphi(n)$, $\widetilde{B}^H$ is an fBm with Hurst index H, not necessarily normalized.

Remark 1.15.10. Condition (1.15.21) is satisfied for $H \in (1/2, 1)$ when $r(k) \sim k^{2\alpha-1}\varphi(k)$, and for $H \in (0, 1/2)$, when $r(k) \sim -k^{2\alpha-1}\varphi(k)$ as $k \to \infty$ with $r(0) + 2\sum_{k=1}^{\infty} r(k) = 0$.

Further, define for a function $f \in L_p[0,1]$ the modulus of continuity in $L_p[0,1]$:

$$\omega_p(f,t) := \sup_{|h| \leq t} \left(\int_{I_h} |f(x+h) - f(x)|^p dx \right)^{1/p},$$

where $I_h := \{x \in [0,1], x+h \in [0,1]\}$. Now, for $0 < \gamma < 1$ and $\beta > 0$, we consider a real function $\omega_\beta^\gamma : (0,1] \to \mathbb{R}$ of the form $\omega_\beta^\gamma(t) := t^\gamma(1 + \log 1/t)^\beta$, $t \in (0,1]$, and denote

$$\|f\|_{p,\omega_\beta^\alpha} := \|f\|_{L_p[0,1]} + \sup_{0 < t \leq 1} \omega_p(f,t)/\omega_\beta^\gamma(t).$$

Recall that the Besov space $Lip_p(\gamma,\beta)$ is the class of functions f in $L_p[0,1]$ such that $\|f\|_{p,\omega_\beta^\gamma} < \infty$; $Lip_p(\gamma,\beta)$ endowed with the norm $\|\cdot\|_{p,\omega_\beta^\gamma}$ is a non-separable Banach space. It is possible to consider a separable subspace $lip_p(\gamma,\beta)$ of $Lip_p(\gamma,\beta)$ of the functions $f \in Lip_p(\gamma,\beta)$ satisfying $\omega_p(f,t) = o(\omega_\beta^\gamma(t))$ as $t \downarrow 0$. According to (BL01), the paths of fBm B^H, $H \in (0,1)$ are a.s. in $lip_p(H,\beta)$ for any $\beta > 0$ and $p \geq 1/H \vee 1/\beta$. The next result is proved in (BL01).

Theorem 1.15.11. *Let $H \in (0,1)$, $\beta > 0$, $p > 1/H \vee 1/\beta$, and let $\{X_n\}_{n\geq 1}$ be a stationary Gaussian sequence with mean 0 and correlations $r(i-j) = EX_iX_j$. Assume that*

$$\sum_{i,j=1}^{n} r(i-j) \sim Cn^{2H} \quad \text{as } n \to \infty, \quad \text{where } C > 0.$$

Then $C^{-1/2}Z_n \to B^H$ as $n \to \infty$ weakly in the space $lip_p(H,\beta)$.

(iv) *Convergence of martingale differences to fBm.* We follow here Nieminen's paper (Nie04), which generalizes the result from (Sot01). Consider the following scheme of series: let $(\Omega, \mathcal{F}, P)$ be a probability space, $(X_{i,n}, \mathcal{F}_{i,n})_{n \geq 1}$, $1 \leq i \leq n$ be a sequence of square integrable martingale-differences, i.e., $X_{i,n}$ is $\mathcal{F}_{i,n}$-adapted, $EX_{i,n}^2 < \infty$, $E(X_{i,n}/\mathcal{F}_{i-1,n}) = 0$, $\mathcal{F}_{0,n} = (\emptyset, \Omega)$, $\mathcal{F}_{i,n} \subset \mathcal{F}_{i+1,n} \subset \mathcal{F}$. Consider the sequence of kernels for $H \in (1/2, 1)$

$$Z^{(n)}(t, s) = n \int_{s-1/n}^{s} m_H\left(\frac{[nt]}{n}, u\right) du$$

for $s \in [1/n, 1]$ and $t \in [0, 1]$, where $[x] = k$ for $k \leq x < k+1$, $k \in \mathbb{Z}$. Define the processes

$$W_t^n := \sum_{i=1}^{[nt]} X_{i,n}, \quad t \in [0, 1],$$

and

$$Z_t^n := \int_0^t Z^{(n)}(t, s) dW_s^n = \sum_{i=1}^{[nt]} n \int_{i-1/n}^{i/n} m_H\left(\frac{[nt]}{n}, u\right) du \cdot \xi_i^{(n)}.$$

Theorem 1.15.12 ((Nie04)). *Let* $\lim_{n \to \infty} n(X_{i,n})^2 = 1$ *a.s.,* $1 \leq i \leq n$ *and* $\max_{1 \leq i \leq n} |X_{i,n}| \leq Cn^{-1/2}$ *a.s. for some* $C \geq 1$.
Then $Z^n \xrightarrow{D} B^H$, $n \to \infty$, *where the convergence is in* $\mathcal{D}[0, 1]$.

In the case when $\xi_i^{(n)}$ are i.i.d. random variables, the corresponding result is proved by Sottinen (Sot01) under weaker conditions.

Theorem 1.15.13 ((Sot01)). *Let* $\xi_i^{(n)} = 0$, $D\xi_i^{(n)} = 1$. *Then* $Z_n \xrightarrow{D} B^H$ *in the Skorohod space* $\mathcal{D}[0, T]$ *for any* $T > 0$.

(v) *Convergence of integral functionals.* Using Theorem 1.15.13, we can prove the result, similar to limit theorems for integral functionals on random walks, established in (SS70) and (Yos78). For example, (Yos78) considers sufficient conditions for

$$\sum_{i=1}^{n-1} f_n\left(\frac{i}{n}, \frac{S_i}{\sqrt{n}}\right) \frac{\xi_{i+1}}{\sqrt{n}} \xrightarrow{D} \int_0^1 f(t, W_t) dW_t,$$

where ξ_i is a sequence of martingale differences, $S_i = \sum_{k=1}^{i} \xi_k$, W_t is a Wiener process. For technical simplicity, we consider i.i.d. random variables and the interval $[0, 1]$. Let $\{f_n\}, n \geq 1$, $f_n : \mathbb{R} \to \mathbb{R}$ be the sequence of functions satisfying the conditions
1) $f_n, f \in C^1(\mathbb{R})$ and $\forall R > 0 \; \exists M_R > 0$ such that

$$\sup_{n\geq 1}\sup_{|x|\leq R}(|f_n(x)|+|f'_n(x)|)\leq M_R;$$

2) $f_n \rightrightarrows f$ uniformly on any $[-R, R]$. Let $\pi_r := \{0 = t_0^{(r)} < t_1^{(r)} < \cdots < t_{p_r}^{(r)} = 1\}$ be the sequence of partitions of $[0,1]$, $|\pi_r| \to 0$ as $r \to \infty$. Denote $\Delta Z_n\left(\frac{i}{n}\right) := Z_n\left(\frac{i+1}{n}\right) - Z_n\left(\frac{i}{n}\right)$, $\Delta Z_{n,j,r} := Z_n\left(t_{j+1}^{(r)}\right) - Z_n\left(t_j^{(r)}\right)$ and define the sequence of integral sums

$$S_n(\pi_r) := \sum_{j=1}^{p_r-1} f_n(Z_n(t_j^{(r)}))\Delta Z_{n,j,r}.$$

Lemma 1.15.14. *Under the conditions of Theorem 1.15.13*

$$P\text{-}\lim_{n\to\infty}\sum_{i=1}^{n-1}\left(\Delta Z_n\left(\frac{i}{n}\right)\right)^2 = P\text{-}\lim_{r\to\infty}\lim_{n\to\infty}\sum_{j=1}^{p_r-1}(\Delta Z_{n,j,r})^2 = 0.$$

Proof. We can prove even the convergence in $L_1(P)$. For this purpose, we can rewrite the difference $Z_n(t_2) - Z_n(t_1)$ for any $0 \leq t_1 < t_2 \leq 1$ in the form

$$Z_n(t_2) - Z_n(t_1) = \sqrt{n}\sum_{k=1}^{[nt_1]}\int_{\frac{k-1}{n}}^{\frac{k}{n}}\left(m_H\left(\frac{[nt_2]}{n},s\right) - m_H\left(\frac{[nt_1]}{n},s\right)\right)ds \cdot \xi_k^{(n)}$$
$$+\sqrt{n}\sum_{k=[nt_1]+1}^{[nt_2]}\int_{\frac{k-1}{n}}^{\frac{k}{n}}m_H\left(\frac{[nt_2]}{n},s\right)ds \cdot \xi_k^{(n)}.$$

Denote $\alpha_n(m,l) := \int_{\frac{m-1}{n}}^{\frac{m}{n}} m_H\left(\frac{l}{n},s\right)ds$, and

$$\beta_n(n,l_1,l_2) := \begin{cases} \alpha_n(m,l_2) - \alpha_n(m,l_1), & m \leq l_1 \leq l_2, \\ \alpha_n(m,l_2), & l_1 \leq m \leq l_2. \end{cases}$$ Then

$$Z_n(t_2) - Z_n(t_1) = \sqrt{n}\sum_{k=1}^{[nt_2]}\beta_n(k,[nt_1],[nt_2])\xi_k^{(n)}, \text{ and}$$

$$E|Z_n(t_2) - Z_n(t_1)|^2 = n\sum_{k=1}^{[nt_2]}\beta_n^2(k,[nt_1],[nt_2])$$
$$= n\sum_{k=1}^{[nt_1]}\left(\int_{\frac{k-1}{n}}^{\frac{k}{n}}\left(m_H\left(\frac{[nt_2]}{n},s\right) - m_H\left(\frac{[nt_1]}{n},s\right)\right)ds\right)^2$$
$$+ n\sum_{k=[nt_1]+1}^{[nt_2]}\left(\int_{\frac{k-1}{n}}^{\frac{k}{n}}m_H\left(\frac{[nt_2]}{n},s\right)ds\right)^2$$
$$\leq \int_0^{\frac{[nt_1]}{n}}\left(m_H\left(\frac{[nt_2]}{n},s\right) - m_H\left(\frac{[nt_1]}{n},s\right)\right)^2 ds + \int_{\frac{[nt_1]}{n}}^{\frac{[nt_2]}{n}}\left(m_H\left(\frac{[nt_2]}{n},s\right)\right)^2 ds$$
$$= (C_H^{(5)})^2\left(\int_0^{\frac{[nt_1]}{n}}s^{-2\alpha}\left(\int_{\frac{[nt_1]}{n}}^{\frac{[nt_2]}{n}}u^\alpha(u-s)^{\alpha-1}du\right)^2 ds\right.$$
$$\left.+ \int_{\frac{[nt_1]}{n}}^{\frac{[nt_2]}{n}}\left(\int_s^{\frac{[nt_2]}{n}}u^\alpha(u-s)^{\alpha-1}du\right)^2 ds\right)$$
$$= E|B_H(t_2) - B_H(t_1)|^2 = \left|\frac{[nt_2]}{n} - \frac{[nt_1]}{n}\right|^{2H} \leq (t_2-t_1)^{2H}.$$

(1.15.22)

From (1.15.22) $E \sum_{i=1}^{n-1} |\Delta Z_n(\frac{i}{n})|^2 \leq \sum_{i=1}^{n-1} n^{-2H} \to 0$, $n \to \infty$ and

$$E \sum_{i=1}^{p_r-1} |\Delta Z_{n,j,r}|^2 \leq \sum_{j=1}^{p_r-1} (t_{j+1}^{(r)} - t_j^{(r)})^{2H} \to 0, \quad r \to \infty. \qquad \square$$

Lemma 1.15.15. *Under conditions 1) and 2)*

$$\lim_{r \to \infty} \limsup_{n \to \infty} P\left(\left|S_n(\pi_r) - \sum_{i=1}^{n-1} f(Z_n(\frac{i}{n}))\Delta Z_n(\frac{i}{n})\right| > \delta\right) = 0 \quad \text{for any } \delta > 0.$$

Proof. Let a function $F : \mathbb{R} \to \mathbb{R}$ be such that $F'(x) = f(x), x \in \mathbb{R}$. Then by the Taylor formula

$$F(Z_n(1)) - F(0) = \sum_{i=0}^{n-1} \left(F\left(Z_n\left(\tfrac{i+1}{n}\right)\right) - F\left(Z_n\left(\tfrac{i}{n}\right)\right)\right)$$
$$= \sum_{i=0}^{n-1} f\left(Z_n\left(\tfrac{i}{n}\right)\right) \Delta Z_n\left(\tfrac{i}{n}\right) + \tfrac{1}{2} \sum_{i=0}^{n-1} f'(\theta_{i,n}) \left(\Delta Z_n\left(\tfrac{i}{n}\right)\right)^2,$$
$$F(Z_n(1)) - F(0) = \sum_{j=0}^{p_r-1} \left(F\left(Z_n\left(t_{j+1}^{(r)}\right)\right) - F\left(Z_n\left(t_j^{(r)}\right)\right)\right)$$
$$= \sum_{j=0}^{p_r-1} f\left(Z_n\left(t_j^{(r)}\right)\right) \Delta Z_{n,j,r} + \tfrac{1}{2} \sum_{j=0}^{p_r-1} f'(\theta_{n,j,r}) (\Delta Z_{n,j,r})^2,$$

where the points $\theta_{i,n}$ are between $Z_n\left(\tfrac{i}{n}\right)$ and $Z_n\left(\tfrac{i+1}{n}\right)$, and the points $\theta_{n,j,r}$ are between $Z_n\left(t_j^{(r)}\right)$ and $Z_n\left(t_{j+1}^{(r)}\right)$. Therefore

$$\left|S_n(\pi_r) - \sum_{i=1}^{n-1} f\left(Z_n\left(\tfrac{i}{n}\right)\right) \Delta Z_n\left(\tfrac{i}{n}\right)\right| \leq \tfrac{1}{2} \sum_{i=0}^{n-1} |f(\theta_{i,n})| \left|\Delta Z_n\left(\tfrac{i}{n}\right)\right|^2$$
$$+ \tfrac{1}{2} \sum_{j=0}^{p_r-1} |f'(\theta_{n,j,r})| |\Delta Z_{n,j,r}|^2,$$

and for any $\delta > 0$

$$P\left(\left|S_n(\pi_r) - \sum_{i=1}^{n-1} f\left(Z_n\left(\tfrac{i}{n}\right)\right) \Delta Z_n\left(\tfrac{i}{n}\right)\right| > \delta\right) \leq P\left(\sup_{0 \leq t \leq 1} |Z_n(t)| \geq R\right)$$
$$+ P\left(\sum_{i=0}^{n-1} \left(\Delta Z_n\left(\tfrac{i}{n}\right)\right)^2 \geq \tfrac{2\delta}{M_R}\right) + P\left(\sum_{i=0}^{n-1} (\Delta Z_{n,j,r})^2 \geq \tfrac{2\delta}{M_R}\right).$$
(1.15.23)

Note that $Z_n \xrightarrow{D} B^H$, and functionals sup and inf are continuous in the Skorohod topology, whence $P\left(\sup_{0 \leq t \leq 1} |Z_n(t)| \geq R\right) \to P\left(\sup_{0 \leq t \leq 1} |B_t^H| \geq R\right)$, and the last probability tends to 0 as $R \to \infty$, according to (Sin97). The proof follows now from Lemma 1.15.14 and (1.15.23). $\qquad \square$

Theorem 1.15.16. *Under the conditions of Lemma 1.15.15*

1.15 Nonsemimartingale Properties of fBm

$$\sum_{i=1}^{n-1} f_n\left(Z_n\left(\frac{i}{n}\right)\right) \Delta Z_n\left(\frac{i}{n}\right) \xrightarrow{d} \int_0^1 f(B_t^H) dB_t^H, \quad n \to \infty,$$

where $\xrightarrow{d}$ denotes here the convergence in distribution.

Remark 1.15.17. The existence of integral $\int_0^1 f(B_t^H) dB_t^H$ for $H \in (1/2, 1)$ and $f \in C^1(\mathbb{R})$ follows from (Zah98) (see also Section 2.1), and this integral is a limit a.s. of Riemann–Stieltjes sums.

Proof. Consider the difference

$$\Delta_n := \int_0^1 f(B_t^H) dB_t^H - \sum_{i=1}^{n-1} f_n\left(Z_n\left(\frac{i}{n}\right)\right) \Delta Z_n\left(\frac{i}{n}\right)$$

and write it in the form $\Delta_n := \sum_{j=1}^{4} \Delta_{n,r}^{(j)}$, where $\Delta_{n,r}^{(1)} = \int_0^1 f(B_t^H) dB_t^H - \sum_{j=1}^{p_r-1} f\left(B_{t_j^{(r)}}^H\right) \Delta B_{j,r}^H$ is independent of n, $\Delta B_{j,r}^H = B_{t_{j+1}^{(r)}}^H - B_{t_j^{(r)}}^H$.

$$\Delta_{n,r}^{(2)} = \sum_{j=1}^{p_r-1} f(B_{t_j^{(r)}}^H) \Delta B_{j,r}^H - \sum_{j=1}^{p_r-1} f\left(Z_{t_j^{(r)}}^{(n)}\right) \Delta Z_{j,r}^{(n)},$$

$$\Delta_{n,r}^{(3)} = \sum_{j=1}^{p_r-1} f(Z_{t_j^{(r)}}^{(n)}) \Delta Z_{j,r}^{(n)} - \sum_{j=1}^{p_r-1} f_n\left(Z_{t_j^{(r)}}^{(n)}\right) \Delta Z_{j,r}^{(n)},$$

$$\Delta_{n,r}^{(4)} = \sum_{j=1}^{p_r-1} f_n(Z_{t_j^{(r)}}^{(n)}) \Delta Z_{j,r}^{(n)} - \sum_{j=1}^{p_r-1} f_n\left(Z_{\frac{i}{n}}^{(n)}\right) \Delta Z_{\frac{i}{n}}^{(n)}.$$

From the result of Zähle (Zah98) cited above, $P\text{-}\lim_{r \to \infty} \Delta_{n,r}^{(1)} = 0$. By Lemma 1.15.15 $P\text{-}\lim_{r \to \infty} \Delta_{n,r}^{(4)} = 0$.

As to $\Delta_{n,r}^{(2)}$, we have from the weak convergence of $Z^{(n)}$ to B^H that

$$\sum_{j=1}^{p_r-1} f\left(Z_{t_j^{(r)}}^{(n)}\right) \Delta Z_{j,r}^{(n)} \xrightarrow{d} \sum_{j=1}^{p_r-1} f(B_{t_j^{(r)}}^H) \Delta B_{j,r}^H,$$

as $n \to \infty$, for any fixed $r \geq 1$. We must estimate now $\Delta_{n,r}^{(3)}$. The technique here is similar to the proof of Lemma 1.15.15.

Let $F(x) = \int_0^x f(t) dt$, $F_n(x) = \int_0^x f_n(t) dt$. Then

$$F(Z_1^{(n)}) = \sum_{j=1}^{p_r-1} \left(F\left(Z_{t_{j+1}^{(r)}}^{(n)}\right) - F\left(Z_{t_j^{(r)}}^{(n)}\right)\right)$$
$$= \sum_{j=1}^{p_r-1} f(Z_{t_j^{(r)}}^{(n)}) \Delta Z_{j,r}^{(n)} + \frac{1}{2} \sum_{j=0}^{p_r-1} f'\left(\theta_{j,r}^{(n)}\right) \left(\Delta Z_{j,r}^{(n)}\right)^2, \quad (1.15.24)$$

and, similarly,

$$F_n(Z_1^{(n)}) = \sum_{j=1}^{p_r-1} f_n(Z_{t_j^{(r)}}^{(n)}) \Delta Z_{j,r}^{(n)} + \frac{1}{2} \sum_{j=0}^{p_r-1} f_n'\left(\widetilde{\theta}_{j,r}^{(n)}\right) \left(\Delta Z_{j,r}^{(n)}\right)^2, \quad (1.15.25)$$

where $\theta_{j,r}^{(n)}$ and $\widetilde{\theta}_{j,r}^{(n)}$ are between $Z_{t_j^{(r)}}^{(n)}$ and $Z_{t_{j+1}^{(r)}}^{(n)}$. Now,

$$|F(Z_1^{(n)}) - F_n(Z_1^{(n)})| \le |Z_1^{(n)}| \sup_{|t| \le |Z_1^{(n)}|} |f_n(t) - f(t)|,$$

whence

$$\begin{aligned} &P\{|F(Z_1^{(n)}) - F_n(Z_1^{(n)})| \ge \delta\} \le P\{|Z_1^{(n)}| \ge R\} \\ &+ P\left\{\sup_{|t| \le R} |f_n(t) - f(t)| \ge \tfrac{\delta}{R}\right\} \end{aligned} \quad (1.15.26)$$

(The last event is not random.) Since f_n uniformly converges to f on $[-R, R]$, the last term in (1.15.26) is zero for all sufficiently large n, and $\lim_{n\to\infty} P\{|Z_1^{(n)}| \ge R\} = P\{|B_1^H| \ge R\} \le \frac{1}{R^2}$. Therefore, from (1.15.24)–(1.15.26) and Lemmas 1.15.14–1.15.15

$$P\text{-}\lim_{r\to\infty} \lim_{n\to\infty} \Delta_{n,r}^{(3)} = 0,$$

and the theorem is proved. □

Remark 1.15.18. The paper (Wang03) contains a result on a weak convergence to fBm in the Brownian scenery.

(vi) *fBm as a weak limit of Poisson shot noise processes.*

Let for all $n \in \mathbb{Z}\setminus\{0\}$ X_n be i.i.d.r.v. with $EX_1 = 0$ and $EX_1^2 \in (0, \infty)$, $g: \mathbb{R}_+ \to \mathbb{R}$ be a continuously differentiable function with $g'(u) = O(u^{-1/2-\varepsilon})$, $u \to \infty$ for some $\varepsilon > 0$. Consider the special model of multiplicative shots: $X_i(u) = g(u) X_i$, $u \ge 0$, and a shot noise model, which is defined as

$$S(t) = \sum_{i=1}^{N(t)} X_i(t - T_i) + \sum_{i \le -1} [X_i(t - T_i) - X_i(-T_i)], \ t \ge 0,$$

where N is a two-sided homogeneous Poisson process with the rate $\alpha > 0$ and points $\cdots < T_{-2} < T_{-1} < 0 < T_1 < T_2 < \cdots$. For $t = 0$ we put $S(0) = 0$.

According to (KK04), the multiplicative process with the above restrictions on g and X_i exists and has the following sample path properties.

Lemma 1.15.19. *The process S possesses a right-continuous version with left limits on $\mathbb{R}_+$ and has a finite variation on any $[0, T]$, $T > 0$. Therefore, it is a semimartingale with respect to its natural filtration.*

Now, suppose that $\lim_{u\to\infty} ug'(u)/g(u) = \gamma$ with $\gamma \in (0, 1/2)$. Introduce the rescaled process

$$S(x,t) = \frac{S(xt)}{\sigma(t)}, \quad x \in [0,\infty), \ t > 0,$$

where $\sigma^2(t) = Var(S(t))$.

Theorem 1.15.20. *Under the above assumptions,*

$$S(\cdot, t) \longrightarrow B^H, \quad t \to \infty$$

when the convergence is in $\mathcal{D}[0, \infty)$ with the metric of uniform convergence on compacts, and $H = 1/2 + \gamma$.

1.16 Hölder Properties of the Trajectories of fBm and of Wiener Integrals w.r.t. fBm

Let $\{\xi_t, t \in [0, T]\}$ be a separable modification of Gaussian process, $\rho_\xi^2(s,t) = E(\xi_s - \xi_t)^2$, $G = G(x) : \mathbb{R}_+ \to \mathbb{R}_+$ be a continuous increasing function, $G(0) = 0$, $D(T, \varepsilon) = \int_0^\varepsilon H(T, u)^{1/2} du$ be the Dudley integral (see Section 1.10), $\rho(s, t)$ be some semi-metric in $[0, T]$.

Definition 1.16.1. A function $\Theta = \Theta(x) : \mathbb{R}_+ \to \mathbb{R}_+$ is called a modulus of continuity if $\Theta(0) = 0$ and for any $x_1, x_2 \geq 0$

$$\Theta(x_1) \leq \Theta(x_1 + x_2) \leq \Theta(x_1) + \Theta(x_2).$$

Definition 1.16.2. Let $g : [0, T] \to \mathbb{R}$ be some function. The function

$$\varepsilon \to \Delta_\rho(g, \varepsilon) := \sup_{\substack{\rho(s,t) \leq \varepsilon \\ s,t \in [0,T]}} |g(s) - g(t)|$$

is called a modulus of uniform continuity of the function g with respect to the semi-metric ρ.

Definition 1.16.3. A modulus $\Theta(\cdot)$ is called a uniform modulus of a Gaussian process ξ with respect to the semi-metric ρ if for a.a. $\omega \in \Omega$

$$\limsup_{\varepsilon \to 0} \Delta_\rho(\xi.(\omega), \varepsilon)/\Theta(\varepsilon) < \infty.$$

The next result is formulated in the book (Lif95).

Theorem 1.16.4. *1. Let for any $s, t \in [0, T]$*

$$\rho_\xi(s,t) \leq G(\rho(s,t)). \tag{1.16.1}$$

Then the function $\Theta(\varepsilon) := D(T, G(\varepsilon))$ is a uniform modulus of the Gaussian process ξ with respect to the semi-metric ρ.

2. Under assumption (1.16.1) with $\rho(s,t) = |s-t|$, the function

$$\Theta(\varepsilon) = \int_0^\varepsilon |\log r|^{1/2} dG(r)$$

is a uniform modulus of the Gaussian process ξ with respect to ρ.

Definition 1.16.5. We say that the function $f : [0,T] \to \mathbb{R}$ belongs to the space $C^{\beta-}[0,T]$ if $f \in C^\gamma[0,T]$ for any $\gamma < \beta$.

Let $\xi_t = B_t^H$ be an fBm with Hurst index $H \in (0,1)$. Then, evidently, we can take $G(x) = x^H$, so from the second statement of previous theorem, the function $\Theta(\varepsilon) \sim \varepsilon^H |\log \varepsilon|^{1/2}$ will be a uniform modulus of B^H on any $[0,T]$. In particular, $|B_t^H - B_s^H| \leq c(\omega)|t-s|^{H-\beta}$ for any $0 < \beta < H$, i.e. $B^H \in C^{H-}[0,T]$ for a.a. ω and any $T > 0$. Now, let $\xi_t = I_t(f) = \int_0^t f(s) dB_s^H$ with $f \in L_2^H[0,t]$ for any $0 \leq t \leq T$, $H \in (1/2, 1)$. We can take $\rho(s,t) = \int_s^t |f(u)|^{\frac{1}{H}} du$, $G(x) = C_H x^H$,

$$\Delta_\rho(I, \varepsilon) = \sup_{\substack{0 \leq s < t \leq T: \\ \int_s^t |f(u)|^{\frac{1}{H}} du < \varepsilon}} |\xi_t - \xi_s|,$$

$D(T, G(\varepsilon)) = \int_0^{C_H \varepsilon^H} H(T,u)^{1/2} du$. Then, according to the first statement of Theorem 1.16.4 and Theorem 1.10.3

$$\limsup_{\varepsilon \to 0} \Delta_\rho(I, \varepsilon)/D(T, G(\varepsilon)) < \infty.$$

Now we simplify the situation supposing that f is essentially bounded on $[0,T]$, $f_T^* := \operatorname{ess\,sup}_{0 \leq t \leq T} |f(t)| < \infty$. Then we can take $\rho(s,t) = |s-t|$, $G(x) = C_H f_T^* \cdot x^H$, and $\Theta(\varepsilon) \sim C_H f_T^* \varepsilon^H |\log \varepsilon|^{1/2}$ will be a uniform modulus of $I(f)$ on $[0,T]$.

Now consider the case $H \in (0, 1/2)$, and f, as before, belongs to $L_2^H[0,t]$ for any $0 \leq t \leq T$. We suppose additionally that $f \in C^\beta[0,T]$ for $H + \beta > 1/2$. Then, according to Remark 1.10.7, we can take $\rho(s,t) = |s-t|$, $G(x) = C_H \|f\|_{C^\beta[0,T]} x^H$, and $\Theta(\varepsilon) \sim C_H \|f\|_{C^\beta[0,T]} \varepsilon^H |\log \varepsilon|^{1/2}$ will be a uniform modulus of $I(f)$ on $[0,T]$.

Remark 1.16.6. Some results related to moduli of continuity for non-Gaussian processes can be found in Subsection 3.5.9.

1.17 Estimates for Fractional Derivatives of fBm and of Wiener Integrals w.r.t. Wiener Process via the Garsia–Rodemich–Rumsey Inequality

The following results are not only of independent interest but also will be used in Chapter 3, devoted to stochastic differential equations involving fBm.

1.17 Estimates for Fractional Derivatives of fBm and Wiener Integrals

Consider for any $T > 0$ the random variable that is the right-sided Riemann–Liouville fractional derivative of order β (in Weyl representation) of fBm B^H, where $1 - H < \beta < 1/2$ and $H \in (1/2, 1)$:

$$G_t := \frac{1}{\Gamma(\beta)} \sup_{0 \le s < z \le t} |D_{z-}^{1-\beta} B_{z-}^H(s)|, t \in [0, T].$$

Lemma 1.17.1. *For any $1 - H < \beta < 1/2$ and any $p > 0$*

$$EG_t^p < \infty.$$

Proof. By the Garsia–Rodemich–Rumsey inequality (GRR71), for any $p \ge 1$ and $\rho > p^{-1}$ there exists a constant $C_{\rho,p} > 0$ such that for any continuous function f on $[0, T]$ and for all $s < z \le t \in [0, T]$

$$|f(z) - f(s)|^p \le C_{\rho,p} |z - s|^{\rho p - 1} \int_0^z \int_0^z \frac{|f(x) - f(y)|^p}{|x - y|^{\rho p + 1}} dx\, dy.$$

Choose $\varepsilon < \beta - (1 - H)$ and put $\rho = H - \frac{\varepsilon}{2}$, $p = \frac{2}{\varepsilon}$ and $f(t) = B_t^H$:

$$|B_z^H - B_s^H| \le C_{H,\varepsilon} |z - s|^{H-\varepsilon} \xi_{t,\varepsilon},$$

where

$$\xi_{t,\varepsilon} = \left(\int_0^t \int_0^t \frac{|B_x^H - B_y^H|^{\frac{2}{\varepsilon}}}{|x - y|^{\frac{2H}{\varepsilon}}} dx\, dy \right)^{\frac{\varepsilon}{2}}, 0 < \varepsilon < H. \qquad (1.17.1)$$

Since $B_x^H - B_y^H$ is a Gaussian random variable, and $E|B_x^H - B_y^H|^2 = |x - y|^{2H}$, we have that for the random variable $\xi_{t,\varepsilon}$ for any $q > 1$

$$E|\xi_{t,\varepsilon}|^q = E \left(\int_0^t \int_0^t \frac{|B_x^H - B_y^H|^{\frac{2}{\varepsilon}}}{|x-y|^{\frac{2H}{\varepsilon}}} dx\, dy \right)^{\frac{q\varepsilon}{2}}$$
$$\le C_{q,H,T} \int_0^T \int_0^T \frac{E|B_x^H - B_y^H|^q}{|x-y|^{Hq}} dx\, dy \le C_{q,H,T},$$

which means that all moments of $\xi_{t,\varepsilon}$ are finite.

Further, for $\varepsilon < \beta - (1 - H)$

$$G_t \le C_\beta \sup_{0 \le s < z \le t} \left(\frac{|B_z^H - B_s^H|}{|z-s|^{1-\beta}} + \int_s^z \frac{|B_s^H - B_y^H|}{|s-y|^{2-\beta}} dy \right)$$
$$\le C_{\beta,H,\varepsilon} \sup_{0 \le s < t} (t-s)^{H - \varepsilon - 1 + \beta} \xi_{t,\varepsilon} \le C_{\beta,H,\varepsilon} \xi_{t,\varepsilon},$$

so, $EG_t^p < \infty$ for any $p > 0$. □

Remark 1.17.2. 1) It is easy to see that the random process $\{G_t, t \in [0, T]\}$ is dominated, up to a constant, by some continuous process with moments of any order, namely, by $\xi_{t,\varepsilon}$.
2) Evidently, all moments of the random variable G_T are finite.
3) It follows immediately from Corollary 1.9.4 that the same conclusions hold for a Wiener integral w.r.t. fBm with a bounded integrand and $H \in (1/2, 1)$.

Now, we establish Hölder properties and estimates, similar to the aforementioned, for the integral $\{\int_0^t b_s dW_s, t \in [0,T]\}$, where b_s is a predictable bounded process. For any $0 < \delta < 1/4$ put $p = \frac{2}{\delta}$, $\theta = 1/2 - \delta/2$ in the Garsia–Rodemich–Rumsey inequality. Then

$$\left| \int_s^t b_u dW_u \right| \leq C_\delta |t-s|^{1/2-\delta} \xi_{t,\delta}^b,$$

where

$$\xi_{t,\delta}^b := \left(\int_0^t \int_0^t \frac{|\int_x^y b_u dW_u|^{2/\delta}}{|x-y|^{1/\delta}} dx\, dy \right)^{\delta/2}, \qquad (1.17.2)$$

and for any $q > 1$ from the Hölder and Burkholder inequalities

$$E|\xi_{t,\delta}^b|^q = E \left(\int_0^t \int_0^t \frac{|\int_x^y b_u dW_u|^{2/\delta}}{|x-y|^{1/\delta}} dx\, dy \right)^{q\delta/2}$$

$$\leq C_{q,t} \int_0^t \int_0^t \frac{E|\int_x^y b_u dW_u|^q dx\, dy}{|x-y|^{q/2}} \leq C_{q,t} \int_0^t \int_0^t \frac{|\int_x^y b_u^2 du|^{q/2} dx\, dy}{|x-y|^{q/2}} \leq C_{q,t},$$

Note that the process $\xi_{t,\delta}^b$ is continuous and strictly increasing, so, our Wiener integral with respect to the Wiener process is dominated by a strictly increasing process with all moments bounded on $[0,T]$.

1.18 Power Variations of fBm and of Wiener Integrals w.r.t. fBm

We start here with the simple result obtained by Rogers in (Rog97). Consider for fBm $\{B_t^H, t \geq 0\}$ with $H \in (0,1)$ and for $p > 0$ the sums

$$S_{n,p}(t) = \sum_{j=1}^{2^n} \left| B_{\frac{jt}{2^n}}^H - B_{\frac{(j-1)t}{2^n}}^H \right|^p \cdot 2^{n(pH-1)}, \qquad (1.18.1)$$

and

$$\tilde{S}_{n,p}(t) = 2^{-n} \sum_{j=1}^{2^n} \left| B_{jt}^H - B_{(j-1)t}^H \right|^p.$$

Then $Law(S_{n,p}(t)) = Law(\tilde{S}_{n,p}(t))$ (i.e., these sums have identical distribution), due to the self-similarity property of B^H: $(Law(B_{ct}^H, t > 0) = Law(c^H B_t^H, t > 0))$.

The sequence $(B_k^H - B_{k-1}^H)_{k \in N}$ is stationary. Therefore, from the ergodic theorem

$$\tilde{S}_{n,p}(t) \to E|B_t^H|^p =: C_p t^{pH} \quad \text{as} \quad n \to \infty$$

1.18 Power Variations of fBm and of Wiener Integrals w.r.t. fBm

with probability 1 and in $L_1(P)$, whence

$$S_{n,p}(t) \xrightarrow{d} C_p t^{pH}, n \to \infty, \tag{1.18.2}$$

so $S_{n,p}(t) \xrightarrow{P} C_p t^{pH}$, $n \to \infty$.
From (1.18.1)–(1.18.2)

$$\sum_{j=1}^{2^n} \left| B^H_{\frac{jt}{2^n}} - B^H_{\frac{(j-1)t}{2^n}} \right|^p \xrightarrow{P} \begin{cases} 0, & p > \frac{1}{H}, \\ +\infty, & p < \frac{1}{H}, \\ E\left|B^H_t\right|^{1/H}, & p = 1/H. \end{cases} \tag{1.18.3}$$

Now, consider the interval $[0,1]$; let $\{\pi_k, k \geq 1\}$ be a sequence of refining partitions and $\Pi(\delta)$ be the set of all partitions π of $[0,1]$ with $|\pi| < \delta$.
Evidently, from (1.18.3) we obtain that

$$\lim_{\delta \to 0} \sup_{\pi \in \Pi(\delta)} S(|x|^p, \pi, B^H) = +\infty$$

with probability 1, where $p < \frac{1}{H}$ and

$$S(\psi(x), \pi, X) := \sum_{t_j \in \pi} \psi(X_{t_j} - X_{t_{j-1}}).$$

Now we use the result of Kawada and Kôno (KK73).

Theorem 1.18.1. *Let $\{X_t, 0 \leq t \leq 1\}$ be a centered Gaussian process with continuous trajectories such that*

$$E|X_t - X_s|^2 \leq \sigma^2(|t-s|),$$

where $\{\sigma(t), 0 \leq t \leq 1\}$ is a continuous function with $\sigma(0) = 0$. Let $\{\psi(t), 0 \leq t \leq 1\}$ be a non-decreasing regular varying function with exponent $\alpha > 0$ satisfying

$$\psi(\sigma(t)) \leq t\gamma(t) \quad \text{for} \quad 0 \leq t \leq 1 \quad \text{and} \quad \lim_{t \downarrow 0} \gamma(t) = 0.$$

Then $\lim_{\delta \to 0} \sup_{\pi \in \Pi(\delta)} S(\psi(x), \pi, X) = \text{constant}$ (including ∞) holds with probability 1.

Put $X_t = B^H_t$, $\sigma^2(t) = t^{2\alpha+1}$, $\psi(t) = t^{\frac{1}{H}+\varepsilon}$ for some $\varepsilon > 0$ (recall that a function is regularly varying if $\frac{\psi(xt)}{\psi(t)} \to \rho(x)$ as $t \to \infty$ and in this case $\rho(x) = x^\beta$ for some $\beta \geq 0$). Then $\psi(\sigma(t)) = t^{1+H\varepsilon}$ and all the assumptions of Theorem 1.18.1 are satisfied. So, $\lim_{\delta \to 0} \sup_{\pi \in \Pi(\delta)} S(|x|^p, \pi, B^H) = \text{const}$ for any $p > \frac{1}{H}$. Evidently, this constant is zero since for any $p' > p > \frac{1}{H}$

$$S(x^{p'}, \pi, B^H) \leq \sup_{0 \leq t < t' \leq t+\delta \leq 1} |B^H_t - B^H_{t'}|^{p'-p} \cdot S(x^p, \pi, B^H),$$

and the first factor tends to zero a.s. as $\delta \to 0$.

Now, let $H \in (0, \frac{1}{2})$. In this case we can use the following theorem for the case $p = \frac{1}{H}$.

1 Wiener Integration with Respect to Fractional Brownian Motion

Theorem 1.18.2 ((KK73)). *1) Let the following assumptions hold:*
(a) $E|X_s - X_t|^2 \leq \sigma^2(|t-s|)$;
(b) $\sigma(t)$ *is a non-decreasing regular varying function;*
(c) *the function* $\sigma(t)\sqrt{2\log\log \frac{1}{t}}$ *is strictly increasing near the origin.*

Let $\tilde{\Pi}(k)$ be the set of all partitions such that $\min|t_j - t_{j-1}| \geq \frac{1}{k}$. Then

$$\limsup_{k\to\infty} \sup_{\pi\in\tilde{\Pi}(k)} \frac{S(\sigma^{-1}(x), \pi, X)}{\Phi(\frac{1}{k})} \leq 1,$$

with probability 1, where

$$\Phi(t) = \sup_{s\geq t} \frac{\sigma^{-1}\left(\sigma(s)\sqrt{2\log\log\frac{1}{s}}\right)}{s}.$$

2) Let the assumption (b) hold and also
(d) $E|X_s - X_t|^2 \leq \sigma^2(|t-s|)$;
(e) $\sigma^2(t) - \sigma^2(t-h) \leq C\sigma^2(h)$ *for some* $C > 0$, *any small* t *and* $0 \leq h \leq t$.
Then $\liminf_{k\to\infty} \sup_{\pi\in\tilde{\Pi}(k)} \frac{S(\sigma^{-1}(x),\pi,X)}{\Phi(\frac{1}{k})} \geq 1$ *with probability 1.*

Put $\sigma(t) = t^H$, $X_t = B_t^H$. Then conditions (a), (b), (c) and (d) hold. Moreover, for $H \in (0, \frac{1}{2})$ $\sigma^2(t) - \sigma^2(t-h) = t^{2\alpha+1} - (t-h)^{2\alpha+1} \leq h^{2\alpha+1}$ for all $0 \leq h \leq t \leq 1$. The function $\Phi(t)$ now has the form $\Phi(t) = (2\log\log\frac{1}{t})^{\frac{1}{H}}$, whence $\lim_{k\to\infty} \sup_{\pi\in\tilde{\Pi}(k)} \frac{S(|x|^{\frac{1}{H}},\pi,B^H)}{(2\log\log k)^{\frac{1}{2H}}} = 1$ or, in other words, $\lim_{k\to\infty} \sup_{\pi\in\tilde{\Pi}(k)} \frac{\sum_{t_j\in\pi}|B_{t_j}^H - B_{t_{j-1}}^H|^{\frac{1}{H}}}{(2\log\log k)^{\frac{1}{2H}}} = 1$.

For $H \in (1/2, 1)$ we have no assumption (e), so, give only upper bounds. Namely, from the first statement of Theorem 1.18.2, we can deduce that

$$\limsup_{k\to\infty} \sup_{\pi\in\tilde{\Pi}(k)} \frac{\sum_{t_j\in\pi}|B_{t_j}^H - B_{t_{j-1}}^H|^{\frac{1}{H}}}{(2\log\log k)^{\frac{1}{2H}}} \leq 1.$$

Moreover, the following result holds.

Theorem 1.18.3. *Under assumptions (a)–(c)*

$$\lim_{\delta\to 0} \sup_{\pi\in\Pi(\delta)} S(\psi(x), \pi, X) \leq 1,$$

with probability 1, where $\psi(x)$ is the inverse function to $\sigma(t)\sqrt{2\log\log\frac{1}{t}}$ near the origin.

In our case it means that

$$\lim_{\delta\to 0} \sup_{\pi\in\Pi(\delta)} \sum_{t_j\in\pi} \psi(|B_{t_j}^H - B_{t_{j-1}}^H|) \leq 1,$$

where $\psi(t)$ is the inverse function to $t^H\sqrt{2\log\log\frac{1}{t}}$.

Let, as before, Π be the set of all partitions of the interval $[0,1]$.

Definition 1.18.4. For any $p > 0$ define p-variation of the function f on the interval $[a,b]$ as
$$v_p(f) = \sup_{\pi \in \Pi} S(|x|^p, \pi, f).$$

Also, let p-variation index of the function f be $v(f) := \inf(p : v_p(f) < \infty)$.

The last relations mean that $v(B_H) = \frac{1}{H}$ with probability 1, and, moreover,
$$v_p(B_H) < \infty \quad \text{for} \quad p > \frac{1}{H} \quad \text{and} \quad = \infty \quad \text{for} \quad p < \frac{1}{H}.$$

This result was obtained in (Nrv99) from another point of view. Let $\{X_t, t \geq 0\}$ be a Gaussian process with stationary increments and $E|X_{t+s} - X_t|^2 = \sigma^2(s)$. Let $\gamma_* := \inf\{\gamma > 0 : \lim_{s\downarrow 0} \frac{s^\gamma}{\sigma(s)} = 0\}$ and $\gamma^* := \sup\{\gamma > 0 : \lim_{s\downarrow 0} \frac{s^\gamma}{\sigma(s)} = \infty\}$. Then $0 \leq \gamma^* \leq \gamma_* \leq +\infty$. If $\gamma^* = \gamma_*$ then we say that the process X_t has the Orey index $\gamma(X) = \gamma^* = \gamma_*$. Let X_t have the Orey index $\gamma(X) \in (0,1)$; then it follows from the results of Berman (Ber69) and also from (JM83) that the p-variation index of X_t equals $v(X) = \frac{1}{\gamma(X)}$. Evidently, the Orey index of the fBm equals its Hurst index and equals H.

Now consider briefly the Gaussian process $X_t = I_t(f) = \int_0^t f(s)dB_s^H$. Let $H \in (\frac{1}{2}, 1)$ and the function f is essentially bounded on $[0,1]$, ess $\sup_{0\leq t \leq 1}|f(t)| = f^*$.

Then, according to Theorem 1.10.3, $E|X_t - X_s|^2 \leq \sigma^2(|t-s|)$, where $\sigma^2(t) = C_H(f^*)^2 t^{2\alpha+1}$, therefore from Theorem 1.18.1 $\lim_{\delta \to 0} \sup_{\pi \in \Pi(\delta)} S(|x|^p, \pi, I) = 0$ for any $p > \frac{1}{H}$ and from Theorems 1.18.2 and 1.18.3

$$\limsup_{k\to\infty} \sup_{\pi \in \tilde{\Pi}(k)} \frac{S(|x|^{\frac{1}{H}}, \pi, I)}{\Phi(\frac{1}{k})} \leq 1 \quad P\text{-a.s.,} \tag{1.18.4}$$

$$\lim_{\delta \to \infty} \sup_{\pi \in \tilde{\Pi}(\delta)} S(\psi(x), \pi, I) \leq 1 \quad P\text{-a.s.} \tag{1.18.5}$$

where $\psi(x)$ is the inverse to $C_H^{1/2} f^* t^H \sqrt{2\log\log\frac{1}{t}}$ near the origin.

Let $f_* := \operatorname{ess\,inf}_{0\leq t\leq 1} f(t) > 0$. Then
$$E|I_t - I_s|^2 = C_H \int_s^t \int_s^t f(u)f(v)|u-v|^{2\alpha-1}du\,dv \geq C_H f_*^2 |t-s|^{2\alpha+1},$$

whence $S(|x|^p, \pi, I) \xrightarrow{P} \infty$ as $|\pi| \to 0$ and $p < \frac{1}{H}$, and together with Theorem 1.18.1 it means that

$$\lim_{\delta \to 0} \sup_{\pi \in \Pi(\delta)} S(|x|^p, \pi, I) = \infty \quad P\text{-a.s.,} \quad p < \frac{1}{H}.$$

For $H \in (0, \frac{1}{2})$ and f with $f_* > 0$ we can immediately conclude from Theorem 1.9.1 that

$$E|I_t - I_s|^2 \geq C_H \|f\|^2_{L_{\frac{1}{H}}[s,t]} \geq C_H f_*^2 |t-s|^{2\alpha+1},$$

whence $S(|x|^p, \pi, I) \xrightarrow{P} \infty$ as $|\pi| \to 0$ and $p < \frac{1}{H}$. Let $f \in C^\beta[0,1]$. Then we can deduce from Remark 1.10.7 that

$$E|I_t - I_s|^2 \leq C_H \|f\|_{C^\beta([0,1])}((t-s)^{2\alpha+1} + (t-s)^{2H+2\beta}),$$

whence (1.18.4)–(1.18.5) follow for $H \in (0, \frac{1}{2})$.

Remark 1.18.5. In the paper (CNW06) the process of the form $\int_0^t u_s dB_s^H$ is considered where u_s is a stochastic process with paths of finite q-variation and the integral is pathwise Riemann–Stieltjes integral (construction of such integrals is described in Section 2.1). The convergence in probability of the normalized power variations of these integrals is established and their deviations are considered.

Remark 1.18.6. Modern results on power variation of the integrals and other processes related to fBm are established in (GuNu05), (Nrv99), (CNW06), (DN99).

1.19 Lévy Theorem for fBm

The idea of this problem belongs to E. Valkeila. The results are published in (MV06). We start with the classical Lévy theorem:

Theorem 1.19.1. *Let $\{\mu(t), t \geq 0\}$ be a continuous local martingale with the angle bracket $\langle \mu \rangle_t = t$. Then μ_t is the Wiener process.*

The natural question is: how can the fBm be characterized in a similar way or by some other properties?

Let $\{\Omega, \mathcal{F}, \{\mathcal{F}_t\}_{t \geq 0}, P\}$ be some stochastic basis, $\{X_t, t \geq 0\}$ be a stochastic process (not necessarily adapted, as for beginning). For any $t > 0$, denote $t_k := t\frac{k}{n}$, $1 \leq k \leq n$. The main result of this section is:

Theorem 1.19.2. *Let the process X_t satisfy the following conditions:*

(a) trajectories of X are Hölder of any order $0 < \beta < H$, where $0 < H < 1$;
(b) $n^{2\alpha} \sum_{k=1}^n (X_{t_k} - X_{t_{k-1}})^2 \to t^{2\alpha+1}$ for any $t > 0$ in the space $L_1(P)$, as $n \to \infty$.
(c) the process $M_t := \int_0^t s^{-\alpha}(t-s)^{-\alpha} dX_s$ is an $\mathcal{F}_t$-adapted continuous square-integrable martingale, where $\alpha = H - 1/2$.

Then X_t is an $\mathcal{F}_t$-adapted fBm with Hurst index H.

Proof. We shall divide the proof into several steps. First, consider the case $H \in (\frac{1}{2}, 1)$. Let the square-integrable martingale $W_t := \int_0^t s^\alpha dM_s$, $t \in [0, T]$, $T > 0$ and the process $Y_t := \int_0^t s^{-\alpha} dX_s$. For convenience we put $T = 1$. We can establish the existence in the pathwise sense of the latter integral using Hölder properties of X and integration by parts.

Evidently,
$$M_t = \int_0^t (t-s)^{-\alpha} dY_s. \tag{1.19.1}$$

Lemma 1.19.3. *The process X_t admits the representation*
$$X_t = \frac{1}{C_H} \int_0^t \left[\int_u^t s^\alpha (s-u)^{\alpha-1} ds \right] u^{-\alpha} dW_u,$$
where $C_H = B(\alpha, 1-\alpha)$.

Proof. Equation (1.19.1) is a generalized Abel integral equation and has the formal solution
$$Y_t = \frac{1}{C_H} \int_0^t (t-s)^{\alpha-1} M_s ds. \tag{1.19.2}$$

It is very easy to check that (1.19.1) becomes an identity, if we substitute (1.19.2) into (1.19.1), rewritten as
$$M_t = t^{-\alpha} Y_t + \alpha \int_0^t (t-s)^{-1-\alpha} (Y_t - Y_s) ds. \tag{1.19.3}$$

Moreover, the corresponding homogeneous equation
$$0 = t^{-\alpha} Y_t + \alpha \int_0^t (t-s)^{-1-\alpha} (Y_t - Y_s) ds,$$
has only a zero solution, whence Y_t admits the representation (1.19.2). Further,
$$X_t = \int_0^t s^\alpha dY_s = t^\alpha Y_t - \alpha \int_0^t s^{\alpha-1} Y_s ds$$
$$= \frac{t^\alpha}{C_H} \int_0^t (t-s)^{\alpha-1} M_s ds - \frac{\alpha}{C_H} \int_0^t s^{\alpha-1} \int_0^s (s-u)^{\alpha-1} M_u du\, ds$$
$$= \frac{1}{C_H} \int_0^t \left[\int_u^t s^\alpha (s-u)^{\alpha-1} ds \right] dM_u.$$
□

Remark 1.19.4. From Lemma 1.19.3, for any $1 \le k \le n$, it follows that
$$X_{t_k} - X_{t_{k-1}} = \frac{1}{C_H} \left(\int_0^{t_k} \left(\int_s^{t_k} u^\alpha (u-s)^{\alpha-1} du \right) dM_s \right.$$

$$-\int_0^{t_{k-1}}\left(\int_s^{t_{k-1}} u^\alpha(u-s)^{\alpha-1}du\right)dM_s\Big)$$

$$=\frac{1}{C_H}\Bigg(\int_0^{t_{k-1}}\left(\int_{t_{k-1}}^{t_k} u^\alpha(u-s)^\alpha du\right)dM_s$$

$$+\int_{t_{k-1}}^{t_k}\left(\int_s^{t_k} u^\alpha(u-s)^{\alpha-1}du\right)dM_s\Bigg). \tag{1.19.4}$$

Denote
$$\varphi_k^t(s):=\int_{t_{k-1}}^{t_k} u^\alpha(u-s)^{\alpha-1}du,$$

and
$$\psi_k^t(s):=\int_s^{t_k} u^\alpha(u-s)^{\alpha-1}du.$$

Then
$$\Delta X_{t_k}:=X_{t_k}-X_{t_{k-1}}=\frac{1}{C_H}\left(\int_0^{t_{k-1}}\varphi_k^t(s)dM_s+\int_{t_{k-1}}^{t_k}\psi_k^t(s)dM_s\right).$$
$$\tag{1.19.5}$$

Now, let $0<s<t$, and let $\frac{s}{t}$ be a rational number, such that $\frac{s}{t}\in\mathbb{Q}$.

Lemma 1.19.5. *Let $\widetilde{n}\in\mathbb{N}$ be an increasing sequence, such that $\widetilde{n}\frac{s}{t}\in\mathbb{N}$, $t_{\widetilde{k}}:=tk/\widetilde{n}$.*

Then $\widetilde{n}^{2\alpha}\sum_{k=\widetilde{n}\frac{s}{t}+1}^{\widetilde{n}}(X_{t_{\widetilde{k}}}-X_{t_{\widetilde{k-1}}})^2 \xrightarrow{L_1(P)} t^{2\alpha}(t-s),\ \widetilde{n}\to\infty.$

Proof. Evidently,

$$\widetilde{n}^{2\alpha}\sum_{k=1}^{\widetilde{n}\frac{s}{t}}(\Delta X_{t_{\widetilde{k}}})^2 = \left(\widetilde{n}\frac{s}{t}\right)^{2\alpha}\cdot\left(\frac{t}{s}\right)^{2\alpha}\sum_{k=1}^{\widetilde{n}\frac{s}{t}}\left(\Delta X_{\frac{sk}{\widetilde{n}\frac{s}{t}}}\right)^2 \to s^{2\alpha+1}\cdot\left(\frac{t}{s}\right)^{2\alpha} = st^{2\alpha}.$$

We know from condition (b) that $\widetilde{n}^{2\alpha}\sum_{k=1}^{\widetilde{n}}\left(\Delta X_{t_{\widetilde{k}}}\right)^2\to t^{2\alpha+1}$, whence the claim follows. □

Now we want to estimate $\widetilde{n}^{2\alpha}\sum_{k=\widetilde{n}\frac{s}{t}}^{\widetilde{n}}\left(\Delta X_{t_{\widetilde{k}}}\right)^2$ in terms of the angle bracket $\langle M\rangle$, by using representations (1.19.4)) and (1.19.5). In order to do this, rewrite the increment of the process X in the form

$$\Delta X_{t\frac{k}{n}} = \frac{1}{C_H}\left(\int_0^{t_{k-2}}\varphi_k^t(s)dM_s + \int_{t_{k-2}}^{t_{k-1}}\varphi_k^t(s)dM_s + \int_{t_{k-1}}^{t_k}\psi_k^t(s)dM_s\right)$$

$$=:\frac{1}{C_H}(I_1^k+I_2^k+I_3^k).$$

Evidently,

$$\varphi_k^t(s) \leq \left(t_k^\alpha (t_{k-1} - s)^{\alpha-1} \cdot \frac{t}{n}\right) \wedge \left(\frac{n^{-\alpha}}{\alpha}\right) \quad (1.19.6)$$

and

$$\widetilde{n}^{2\alpha} \sum_{k=\widetilde{n}\frac{s}{t}+1}^{\widetilde{n}} \left(\Delta X_{t\frac{k}{n}}\right)^2$$

$$= \frac{\widetilde{n}^{2\alpha}}{C_H^2} \left(\sum_{k=\widetilde{n}\frac{s}{t}+1}^{\widetilde{n}} \left((I_1^k)^2 + (I_2^k)^2 + (I_3^k)^2 + 2I_1^k \cdot I_2^k + 2I_1^k \cdot I_3^k + 2I_2^k \cdot I_3^k\right)\right). \quad (1.19.7)$$

Now we shall estimate the terms on the right-hand side of (1.19.7).

Lemma 1.19.6. *There exist two constants $C_1 > 0$, $C_2 > 0$ such that*

$$C_1 t^{2\alpha} \int_s^t u^{2\alpha} d\langle M \rangle_u \leq \underset{\widetilde{n}\to\infty}{P\text{-lim}}(\widetilde{n}^{2\alpha} \sum_{k=\frac{\widetilde{n}s}{t}+1}^{\widetilde{n}} (I_1^k)^2) \leq C_2 t^{4\alpha}(\langle M \rangle_t - \langle M \rangle_s).$$

Proof. For simplicity, we shall omit $\sim$, and consider only such n that $n\frac{s}{t} \in \mathbb{N}$. From the Itô formula for square-integrable martingales, it follows that

$$(I_1^k)^2 = \left(\int_0^{t_{k-2}} \varphi_k^t(u) dM_u\right)^2 = \int_0^{t_{k-2}} (\varphi_k^t(u))^2 d\langle M \rangle_u$$

$$+ 2\int_0^{t_{k-2}} \int_0^u \varphi_k^t(v) dM_v \cdot \varphi_k^t(u) dM_u.$$

First, we estimate

$$S_1^n := n^{2\alpha} \sum_{k=n\frac{s}{t}+2}^{n} \int_0^{t_{k-2}} (\varphi_k^t(u))^2 d\langle M \rangle_u.$$

From (1.19.6), we obtain that

$$\int_0^{t_{k-2}} (\varphi_k^t(u))^2 d\langle M \rangle_u \leq t^{2\alpha} \left(\frac{k}{n}\right)^{2\alpha} \frac{t^2}{n^2} \int_0^{t_{k-2}} (t_{k-1} - u)^{2\alpha-2} d\langle M \rangle_u.$$

So, the estimate of S_1^n from above has the form

$$S_1^n \leq n^{2\alpha-2} t^{2\alpha+2} \sum_{k=n\frac{s}{t}+2}^{n} \left(\frac{k}{n}\right)^{2\alpha} \int_0^{t_{k-2}} (t_{k-1} - u)^{2\alpha-2} d\langle M \rangle_u. \quad (1.19.8)$$

Now, we rewrite the sum in (1.19.8) for $0 < s < t$ and $2 \leq n\frac{s}{t} \leq n-3$:

$$S_{11}^n := \sum_{k=n\frac{s}{t}+2}^{n} \int_0^{t_{k-2}} (t_{k-1} - u)^{2\alpha-2} d\langle M \rangle_u$$

$$= \left(\sum_{i=1}^{n\frac{s}{t}} \sum_{k=n\frac{s}{t}+2}^{n} + \sum_{i=n\frac{s}{t}+1}^{n-2} \sum_{k=i+2}^{n} \right) \int_{t_{i-1}}^{t_i} (t_{k-1} - u)^{2\alpha-2} d\langle M \rangle_u$$

$$+ \sum_{i=n\frac{s}{t}+1}^{n-2} \int_{t_{i-1}}^{t_i} \left(\sum_{k=i+2}^{n} (t_{k-1} - u)^{2\alpha-2} \right) d\langle M \rangle_u. \tag{1.19.9}$$

Evidently,

$$\frac{1}{n} \sum_{k=n\frac{s}{t}+2}^{n} (t_{k-1} - u)^{2\alpha-2} \leq \int_{s-u}^{s+t-u} x^{2\alpha-2} dx \cdot \frac{1}{t} \leq (s-u)^{2\alpha-1} \frac{1}{t} \cdot (1 - 2\alpha)^{-1},$$

and

$$\frac{1}{n} \sum_{k=i+2}^{n} (t_{k-1} - u)^{2\alpha-2} \leq \frac{1}{n} (t_{i+1} - u)^{2\alpha-2} + \frac{1}{(1-2\alpha)t} (t_{i+1} - u)^{2\alpha-1}.$$

We substitute these estimates into (1.19.9):

$$S_{11}^n \leq \frac{n}{1-2\alpha} \sum_{i=1}^{n\frac{s}{t}} \int_{t_{i-1}}^{t_i} (s-u)^{2\alpha-1} \frac{1}{t} d\langle M \rangle_u$$

$$+ n \sum_{i=n\frac{s}{t}+1}^{n} \int_{t_{i-1}}^{t_i} \left[\frac{1}{n} (t_{i+1} - u)^{2\alpha-2} + \frac{1}{(1-2\alpha)t} (t_{i+1} - u)^{2\alpha-1} \right] d\langle M \rangle_u$$

$$\leq \frac{n}{t} \int_0^s (s-u)^{2\alpha-1} d\langle M \rangle_u + t^{2\alpha-2} n^{2-2\alpha} \left(1 + \frac{1}{1-2\alpha} \right) (\langle M \rangle_t - \langle M \rangle_s).$$

We return to (1.19.8) and obtain that

$$S_1^n \leq n^{2\alpha-1} t^{2\alpha+1} \int_0^s (s-u)^{2\alpha-1} d\langle M \rangle_u + t^{4\alpha} \left(\frac{1}{1-2\alpha} + 1 \right) (\langle M \rangle_t - \langle M \rangle_s).$$

Note that the martingale M is Hölder continuous up to order $\frac{1}{2}$, so W is Hölder continuous up to order $\frac{1}{2}$, $\langle W \rangle$ is Hölder continuous up to 1, and the integral

$$\int_0^s (s-u)^{2\alpha-1} d\langle M \rangle_u = \int_0^s (s-u)^{2\alpha-1} u^{-2\alpha} d\langle W \rangle_u$$

exists. Therefore, $n^{2\alpha-1} \int_0^s (s-u)^{2\alpha-1} d\langle M \rangle_u \to 0$, $n \to \infty$. We obtain that $\lim_{n \to \infty} S_1^n \leq C_2 t^{4\alpha} (\langle M \rangle_t - \langle M \rangle_s)$. Now we estimate S_1^n from below: first,

$$(\varphi_k^t(u))^2 \geq (t_{k-1})^{2\alpha}(t_k - u)^{2\alpha-2} \cdot \frac{t^2}{n^2}.$$

Then,

$$S_1^n \geq n^{2\alpha-2} t^2 \sum_{k=n\frac{s}{t}+2}^{n} \int_0^{t_{k-2}} (t_{k-1})^{2\alpha} (t_k - u)^{2\alpha-2} d\langle M \rangle_u$$

$$= n^{2\alpha-2} t^2 \sum_{i=1}^{n\frac{s}{t}} \int_{t_{i-1}}^{t_i} \left(\sum_{k=n\frac{s}{t}+2}^{n} (t_{k-1})^{2\alpha}(t_k - u)^{2\alpha-2} \right) d\langle M \rangle_u$$

$$+ n^{2\alpha-2} t^2 \sum_{i=n\frac{s}{t}+1}^{n-2} \int_{t_{i-1}}^{t_i} \left(\sum_{k=i+2}^{n} (t_{k-1})^{2\alpha}(t_k - u)^{2\alpha-2} \right) d\langle M \rangle_u. \quad (1.19.10)$$

Consider the interior sum of the second term:

$$\frac{1}{n} \sum_{k=i+2}^{n} (t_{k-1})^{2\alpha} (t_k - u)^{2\alpha-2} \geq \frac{1}{t} \int_{t_{i+2}-u}^{t-u} x^{2\alpha-2} \left(x + u - \frac{1}{n} \right)^{2\alpha} dx$$

$$\geq \frac{1}{t} (t_{i+1})^{2\alpha} \int_{t_{i+2}-u}^{t-u} x^{2\alpha-2} dx$$

$$\geq t^{2\alpha-1} \left(\frac{i+1}{n} \right)^{2\alpha} \cdot \frac{(t_{i+2} - u)^{2\alpha-1} - (t - u)^{2\alpha-1}}{1 - 2\alpha}.$$

So,

$$S_1^n \geq \frac{t^{2\alpha+1} n^{2\alpha-1}}{1 - 2\alpha} \sum_{i=n\frac{s}{t}+1}^{n-3} \left(\frac{i+1}{n} \right)^{2\alpha} \int_{t_{i-1}}^{t_i} \left[(t_{i+2} - u)^{2\alpha-1} \right.$$

$$\left. - (t - u)^{2\alpha-1} \right] d\langle M \rangle_u.$$

Consider the function $f(u) := (t_{i+2} - u)^{2\alpha-1} - (t - u)^{2\alpha-1}$ on the interval $[t_{i-1}, t_i]$:

$$f(u) \geq (t_{i+2} - t_{i-1})^{2\alpha-1} - (t - t_{i-1})^{2\alpha-1} = \frac{3^{2\alpha-1} - 4^{2\alpha-1}}{n^{2\alpha-1}} t^{2\alpha-1}.$$

Therefore,

$$S_1^n \geq \frac{t^{2\alpha+1} n^{2\alpha-1}}{1 - 2\alpha} \sum_{i=n\frac{s}{t}+1}^{n-3} \int_{t_{i-1}}^{t_i} \left(\frac{i+1}{n} \right)^{2\alpha} \frac{t^{2\alpha-1}}{n^{2\alpha-1}} d\langle M \rangle_u$$

$$\times(3^{2\alpha-1} - 4^{2\alpha-1}) \geq C_1 t^{2\alpha} \sum_{i=n\frac{s}{t}+1}^{n-3} \int_{t_{i-1}}^{t_i} \left(u + \frac{2t}{n}\right)^{2\alpha} d\langle M\rangle_u,$$

and

$$\lim_{n\to\infty} S_1^n \geq C_1 t^{2\alpha} \int_s^t u^{2\alpha} d\langle M\rangle_u,$$

or, in terms of $\langle W\rangle$,

$$\lim_{n\to\infty} S_1^n \geq C_1 t^{2\alpha} \left(\langle W\rangle_t - \langle W\rangle_s\right).$$

Now, we try to prove that $S_2^n \to 0$ in probability, where

$$S_2^n = n^{2\alpha} \sum_{k=n\frac{s}{t}+2}^{n} \int_0^{t_{k-2}} \left(\int_0^u \varphi_k^t(s) dM_s\right) \varphi_k^t(u) dM_u.$$

Evidently, it is sufficient to consider the sums of the form

$$S_3^n = n^{2\alpha} \sum_{k=2}^{n} \int_0^{l_{k-2}} \left(\int_0^u \varphi_k^t(s) dM_s\right) \varphi_k^t(u) dM_u,$$

because the sums

$$\sum_{k=2}^{n\frac{s}{t}+2} \int_0^{t_{k-2}} \left(\int_0^u \varphi_k^t(s) dM_s\right) \varphi_k^t(u) dM_u$$

can be considered in a similar way.

We use a very weak version of the Lenglart inequality: if N is a locally square integrable martingale on $\mathbb{R}$, then for any $\varepsilon > 0$, $A > 0$ and $T > 0$ we have that

$$P\{\sup_{0 \leq t \leq T} |N(t)| \geq \varepsilon\} \leq \frac{A}{\varepsilon^2} + P\{\langle N\rangle_T \geq A\}. \tag{1.19.11}$$

Rewrite S_3^n as

$$S_3^n = n^{2\alpha} \sum_{i=1}^{n-2} \int_{t_{i-1}}^{t_i} \left(\sum_{k=i+2}^{n} \varphi_k^t(u) \int_0^u \varphi_k^t(s) dM_s\right) dM_u = n^{2\alpha} \int_0^{1-\frac{2}{n}} \psi_u^M dM_u,$$

where

$$\psi_u^M = \sum_{k=i+2}^{n} \varphi_k^t(u) \int_0^u \varphi_k^t(s) dM_s, \quad u \in \left[\frac{i-1}{n}, \frac{i}{n}\right).$$

Since the martingale M is continuous (and square integrable), we can localize it: let for some $L > 1$

$$\tau_L = \inf\{t > 0 : |M_t| \vee \langle M\rangle_t \geq L\},$$

$\overline{M}_t = M_{t\wedge\tau_L}$, $\langle\overline{M}\rangle_t = \langle M\rangle_{t\wedge\tau_L}$, $\overline{\psi}_u = \psi_u^{\overline{M}}$, $\tau_L = \infty$ if $|M_t| \vee \langle M\rangle_\infty < L$ for all $t > 0$.

By (1.19.11), it is sufficient to prove that for any $L > 0$,

$$n^{4\alpha} \int_0^{t(1-\frac{2}{n})} \overline{\psi}_u^2 d\langle\overline{M}\rangle_u$$

$$= n^{4\alpha} \sum_{i=1}^{n-2} \int_{t_{i-1}}^{t_i} \left(\sum_{k=i+2}^n \varphi_k^t(u) \int_0^u \varphi_k^t(s) d\overline{M}_s \right)^2 d\langle\overline{M}\rangle_u \xrightarrow{P} 0, \quad n \to \infty.$$

(1.19.12)

First, we estimate the function $\psi_u := \sum_{k=i+2}^n \varphi_k^t(u) \int_0^u \varphi_k^t(s) d\overline{M}_s = \sum_{k=i+2}^n \varphi_k^t(u)(\varphi_k^t(u)\overline{M}_u - \int_0^u \overline{M}_s(\varphi_k^t(s))'_s ds)$. Evidently,

$$(\varphi_k^t(u))'_u = (1-\alpha) \int_{t_{k-1}}^{t_k} v^\alpha (v-u)^{\alpha-2} dv.$$

Therefore,

$$|\psi_u| \le L \sum_{k=i+2}^n (\varphi_k^t(u))^2 + L(1-\alpha) \sum_{k=i+2}^n \varphi_k^t(u) \int_0^u \int_{t_{k-1}}^{t_k} v^\alpha (v-s)^{\alpha-2} dv\, ds.$$

Estimate the terms separately:

$$\varphi_k^t(u) \le \frac{t^{\alpha+1}}{n} (t_{k-1}-u)^{\alpha-1},$$

whence,

$$\sum_{k=i+2}^n (\varphi_k^t(u))^2 \le \frac{t^{2+2\alpha}}{n^2} \sum_{k=i+2}^n (t_{k-1}-u)^{2\alpha-2} \le \frac{t^{2+2\alpha}}{n^2}(t_{i+1}-u)^{2\alpha-2}$$

$$+ \frac{t^{2+2\alpha}}{n} \int_{\frac{i+1}{n}}^1 (tx-u)^{2\alpha-2} dx = \frac{t^{4\alpha}}{n^{2\alpha}} + \frac{t^{2\alpha+1}(t_{i+1}-u)^{2\alpha-1}}{n\,1-2\alpha} \le Cn^{-2\alpha},$$

and

$$\sum_{k=i+2}^n \varphi_k^t(u) \int_0^u \int_{t_{k-1}}^{t_k} v^\alpha(v-s)^{\alpha-2} dv\,ds \le C \sum_{k=i+2}^n \varphi_k^t(u) \int_{t_{k-1}}^{t_k} v^\alpha(v-u)^{\alpha-1} dv$$

$$\le C \sum_{k=i+2}^n (\varphi_k^t(u))^2 \le Cn^{-2\alpha}.$$

From these estimates, it follows that $\overline{\psi}_u^2 n^{4\alpha} \le C$. Therefore, there exists the bounded dominant. In order to establish (1.19.12), it is sufficient to prove that $\overline{\psi}_u n^{2\alpha} \xrightarrow{P} 0$, $0 < u < 1$. We have that

$$\mathbb{E}(\overline{\psi}_u n^{2\alpha})^2 = n^{4\alpha}\mathbb{E}\left(\sum_{k=i+2}^{n} \varphi_k^t(u)\int_0^u \varphi_k^t(s) d\overline{M}_s\right)^2$$

$$= n^{4\alpha}\mathbb{E}\int_0^u \left(\sum_{k=i+2}^{n} \varphi_k^t(u)\varphi_k^t(s)\right)^2 d\langle M\rangle_s.$$

Similarly to previous estimates, we obtain that

$$n^{4\alpha}\left(\sum_{k=i+2}^{n} \varphi_k^t(u)\varphi_k^t(s)\right)^2 \leq C n^{4\alpha}\left(\sum_{k=i+2}^{n} \frac{1}{n^2}(t_{k-1}-u)^{\alpha-1}\right.$$

$$\left.\times (t_{k-1}-s)^{\alpha-1}\right)^2 \leq Cn^{4\alpha-2}\left(\frac{1}{n}\sum_{k=i+2}^{n}(t_{k-1}-u)^{2\alpha-2}\right)^2$$

$$\leq Cn^{4\alpha-2}\left(\frac{n^{2-2\alpha}}{n} + n^{1-2\alpha}\right)^2 \leq C, \text{ for some } C > 0.$$

This means that the bounded dominant exists. Moreover,

$$n^{2\alpha}\sum_{k=i+2}^{n} \varphi_k^t(u)\varphi_k^t(s) \leq Cn^{2\alpha}\sum_{k=i+2}^{n} \varphi_k^t(u) \cdot \frac{1}{n}(u-s)^{\alpha-1}$$

$$\leq Cn^{2\alpha} \cdot \frac{1}{n}\int_{\frac{i+1}{n}}^{1} v^\alpha(v-u)^{\alpha-1}du \cdot (u-s)^{\alpha-1} \to 0$$

for any $s < u$. This means that $S_3^n \xrightarrow{P} 0$, and the lemma is proved. $\square$

Lemma 1.19.7. *There exists a constant $C_3 > 0$, such that*

$$P\text{-}\lim_{n\to\infty} n^{2\alpha}\sum_{k=n\frac{s}{t}+2}^{n}(I_2^k)^2 \leq C_3 t^{4\alpha}(\langle M\rangle_t - \langle M\rangle_s).$$

Proof. We apply the Itô formula to $(I_2^k)^2$ and obtain that

$$(I_2^k)^2 = \int_{t_{k-2}}^{t_{k-1}} (\varphi_k^t(s))^2 d\langle M\rangle_s + \int_{t_{k-2}}^{t_{k-1}}\left(\int_{t_{k-2}}^{s} \varphi_k^t(u) dM_u\right)\varphi_t^k(s) dM_s.$$

From (1.19.6) it follows that

$$\sum_{k=n\frac{s}{t}+2}^{n}\int_{t_{k-2}}^{t_{k-1}}(\varphi_k^t(s))^2 d\langle M\rangle_s \cdot n^{2\alpha} \leq t^{4\alpha}C(\langle M\rangle_t - \langle M\rangle_s).$$

Similarly to the estimates from Lemma 1.19.6, we obtain that for any $A > 0$ and $\varepsilon > 0$

$$P\left\{n^{2\alpha}\left|\sum_{k=n\frac{s}{t}+2}^{n}\int_{t_{k-2}}^{t_{k-1}}\left(\int_{t_{k-2}}^{s}\varphi_k^t(u)d\overline{M}_u\right)\varphi_k^t(s)d\overline{M}_s\right| \geq \varepsilon\right\}$$

$$\leq \frac{A}{\varepsilon^2} + P\left\{n^{4\alpha}\sum_{k=n\frac{s}{t}+2}^{n}\int_{t_{k-2}}^{t_{k-1}}\left(\int_{t_{k-2}}^{s}\varphi_k^t(u)d\overline{M}_u\right)^2\left(\varphi_k^t(s)\right)^2 d\langle\overline{M}\rangle_s \geq A\right\}.$$

So, it is sufficient to prove that

$$n^{4\alpha}\sum_{k=n\frac{s}{t}+2}^{n}\int_{t_{k-2}}^{t_{k-1}}\left(\int_{t_{k-2}}^{s}\varphi_k^t(u)d\overline{M}_u\right)^2(\varphi_k^t(v))^2 d\langle\overline{M}\rangle_v \xrightarrow{P} 0.$$

The existence of the bounded dominant is established by the estimates:

$$n^{4\alpha}\cdot\left(\int_{t_{k-2}}^{s}\varphi_k^t(u)d\overline{M}_u\right)^2(\varphi_k^t(s))^2$$

$$\leq n^{4\alpha}\left(\varphi_k^t(s)\overline{M}_s - \varphi_k^t(t_{k-2})\cdot\overline{M}_{t_{k-2}} - \left(\int_{t_{k-2}}^{s}(\varphi_k^t(u))'_u\overline{M}_u du\right)^2\right)\cdot(\varphi_k^t(s))^2$$

$$\leq CL^2 n^{4\alpha}\left(\varphi_k^t(s) + \varphi_k^t(t_{k-2}) + \int_{t_{k-2}}^{s}\int_{t_{k-1}}^{t_k}v^\alpha(v-u)^{\alpha-2}dv\,du\right)^2\cdot(\varphi_k^t(s))^2$$

$$\leq 9CL^2 n^{4\alpha}(1/n)^{4\alpha}\cdot t^{4\alpha} \leq CL^2 t^{4\alpha}.$$

Therefore, we must prove, that for any $s < v < t$

$$n^{2\alpha}\left(\int_{t_{k-2}}^{s}\varphi_k^t(u)d\overline{M}_u\right)(\varphi_k^t(v)) \xrightarrow{P} 0, \quad n \to \infty.$$

Here $(\varphi_k^t(v))^2 \leq \frac{t^{4\alpha}}{n^{2\alpha}}$. Taking into account that $\langle\overline{M}\rangle$ is bounded and continuous, and by using the relation

$$n^{4\alpha}(\varphi_k^t(v))^2 E\int_{t_{k-2}}^{s}(\varphi_k^t(u))^2 d\langle\overline{M}\rangle_u d\langle\overline{M}\rangle_u \leq CE(\langle\overline{M}\rangle_s - \langle\overline{M}\rangle_{t_{k-2}}) \to 0,$$

for $s < t_{k-1}$, we obtain the necessary estimates, whence the proof follows. □

Lemma 1.19.8. *There exists a constant $C_4 > 0$ such that*

$$P\text{-}\lim_{n\to\infty} n^{2\alpha}\sum_{k=\frac{ns}{t}+1}^{n}(I_3^k)^2 \leq C_4(\langle\overline{M}\rangle_t - \langle\overline{M}\rangle_s)\cdot t^{4\alpha}.$$

The proof is similar to Lemma 1.19.7.

Lemma 1.19.9. *We have that*

$$\lim_{n\to\infty} n^{2\alpha} \sum_{k}^{n} I_i^k I_j^k = 0$$

in probability.

Proof. Consider, for example, $n^{2\alpha}\sum_{k=1}^n I_1^k I_2^k$, where we substitute $\overline{M}$ instead of M. But in this case,

$$n^{4\alpha} E\left(\sum_{k=1}^n I_1^k I_2^k\right)^2 = n^{4\alpha} E \sum_{k=1}^n (I_1^k)^2 (I_2^k)^2,$$

where $I_2^k = \int_{t_{k-2}}^{t_{k-1}} (\varphi_k^t(s))^2 d\langle \overline{M} \rangle_s$, since I_1^k, I_2^k, I_3^k are pairwise orthogonal. Moreover, from inequality (1.19.11), it follows that we must only prove the relation

$$n^{4\alpha} \sum_{k=1}^n (I_1^k)^2 (I_2^k)^2 \xrightarrow{P} 0.$$

According to Lemma 1.19.6, we have that

$$P\text{-}\lim_{n\to\infty} n^{2\alpha} \sum_{k=1}^n (I_1^k)^2 \leq C_2 t^{4\alpha} \langle M \rangle_t$$

and

$$n^{2\alpha} \max_{1\leq k\leq n} \int_{t_{k-2}}^{t_{k-1}} (\varphi_k^t(s))^2 d\langle \overline{M} \rangle_s \leq \alpha^{-2} \max_{1\leq k\leq n} (\langle \overline{M} \rangle_{t_{k-1}} - \langle \overline{M} \rangle_{t_{k-2}}) \xrightarrow{P} 0.$$

All other terms can be estimated similarly, whence the claim follows. □

By using our estimates, we can conclude that for rational s, consequently for any $s < t$, the following claims hold:

(a) there exist two constants, $C_1 > 0$ and $C_2 > 0$ such that

$$C_1 \int_s^t u^{2\alpha} d\langle M \rangle_u \leq (t-s) \leq C_2 t^{2\alpha} (\langle \overline{M} \rangle_t - \langle \overline{M} \rangle_s).$$

This estimate can be rewritten in terms of W and $\langle W \rangle$:

$$C_1(\langle W \rangle_t - \langle W \rangle_s) \leq (t-s) \leq C_2 t^{2\alpha} \int_s^t u^{-2\alpha} d\langle W \rangle_u.$$

(b)

$$P\text{-}\lim_{n\to\infty} n^{2\alpha} \sum_{k=n\frac{s}{t}+1}^n (\Delta X_{t_k})^2 = P\text{-}\lim_{n\to\infty} \int_s^t \varphi_s^n d\langle M \rangle_s,$$

where φ_s^n is a positive, bounded, nonrandom function, separated from 0 by some constant.

From the left-hand side of (a), it follows that $\langle W \rangle_t$ is absolutely continuous w.r.t. the Lebesgue measure, so $\langle W \rangle_t = \int_0^t \theta_s ds$, where θ_s is a bounded, possibly, random variable. From the right-hand side of (a), it follows that

$$\int_s^t u^{-2\alpha} \theta_u du \geq \frac{1}{C_2}(t^{1-2\alpha} - st^{-2\alpha}) \geq C_3(t^{1-2\alpha} - s^{1-2\alpha}) = C_3 \int_s^t u^{-2\alpha} du.$$

This means that
$$\int_s^t u^{-2\alpha}(\theta_u - C_3) du \geq 0.$$

Evidently, for any set $A \in \mathcal{F}$
$$\int_A \int_s^t u^{-2\alpha}(\theta_u - C_3) du\, dP \geq 0.$$

Now, let the set $D \in \sigma\{\mathcal{F} \times \mathcal{B}[\delta, 1]\}$, and let $\delta > 0$ be fixed. Then $\mu(D) < \infty$, where $\mu = P \times \lambda$, λ is the Lebesgue measure on $[0, 1]$.

By the theorem of approximation of measurable sets, for any $\varepsilon > 0$ there exists a collection of the sets

$$\{D_i = B_i \times [s_i, t_i], B_i \in \mathcal{F}, [s_i, t_i] \in \mathcal{B}[\delta, 1]\},$$

such that
$$\mu\left(\left(D \backslash \bigcup_{i=1}^k D_i\right) \bigcup \left(\bigcup_{i=1}^k D_i \backslash D\right)\right) < \varepsilon.$$

Therefore, since $u^{-2\alpha}(\theta_u - C_3)$ is bounded on D,

$$\int_D u^{-2\alpha}(\theta_u - C_3) d\mu \geq 0. \tag{1.19.13}$$

Now, set
$$D = \{(\omega, u) : \theta_u - C_3 < 0, \text{ and } u \geq \delta\}$$

and we immediately obtain that $\mu(D) = 0$. From here we conclude that $\langle W \rangle$ is equivalent to the Lebesgue measure, and $W_t = \int_0^t \theta_s^{\frac{1}{2}} dV_s$, where $\{V_s, \mathcal{F}_s, s \geq 0\}$ is some Wiener process.

Now, if we do all the same calculations as before, but for "true" fractional Brownian motion B_t^H, we obtain that

$$P\text{-}\lim_{n \to \infty} n^{2\alpha} \sum_{k=n\frac{s}{t}+1}^n (\triangle B_{t_k}^H)^2 = P\text{-}\lim_{n \to \infty} \int_s^t \varphi_s^n s^{-2\alpha} ds$$

$$= P\text{-}\lim_{n \to \infty} n^{2\alpha} \sum_{k=n\frac{s}{t}+1}^n (\triangle B_{t_k}^H)^2.$$

(It is sufficient to take $s = 0$.) Therefore, $P\text{-}\lim_{n\to\infty} \int_s^t \psi_u^n du = 0$, where $\psi_u^n = u^{-2\alpha}\varphi_u^n(\theta_u - 1)$.

Consider any set $D \in \sigma\{\mathcal{F} \times \mathcal{B}[\delta, 1]\}$, repeat all the previous reasonings and obtain that $\theta_u \equiv 1$ (otherwise, put $D = \{(\omega, u): \theta_u > 1 + \alpha$, or $\theta_u < 1 - \alpha\}$).

We proved Theorem 1.19.1 for $H \in (1/2, 1)$. Now we consider the case $H \in (0, 1/2)$. Similarly to Lemma 1.19.3, we can present the process X_t as

$$X_t = \int_0^t z(t,s) dW_s, \text{ where } W_t = \int_0^t s^\alpha dM_s,$$

and

$$z(t,s) := (C_H^{(6)})^{-1} m_H(t,s) = \left(\frac{t}{s}\right)^\alpha (t-s)^\alpha - \alpha s^{-\alpha} \int_s^t u^{\alpha-1}(u-s)^\alpha du.$$

Therefore,

$$\begin{aligned}
X_{t_k} - X_{t_{k-1}} &= -\alpha \int_0^{t_{k-2}} \int_{t_{k-1}}^{t_k} \left(\frac{s}{u}\right)^{-\alpha} (u-s)^{\alpha-1} du\, dW_s \\
&\quad - \alpha \int_{t_{k-2}}^{t_{k-1}} \int_{t_{k-1}}^{t_k} \left(\frac{s}{u}\right)^{-\alpha} (u-s)^{\alpha-1} du\, dW_s \\
&\quad + \int_{t_{k-1}}^{t_k} \left(\frac{s}{t_k}\right)^{-\alpha} (t_k - s)^\alpha\, dW_s \\
&\quad - \alpha \int_{t_{k-1}}^{t_k} s^{-\alpha} \int_s^{t_k} u^{\alpha-1}(u-s)^\alpha du\, dW_s \\
&= J_1^k + J_2^k + J_3^k + J_4^k.
\end{aligned}$$

For $H \in (0, 1/2)$ it is more convenient to deal with W_t, not M_t.

Evidently,

$$\lim_{n\to\infty} n^{2\alpha} \sum_{k=n\frac{s}{t}+2}^n (\Delta X_{t_k})^2 = \lim_{n\to\infty} n^{2\alpha} \left(\sum_{k=n\frac{s}{t}+2}^n (J_1^k)^2 \right.$$

$$\left. + \sum_{k=n\frac{s}{t}+2}^n (J_2^k + J_3^k + J_4^k)^2 + \sum_{k=n\frac{s}{t}+2}^n J_1^k (J_2^k + J_3^k + J_4^k) \right).$$

First, estimate

$$\lim_{n\to\infty} n^{2\alpha} \sum_{k=n\frac{s}{t}+2}^n (J_1^k)^2$$

from below and from above. As before,

1.19 Lévy Theorem for fBm   107

$$\lim_{n\to\infty} n^{2\alpha} \sum_{k=n\frac{s}{t}+2}^{n} (\Delta X_{t_k})^2 \to t^{2\alpha}(t-s).$$

First, we obtain upper bound for the sum

$$S_4^n := n^{2\alpha} \sum_{k=n\frac{s}{t}+2}^{n} \int_0^{t_{k-2}} \left(\theta_k^t(s)\right)^2 d\langle W\rangle_s,$$

where $\theta_k^t(s) = \int_{t_{k-1}}^{t_k} \left(\frac{s}{u}\right)^{-\alpha}(u-s)^{\alpha-1}du$, $s \le t_{k-1}$. Evidently, for $s \le t_{k-2}$

$$\theta_k^t(s) \le \left((t_{k-1}-s)^{\alpha-1}\frac{t}{n}\right) \wedge \left(\frac{1}{-\alpha}\left(\frac{t}{n}\right)^{\alpha}\right). \tag{1.19.14}$$

Therefore, for such n, that $n\frac{s}{t} \in \mathbb{N}$ we have that

$$S_4^n = n^{2\alpha} \sum_{k=n\frac{s}{t}+2}^{n} \sum_{i=1}^{k-2} \int_{t_{i-1}}^{t_i} \left(\theta_k^t(u)\right)^2 d\langle W\rangle_u$$

$$= n^{2\alpha} \left(\sum_{i=1}^{n\frac{s}{t}} \sum_{k=n\frac{s}{t}+2}^{n} + \sum_{i=n\frac{s}{t}+1}^{n-2} \sum_{k=i+2}^{n} \right) \int_{t_{i-1}}^{t_i} \left(\theta_k^t(u)\right)^2 d\langle W\rangle_u$$

$$\le n^{2\alpha-1}t \int_0^s \left(s+\frac{t}{n}-u\right)^{2\alpha-1} d\langle W\rangle_u$$

$$+ n^{2\alpha-2}t^2 \int_0^s \left(s+\frac{t}{n}-u\right)^{2\alpha-2} d\langle W\rangle_u$$

$$+ \frac{t}{1-2\alpha}\left(\frac{t}{n}\right)^{2\alpha-1} n^{2\alpha-1} \sum_{i=n\frac{s}{t}+1}^{n-2} \int_{t_{i-1}}^{t_i} d\langle W\rangle_u$$

$$+ t^2 n^{2\alpha-2} \sum_{i=n\frac{s}{t}+1}^{n-2} \int_{t_{i-1}}^{t_i} d\langle W\rangle_u \left(\frac{t}{n}\right)^{2\alpha-2}. \tag{1.19.15}$$

The integral $\int_0^s \left(s+\frac{t}{n}-u\right)^{2\alpha-1} d\langle W\rangle_u$, according to Lemma 2.1 (NVV99), can be estimated as

$$\left|\int_0^s \left(s+\frac{t}{n}-u\right)^{2\alpha-1} d\langle W\rangle_u\right| \le C(\omega)\left(s+\frac{t}{n}-s\right)^{2\alpha-1+\beta},$$

for some random variable $0 < C(\omega) < \infty$, where β is Hölder index of $\langle W\rangle_u$. Evidently, $\beta > 0$, and it holds that

$$\int_0^s \left(s+\frac{t}{n}-u\right)^{2\alpha-1} d\langle W\rangle_u \cdot n^{2\alpha-1} \sim n^{2\alpha-1}\left(\frac{1}{n}\right)^{2\alpha-1+\beta} \to 0.$$

The same is true for

$$\int_0^s \left(s + \frac{t}{n} - u\right)^{2\alpha-2} d\langle W\rangle_u \cdot n^{2\alpha-2}.$$

The last two integrals from (1.19.14) admit the estimate:

$$\sum_{i=n\frac{s}{t}+1}^{n-2} \int_{t_{i-1}}^{t_i} d\langle W\rangle_u \frac{t}{1-2\alpha} \left(\frac{t}{n}\right)^{2\alpha-1} n^{2\alpha-1}$$

$$+ \sum_{i=n\frac{s}{t}+1}^{n-2} \int_{t_{i-1}}^{t_i} d\langle W\rangle_u \left(\frac{t}{n}\right)^{2\alpha-2} t^2 n^{2\alpha-2} \le t^{2\alpha} C_2 \left(\langle W\rangle_t - \langle W\rangle_s\right).$$

Now we obtain the lower bound for S_4^n. Return to $\langle M\rangle$ instead of $\langle W\rangle$.

$$S_4^n = n^{2\alpha} \sum_{k=n\frac{s}{t}+2}^{n} \int_0^{t_{k-2}} \left(\varphi_t^k(u)\right)^2 d\langle M\rangle_u$$

$$\ge t^2 n^{2\alpha-2} \sum_{k=n\frac{s}{t}}^{n} (t_k)^{2\alpha} \int_0^{t_{k-2}} (t_k - u)^{2\alpha-2} d\langle M\rangle_u$$

$$\ge t^2 n^{2\alpha-2} \Bigg(\sum_{i=1}^{n\frac{s}{t}-1} \sum_{k=n\frac{s}{t}+2}^{n} + \sum_{i=n\frac{s}{t}+1}^{n-2} \sum_{k=i+2}^{n} \Bigg) (t_k)^{2\alpha} \int_{t_{i-1}}^{t_i} (t_k - u)^{2\alpha-2} d\langle M\rangle_u$$

$$= C t^{2\alpha+2} n^{2\alpha-1} \sum_{i=n\frac{s}{t}+1}^{n-2} \int_{t_{i-1}}^{t_i} \frac{1}{t} \left((t_{i+2} - u)^{2\alpha-1} - (t-u)^{2\alpha-1} \right) d\langle M\rangle_u.$$

Note that

$$n^{2\alpha-1} \sum_{i=n\frac{s}{t}+1}^{n-2} \int_{t_{i-1}}^{t_i} (t-u)^{2\alpha-1} d\langle M\rangle_u$$

$$\sim \left(t - t + \frac{2}{n}\right)^{2\alpha-1+\beta} \cdot n^{2\alpha-1} \to 0, \quad n \to \infty.$$

Therefore,

$$\lim_{n\to\infty} n^{2\alpha} \sum_{k=n\frac{s}{t}+1}^{n} \left(J_1^k\right)^2$$

$$\ge C t^{2\alpha+1} n^{2\alpha-1} \sum_{i=n\frac{s}{t}+1}^{n-2} \int_{t_{i-1}}^{t_i} (t_{i+2} - u)^{2\alpha-1} d\langle M\rangle_u$$

$$\ge C t^{2\alpha+1} n^{2\alpha-1} \sum_{i=n\frac{s}{t}+1}^{n-2} (t_{i+2} - t_{i-1})^{2\alpha-1} \int_{t_{i-1}}^{t_i} d\langle M\rangle_u.$$

1.19 Lévy Theorem for fBm

The "remainder" term for $\sum \left(J_1^k\right)^2$ equals

$$R_n := n^{2\alpha} \sum_{k=n\frac{s}{t}+2}^{n} \int_0^{t_{k-2}} \left(\int_0^z \theta_k^t(v) dW_v \right) \theta_k^t(u) dW_u.$$

For technical simplicity, it is enough to consider the stopped process $\overline{W}_t$, instead of W_t, and $\sum_{k=3}^{nr}$ for any $r \in \mathbb{N}$, instead of $\sum_{k=n\frac{s}{t}+2}^{n} = -\sum_{k=3}^{n\frac{s}{t}+1} + \sum_{k=3}^{n}$. We obtain that

$$\mathbb{E}(R_n)^2 = n^{4\alpha} \mathbb{E} \left(\sum_{k=3}^{nr} \sum_{i=1}^{k-2} \int_{t_{i-1}}^{t_i} \int_0^u \theta_k^t(v) d\overline{W}_v \cdot \theta_k^t(u) d\overline{W}_u \right)^2$$

$$= n^{4\alpha} \mathbb{E} \left(\sum_{i=1}^{nr-2} \sum_{k=i+3}^{nr} \int_{t_{\frac{i-1}{n}}}^{t_i} \int_0^u \theta_k^t(v) d\overline{W}_v \cdot \theta_k^t(u) d\overline{W}_u \right)^2$$

$$= n^{4\alpha} \sum_{i=1}^{nr-2} \mathbb{E} \int_{t_{\frac{i-1}{n}}}^{t_i} \left(\sum_{k=i+3}^{nr} \int_0^u \theta_k^t(v) d\overline{W}_v \cdot \theta_k^t(u) \right)^2 d\langle \overline{W} \rangle_u.$$

Let us estimate

$$\left| \int_0^u \theta_k^t(v) d\overline{W}_v \right| = \left| \theta_k^t(u) \overline{W}_u - \int_0^u \overline{W}_v \left(\theta_k^t(v) \right)'_v dv \right|$$

$$\leq L \left| \theta_k^t(u) \right| + L \left| \int_0^u \left(\theta_k^t(v) \right)'_v dv \right|.$$

It follows from (1.19.14), that

$$\left| \int_0^u \left(\theta_k^t(v) \right)'_v dv \right| = \left| \theta_k^t(u) - \theta_k^t(0) \right| \leq C \left(\frac{t}{n} \right)^\alpha \quad \text{for some} \quad C > 0.$$

Moreover,

$$n^{2\alpha} \left(\sum_{k=i+3}^{nr} \theta_k^t(u) \right)^2 \leq n^{2\alpha} \left(\int_{t_{i+1}}^{tr} (v-u)^{\alpha-1} dv \right)^2$$

$$= Cn^{2\alpha} \left[-(tr-u)^\alpha + (t_{i+1} - u)^\alpha \right]^2 \leq C,$$

and the integrand

$$n^{4\alpha} \left(\sum_{k=i+2}^{nr} \int_0^u \theta_k^t(v) d\overline{W}_v \cdot \theta_k^t(u) \right)^2 \leq C,$$

i.e. there exists the integrable dominant. Therefore, it is sufficient to establish that for any u

$$n^{2\alpha} \sum_{k=i+3}^{nr} \int_0^u \theta_k^t(v) d\overline{W}_v \cdot \theta_k^t(u) \xrightarrow{P} 0.$$

We take the mathematical expectation and obtain that

$$n^{4\alpha} \mathbb{E} \int_0^u \left(\sum_{k=i+3}^{nr} \theta_k^t(v) \theta_k^t(u) \right)^2 d\langle \overline{W} \rangle_v.$$

The bounded dominant exists. Indeed,

$$n^{4\alpha} \left(\sum_{k=i+3}^{nr} \theta_k^t(v) \theta_k^t(u) \right)^2 \leq n^{2\alpha} \left(\sum_{k=i+2}^{nr} \theta_k^t(v) \right)^2 \leq C,$$

as before. Further, we must prove that

$$n^{2\alpha} \sum_{k=i+3}^{nr} \theta_k^t(v) \theta_k^t(u) \to 0$$

for all fixed $0 < v < u$. We have that

$$n^{2\alpha} \sum_{k=i+2}^{nr} \theta_k^t(v) \theta_k^t(u) \leq n^{2\alpha} \sum_{k=i+3}^{nr} \int_{t_{k-1}}^{t_k} (s-u)^{\alpha-1} ds$$

$$\times \int_{t_{k-1}}^{t_k} (s-v)^{\alpha-1} ds \leq n^{2\alpha} \sum_{k=i+3}^{nr} (t_{k-1}-u)^{\alpha-1} \frac{1}{n} \int_{t_{k-1}}^{t_k} (s-v)^{\alpha-1} ds$$

$$\leq n^{2\alpha-1} (t_{i+2}-u)^{\alpha-1} \int_{t_{i+2}}^{tr} (s-v)^{\alpha-1} ds$$

$$\leq C n^{\alpha-1} (u-v)^{\alpha-1} \to 0, \quad n \to \infty \quad \text{for any} \quad 0 < v < u.$$

From all these estimates, the remainder term $R_n \xrightarrow{P} 0$, $n \to \infty$, and we have established that

$$C_1 t^{4\alpha} \left(\langle M \rangle_t - \langle M \rangle_s \right) \leq \lim_{n \to \infty} n^{2\alpha} \sum_{k=n\frac{s}{t}+2}^{n} \left(J_1^k \right)^2 \leq C_2 t^{2\alpha} \left(\langle W \rangle_t - \langle W \rangle_s \right).$$

(Note, that for $H \in (1/2, 1)$, we obtained opposite estimates.) Note also that we cannot estimate $\sum \left(J_i^k \right)^2$, $i > 1$, from above. Indeed, the integrand of the form $(t \frac{l}{n} - u)^{\alpha}$ that admits the estimate $< \left(\frac{1}{n} \right)^{\alpha} \to 0$ for $H \in (1/2, 1)$, now, for $H \in (0, 1/2)$, tends to ∞. So, we mention that $\sum_{k=n\frac{s}{t}+2}^{n} \left(J_2^k + J_3^k + J_4^k \right)^2 \geq 0$, prove that $\sum J_1^k \left(J_2^k + J_3^k + J_4^k \right) \to 0$, and obtain the estimate from above:

$$C_1 t^{2\alpha} \left(\langle M \rangle_t - \langle M \rangle_s \right) \leq (t-s).$$

In the sequel, we realize this plan.

It is sufficient to estimate the sums from $k = 2$ till $k = n$. By applying the Lenglart inequality to $n^{2\alpha} \sum_{k=2}^n J_1^k J_2^k$, we obtain that it is sufficient to prove that

$$n^{4\alpha} \sum_{k=2}^n \left(\int_0^{t_{k-2}} \int_{t_{k-1}}^{t_k} \left(\frac{s}{u}\right)^{-\alpha} (u-s)^{\alpha-1} du\, d\overline{W}_s \right)^2$$

$$\times \left(\int_{t_{k-2}}^{t_{k-1}} \left(\int_{t_{k-1}}^{t_k} \left(\frac{s}{u}\right)^{-\alpha} (u-s)^{\alpha-1} du \right)^2 d\langle \overline{W} \rangle_s \right)$$

$$\leq C n^{4\alpha} \sum_{k=2}^n \left(\int_0^{t_{k-2}} \theta_k^t(s) d\overline{W}_s \right)^2 \int_{t_{k-2}}^{t_{k-1}} (t_{k-1}-s)^{2\alpha} d\langle \overline{W} \rangle_s \xrightarrow{P} 0.$$

Integrate the last integral by parts:

$$\int_{t_{k-2}}^{t_{k-1}} (t_{k-1}-s)^{2\alpha} d\langle \overline{W} \rangle_s = (t_{k-1} - t_{k-2})^{2\alpha} \left(\langle \overline{W} \rangle_{t_{k-1}} - \langle \overline{W} \rangle_{t_{k-2}} \right)$$

$$- 2\alpha \int_{t_{k-2}}^{t_{k-1}} (t_{k-1}-s)^{2\alpha-1} \left(\langle \overline{W} \rangle_{t_{k-1}} - \langle \overline{W} \rangle_s \right) ds$$

$$\leq C n^{-2\alpha} \Delta \langle \overline{W} \rangle_{t_{k-1}} + C \int_{t_{k-2}}^{t_{k-1}} (t_{k-1}-s)^{2\alpha-1} \left(\langle \overline{W} \rangle_{t_{k-1}} - \langle \overline{W} \rangle_s \right) ds.$$

Now recall that

$$\left(\int_0^{t_{k-2}} \theta_k^t(s) d\overline{W}_s \right)^2$$

$$= \int_0^{t_{k-2}} \left(\theta_k^t(s) \right)^2 d\langle \overline{W} \rangle_s + 2 \int_0^{t_{k-2}} \int_0^s \theta_k^t(v) d\overline{W}_v \theta_k^t(s) d\overline{W}_s.$$

It was proved that

$$\sigma_1^n := n^{2\alpha} \sum_{k=2}^n \int_0^{t_{k-2}} \left(\theta_k^t(s) \right)^2 d\langle \overline{W} \rangle_s$$

is bounded in probability, and

$$\sigma_2^n := n^{2\alpha} \sum_{k=2}^n \int_0^{t_{k-2}} \int_0^s \theta_k^t(v) d\overline{W}_v \theta_k^t(s) d\overline{W}_s \xrightarrow{P} 0, \quad n \to \infty.$$

Therefore,

$$n^{4\alpha} \sum_{k=2}^n \left(\int_0^{t_{k-2}} \theta_k^t(s) d\overline{W}_s \right)^2 \cdot C n^{-2\alpha} \Delta \langle \overline{W} \rangle_{t_{k-1}}$$

$$\leq C\sigma_1^n \cdot \max_k \Delta \langle \overline{W} \rangle_{t_{k-1}} + C\sigma_2^n \cdot \max_k \Delta \langle \overline{W} \rangle_{t_{k-1}} \xrightarrow{P} 0, \quad n \to \infty.$$

Also,

$$n^{4\alpha} \sum_{k=2}^n \left(\int_0^{t_{k-2}} \theta_k^t(s) d\overline{W}_s \right)^2 \cdot \int_{t_{k-2}}^{t_{k-1}} (t_{k-1} - s)^{2\alpha-1}$$

$$\times \left(\langle \overline{W} \rangle_{t_{k-1}} - \langle \overline{W} \rangle_s \right) ds \leq C(\omega) \left(\sigma_1^n + \sigma_2^n \right) n^{2\alpha} \int_{t_{k-2}}^{t_{k-1}} (t_{k-1} - s)^{2\alpha-\varepsilon} ds$$

$$\leq C(\omega) \left(\sigma_1^n + \sigma_2^n \right) n^{2\alpha} (t_{k-1} - t_{k-2})^{2H-\varepsilon} \sim \left(\frac{1}{n} \right)^{1-\varepsilon} \to 0, \quad n \to \infty.$$

Consider $n^{2\alpha} \sum_{k=2}^n J_1^k J_3^k$:

$$n^{2\alpha} \sum_{k=1}^n \int_0^{t_{k-2}} \theta_k^t(s) dW_s \cdot \int_{t_{k-1}}^{t_k} \left(\frac{s}{t_k} \right)^{-\alpha} (t_k - s)^\alpha dW_s.$$

As before, it is sufficient to prove that

$$n^{4\alpha} \sum_{k=1}^n \left(\int_0^{t_{k-2}} \theta_k^t(s) d\overline{W}_s \right)^2 \cdot \int_{t_{k-1}}^{t_k} \left(\frac{s}{t_k} \right)^{-2\alpha} (t_k - s)^{2\alpha} d\langle \overline{W} \rangle_s \xrightarrow{P} 0,$$

$$n \to \infty,$$

or, equivalently,

$$n^{2\alpha} \max_k \int_{t_{k-1}}^{t_k} (t_k - s)^{2\alpha} d\langle \overline{W} \rangle_s \cdot (\sigma_1^n + \sigma_2^n) \xrightarrow{P} 0. \qquad (1.19.16)$$

Note that by (NVV99, Lemma 2.1) and due to Hölder properties of $\langle \overline{W} \rangle$,

$$\int_{t_{k-1}}^{t_k} (t_k - s)^{2\alpha} d\langle \overline{W} \rangle_s \leq C(\omega) (t_k - t_{k-1})^{2\alpha+1-\varepsilon} \sim \left(\frac{1}{n} \right)^{2\alpha+1-\varepsilon},$$

whence we obtain (1.19.16).

Now, consider $n^{2\alpha} \sum J_1^k J_4^k$; other sums can be estimated similarly. After some transformations,

$$n^{4\alpha} \sum_{k=1}^n \left(\int_0^{t_{k-2}} \theta_k^t(v) d\overline{W}_u \right)^2 \cdot \int_{t_{k-1}}^{t_k} s^{-2\alpha} \left(\int_s^{t_k} u^{\alpha-1} (u-s)^\alpha du \right)^2 d\langle \overline{W} \rangle_s$$

$$\leq n^{2\alpha} \max_k \int_{t_{k-1}}^{t_k} \left(\int_s^{t_k} u^{\alpha-1} (u-s)^\alpha du \right)^2 d\langle \overline{W} \rangle_s \cdot (\sigma_1^n + \sigma_2^n)$$

1.19 Lévy Theorem for fBm   113

$$\leq n^{2\alpha} \max_k \int_{t_{k-1}}^{t_k} \left(\int_s^{t_k} u^{2\alpha-2} du \cdot \int_s^{t_k} (u-s)^{2\alpha} du \right) d\langle \overline{W} \rangle_s \cdot (\sigma_1^n + \sigma_2^n)$$

$$\leq C n^{2\alpha} \max_k \int_{t_{k-1}}^{t_k} s^{2\alpha-1} (t_k - s)^{2\alpha+1} d\langle \overline{W} \rangle_s \cdot (\sigma_1^n + \sigma_2^n)$$

$$\leq C n \cdot \frac{1}{n} \max_k \int_{t_{k-1}}^{t_k} (t_k - s)^{2\alpha} d\langle \overline{W} \rangle_s \cdot (\sigma_1^n + \sigma_2^n)$$

$$\leq C \max_k (t_k - t_{k-1})^{2\alpha+1-\varepsilon} \cdot (\sigma_1^n + \sigma_2^n) \to 0, \quad n \to \infty.$$

Due to all these estimates we have proved that

$$t^{2\alpha}(t-s) = \lim_{n \to \infty} n^{2\alpha} \sum_{k=n\frac{s}{t}+2}^{n} (\Delta X_{t_k})^2 \geq C_1 t^{4\alpha} \left(\langle M \rangle_t - \langle M \rangle_s \right),$$

i.e.

$$\langle M \rangle_t - \langle M \rangle_s \leq C_2 t^{-2\alpha}(t-s) = C_2 \left(t^{1-2\alpha} - st^{-2\alpha} \right) \leq C_2 \left(t^{1-2\alpha} - s^{1-2\alpha} \right),$$

or

$$\int_s^t u^{-2\alpha} d\langle W \rangle_u \leq C_2 \int_s^t u^{-2\alpha} du.$$

As before, it follows that $\langle W \rangle_t$ is absolutely continuous w.r.t. Lebesgue measure,

$$\langle W \rangle_t = \int_0^t \theta_s ds, \qquad (1.19.17)$$

$0 \leq \theta_s \leq C$, C is some constant, θ_s possibly is random.

Taking this into account, we can continue estimates from above: for example, if we take for simplicity the sums over $k=2$ till $k=n$, then

$$n^{2\alpha} \sum_{k=1}^{n} \left(J_2^k \right)^2 = \tilde{\sigma}_1^n + \tilde{\sigma}_2^n := C n^{2\alpha} \sum_{k=1}^{n} \int_{t_{k-2}}^{t_{k-1}} \left(\int_{t_{k-1}}^{t_k} \left(\frac{s}{u} \right)^{-\alpha} (u-s)^{\alpha-1} du \right)^2$$

$$\times d\langle W \rangle_s + C n^{2\alpha} \sum_{k=1}^{n} \int_{t_{k-2}}^{t_{k-1}} \left(\int_{t_{k-2}}^{u} \theta_k^t(v) dW_v \right) \theta_k^t(u) dW_u,$$

where

$$\theta_k^t(s) = \int_{t_{k-1}}^{t_k} \left(\frac{s}{u} \right)^{-\alpha} (u-s)^{\alpha-1} du \leq (t_{k-1} - s)^{\alpha} C.$$

Therefore,

$$\tilde{\sigma}_1^n \leq C n^{2\alpha} \sum_{k=1}^{n} \int_{t_{k-2}}^{t_{k-1}} (t_{k-1} - s)^{2\alpha} d\langle W \rangle_s.$$

Direct estimates give nothing (because of singularity at t_{k-1}). So, we go by an indirect way: for some $A > 0$,

$$\int_{t_{k-2}}^{t_{k-1}} (t_{k-1}-s)^{2\alpha} d\langle W\rangle_s \leq \int_{t_{k-2}}^{t_{k-1}-\frac{t}{nA}} + \int_{t_{k-1}-\frac{t}{nA}}^{t_{k-1}}$$

$$\leq \left(t_{k-1} - \left(t_{k-1} - \frac{t}{nA}\right)\right)^{2\alpha} \cdot \Delta\langle W\rangle_{t_k}$$

$$+ \text{(thanks to (1.19.17))} \ C \int_{t_{k-1}-\frac{t}{nA}}^{t_{k-1}} (t_{k-1}-s)^{2\alpha} ds$$

$$\leq \left(\frac{t}{nA}\right)^{2\alpha} \Delta\langle W\rangle_{t_k} + C\left(\frac{t}{nA}\right)^{2\alpha+1}.$$

Taking the sum, we obtain:

$$\widetilde{\sigma}_1^n \leq Cn^{2\alpha} \sum_{k=1}^{n} \left(\frac{t}{nA}\right)^{2\alpha} \Delta\langle W\rangle_{t_k} + Cn^{2\alpha} n \left(\frac{t}{nA}\right)^{2\alpha+1}$$

$$\leq CA^{-2\alpha} t^{2\alpha} \langle W\rangle_t + C\frac{1}{A^{2\alpha+1}} t^{2\alpha+1}.$$

If we estimate the sum from $k = n\frac{s}{t} + 1$ to $k = n$, then

$$\widetilde{\sigma}_1^n \leq CA^{-2\alpha} t^{2\alpha} \left(\langle W\rangle_t - \langle W\rangle_s\right) + C\frac{1}{A^{2\alpha+1}} t^{2\alpha+1}\left(1 - \frac{s}{t}\right)$$

$$= CA^{-2\alpha} t^{2\alpha} \left(\langle W\rangle_t - \langle W\rangle_s\right) + C\frac{1}{A^{2\alpha+1}} t^{2\alpha}(t-s).$$

Now we want to prove that

$$n^{2\alpha} \sum_{k=1}^{n} \int_{t_{k-2}}^{t_{k-1}} \left(\int_{t_{k-2}}^{u} \theta_k^t(v) dW_v\right) \theta_k^t(u) dW_u \xrightarrow{P} 0, \quad n \to \infty.$$

As usual, it is enough to establish that

$$n^{4\alpha} \sum_{k=1}^{n} \int_{t_{k-2}}^{t_{k-1}} \left(\int_{t_{k-2}}^{u} \theta_k^t(v) d\overline{W}_v\right)^2 \left(\theta_k^t(u)\right)^2 d\langle \overline{W}\rangle_u \xrightarrow{P} 0.$$

But we can bound $\langle \overline{W}\rangle_u$ by Cdu, so, it is enough to prove that

$$n^{4\alpha} \sum_{k=1}^{n} \int_{t_{k-2}}^{t_{k-1}} \left(\int_{t_{k-2}}^{u} \theta_k^t(v) d\overline{W}_v\right)^2 \left(\theta_k^t(u)\right)^2 du \xrightarrow{P} 0.$$

By taking the mathematical expectation, we see that it is sufficient to establish that

1.19 Lévy Theorem for fBm 115

$$n^{4\alpha} \sum_{k=1}^{n} \int_{t_{k-2}}^{t_{k-1}} \int_{t_{k-2}}^{u} \left(\theta_k^t(v)\right)^2 d\langle \overline{W}\rangle_v \left(\theta_k^t(u)\right)^2 du \xrightarrow{P} 0.$$

By substituting Cdv instead of $d\langle \overline{W}\rangle_v$, we see that it is enough to establish that

$$\sigma_3^n := n^{4\alpha} \sum_{k=1}^{n} \int_{t_{k-2}}^{t_{k-1}} \left(\int_{t_{k-2}}^{u} \left(\theta_k^t(v)\right)^2 dv \right) \left(\theta_k^t(u)\right)^2 du \to 0.$$

We have that $\left(\theta_k^t(u)\right)^2 \leq Cn^{-2\alpha}$, and

$$\sigma_3^n \leq \sum_{k=1}^{n} \int_{t_{k-2}}^{t_{k-1}} \left(\int_{t_{k-2}}^{u} dv \right) du \leq \frac{1}{n}C \to 0, \quad n \to \infty.$$

Finally,

$$n^{2\alpha} \sum_{k=n\frac{s}{t}+2}^{n} \left(J_2^k\right)^2 \leq CA^{-2\alpha}t^{2\alpha}\left(\langle W\rangle_t - \langle W\rangle_s\right) + C\frac{1}{A^{2\alpha+1}}t^{2\alpha}(t-s).$$

Now, proceed with J_3^k:

$$n^{2\alpha} \sum_{k=1}^{n} \left(J_3^k\right)^2 = n^{2\alpha} \sum_{k=1}^{n} \int_{t_{k-1}}^{t_k} \left(\left(\frac{s}{t_k}\right)^{-\alpha} (t_k - s)^\alpha \right)^2 d\langle W\rangle_s$$

$$+ n^{2\alpha} \sum_{k=1}^{n} \int_{t_{k-1}}^{t_k} \left(\int_{t_{k-1}}^{u} \left(\frac{s}{t_k}\right)^{-\alpha} (t_k - s)^\alpha dW_s \right) \left(\frac{u}{t_k}\right)^{-\alpha} (t_k - u)^\alpha dW_u.$$

The first term can be estimated as

$$n^{2\alpha} \sum_{k=1}^{n} \int_{t_{k-1}}^{t_k} (t_k - s)^{2\alpha} d\langle W\rangle_s \leq C \left(\frac{t}{A}\right)^{2\alpha} \left(\langle W\rangle_t - \langle W\rangle_s\right) + \frac{C}{A^{2\alpha+1}}t^{2\alpha}(t-s),$$

as before.

And with the bound $d\langle W\rangle_s \leq Cds$, the second term can be estimated as $n^{4\alpha} \sum_{k=1}^{n} \int_{t_{k-1}}^{t_k} \int_{t_{k-1}}^{u} (t_k - s)^{2\alpha} ds \cdot (t_k - u)^{2\alpha} du \leq \frac{Cn^{4\alpha}}{n^{4\alpha+2}} \to 0$. Therefore, for $\sum \left(J_3^k\right)^2$ we have the same estimate as for $\sum \left(J_2^k\right)^2$. Finally, estimate

$$n^{2\alpha} \sum_{k=1}^{n} \left(J_4^k\right)^2 = Cn^{2\alpha} \sum_{k=1}^{n} \left(\int_{t_{k-1}}^{t_k} s^{-\alpha} \int_s^{t_k} u^{\alpha-1}(u-s)^\alpha du \, dW_s \right)^2$$

$$= Cn^{2\alpha} \sum_{k=1}^{n} \int_{t_{k-1}}^{t_k} s^{-2\alpha} \left(\int_s^{t_k} u^{\alpha-1}(u-s)^\alpha du \right)^2 d\langle W\rangle_s$$

$$+ Cn^{2\alpha} \sum_{k=1}^{n} \int_{t_{k-1}}^{t_k} \int_{t_{k-1}}^{u} s^{-\alpha} \int_s^{t_k} v^{\alpha-1}(v-s)^\alpha dv \, dW_s$$

$$\times u^{-\alpha} \int_u^{t_k} v^{\alpha-1}(v-u)^\alpha dv dW_u.$$

The first term can be estimated with the help of (1.19.17) as

$$n^{2\alpha} t^{-2\alpha} \sum_{k=2}^n \int_{t_{k-1}}^{t_k} \left(\int_s^{t_k} u^{\alpha-1}(u-s)^\alpha du \right)^2 d\langle W \rangle_s \le C n^{-2H} \to 0 \quad n \to \infty.$$

If $k = 1$, then for $\frac{1}{p} + \frac{1}{q} = 1$, $p, q > 1$

$$n^{2\alpha} t^{-2\alpha} \int_0^{t/n} \left(\int_s^{t/n} u^{\alpha-1}(u-s)^\alpha du \right)^2 ds$$

$$\le n^{2\alpha} t^{-2\alpha} \int_0^{t/n} \left(\int_s^{t/n} u^{p(\alpha-1)} du \right)^{2/p} \left(\int_s^{t/n} (u-s)^{\alpha q} du \right)^{2/q} ds$$

$$\le n^{2\alpha} t^{-2\alpha} \int_0^{t/n} s^{(pH - \frac{3p}{2} + 1)\frac{2}{p}} \left(\frac{t}{n} - s \right)^{(Hq - \frac{q}{2} + 1)\frac{2}{q}} ds$$

$$= n^{2\alpha} t^{-2\alpha} \int_0^{t/n} s^{2\alpha - 2 + \frac{2}{p}} \left(\frac{t}{n} - s \right)^{2\alpha + \frac{2}{q}} ds \sim n^{2\alpha} t^{-2\alpha} \left(\frac{t}{n} \right)^{4\alpha+1} \to 0,$$

i.e. the "main term" of $n^{2\alpha} \sum_{k=1}^n (J_4^k)^2$ tends to 0. For the remainder term of $n^{2\alpha} \sum_{k=1}^n (J_4^k)^2$ it is sufficient to prove that for any $\varepsilon > 0$

$$\sigma_4^n := n^{4\alpha} \sum_{k=\frac{n\varepsilon}{t}}^n \int_{t_{k-1}}^{t_k} \int_{t_{k-1}}^u \left(s^{-\alpha} \int_s^{t_k} v^{\alpha-1}(v-s)^\alpha dv \right)^2 ds$$

$$\times u^{-2\alpha} \left(\int_u^{t_k} v^{\alpha-1}(v-u)^\alpha dv \right)^2 du \to 0 \quad n \to \infty.$$

But

$$\sigma_4^n \le n^{4\alpha} \sum_{k=\frac{n\varepsilon}{t}}^n \int_{t_{k-1}}^{t_k} \int_{t_{k-1}}^u \left(\int_s^{t_k} v^{\alpha-1}(v-s)^\alpha dv \right)^2 ds$$

$$\times \left(\int_u^{t_k} v^{\alpha-1}(v-u)^\alpha dv \right)^2 du \le n^{-6} \sum_{k=\frac{n\varepsilon}{t}}^n (t_{k-1})^{-4} \sim n^{-2} \to 0, \quad n \to \infty.$$

After all estimates, for $s > 0$

$$\lim_{n \to \infty} n^{2\alpha} \sum_{k=n\frac{s}{t}+2}^n (\Delta X_{t_k})^2 \le C_2 A^{-2\alpha} t^{2\alpha} (\langle W \rangle_t - \langle W \rangle_s) + C_2 \frac{1}{A^{2\alpha+1}} t^{2\alpha} (t-s).$$

We have the opposite estimate,

$$C_1 t^{2\alpha}(t-s) \le \lim_{n\to\infty} n^{2\alpha} \sum_{k=n\frac{s}{t}+2}^{n} (\Delta X_{t_k})^2$$

$$\le C_2 A^{-2\alpha} t^{2\alpha} \left(\langle W\rangle_t - \langle W\rangle_s\right) + C_2 \frac{1}{A^{2\alpha+1}} t^{2\alpha}(t-s).$$

So, for A sufficiently large, $C_3 := C_1 - C_2 \frac{1}{A^{2\alpha+1}} > 0$, and we obtain that

$$C_3 t^{2\alpha}(t-s) \le C_2 A^{-2\alpha} t^{2\alpha} \left(\langle W\rangle_t - \langle W\rangle_s\right),$$

whence $\langle W\rangle_t - \langle W\rangle_s \ge \frac{C_3}{C_2} A^{2\alpha}(t-s)$, and constants do not depend on s and t. Therefore, if we write $\langle W\rangle_t = \int_0^t \theta_s ds$, then $\varepsilon_1 \le \theta_s \le \varepsilon_2$, $\varepsilon_i > 0$, and $W_t = \int_0^t \theta_s^{1/2} dV_s$ with some Wiener process V. Then we can conclude the proof of the theorem by the same arguments as for $H \in (1/2, 1)$. □

1.20 Multi-parameter Fractional Brownian Motion

1.20.1 The Main Definition

There can be at least two approaches to the definition of multi-parameter fBm. We consider the process which has a "fractional Brownian" property in each coordinate, but also it is possible to consider this property, for example, along any ray with its origin at zero (MY67).

For technical simplicity we consider two-parameter fBm (fBm-field) $\{B_t^H, t \in \mathbb{R}_+^2\}$, where $t = (t_1, t_2)$. We suppose that $s \le t$ if $s = (s_1, s_2)$, $t = (t_1, t_2)$ and $s_i \le t_i$, $i = 1, 2$.

Definition 1.20.1. The two-parameter process $\{B_t^H, t \in \mathbb{R}_+^2\}$ is called a (normalized) two-parameter fBm with Hurst index $H = (H_1, H_2) \in (0, 1)^2$, if it satisfies the assumptions
  (a) B^H is a Gaussian field, $B_t = 0$ for $t \in \partial \mathbb{R}_+^2$;
  (b) $E B_t^H = 0$, $E B_t^H B_s^H = \frac{1}{4} \prod_{i=1,2} (t_i^{2H_i} + s_i^{2H_i} - |t_i - s_i|^{2H_i})$.

Evidently, such a process has the modification with continuous trajectories, and we will always consider such a modification. Moreover, consider "two-parameter" increments: $\Delta_s B_t^H := B_t^H - B_{s_1 t_2}^H - B_{t_1 s_2}^H + B_s^H$ for $s \le t$. Then they are stationary. Note, that for any fixed $t_i > 0$ the process $B_{(t_i, \cdot)}^H$ will be the fBm with Hurst index H_j, $i = 1, 2, j = 3 - i$, evidently, nonnormalized.

1.20.2 Hölder Properties of Two-parameter fBm

Denote $\mathcal{P}_T := [0, T_1] \times [0, T_2]$.

118   1 Wiener Integration with Respect to Fractional Brownian Motion

Definition 1.20.2. The function $f : \mathbb{R}_+^2 \to \mathbb{R}$ belongs to the class $C^{\lambda_1,\lambda_2}(\mathcal{P}_T)$ for $0 < \lambda_i \leq 1$ (f is Hölder of orders λ_1 and λ_2 on $\mathcal{P}_T$), if there exists a constant $C > 0$, such that for all $s \leq t, s, t \in \mathcal{P}_T$

$$|\Delta_s f_t| \leq C \prod_{i=1,2} (t_i - s_i)^{\lambda_i}, \qquad (1.20.1)$$

$$|f(t) - f(s_1,t_2)| \leq C|t_1 - s_1|^{\lambda_1}, |f(t) - f(t_1,s_2)| \leq C|t_2 - s_2|^{\lambda_2}. \qquad (1.20.2)$$

The norm in the space $C^{\lambda_1,\lambda_2}(\mathcal{P}_T)$ is denoted as

$$\|f\|_{\lambda_1,\lambda_2} := \sup_{0 \leq s < t \leq T} \left(|f(t)| + \frac{|f(t) - f(s_1,t_2)|}{(t_1 - s_1)^{\lambda_1}} \right.$$
$$\left. + \frac{|f(t) - f(t_1,s_2)|}{(t_2 - s_2)^{\lambda_2}} + \frac{|\Delta_s f(t)|}{\prod_{i=1,2}(t_i - s_i)^{\lambda_i}} \right).$$

Evidently, inequalities (1.20.2) hold for B^H with $\lambda_1 < H, \lambda_2 < H$ and any $\mathcal{P}_T \subset \mathbb{R}_+^2$. It was proved by Kamont (Kam96), that (1.20.1) holds for B^H with any $\lambda_1 < H, \lambda_2 < H$ and on any $\mathcal{P}_T \subset \mathbb{R}_+^2$. Therefore, $B^H \in C^{H_1-\varepsilon,H_2-\varepsilon}(\mathcal{P}_T)$ for any $T \geq 0$ and any $0 < \varepsilon_i < H_i$. Moreover, according to (Kam96), for any $T > 0$ there exists the random variable $0 < c(\omega) < \infty$ P-a.s. such that

$$|\Delta_s B_t^H| \leq c(\omega) \prod_{i=1,2} (t_i - s_i)^{H_i} \left(1 + \log \tfrac{1}{t_i - s_i} \right)^{1/2}.$$

1.20.3 Fractional Integrals and Fractional Derivatives of Two-parameter Functions

For $\overline{\alpha} = (\alpha_1, \alpha_2)$ denote $\overline{\Gamma}(\overline{\alpha}) = \frac{1}{\Gamma(\alpha_1)\Gamma(\alpha_2)}$.

Definition 1.20.3. (SKM93) Let $f \in \mathcal{P} := [a,b] := \prod_{i=1,2}[a_i,b_i], a = (a_1,a_2), b = (b_1,b_2)$. Forward and backward Riemann–Liouville fractional integrals of orders $0 < \alpha_i < 1$ are defined as

$$(I_{a+}^{\alpha_1\alpha_2} f)(x) := \overline{\Gamma}(\overline{\alpha}) \int_{[a,x]} \frac{f(u)}{\varphi(x,u,1-\alpha)} du,$$

and

$$(I_{b-}^{\alpha_1\alpha_2} f)(x) := \overline{\Gamma}(\overline{\alpha}) \int_{[x,b]} \frac{f(u)}{\varphi(x,u,1-\alpha)} du,$$

correspondingly, where $[a,x] = \prod_{i=1,2}[a_i,x_i], [x,b] = \prod_{i=1,2}[x_i,b_i], du = du_1 du_2$, $\varphi(u,x,\alpha) = |u_1 - x_1|^{\alpha_1}|u_2 - x_2|^{\alpha_2}, u,x \in [a,b]$.

1.20 Multi-parameter Fractional Brownian Motion

Definition 1.20.4. Forward and backward fractional Liouville derivatives of orders $0 < \alpha_i < 1$ are defined as

$$(D_{a+}^{\alpha_1\alpha_2} f)(x) := \overline{\Gamma(1-\alpha)} \frac{\partial^2}{\partial x_1 \partial x_2} \int_{[a,x]} \frac{f(u)}{\varphi(x,u,\alpha)} du,$$

and

$$(D_{b-}^{\alpha_1\alpha_2} f)(x) := \overline{\Gamma(1-\alpha)} \frac{\partial^2}{\partial x_1 \partial x_2} \int_{[x,b]} \frac{f(u)}{\varphi(x,u,\alpha)} du, \quad x \in [a,b].$$

Definition 1.20.5. Forward fractional Marchaud derivatives of orders $0 < \alpha_i < 1$ are defined as

$$(\widetilde{D}_{a+}^{\alpha_1\alpha_2} f)(x) := \overline{\Gamma(1-\alpha)} \left(\frac{f(x)}{\varphi(x,u,\alpha)} + \alpha_1\alpha_2 \int_{[a,x]} \frac{\Delta_u f(x) du}{\varphi(x,u,1+\alpha)} \right.$$

$$\left. + \sum_{i=1,2, j=3-i} \frac{\alpha_i}{(x_j - a_j)^{\alpha_j}} \alpha_j \int_{a_i}^{x_i} \frac{f(x) - f(u_i, x_j)}{(x_i - u_i)^{1+\alpha_i}} du_i \right),$$

and the backward derivatives can be defined in a similar way.

Let $1 \leq p \leq \infty$, the classes $I_+^{\alpha_1\alpha_2}(L_p(\mathcal{P})) := \{f \mid f = I_{a+}^{\alpha_1\alpha_2}\varphi, \varphi \in L_p(\mathcal{P})\}$, $I_-^{\alpha_1\alpha_2}(L_p(\mathcal{P})) := \{f \mid f = I_{b-}^{\alpha_1\alpha_2}\varphi, \varphi \in L_p(\mathcal{P})\}$. Similarly to Theorem 13.1 (SKM93), the following result can be proved.

Theorem 1.20.6. *Liouville and Marchaud derivatives coincide on the classes $I_\pm^{\alpha_1\alpha_2}(L_p(\mathcal{P}))$.*

Further we denote $D_{a+}^{\alpha_1\alpha_2} =: I_{a+}^{-(\alpha_1\alpha_2)}$. Of course, we can introduce the notions of fractional integrals and fractional derivatives on $\mathbb{R}_+^2$. For example, the Riemann–Liouville fractional integrals and derivatives on $\mathbb{R}_+^2$ are defined by the formulas $(I_+^{\alpha_1\alpha_2} f)(x) := \overline{\Gamma(\alpha)} \int_{(-\infty,x]} \frac{f(t)}{\varphi(x,u,\alpha)} dt$,
$(I_-^{\alpha_1\alpha_2} f)(x) := \overline{\Gamma(\alpha)} \int_{[x,\infty)} \frac{f(t)}{\varphi(x,u,\alpha)} dt$,
$(I_+^{-(\alpha_1\alpha_2)} f)(x) = (D_+^{\alpha_1\alpha_2} f)(x) := \overline{\Gamma(1-\alpha)} \frac{\partial^2}{\partial x_1 \partial x_2} \int_{(-\infty,x]} \frac{f(t)}{\varphi(x,t,\alpha)} dt$, and
$(I_-^{-(\alpha_1\alpha_2)} f)(x) = (D_-^{\alpha_1\alpha_2} f)(x) := \overline{\Gamma(1-\alpha)} \frac{\partial^2}{\partial x_1 \partial x_2} \int_{[x,\infty)} \frac{f(t)}{\varphi(x,t,\alpha)} dt$,
$0 < \alpha_i < 1$. Evidently, all these operators can be expanded into the product of the form $I_+^{\alpha_1\alpha_2} = I_+^{\alpha_1} \otimes I_+^{\alpha_2}$, and so on. In what follows we shall consider only the case $H_i \in (1/2, 1)$. Define the operator

$$M_\pm^{H_1 H_2} f := \prod_{i=1,2} C_{H_i}^{(3)} I_\pm^{\alpha_1\alpha_2} f.$$

Definition 1.20.7. A random field $\{X_t, t \in \mathbb{R}_+^2\}$ is a field with independent increments if its increments $\{\Delta_{s_i} X_{t_i}, i = \overline{1,n}\}$ for any family of disjoint rectangles $\{(s_i, t_i], i = \overline{1,n}\}$ are independent.

Definition 1.20.8. The random field $\{W_t, t \in \mathbb{R}_+^2\}$ is called the Wiener field if $W = 0$ on $\partial \mathbb{R}_+^2$, W is the field with the independent increments and

$$E(\Delta_s W_t)^2 = area((s,t]) = \prod_{i=1,2} (t_i - s_i).$$

Let we have a probability space $(\Omega, \mathcal{F}, P)$ with two-parameter filtration $\{\mathcal{F}_t, t \in \mathbb{R}_+^2\}$ on it. It means that $\mathcal{F}_s \subset \mathcal{F}_t \subset \mathcal{F}$ for $s \leq t$. Denote $\mathcal{F}_s^* := \sigma\{\mathcal{F}_u, s \not< u\}$.

Definition 1.20.9. An adapted random field $\{X_t, \mathcal{F}_t, t \in \mathbb{R}_+^2\}$ is a strong martingale if X vanishes on $\partial \mathbb{R}_+^2$, $E|X_t| < \infty$ for all $t \in \mathbb{R}_+^2$ and for any $s \leq t$ $E(\Delta_s X_t \mid \mathcal{F}_s^*) = 0$.

Evidently, any random field with constant expectation and independent increments is a strong martingale, in particular, the Wiener field is a strong martingale.

It is not difficult to prove the following fact.

Lemma 1.20.10. Let $\{W_t, t \in \mathbb{R}_+^2\}$ be a Wiener field. Then the field

$$B_t^{H_1 H_2} := \int_{\mathbb{R}^2} (M_-^{H_1 H_2} \mathbf{1}_{(0,t)})(x) dW_x \quad (1.20.3)$$

is two-parameter fBm (not necessarily normalized).

Similarly to the one-parameter case, it is easy to show that any two-parameter fBm can be represented by (1.20.3) via underlying random field W.

Introduce the notion of Wiener integral w.r.t. two-parameter fBm.

Definition 1.20.11. Let

$$f \in L_2^{H_1 H_2} := \left\{ f : \mathbb{R}^2 \to \mathbb{R} : \int_{\mathbb{R}^2} ((M_-^{H_1 H_2} f)(t))^2 dt < \infty \right\}.$$

Then we denote $\int_{\mathbb{R}^2} f(t) dB_t^{H_1 H_2}$ as $\int_{\mathbb{R}^2} (M_-^{H_1 H_2} f)(t) dW_t$ for the underlying Wiener process W.

The following facts are proved similarly to the one-parameter case.

Theorem 1.20.12. Let the kernel $l_H^{(2)}(t,s) = \prod_{i=1,2} l_{H_i}(t_i, s_i) \cdot \mathbf{1}_{\{0 < s < t\}}$ for $H = (H_1, H_2)$, $t = (t_1, t_2)$ and $s = (s_1, s_2)$.

Then the field

$$I_t^H(l_H) := \int_{\mathbb{R}^2} l_H^{(2)}(t,s) dB_s^{H_1 H_2}$$

is a strong square integrable Gaussian martingale with independent increments and $E(I_t^H(l_H))^2 = \prod_{i=1,2} t_i^{1-2\alpha_i}$.

1.20 Multi-parameter Fractional Brownian Motion

Similarly to the one-parameter case, we call $I^H(B_H)$ the strong Molchan martingale. It can be presented as

$$I_t^H(l_H) = \prod_{i=1,2}(1-2\alpha_i)^{1/2} \int_{[0,t]} \prod_{i=1,2} s_i^{-\alpha_i} dB_s, \quad \alpha_i = H_i - 1/2,$$

where $\{B_t, \mathcal{F}_t, t \in \mathbb{R}_+^2\}$ is some Wiener field.

In turn, the two-parameter fBm can be presented via some Wiener field B by the integral

$$B_t^{H_1 H_2} = \int_{[0,t]} m_H^{(2)}(t,s) dB_s,$$

where $H_i \in (1/2, 1)$, and $m_H^{(2)}(t,s) = \prod_{i=1,2} m_{H_i}(t_i, s_i) \mathbf{1}_{\{0 < s_i < t_i\}}$

$$= \prod_{i=1,2} C_{H_i}^{(6)} \alpha_i s_i^{-\alpha_i} \int_{s_i}^{t_i} u_i^{\alpha_i}(u_i - s_i)^{\alpha_i - 1} du_i.$$

2
Stochastic Integration with Respect to fBm and Related Topics

2.1 Pathwise Stochastic Integration

2.1.1 Pathwise Stochastic Integration in the Fractional Sobolev-type Spaces

In this subsection we consider pathwise integrals $\int_0^T f(t)dB_t^H$ for processes f from the fractional Sobolev type spaces $I_{a+}^\alpha(L^p)$ for some $p > 1$. This approach was developed by Zähle (Zah98), (Zah99), (Zah01).

Consider two nonrandom functions f and g defined on some interval $[a,b] \subset \mathbb{R}$ and suppose that the limits $f(u+) := \lim_{\delta \downarrow 0} f(u + \delta)$ and $g(u-) := \lim_{\delta \downarrow 0} g(u - \delta)$, $a \leq u \leq b$, exist. Put $f_{a+}(x) := (f(x) - f(a+))\mathbf{1}_{(a,b)}(x)$, $g_{b-}(x) := (g(b-) - g(x))\mathbf{1}_{(a,b)}(x)$. Suppose also that $f_{a+} \in I_{a+}^\alpha(L_p[a,b])$, $g_{b-} \in I_{b-}^{1-\alpha}(L_p[a,b])$ for some $p \geq 1, q \geq 1, 1/p + 1/q \leq 1, 0 \leq \alpha \leq 1$. Then, evidently, $D_{a+}^\alpha f_{a+} \in L_p[a,b]$, $D_{b-}^{1-\alpha} g_{b-} \in L_q[a,b]$.

Definition 2.1.1. *The generalized (fractional) Lebesgue–Stieltjes integral $\int_a^b f(x)dg(x)$ is defined as*

$$\int_a^b f(x)dg(x) := \int_a^b (D_{a+}^\alpha f_{a+})(x)(D_{b-}^{1-\alpha} g_{b-})(x)dx + f(a+)(g(b-) - g(a+)).$$

Lemma 2.1.2. *Definition 2.1.1 does not depend on the possible choice of α.*

Proof. Let $f_{a+} \in (I_{a+}^\alpha \cap I_{a+}^{\alpha+\beta})(L_p[a,b])$, $g_{b-} \in (I_{b-}^{1-\alpha} \cap I_{b-}^{1-\alpha-\beta})(L_q[a,b])$ for some α, β such that $0 \leq \alpha \leq 1$, $0 \leq \alpha + \beta \leq 1$, $1/p + 1/q \leq 1$. Then, according to (1.1.5) (composition formula for fractional derivatives) and (1.1.6) (integration-by-parts formula),

$$\int_a^b (D_{a+}^{\alpha+\beta} f_{a+})(x)(D_{b-}^{1-\alpha-\beta} g_{b-})(x)dx$$

$$= \int_a^b (D_{a+}^\beta D_{a+}^\alpha f_{a+})(x)(D_{b-}^{1-\alpha-\beta} g_{b-})(x)dx$$

$$= \int_a^b (D_{a+}^\alpha f_{a+})(x)(D_{b-}^\beta D_{b-}^{1-\alpha-\beta} g_{b-})(x)dx$$

$$= \int_a^b (D_{a+}^\alpha f_{a+})(x)(D_{b-}^{1-\alpha} g_{b-})(x)dx.$$

□

Let $\alpha p < 1$. Then $f_{a+} \in I_{a+}^\alpha(L_p[a,b])$ if and only if $f \in I_{a+}^\alpha(L_p[a,b])$ and in this case we can simplify the formula for the generalized integral:

$$\int_a^b f(x)dg(x) = \int_a^b \left((D_{a+}^\alpha f)(x) - \frac{1}{\Gamma(1-\alpha)} \cdot \frac{f(a+)}{(x-a)^\alpha}\right)(D_{b-}^{1-\alpha} g_{b-})(x)dx$$
$$+ f(a+)(g(b-) - g(a+)) = \int_a^b (D_{a+}^\alpha f)(x)(D_{b-}^{1-\alpha} g_{b-})(x)dx \quad (2.1.1)$$
$$- f(a+)I_{b-}^{1-\alpha}(D_{b-}^{1-\alpha} g)(a) + f(a+)(g(b-) - g(a+))$$
$$= \int_a^b (D_{a+}^\alpha f)(x)(D_{b-}^{1-\alpha} g_{b-})(x)dx.$$

Lemma 2.1.3. *Let $g_{b-} \in I_{b-}^{1-\alpha}(L_q[a,b]) \cap C[a,b]$ for some $q > \frac{1}{1-\alpha}$ and $0 < \alpha < 1$. Then for any $a < c < d < b$*

$$\int_a^b (D_{a+}^\alpha \mathbf{1}_{(c,d)})(x)(D_{b-}^{1-\alpha} g_{b-})(x)dx = g(d) - g(c). \quad (2.1.2)$$

Proof. We have that

$$(D_{a+}^\alpha \mathbf{1}_{(c,d)})(x) = \begin{cases} 0, & x \leq c, \\ \frac{(x-c)^{-\alpha}}{\Gamma(1-\alpha)}, & c < x \leq d, \\ \frac{(x-c)^{-\alpha} - (x-d)^{-\alpha}}{\Gamma(1-\alpha)}, & d \leq x \leq b. \end{cases}$$

Therefore, by using (2.1.1), we obtain for $\alpha p < 1$, or $q > \frac{1}{1-\alpha}$, that

$$\int_a^b (D_{a+}^\alpha \mathbf{1}_{(c,d)})(x)(D_{b-}^{1-\alpha} g_{b-})(x)dx = \frac{1}{\Gamma(1-\alpha)} \int_c^b (x-c)^{-\alpha}(D_{b-}^{1-\alpha} g_{b-})(x)dx$$
$$- \frac{1}{\Gamma(1-\alpha)} \int_d^b (x-d)^{-\alpha}(D_{b-}^{1-\alpha} g_{b-})(x)dx = I_{b-}^{1-\alpha}(D_{b-}^{1-\alpha} g_{b-})(c)$$
$$- I_{b-}^{1-\alpha}(D_{b-}^{1-\alpha} g_{b-})(d) = g(d) - g(c).$$

□

Corollary 2.1.4. *Let the function $g \in C^\lambda[a,b]$ for some $\lambda \leq 1$, then $g_{b-} \in I_{b-}^{1-\alpha}(L_p[a,b])$ for any $p \geq 1$ and $1 - \alpha < \lambda$. So, we can put $p > 2/\lambda$, $\alpha = 1 - \lambda/2$ and obtain for g (2.1.2).*

Corollary 2.1.5. *For any step function $f_\pi(x) = \sum_{k=0}^{n-1} c_k \mathbf{1}_{[x_k, x_{k+1})}(x)$ with $a = x_0 < \cdots < x_n = b$ and g satisfying the conditions of Lemma 2.1.3, we have that $\int_a^b f(x)dg(x) = \sum_{k=0}^{n-1} c_k(g(x_{k+1}) - g(x_k))$.*

Further we suppose that $g(b-) = g(b)$ and $g(a+) = g(a)$.
Denote by $BV[a,b]$ the class of functions of bounded variation on $[a,b]$.

Lemma 2.1.6. *Let the functions* $f_{a+} \in I^{\alpha}_{a+}(L_p[a,b])$, $g_{b-} \in I^{1-\alpha}_{b-}(L_q[a,b]) \cap BV[a,b]$ *with* $p \geq 1, q \geq 1, 1/p + 1/q \leq 1$ *and*

$$\int_a^b I^{\alpha}_{a+}(|(D^{\alpha}_{a+}f)|)(x)|g|(dx) < \infty. \qquad (2.1.3)$$

Then

$$\int_a^b f(x)dg(x) = \text{(L-S)} \int_a^b f(x)dg(x).$$

Proof. We have that

$$\begin{aligned}\text{(L-S)} \int_a^b f(x)dg(x) &= \text{(L-S)} \int_a^b I^{\alpha}_{a+}(D^{\alpha}_{a+}f)(x)dg(x) \\ &= \tfrac{1}{\Gamma(1-\alpha)}\text{(L-S)} \int_a^b (\int_a^x (x-y)^{\alpha-1}(D^{\alpha}_{a+}f)(y)dy)dg(x).\end{aligned} \qquad (2.1.4)$$

Condition (2.1.3) together with Fubini theorem permits us to change the order of integration:

$$\begin{aligned}&\text{(L-S)} \int_a^b (\int_a^x (x-y)^{\alpha-1}(D^{\alpha}_{a+}f)(y)dy)dg(x) \\ &= \int_a^b (D^{\alpha}_{a+}f)(y)(\int_y^b (x-y)^{\alpha-1}dg(x))dy \\ &= (\alpha - 1)\int_a^b (D^{\alpha}_{a+}f)(y)(\int_y^b (\int_x^{\infty} (z-y)^{\alpha-2}dz)dg(x))dy.\end{aligned} \qquad (2.1.5)$$

Further, if $y \in (a,b)$ is the point of continuity of function g, then

$$\begin{aligned}&\int_y^b (\int_x^{\infty}(z-y)^{\alpha-2}dz)dg(x) = \int_y^b(\int_y^z dg(x))(z-y)^{\alpha-2}dz \\ &+ \int_b^{\infty}(\int_y^b dg(x))(z-y)^{\alpha-2}dz = \int_y^b \tfrac{g(z)-g(y)}{(z-y)^{2-\alpha}}dz \\ &+ \tfrac{g(b)-g(y)}{(\alpha-1)(b-y)^{\alpha-1}} = \tfrac{\Gamma(\alpha)}{\alpha-1}(D^{1-\alpha}_{b-}g_{b-})(y).\end{aligned} \qquad (2.1.6)$$

Since set of discontinuity points of g is at most countable, and taking (2.1.4)–(2.1.6) together, we obtain the proof. $\square$

Now we consider the case of Hölder functions f and g. The existence of (R-S) $\int_a^b f dg$ for $f \in C^{\lambda}[a,b]$, $g \in C^{\mu}[a,b]$ with $\lambda + \mu > 1$ was established by Kondurar (Kon37). Moreover, this integral coincides with $\int_a^b f dg$, as the next theorem states.

Let $f \in C^{\lambda}[a,b]$ for some $0 < \lambda \leq 1$ and $|f(x) - f(y)| \leq c(\lambda)|x-y|^{\lambda}$, $x, y \in [a,b]$. Consider the following step function:

$$f_{\pi}(x) = \sum_{k=0}^{n-1} f(x_k)\mathbf{1}_{[x_k, x_{k+1})}(x),$$

where the partition $\pi = \{a = x_0 < x_1 < \cdots < x_n = b\}$.
Evidently, $\lim_{|\pi| \to 0} \sup_{\pi} \|f_{\pi} - f\|_{L_{\infty}[a,b]} = 0$.

Theorem 2.1.7. *1) For any $0 < \alpha < \lambda$*

$$\lim_{|\pi| \to 0} \sup_{\pi} \|(D_{a+}^\alpha f_\pi) - (D_{a+}^\alpha f)\|_{L_1[a,b]} = 0.$$

2) Let $f \in C^\lambda([a,b])$, $g \in C^\mu[a,b]$ with $\lambda + \mu > 1$, then (R-S) $\int_a^b f dg$ exists and

$$\int_a^b f dg = (\text{R-S}) \int_a^b f dg.$$

Proof. 1) It is sufficient to prove that $\int_a^b \frac{|f_\pi(x) - f(x)|}{(x-a)^\alpha} dx \to 0$ and $\int_a^b \int_a^x (x-y)^{-\alpha-1} |f_\pi(x) - f(x) - f_\pi(y) + f(y)| dy\, dx \to 0$ as $|\pi| \to 0$. But $|f_\pi(x) - f(x)| \le |f(x_k) - f(x)| \le c(\lambda)|\pi|^\lambda$ for $x \in [x_k, x_{k+1})$, therefore $\int_a^b \frac{|f_\pi(x) - f(x)|}{(x-a)^\alpha} dx \le c(\lambda)|\pi|^\lambda \frac{(b-a)^{1-\alpha}}{1-\alpha} \to 0$ as $|\pi| \to 0$. Also, for $x \in [x_k, x_{k+1})$

$$A(x) := \int_a^x (x-y)^{-\alpha-1} |f_\pi(x) - f(x) - f_\pi(y) + f(y)| dy$$

$$= \sum_{i=0}^{k-1} \int_{x_i}^{x_{i+1}} (x-y)^{-\alpha-1} |f(x_k) - f(x) - f(x_i) + f(y)| dy$$

$$+ \int_{x_k}^x (x-y)^{-\alpha-1} |f(y) - f(x)| dy \le 2c(\lambda) \sum_{i=0}^{k-1} \int_{x_i}^{x_{i+1}} (x-y)^{-\alpha-1} dy \cdot |\pi|^\lambda$$

$$+ c(\lambda) \int_{x_k}^x (x-y)^{\lambda-\alpha-1} dy \le 2c(\lambda)|\pi|^\lambda \frac{(x-x_k)^{-\alpha}}{1-\alpha} + c(\lambda) \frac{(x-x_k)^{\lambda-\alpha}}{\lambda-\alpha}$$

$$\le 3c(\lambda) \frac{|\pi|^{\lambda-\alpha}}{\lambda-\alpha},$$

which means that $\int_a^b A(x) dx \to 0$ as $|\pi| \to 0$.

2) We take $1 - \mu < \alpha < \lambda$, then the fractional derivatives $D_{a+}^\alpha f(x)$ and $(D_{b-}^{1-\alpha} g)_{b-}(x)$ exist, and, moreover,

$$|(D_{b-}^{1-\alpha} g)_{b-}(x)| \le \frac{1}{\Gamma(1-\alpha)} \left(\frac{|g(b) - g(x)|}{(b-x)^{1-\alpha}} + (1-\alpha) \int_x^b \frac{|g(y) - g(x)|}{(y-x)^{2-\alpha}} dy \right)$$

$$\le \frac{1}{\Gamma(1-\alpha)} \cdot c(\lambda)(b-x)^{\mu+\alpha-1} \left(1 + \frac{1-\alpha}{\mu+\alpha-1} \right) \le C$$

for some constant C. Therefore, according to part 1) of the proof,

$$|\int_a^b f_\pi dg - \int_a^b f dg| \le \int_a^b |(D_{a+}^\alpha f_\pi)(x) - (D_{a+}^\alpha f)(x)||(D_{b-}^{1-\alpha} g)_{b-}(x)| dx$$

$$\le C \int_a^b |(D_{a+}^\alpha f_\pi)(x) - (D_{a+}^\alpha f)(x)| dx \to 0,$$

(2.1.7)

as $|\pi| \to 0$.

Furthermore, according to Corollary 2.1.5,

$$\int_a^b f_\pi dg = \sum_{k=0}^{n-1} f(x_k)(g(x_{k+1}) - g(x_k)) \to (\text{R-S}) \int_a^b f dg, \qquad (2.1.8)$$

and from (2.1.7)–(2.1.8) we obtain the desired equality. □

Now we establish the properties of generalized integral $\int_s^t f dg$ as the function of upper and lower boundaries.

Lemma 2.1.8 ((Zah98)). *1) Let $a \le s < t \le b$ and the functions f and g satisfy the assumptions*
(i) $(f \cdot \mathbf{1}_{(s,t)}) \in I_+^\alpha(L_p[a,b])$, $g_{b-} \in I_-^{1-\alpha}(L_q[a,b])$ for some $0 < \alpha < 1$, $p \ge 1, q \ge 1, 1/p + 1/q \le 1$,
(ii) $f_{s+} \in I_+^{\alpha'}(L_{p'}[s,t])$, $g_{t-} \in I_-^{1-\alpha'}(L_{q'}[s,t])$ for some $0 < \alpha' < 1$, $p' \ge 1, q' \ge 1, 1/p' + 1/q' \le 1$. Then

$$\int_a^b \mathbf{1}_{(s,t)} f dg = \int_s^t f dg.$$

2) The equality

$$\int_s^t f dg + \int_t^u f dg = \int_s^u f dg$$

holds for $a \le s < t < u \le b$, if all the integrals exist as generalized Lebesgue-Stieltjes integrals.

Proof. 1) Let $\{\varphi_n(x), x \in \mathbb{R}\}$ be a sequence of smooth kernels, i.e. $\varphi_n \in C^\infty(\mathbb{R})$, $\varphi_n \ge 0$, $\varphi_n = 0$ outside $[-1/n, 0]$ and $\int_{-1/n}^0 \varphi_n(x)dx = 1$. More exactly, let $\varphi_n(x) = n\varphi(nx)$ for $\varphi \in C^\infty(\mathbb{R})$, $\varphi = 0$ outside of $[-1, 0]$. Then we can approximate the function g_{b-} by smooth functions $g_n := g_{b-} * \varphi_n$, and the following properties hold:

$$g_n(b-) = n \int_{[x-b, x-a] \cap [-1/n, 0]} (g(b-) - g(x-t))\varphi(nt)dt \,|_{x=b-} = 0;$$
$$(D_{b-}^{1-\alpha} g_n)(x) = D_{b-}^{1-\alpha}(\int_{\mathbb{R}} g_{b-}(x-t)\varphi_n(t)dt)$$
$$= \mathbf{1}_{(a,b)}(x)(\Gamma(1-\alpha))^{-1} \Big(\int_{\mathbb{R}} g_{b-}(x-t)\varphi_n(t)dt(b-x)^{\alpha-1}$$
$$+ \alpha \int_x^b (y-x)^{-\alpha}(\int_{\mathbb{R}}(g_{b-}(x-t) - g_{b-}(y-t))\varphi_n(t)dt)dy\Big) \quad (2.1.9)$$
$$= \frac{\mathbf{1}_{(a,b)}(x)}{\Gamma(1-\alpha)} \int_{\mathbb{R}} \varphi_n(t) \Big(\frac{g_{b-}(x-t)}{(b-x)^{1-\alpha}} + \alpha \int_x^b \frac{g_{b-}(x-t) - g_{b-}(y-t)}{(y-x)^{2-\alpha}} dy\Big) dt$$
$$= \mathbf{1}_{(a,b)}(x)((D_{b-}^{1-\alpha} g_{b-}) * \varphi_n)(x);$$

$$\|(D_{b-}^{1-\alpha} g_n) - (D_{b-}^{1-\alpha} g_{b-})\|_{L_q[a,b]}^q$$
$$\|(D_{b-}^{1-\alpha} g_{b-}) * \varphi_n - (D_{b-}^{1-\alpha} g_{b-})\|_{L_q[a,b]}^q$$
$$= \int_a^b |\int_{-1}^0 ((D_{b-}^{1-\alpha} g_{b-})(x - \tfrac{t}{n}) - (D_{b-}^{1-\alpha} g_{b-})(x))\varphi(t)dt|^q dx$$
$$\le C \int_a^b \int_{-1}^0 |(D_{b-}^{1-\alpha} g_{b-})(\cdot - \tfrac{t}{n}) - (D_{b-}^{1-\alpha} g_{b-})(\cdot)|^q dt\, dx \to 0, \quad n \to \infty. \quad (2.1.10)$$

Therefore, from this L_q-convergence, from Lemma 2.1.2 and the properties of convolutions,

$$\int_a^b \mathbf{1}_{(s,t)} f dg = \int_a^b (D_{a+}^\alpha \mathbf{1}_{(s,t)} f)(u)(D_{b-}^{1-\alpha} g_{b-})(u)du$$
$$= \lim_{n \to \infty} \int_a^b (D_{a+}^\alpha \mathbf{1}_{(s,t)} f)(u)(D_{b-}^{1-\alpha} g_n)(u)du$$
$$= \lim_{n \to \infty} \int_a^b (\mathbf{1}_{(s,t)} f)(u) g_n'(u) du = \lim_{n \to \infty} \int_s^t f(u)(g_{b-} * \varphi_n')(u)du.$$

Further, for any $c > 0$ $(c * \varphi_n')(u) = 0$, therefore

$$\int_s^t f(u)(g_{b-} * \varphi_n')(u)du = \int_s^t f(u)(g * \varphi_n')(u)du$$
$$= \int_s^t f(u)(g_{t-} * \varphi_n')(u)du, \qquad (2.1.11)$$

and

$$\lim_{n\to\infty} \int_s^t f(u)(g_{b-} * \varphi_n')(u)du = \lim_{n\to\infty} \int_s^t f(u)(g_{t-} * \varphi_n')(u)du$$
$$= \lim_{n\to\infty} \int_s^t f(u)(g_{t-} * \varphi_n)'(u)du. \qquad (2.1.12)$$

Thanks to Lemma 2.1.2, assumption (ii), (2.1.9) and (2.1.10), applied to t instead of b,

$$\begin{aligned}
&\lim_{n\to\infty} \int_s^t f(u)(g_{t-} * \varphi_n)'(u)du \\
&= \lim_{n\to\infty} \int_s^t (D_{s+}^{\alpha'} f_{s+})(u)(D_{t-}^{1-\alpha'}(g_{t-} * \varphi_n))(u)du \\
&= \lim_{n\to\infty} \int_s^t (D_{s+}^{\alpha'} f_{s+})(u)((D_{t-}^{1-\alpha'} g_{t-}) * \varphi_n)(u)du \\
&= \int_s^t (D_{s+}^{\alpha'} f_{s+})(u)(D_{t-}^{1-\alpha'} g_{t-})(u)du = \int_s^t f dg,
\end{aligned} \qquad (2.1.13)$$

and we obtain the first statement. The second one we obtain by using some of the equalities from (2.1.11):

$$\int_s^t f dg + \int_t^u f dg = \lim_{n\to\infty} \int_s^t f(r)(g * \varphi_n')(r)dr$$
$$+ \lim_{n\to\infty} \int_t^u f(r)(g * \varphi_n')(r)dr = \lim_{n\to\infty} \int_s^u f(r)(g * \varphi_n')(r)dr$$
$$= \int_s^u f dg.$$

□

2.1.2 Pathwise Stochastic Integration in Fractional Besov-type Spaces

In this subsection we consider the approach to pathwise stochastic integration in fractional Besov-type spaces, introduced by Nualart and Răşcanu (NR00) (see also (CKR93) and (NO03a)).

Consider the following functional spaces. Let for $0 < \beta < 1$
$\varphi_f^\beta(t) := |f(t)| + \int_0^t |f(t) - f(s)|(t-s)^{-\beta-1}ds$, and $W_0^\beta = W_0^\beta[0T]$ be the space of real-valued measurable functions $f : [0,T] \to \mathbb{R}$ such that

$$\|f\|_{0,\beta} := \sup_{t\in[0,T]} \varphi_f^\beta(t) < \infty.$$

Furthermore, let $W_1^\beta = W_1^\beta[0,T]$ be the space of real-valued measurable functions $f : [0,T] \to \mathbb{R}$ such that

$$\|f\|_{1,\beta} := \sup_{0\le s<t\le T} \left(\frac{|f(t) - f(s)|}{(t-s)^\beta} + \int_s^t \frac{|f(u) - f(s)|}{(u-s)^{1+\beta}} du \right) < \infty$$

and $W_2^\beta = W_2^\beta[0,T]$ be the space of real-valued measurable functions $f : [0,T] \to \mathbb{R}$ such that

$$\|f\|_{2,\beta} := \int_0^T \frac{|f(s)|}{s^\beta} ds + \int_0^T \int_0^s \frac{|f(s)-f(u)|}{(s-u)^{\beta+1}} du < \infty.$$

Note that the spaces $W_i^\beta, i=0,2$ are Banach spaces with respect to corresponding norms and $\|f\|_{1,\beta}$ is not the norm in a usual sense.

Moreover, for any $0 < \varepsilon < \beta \wedge (1-\beta)$

$$C^{\beta+\varepsilon}[0,T] \subset W_i^\beta[0,T] \subset C^{\beta-\varepsilon}[0,T],\; i=0,1,\; C^{\beta+\varepsilon}[0,T] \subset W_2^\beta[0,T].$$

Therefore, the trajectories of fBm B^H for a.a. $\omega \in \Omega$, any $T > 0$ and any $0 < \beta < H$ belong to $W_1^\beta[0,T]$.

Let $f \in W_1^\beta[0,T]$. Then its restriction to $[0,t] \subset [0,T]$ belongs to $I_-^\beta(L_\infty[0,t])$ and

$$\Lambda_\beta(f) := \sup_{0 \leq s < t \leq T} |(D_{t-}^\beta f_{t-})(s)| \leq \frac{1}{\Gamma(1-\beta)} \|f\|_{1,\beta} < \infty.$$

The restriction of $f \in W_2^\beta[0,T]$ to $[0,t] \subset [0,T]$ belongs to $I_+^\beta(L_1[0,t])$.

Now, let $f \in W_2^\beta[0,T]$, $g \in W_1^{1-\beta}[0,T]$. Then for any $0 < t \leq T$ there exists the Lebesgue integral $\int_0^t (D_{0+}^\beta f)(x)(D_{t-}^{1-\beta} g_{t-})(x)dx$, so we can define $\int_0^t fdg$ according to Definition 2.1.1 and formula (2.1.2). Moreover, for any $0 < t \leq T$ $\int_0^t fdg = \int_0^T 1_{(0,t)} fdg$, and the integral $\int_0^t fdg$ admits an estimate

$$|\int_0^t fdg| \leq \int_0^t |(D_{0+}^\beta f)(x)||(D_{t-}^{1-\beta} g_{t-})(x)|dx$$
$$\leq \Lambda_{1-\beta}(g) \|f\|_{2,\beta} \leq (\Gamma(\beta))^{-1} \|g\|_{1,1-\beta} \|f\|_{2,\beta}.$$

Further we fix some $0 < \beta < 1/2$.

Lemma 2.1.9 ((NR00)). *1. Let $f \in W_0^\beta[0,T]$, $g \in W_1^{1-\beta}[0,T]$, $G_t(f) := \int_0^t fdg$, $t \in [0,T]$. Then*

$$\varphi_{G.(f)}^\beta(t) \leq C_{\beta,T}^1 \Lambda_{1-\beta}(g) \int_0^t ((t-s)^{-2\beta} + s^{-\beta}) \varphi_f^\beta(s)ds.$$

2. Let $f \in W_0^\beta[0,T]$, $g \in W_1^{1-\beta}[0,T]$. Then $G.(f) \in C^{1-\beta}[0,T]$ and

$$\|G(f)\|_{1,1-\beta} \leq C_{\beta,T}^2 \Lambda_{1-\beta}(g) \|f\|_{0,\beta}.$$

Here $C_{\beta,T}^i, i=1,2$ depend only on T and β.

Proof. 1. It is not hard to check that for $f \in W_0^\beta[0,T]$ and $g \in W_1^{1-\beta}[0,T]$ condition 1) of Lemma 2.1.8 holds. Therefore, evidently,

$$|G_t(f) - G_s(f)| = |\int_s^t f\,dg| \leq \int_s^t |(D_{s+}^\beta f)(u)||(D_{t-}^{1-\beta} g_{t-})(u)|du$$
$$\leq \Lambda_{1-\beta}(g) \int_s^t \left(\frac{|f(u)|}{(u-s)^\beta} + \beta \int_s^u \frac{|f(u)-f(v)|}{(u-v)^{\beta+1}} dv\right) du. \tag{2.1.14}$$

From (2.1.14) it follows that

$$\int_0^t \frac{|G_t(f)-G_u(f)|}{(t-u)^{\beta+1}} du \leq \Lambda_{1-\beta}(g)\left(\int_0^t |f(u)|(\int_0^u (t-s)^{-\beta-1}(u-s)^{-\beta}ds)du \right.$$
$$\left. + \int_0^t \int_0^u \frac{|f(u)-f(v)|}{(u-v)^{\beta+1}}(t-v)^{-\beta}dv\,du\right). \tag{2.1.15}$$

The first integral on the right-hand side of (2.1.15) can be estimated as $C\int_0^t |f(u)|(t-u)^{-2\beta}du$ with $C = \int_0^\infty (1+u)^{-\beta-1} u^{-\beta} du$, and the second one can be estimated as $\int_0^t (t-u)^{-\beta} \int_0^u \frac{|f(u)-f(v)|}{(u-v)^{\beta+1}} dv\,du$.

Since $(t-u)^{-2\beta} \geq (t-u)^{-\beta} T^{-\beta}$, we obtain from (2.1.15) that

$$\int_0^t \frac{|G_t(f) - G_u(f)|}{(t-u)^{\beta+1}} du \leq \Lambda_{1-\beta}(g)(C+T^\beta)\int_0^t (t-u)^{-2\beta}\varphi_f^\beta(u)du. \tag{2.1.16}$$

Further, from (2.1.14) it follows that

$$|G_t(f)| \leq \Lambda_{1-\beta}(g) \int_0^t \left(\frac{|f(u)|}{u^\beta} + \beta \int_0^u \frac{|f(u)-f(v)|}{(u-v)^{\beta+1}}dv\right) du$$
$$\leq \Lambda_{1-\beta}(g)(1+\beta T^\beta)\int_0^t u^{-\beta}\varphi_f^\beta(u)du, \tag{2.1.17}$$

and the proof follows from (2.1.16)–(2.1.17).

2. It follows from (2.1.14) that

$$|G_t(f) - G_s(f)| \leq \Lambda_{1-\beta}(g)\frac{1+\beta T^\beta}{1-\beta}\|f\|_{0,\beta}(t-s)^{1-\beta},$$

and from (2.1.17) we obtain that

$$|G_t(f)| \leq \Lambda_{1-\beta}(g)\frac{1+\beta T^\beta}{1-\beta}T^{1-\beta}\|f\|_{0,\beta},$$

whence the proof follows with $C_{\beta,T}^2 = (1 \vee T^{1-\beta})\frac{1+\beta T^\beta}{1-\beta}$. □

Similar but more simple estimates hold for the Lebesgue integral $F_t(f) = \int_0^t f(s)ds$, so we omit the proof of the following lemma.

Lemma 2.1.10 ((NR00)). 1. Let $0 < \beta < 1$ and $f : [0,T] \to \mathbb{R}$ be a measurable function with $\sup_{t\in[0,T]} \int_0^t |f(s)|(t-s)^{-\beta}ds < \infty$.
Then

$$\varphi_{F.(f)}^\beta(t) \leq C_{\beta,T}^3 \int_0^t |f(s)|(t-s)^{-\beta}ds,$$

with $C_{\beta,T}^3 = T^\beta + 1/\beta$.

2. Let f be bounded on $[0,T]$. Then $F(f) \in C^1[0,T]$ and $\|F(f)\|_{0,\beta} \leq C_{\beta,T}^4 f_T^*$, where $f_T^* := \sup_{t\in[0,T]}|f(t)|$, $C_{\beta,T}^4$ depends on β and T.

2.2 Pathwise Stochastic Integration w.r.t. Multi-parameter fBm

2.2.1 Some Additional Properties of Two-parameter Fractional Integrals and Derivatives

Throughout this section we consider two-parameter functions and fields. The first result can be proved similarly to the one-parameter case. Let the rectangle $\mathcal{P} = [a, b]$ be fixed.

Lemma 2.2.1. *1. Let* $f \in I_{\pm}^{\beta_1 \beta_2}(L_p(\mathcal{P}))$ *for some* $p > 1$. *Then* $\lim_{\beta_1 \to 0, \beta_2 \to 0} D_{a+(b-)}^{\beta_1 \beta_2} f(x) = f(x)$, *where the limit is in* $L_p(\mathcal{P})$. *2. Let, in addition, the function* f *be twice continuously differentiable in the neighborhood of the point* x. *Then* $\lim_{\beta_1 \to 1, \beta_2 \to 1} D_{a+(b-)}^{\beta_1 \beta_2} f(x) = \frac{\partial^2 f}{\partial x_1 \partial x_2}(x)$. *So, we can put* $D_{a+(b-)}^{00} f := f$, $D_{a+(b-)}^{11} f := f$.

Theorem 2.2.2. *Let* $0 < \beta_i < 1$ *and* $1 < p < \beta_1^{-1} \vee \beta_2^{-1}$. *Then the operator* $I_{a+}^{\beta_1 \beta_2}$ *is bounded from* $L_p(\mathcal{P})$ *into* $L_q(\mathcal{P})$, *where* $1 < q < p((1 - \beta_1 p)^{-1} \wedge (1 - \beta_2 p)^{-1})$.

Proof. Denote $r := p((1 - \beta_1 p)^{-1} \vee (1 - \beta_2 p)^{-1})$. Since $r > p$, it is sufficient to consider $q \in (p, r)$. Then for $\frac{1}{p'} + \frac{1}{p} = 1$, $\frac{1}{p_i'} + \frac{1}{r} = 1 - \beta_i$, from the generalized Hölder inequality, it holds that

$$|(I_{a+}^{\beta_1 \beta_2} f)(x)| \leq C \Big(\int_{[a,x]} |f(u)|^p \prod_{i=1,2} (x_i - u_i)^{(\beta_i - 1)\gamma q} du \Big)^{\frac{1}{q}}$$

$$\times \Big(\int_{[a,x]} |f(u)|^p du \Big)^{\frac{1}{p} - \frac{1}{q}} \Big(\int_{[a,x]} \prod_{i=1,2} (x_i - u_i)^{(\beta_i - 1)(1 - \gamma) p'} du_i \Big)^{\frac{1}{p'}}$$

$$\leq C \|f\|_{L_p(\mathcal{P})}^{1 - \frac{p}{q}} \Big(\int_{[a,x]} |f(u)|^p \prod_{i=1,2} (x_i - u_i)^{(\beta_i - 1)\gamma q} du \Big)^{\frac{1}{q}}.$$

Here we choose γ satisfying the inequalities $(1 - \beta_i)\gamma q < 1$ and $(1 - \beta_i)(1 - \gamma) p' < 1$, which is equivalent to $1 - (p'(1 - \beta_i))^{-1} < \gamma < (q(1 - \beta_i))^{-1}$. Such a choice is possible, since the inequality $1 - (p'(1 - \beta_i))^{-1} < (q(1 - \beta_i))^{-1}$ is equivalent to $q < p(1 - \beta_i p)^{-1}$, and this is evident under our suppositions. By integration over $\mathcal{P}$ we obtain that

$$\|I_{a+}^{\beta_1 \beta_2} f\|_{L_q(\mathcal{P})} \leq C \|f\|_{L_p(\mathcal{P})}^{1 - \frac{p}{q}} \Big(\int_{\mathcal{P}} |f(u)|^p du \cdot \int_{\mathcal{P}} \prod_{i=1,2} (x_i - u_i)^{(\beta_i - 1)\gamma q} dx \Big)^{\frac{1}{q}}$$

$$\leq C \|f\|_{L_p(\mathcal{P})}.$$

□

Corollary 2.2.3. *Let* $f \in L_p(\mathcal{P}), g \in L_q(\mathcal{P}), I_{a+}^{\beta_1\beta_2}g \in L_r(\mathcal{P})$ *for* $1/p + 1/r = 1$ *and* $r < q((1-\beta_1 q)^{-1} \wedge (1-\beta_2 q)^{-1})$, *i.e.* $1/p + 1/q < 1 + \beta_1 \wedge \beta_2$. *Then*

$$\int_{\mathcal{P}} f(u) I_{a+}^{\beta_1\beta_2} g(u) du = \int_{\mathcal{P}} g(u) I_{b-}^{\beta_1\beta_2} f(u) du.$$

Evidently,

$$I_{\pm}^{\rho_1\rho_2} I_{\pm}^{\beta_1\beta_2} = I_{\pm}^{\rho_1+\beta_1,\rho_2+\beta_2} \quad \text{on} \quad L_1(\mathcal{P});$$

for $f \in I_{\pm}^{\rho_1+\beta_1,\rho_2+\beta_2}(L_1(\mathcal{P}))$, $\rho_i, \beta_i \geq 0$, $\rho_i + \beta_i \leq 1$

$$D_{a+(b-)}^{\rho_1\rho_2} D_{a+(b-)}^{\beta_1\beta_2} f = D_{a+(b-)}^{\rho_1+\beta_1,\rho_2+\beta_2} f;$$

for $f \in I_{a+(b-)}^{\rho_1\rho_2}(L_p(\mathcal{P}))$, $g \in I_{b-}^{\rho_1\rho_2}(L_q(\mathcal{P}))$, $p, q > 1$, $1/p + 1/q < 1 + \rho_1 \wedge \rho_2$

$$\int_{\mathcal{P}} D_{a+}^{\rho_1\rho_2} f(u) g(u) du = \int_{\mathcal{P}} f(u) D_{b-}^{\rho_1\rho_2} g(u) du.$$

2.2.2 Generalized Two-parameter Lebesgue–Stieltjes Integrals

We suppose that all the functions, considered on some rectangle $\mathcal{P} = [a, b]$, belong to the space $D(\mathcal{P})$, i.e. they have the limits in all the quadrants,

$$Q^{++}(x) = \{s \in \mathcal{P} | s \geq x\}, \qquad Q^{+-}(x) = \{s \in \mathcal{P} | s_1 \geq x_1, s_2 < x_2\},$$
$$Q^{-+}(x) = \{s \in \mathcal{P} | s_1 < x_1, s_2 \geq x_2\}, \ Q^{--}(x) = \{s \in \mathcal{P} | s < x\},$$

$f(x) = \lim_{s \to x, s \geq x} f(s)$, and on the sides of rectangle the limits that can be defined are supposed to exist and denoted as $f(x_1, b_2-), f(b_1-, x_2), f(b-)$. Denote $f_{a+}(x) = \Delta_a f(x)$, $x \in \mathcal{P}$, and $f_{b-}(x) := f(x) - f(x_1, b_2-) - f(b_1-, x_2) + f(b-)$.

Definition 2.2.4. *Let* $f, g : \mathcal{P} \to \mathbb{R}$. *The generalized two-parameter Lebesgue–Stieltjes integral of f w.r.t. g is defined by*

$$\int_{\mathcal{P}} f dg := \int_{\mathcal{P}} (D_{a+}^{\beta_1\beta_2} f_{a+})(u)(D_{b-}^{1-\beta_1,1-\beta_2} g_{b-})(u) du$$
$$+ \sum_{i=1,2} \int_{a_i}^{b_i} (D_{a_i+}^{\beta_i} f_{a_i+})(u, a_i)(D_{b_i-}^{1-\beta_i})(g_{b_i-}(u, b_i-) - g_{b_i-}(u, b_i-)) du$$
$$+ f(a) \Delta_a g(b), \quad (2.2.1)$$

under the assumption that all the integrals on the right-hand side exist.

A more convenient formula for $\int_{\mathcal{P}} f dg$ has a form

$$\int_{\mathcal{P}} f dg = \int_{\mathcal{P}} (D_{a+}^{\beta_1\beta_2} f)(u)(D_{b-}^{1-\beta_1,1-\beta_2} g_{b-})(u) du.$$

(We do not specify here the conditions ensuring the latter equality but it is very easy to do it, similarly to the one-parameter case.) The next results also can be proved similarly to the one-parameter case ((SKM93) and (Zah98)).

Theorem 2.2.5. *Definition 2.2.4 is correct, i.e. the right-hand side of (2.2.1) does not depend on the choice of $\beta_i, i = 1, 2$.*

Theorem 2.2.6. *Let $f : \mathcal{P} \to \mathbb{R}, f \in C^{\lambda_1 \lambda_2}(\mathcal{P})$ and $\lambda_i + \beta_i < 1, i = 1, 2$, $0 < \beta_i < 1$. Then $I_{a+(b-)}^{\beta_1 \beta_2}(f_{a+(b-)}) \in C^{\lambda_1 + \beta_1 \lambda_2 + \beta_2}(\mathcal{P})$.*

Theorem 2.2.7. *Let the function $f \in C^{\lambda_1 \lambda_2}(\mathcal{P})$. Then for any $p \geq 1$ and $0 < \varepsilon_i < \lambda_i, i = 1, 2$*

$$f_{a+(b-)} \in I_{\pm}^{\varepsilon_1 \varepsilon_2}(L_p(\mathcal{P}))$$

and

$$D_{a+(b-)}^{\varepsilon_1 \varepsilon_2} f_{a+(b-)} \in C^{\lambda_1 - \varepsilon_1 \lambda_2 - \varepsilon_2}(\mathcal{P}).$$

Theorem 2.2.8. *Let $f \in C(\mathcal{P})$, $g \in BV(\mathcal{P})$, $f \in I_+^{\beta_1 \beta_2}(L_p(\mathcal{P}))$, $g_{b-} \in I_-^{1-\beta_1 1-\beta_2}(L_q(\mathcal{P}))$, $i = 1, 2$, $j = 3 - i$, $\frac{1}{p} + \frac{1}{q} \leq 1$, $0 \leq \beta_i \leq 1, i = 1, 2$. Then the generalized two-parameter Lebesgue–Stieltjes integral $\int_{\mathcal{P}} f dg$ equals the Riemann–Stieltjes integral $\int_{\mathcal{P}} f(x) dg(x)$.*

Theorem 2.2.9. *1. Let $g \in C^{\lambda_1 \lambda_2}(\mathcal{P})$ for some $0 < \lambda_i \leq 1, i = 1, 2$. Then for any $\mathcal{P}_1 = [c, d) \subset \mathcal{P}$*

$$\int_{\mathcal{P}} \mathbf{1}_{\mathcal{P}_1} dg = \Delta_c g(d).$$

2. Let $g \in C^{\lambda_1 \lambda_2}(\mathcal{P})$ and let the partition $\pi = \pi^1 \times \pi^2$, where $\pi^i = \{a_i = x_0^i < \cdots < x_{n_i}^i = b_i\}$ be the partition of $[a_i, b_i]$. Also, let $f_{\pi}(x) = \sum_{i=1,2} \sum_{j_i=0}^{n_i-1} f_{j_1 j_2} \mathbf{1}_{\mathcal{P}_{j_1 j_2}}(x)$, where $\mathcal{P}_{j_1 j_2} = \prod_{i=1,2} [x_{j_i}^i, x_{j_i+1}^i)$. Then $\int_{\mathcal{P}} f_{\pi} dg = \sum_{i=1,2} \sum_{j_i=0}^{n_i-1} f_{j_1 j_2} \Delta_{x_j} g(x_{j+1})$, where $x_j = (x_{j_1}^1, x_{j_2}^2)$.

Now, let π_n be the sequence of partitions of rectangle $\mathcal{P}$, $\pi_n \subset \pi_{n+1}$ and $|\pi_n| = \max_{i=1,2} \max_{0 \leq j_i \leq n_{i,n}-1}(x_{j_i+1}^{i,n} - x_{j_i}^{i,n})$. Let $f : \mathcal{P} \to \mathbb{R}$, $f_{j_1 j_2} = f(x_{j+1}^n)$. We say that the partitions π_n are uniform, if $n_1^{(n)} = n_2^{(n)}$ and $x_{j_i+1}^{i,n} - x_{j_i}^{i,n} = \frac{b_i - a_i}{n_1^{(n)}}, i = 1, 2$.

Theorem 2.2.10. *1. Let $f \in C^{\lambda_1 \lambda_2}(\mathcal{P})$ for some $0 < \lambda_i \leq 1, i = 1, 2$. Then*

$$\lim_{n \to \infty} \sup_{\pi_n} \|f_{\pi_n} - f\|_{L_\infty(\mathcal{P})} = 0,$$

where $\sup_{\pi_n}$ is taken over all the sequences of partitions mentioned above.

2. $\lim_{n \to \infty} \sup_{\pi'_n} \left\| D_{a+}^{\beta_1 \beta_2}(f_{\pi'_n})_{a+} - D_{a+}^{\beta_1 \beta_2} f_{a+} \right\|_{L_1(\mathcal{P})} = 0$, for any $\beta_1 \vee \beta_2 < \lambda_1 \wedge \lambda_2$ and all the sequences of uniform partitions of $\mathcal{P}$.

Proof. The first statement is a direct consequence of uniform continuity f on $\mathcal{P}$. Further, let $g_n(x) = f_{\pi'_n}(x) - f(x)$. For the second statement it is sufficient to prove that any of the following functions

$$G_1^n(x) := g_n(x)(x - a_1)^{-\beta_1}(y - a_2)^{-\beta_2},$$

$$G_2^n(x) := (x_2 - a_2)^{-\beta_2} \int_{a_1}^{x_1} (g_n(x) - g_n(s_1, x_2))(x_1 - s_1)^{-1-\beta_1} ds_1,$$

$$G_3^n(x) := (x_1 - a_1)^{-\beta_1} \int_{a_2}^{x_2} (g_n(x) - g_n(x_1, s_2))(x_2 - s_2)^{-1-\beta_2} ds_2,$$

$$G_4^n(x) := \int_{[a,x]} \Delta_s g_n(x) \prod_{i=1,2} (x_i - s_i)^{-1-\beta_i} ds$$

tends to zero in $L_1(\mathcal{P})$. First, note that $|g_n(x)| \leq C(|\pi_n|^{\lambda_1} + |\pi_n|^{\lambda_2})$, whence $\|G_1^n\|_{L_1(\mathcal{P})} \leq C(|\pi_n|^{\lambda_1} + |\pi_n|^{\lambda_2}) \prod_{i=1,2}(b_i - a_i)^{1-\beta_i} \to 0$, $n \to \infty$. Further, let the point $x \in \mathcal{P}_j^n := \prod_{i=1,2}[x_{j_i}^{i,n}, x_{j_i+1}^{i,n}) =: [x_j^n, x_{j+1}^n)$. Then it holds that

$$G_2^n(x) = (x_2 - a_2)^{-\beta_2} \left(\sum_{k=0}^{j_1-1} \int_{x_k^{1,n}}^{x_{k+1}^{1,n}} (x_1 - s_1)^{-1-\beta_1} ds_1 \right. $$

$$\left. + \int_{x_{j_1}}^{x_1} g_n(x, x_j^n, s_1)(x_1 - s_1)^{-1-\beta_1} ds_1 \right),$$

where $g_n(x, x_j^n, s_1) = f(x_j^n) - f(x) - f(x_k^{1,n}, x_{j_2}^{2,n}) + f(s_1, x_2)$. Therefore,

$$|G_2^n(x)|I\{x \in \mathcal{P}_j^n\}$$

$$\leq C(x_2 - a_2)^{-\beta_2} \left[\left(\sum_{i=1,2} |x_{j_i}^{i,n} - x_i|^{\lambda_i} \right) \int_{a_1}^{x_{j_1}^{1,n}} (x_1 - s_1)^{-1-\beta_1} ds_1 \right.$$

$$+ \sum_{k=0}^{j_1-1} ((x_{k+1}^{1,n} - x_k^{1,n})^{\lambda_1} + (x_{j_2+1}^{2,n} - x_{j_2}^{2,n})^{\lambda_2}) \int_{x_k^{1,n}}^{x_{k+1}^{1,n}} (x_1 - s_1)^{-1-\beta_1} ds_1$$

$$\left. + \int_{x_{j_1}}^{x} (x_1 - s_1)^{\lambda_1 - 1 - \beta_1} ds_1 \right]$$

$$\leq C(x_2 - a_2)^{-\beta_2} \left[\sum_{i=1,2} (x_1 - x_{j_1}^{1,n})^{\lambda_i}(x_1 - x_{j_1}^{1,n})^{-\beta_1} \right.$$

$$+ \sum_{k=0}^{j_1-1}((x_{k+1}^{1,n} - x_k^{1,n})^{\lambda_1} + (x_{j_2+1}^{2,n} - x_{j_2}^{2,n})^{\lambda_2}) \int_{x_k^{1,n}}^{x_{k+1}^{1,n}} (x_1 - s_1)^{-1-\beta_1} ds_1$$

$$\left. + (x_1 - x_{j_1}^{1,n})^{\lambda_1 - \beta_1} \right],$$

2.2 Pathwise Stochastic Integration w.r.t. Multi-parameter fBm   135

and

$$\|G_1^n\|_{L_1(\mathcal{P})} \le \sum_{j_1,j_2} \|G_1^n\|_{L_1(\mathcal{P}_j^n)} \le C \sum_{j_1,j_2} \left(\int_{\mathcal{P}_j^n} \left((x_2-a_2)^{-\beta_2}(x_1-x_{j_1}^{1,n})^{\lambda_1-\beta_1} \right. \right.$$
$$+ (x_2-a_2)^{-\beta_2}(x_2-x_{j_2}^{2,n})^{\lambda_2}(x_1-x_{j_1}^{1,n})^{-\beta_1}$$
$$+ \sum_{k=0}^{j_1-1} (x_{k+1}^{1,n}-x_k^{1,n})^{\lambda_1}(x_2-a_2)^{-\beta_2} \int_{x_k^{1,n}}^{x_{k+1}^{1,n}} (x_1-s_1)^{-1-\beta_1} ds_1$$
$$+ (x_2-a_2)^{-\beta_2}(x_{j_2+1}^{2,n}-x_{j_2}^{2,n})^{\lambda_1} \sum_{k=0}^{j_1-1} \int_{x_k^{1,n}}^{x_{k+1}^{1,n}} (x_1-s_1)^{-1-\beta_1} ds_1$$
$$\left. \left. + (x_2-a_2)^{-\beta_2}(x_1-x_{j_1}^{1,n})^{\lambda_1-\beta_1} \right) dx \right)$$
$$\le C(b_2-a_2)^{1-\beta_2} \left(|\pi_n|^{\lambda_1-\beta_1} + |\pi_n|^{\lambda_2} \sum_{j_1=1}^{n_1^{(n)}} (x_{j_1+1}^{1,n}-x_{j_1}^{1,n})^{1-\beta_1} \right.$$
$$+ \sum_{j_1=0}^{n_1^{(n)}-1} (x_{k+1}^{1,n}-x_k^{1,n})^{\lambda_1} \int_{x_k^{1,n}}^{x_{k+1}^{1,n}} \left(\int_{x_{k+1}^{1,n}}^{b_1} (x_1-s_1)^{-1-\beta_1} dx_1 \right) ds_1$$
$$\left. + |\pi_n|^{\lambda_2} \sum_{j_1=0}^{n_1^{(n)}-1} \int_{x_k^{1,n}}^{x_{k+1}^{1,n}} \int_{a_1}^{x_k^{1,n}} (x_1-s_1)^{-1-\beta_1} ds_1 dx_1 + |\pi_n|^{\lambda_1-\beta_1} \right).$$

(2.2.2)

The first, third and fifth terms on the right-hand side of (2.2.2) are bounded from above by $C|\pi_n|^{\lambda_1-\beta_1} \to 0$, $n \to \infty$, and it is true for any π_n. The second and fourth terms can be effectively estimated when $\pi_n = \pi_n'$ is uniform. In this case

$$|\pi_n'|^{\lambda_2} \sum_{j_1=1}^{n_1^{(n)}} (x_{j_1+1}^{1,n}-x_{j_1}^{1,n})^{1-\beta_1} \le \frac{C}{(n_1^{(n)})^{\lambda_2-\beta_1}} \to 0, \quad n \to \infty,$$

and

$$|\pi_n'|^{\lambda_2} \sum_{j_1=0}^{n_1^{(n)}-1} \int_{x_k^{1,n}}^{x_{k+1}^{1,n}} \int_{a_1}^{x_k^{1,n}} (x_1-s_1)^{-1-\beta_1} ds_1 dx_1$$
$$\le |\pi_n'|^{\lambda_2} \sum_{j_1=1}^{n_1^{(n)}} (x_{j_1+1}^{1,n}-x_{j_1}^{1,n})^{1-\beta_1} \to 0, \quad n \to \infty.$$

G_3^n and G_4^n can be estimated in a similar way.   □

Definition 2.2.11. We say that the two-parameter left Riemann–Stieltjes integral $l\text{-}\int_{\mathcal{P}} f dg$ exists if the sums S_n have the limit for all sequences of uniform partitions of $\mathcal{P}$ with vanishing diameter.

Theorem 2.2.12. *Let $f \in C^{\lambda_1 \lambda_2}(\mathcal{P})$, $g \in C^{\mu_1 \mu_2}(\mathcal{P})$ and $\lambda_i + \mu_i > 1, i = 1, 2$. Then the generalized two-parameter Lebesgue–Stieltjes integrals $\int_\mathcal{P} f dg$ and l-$\int_\mathcal{P} f dg$ exist and coincide.*

Proof. It is sufficient to prove that $S_n \to \int_\mathcal{P} f dg$. But the sums S_n equal $S_n = \int_\mathcal{P} f_{\pi_n} dg$. Denote $f^{(n)} := f_{\pi_n}$. Then

$$\int_\mathcal{P} f^{(n)} dg = \int_\mathcal{P} D_{a+}^{\beta_1 \beta_2} f^{(n)}(x) D_{b-}^{1-\beta_1, 1-\beta_2} g_{b-}(x) dx$$

for any $1 - \mu_i < \beta_i < \lambda_i$. According to previous theorem, $D_{a+}^{\beta_1 \beta_2} f^{(n)} \to D_{a+}^{\beta_1 \beta_2} f$ in $L_1(\mathcal{P})$, whence the proof follows. □

Remark 2.2.13. We can use the Hölder properties of f in order to establish that $\int_\mathcal{P} f dg = \lim \widetilde{S}_n$, where

$$\widetilde{S}_n = \sum_{j_1 j_2} (f(x_{j_1}^{1,n}, \xi_{j_2}^{2,n}) + f(\xi_{j_1}^{1,n}, x_{j_2}^{2,n}) - f(\xi_j^n)) \Delta_{x_j^n} g(x_{j+1}^n)$$

and ξ_j^n is any point of $\mathcal{P}_j^n$.

2.2.3 Generalized Integrals of Two-parameter fBm in the Case of the Integrand Depending on fBm

Since the trajectories of two-parameter fBm $B^{H_1 H_2}$ a.s. belong to $C^{H_1 - \varepsilon_1, H_2 - \varepsilon_2}(\mathcal{P})$ for any rectangle $\mathcal{P} \subset \mathbb{R}_+^2$ and any $0 < \varepsilon_i < H_i$, the next result is a direct consequence of Theorem 2.2.12.

Theorem 2.2.14. *Let $B^{H_1 H_2}$ be a two-parameter fBm with $H_i \in (1/2, 1)$, and the function $F : \mathbb{R}_+ \times \mathbb{R} \to \mathbb{R}$, $F \in C^1(\mathbb{R}_+ \times \mathbb{R})$. Then there exists the generalized two-parameter Lebesgue–Stieltjes integral $\int_\mathcal{P} F(\cdot, B^{H_1 H_2}) dB^{H_1 H_2}$ which coincides with the left Riemann–Stieltjes integral l-$\int_\mathcal{P} F(\cdot, B^{H_1 H_2}) dB^{H_1 H_2}$.*

Remark 2.2.15. Theorem 2.2.14 holds if we replace $F(\cdot, B^{H_1 H_2})$ with any Hölder field $f \in C^{\lambda_1 \lambda_2}(\mathcal{P})$, such that $\lambda_i + H_i > 1$. It means that for such an f, we can consider the integral $\int_\mathcal{P} f dB^{H_1 H_2}$ for any $\omega \in \Omega', P(\Omega') = 1$ as the limit of corresponding integral sums.

2.2.4 Pathwise Integration in Two-parameter Besov Spaces

According to the form of two-parameter forward and backward fractional Marchaud derivatives (Definition 1.20.8), the Besov type spaces in this case receive the following form.

Let $\mathcal{P}_t := [0, t] = \prod_{i=1,2}[0, t_i]$,
$\varphi_1^{\beta_1}(f)(t) := \int_0^{t_1} |f(t) - f(s_1, t_2)|(t_1 - s_1)^{-\beta_1 - 1} ds_1$,
$\varphi_2^{\beta_2}(f)(t) := \int_0^{t_2} |f(t) - f(t_1, s_2)|(t_2 - s_2)^{-\beta_2 - 1} ds_2$,

$\varphi_3^{\beta_1\beta_2}(f)(t) := \int_{\mathcal{P}_t} |\Delta_s f(t)|(\varphi(t,s,1+\beta))^{-1}ds,\ 0 < \beta_i < 1,$

and $\varphi_f^{\beta_1\beta_2}(t) := |f(t)| + \sum_{i=1,2} \varphi_i^{\beta_i}(f)(t) + \varphi_3^{\beta_1\beta_2}(f)(t).$

Denote by $W_0^{\beta_1,\beta_2}(\mathcal{P}_T)$ the Banach space of measurable functions $f : \mathcal{P}_T \to \mathbb{R}$, such that

$$\|f\|_{0,\beta_1,\beta_2} := \sup_{t \in \mathcal{P}_T} \varphi_f^{\beta_1\beta_2}(t) < \infty,$$

$W_1^{\beta_1,\beta_2}(\mathcal{P}_T)$ the Banach space of measurable functions $f : \mathcal{P}_T \to \mathbb{R}$, such that

$$\|f\|_{1,\beta_1,\beta_2} := \sup_{0<s\leq t<T} \Big(|\Delta_s f(t)| \prod_{i=1,2} (t_i - s_i)^{-\beta_i}$$
$$+ (t_2 - s_2)^{-\beta_2} \int_{s_1}^{t_1} |f_{t-}(u,s_2) - f_{t-}(s)|(u-s_1)^{-1-\beta_1} du$$
$$+ (t_1 - s_1)^{-\beta_1} \int_{s_2}^{t_2} |f_{t-}(s_1,v) - f_{t-}(s)|(v-s_2)^{-1-\beta_2} dv$$
$$+ \int_{[s,t]} |\Delta_s f(r)|(\varphi(r,s,1+\beta))^{-1} dr \Big) < \infty$$

(for the notation of $\varphi(r,s,\beta)$ see Definition 1.20.3) and $W_2^{\beta_1,\beta_2}(\mathcal{P}_T)$ the Banach space of measurable functions $f : \mathcal{P}_T \to \mathbb{R}$, such that

$$\|f\|_{2,\beta_1,\beta_2} := \int_{\mathcal{P}_T} \Big(|f(s)| \prod_{i=1,2} s_i^{-\beta_i} + s_2^{-\beta_2} \varphi_1^{\beta_1}(f)(s)$$
$$+ s_1^{-\beta_1} \varphi_2^{\beta_2}(f)(s) + \varphi_3^{\beta_1\beta_2}(f)(s) \Big) ds < \infty.$$

Similarly to Lemmas 2.1.9 and 2.1.10, the following bounds can be established. Let $0 < \beta_i < 1/2,\ i = 1,2,\ G_t(f) = \int_{\mathcal{P}_t} f dg,\ F_t(f) = \int_{\mathcal{P}_t} f ds.$

Lemma 2.2.16. *1. Let $f \in W_2^{\beta_1\beta_2}(\mathcal{P}_T),\ g \in W_1^{1-\beta_1,1-\beta_2}(\mathcal{P}_T)$. Then*

$$\varphi_{G.(f)}^{\beta_1\beta_2}(t) \leq C_{\beta_1,\beta_2,T}^1 \Lambda_{1-\beta_1,1-\beta_2}(g) \int_{\mathcal{P}_t} \prod_{i=1,2}(r_i^{-\beta_i} + (t_i - r_i)^{-2\beta_i})\varphi_f^{\beta_1\beta_2}(r) dr.$$

2. Let $f \in W_0^{\beta_1\beta_2}(\mathcal{P}_T),\ g \in W_1^{1-\beta_1,1-\beta_2}(\mathcal{P}_T)$. Then $G.(f) \in C^{1-\beta_1,1-\beta_2}(\mathcal{P}_T)$ and

$$\|G(f)\|_{1-\beta_1,1-\beta_2} \leq C_{\beta_1,\beta_2,T}^2 \Lambda_{1-\beta_1,1-\beta_2}(g) \|f\|_{0,\beta_1,\beta_2}.$$

3. Let $0 < \beta_i < 1$ and $f_T^* := \sup_{t \in \mathcal{P}_T} |f(t)| < \infty$. Then $F.(f) \in W_0^{\beta_1\beta_2}(\mathcal{P}_T) \cap C^2(\mathcal{P}_T)$ and

$$\|F(f)\|_{0,\beta_1,\beta_2} \leq C_{\beta_1,\beta_2,T}^3 f_T^*.$$

2.2.5 The Existence of the Integrals of the Second Kind of a Two-parameter fBm

We fix the rectangle $\mathcal{P} = [0,T] \subset \mathbb{R}_+^2$ and consider the sequence of uniform partitions

$$\pi_n = \{t_j^n = (T_1 j_1 \cdot 2^{-n}, T_2 j_2 \cdot 2^{-n}), 0 \le j_i \le 2^n\}.$$

Let the functions $f, g : \mathcal{P} \to \mathbb{R}$, $f|_{\partial \mathbb{R}_+^2} = f_0 \in \mathbb{R}$, $g|_{\partial \mathbb{R}_+^2} = g_0 \in \mathbb{R}$, $f \in C^{\lambda_1 \lambda_2}(\mathcal{P})$ and $g \in C^{\mu_1 \mu_2}(\mathcal{P})$.

Consider the sequence of integral sums of the second kind, i.e.

$$\widetilde{S}_n := \sum_{j_1, j_2 = 0}^{2^n - 1} f(t_j^n) \Delta_j^1 g \Delta_j^2 g,$$

where $\Delta_j^1 g = g(t_{j_1+1 j_2}^n) - g(t_j^n)$, $\Delta_j^2 g = g(t_{j_1 j_2+1}^n) - g(t_j^n)$.

Theorem 2.2.17. *Let $\lambda_i, \mu_i > \frac{1}{2}$, $\lambda_i + \mu_1 + \mu_2 > 2$, $i = 1, 2$. Then there exists $\lim_{n \to \infty} \widetilde{S}_n =: \widetilde{S}$. This limit will be called the integral of the second kind of f w.r.t. g and denoted as $\widetilde{S} = \int_{\mathcal{P}} f d_1 g d_2 g$.*

Proof. Let, for technical simplicity, $T_1 = T_2 = 1$. Also, let $m > n$. Consider the difference $S_n - S_m = S_n - S_{mn} + S_{mn} - S_m$, where

$$S_{mn} = \sum_{j_1, j_2 = 0}^{2^n - 1} \sum_{r \in A_{j_1}} f(r 2^{-m}, j_2 2^{-n})(g((r+1) 2^{-m}, j_2 2^{-n}) - g(r 2^{-m}, j_2 2^{-n}))$$
$$\times (g(r 2^{-m}, (j_2 + 1) 2^{-n}) - g(r 2^{-m}, j_2 2^{-n})),$$
$$A_{j_1} = \{r : j_1 2^{m-n} \le r < (j_1 + 1) 2^{m-n}\}.$$

It is sufficient to estimate only $S_n - S_{mn}$, because $S_{mn} - S_m$ can be estimated similarly. We have that

$$|S_n - S_{mn}| \le |\Delta_{mn}^1| + |\Delta_{mn}^2|,$$

where

$$\Delta_{mn}^1 = \sum_{j_1, j_2 = 0}^{2^n - 1} \sum_{r \in A_{j_1}} f(t_j^n) \Delta_{jr} g \Delta_{j_2 r}^1 g, \quad \Delta_{mn}^2 = \sum_{j_1, j_2 = 0}^{2^n - 1} \sum_{r \in A_{j_1}} \Delta_{jr}^1 f \Delta_{j_2 r}^1 g \Delta_{j_2 r}^2 g,$$
$$\Delta_{jr} g = \Delta_{t_j^n} g(r 2^{-m}, (j_2 + 1) 2^{-n}),$$
$$\Delta_{j_2 r}^1 g = \Delta_{(r 2^{-m}, j_2 2^{-n})}^1 g((r + 1) 2^{-m}, (j_2 + 1) 2^{-n}),$$
$$\Delta_{jr}^1 f = \Delta_{t_j^n}^1 f(r 2^{-m}, j_2 2^{-n}), (j_2 + 1) 2^{-n}),$$
$$\Delta_{j_2 r}^2 g = \Delta_{(r 2^{-m}, j_2 2^{-n})}^2 g(r 2^{-m}, (j_2 + 1) 2^{-n}).$$

Transform Δ_{mn}^1 into the sum

$$\Delta_{mn}^1 = \sum_{j_1, j_2 = 0}^{2^n - 1} \sum_{r \in A_{j_1}} f(t_j^n) \Delta_{j_2 r} g \Delta_{jr}^1 g,$$

where $\Delta_{j_2 r} g = \Delta_{(r 2^{-m}, j_2 2^{-n})}(g((r + 1) 2^{-m}(j_2 + 1) 2^{-n}))$, and $\Delta_{jr}^1 g = \Delta_{(r 2^{-m}, j_2 2^{-n})}^1 g(t_{j_1+1 j_2}^n)$. The increments $\Delta_{j_2 r} g$ correspond to the rectangles $\Delta_{j_2r} = (r2^{-m}, (r+1)2^{-m}] \times (j_2 2^{-n}, (j_2+1)2^{-n}]$, that do not intersect, and $\cup \Delta_{j_2r} = (0,1]^2$. Therefore the sum $\Delta^1_{n,m}$ can be presented as a two-parameter generalized Lebesgue–Stieltjes integral $\int_{\mathcal{P}} \widetilde{f}_{mn} dg$, where

$$\widetilde{f}_{mn}(s) = f(t_j^n)\Delta^1_{jr} g \cdot \mathbf{1}_{\{s \in \Delta_{j_2r}\}}.$$

In turn,

$$\int_{\mathcal{P}} \widetilde{f}_{mn} dg = \int_{\mathcal{P}} (D^{\beta_1 \beta_2}_{0+} \widetilde{f}_{mn})(s)(D^{1-\beta_1 1-\beta_2}_{1-} g_{1-})(s) ds,$$

where $1 = (1,1), 0 = (0,0), 1 - \mu_i < \beta_i < \lambda_i, i = 1,2$. With such a choice of β_i $D^{1-\beta_1 1-\beta_2}_{1-} g_{1-} \in C^{\mu_1 + \beta_1 - 1, \mu_2 + \beta_2 - 1}(\mathcal{P})$, in particular, there exists such a $C > 0$ that $|(D^{1-\beta_1 1-\beta_2}_{1-} g_{1-})(s)| \leq C$, $s \in \mathcal{P}$. Therefore, it is sufficient to prove that $\int_{\mathcal{P}} |(D^{\beta_1 \beta_2}_{0+} \widetilde{f}_{mn})(s)| ds \to 0$, $n, m \to \infty$. Since $D^{\beta_1 \beta_2}_{0+} \widetilde{f}_{mn}$ consists of four terms, we must consider them separately. Estimate only $\int_{\mathcal{P}} |\varphi_{mni}(s)| ds$, where
$\varphi_{mn1}(s) = s_2^{-\beta_2} \int_0^{s_1} (\widetilde{f}_{mn}(s) - \widetilde{f}_{mn}(u_1, s_2))(s_1 - u_1)^{-1-\beta_1} du_1$,
and
$\varphi_{mn2}(s) = \int_{[0,s]} \Delta_u \widetilde{f}_{mn}(s) \prod_{i=1,2}(s_i - u_i)^{-1-\beta_i} du_1$;
the other two terms can be considered similarly.

Let $s \in \Delta_{j_2r}$. Then, taking into account that $|f(s)| \leq C$ for some $C > 0$, we obtain that

$$|\varphi_{mn1}(s)| \leq s_2^{-\beta_2}(\int_0^{j_1 2^{-n}} + \int_{j_1 2^{-n}}^{r2^{-m}})|\widetilde{f}_{mn}(s) - \widetilde{f}_{mn}(u_1, s_2)|(s_1 - u_1)^{-1-\beta_1} du_1$$
$$\leq s_2^{-\beta_2} \int_0^{j_1 2^{-n}} (|\widetilde{f}_{mn}(s)| + |\widetilde{f}_{mn}(u_1, s_2)|)(s_1 - u_1)^{-1-\beta_1} du_1$$
$$+ Cs_2^{-\beta_2} \int_{j_1 2^{-n}}^{r2^{-m}} |f(t_j^n)|(s_1 - u_1 + 2^{-m})^{\mu_1}(s_1 - u_1)^{-1-\beta_1} du_1 \leq Cs_2^{-\beta_2}$$
$$\times (2^{-n\mu_1}(s_1 - j_1 2^{-n})^{-\beta_1} + (s_1 - r2^{-m})^{\mu_1 - \beta_1} + 2^{-m\mu_1}(s_1 - r2^{-m})^{-\beta_1}),$$

whence

$$\int_{\mathcal{P}} |\varphi_{mn1}(s)| ds \leq C \sum_{j_1, j_2 = 0}^{2^n - 1} \sum_{r \in A_{j_1}} \left(2^{-n\mu_1} \int_{\Delta_{j_2r}} s_2^{-\beta_2}(s_1 - j_1 2^{-n}) ds \right.$$
$$+ \int_{\Delta_{j_2r}} s_2^{-\beta_2}(s_1 - r2^{-m})^{\mu_1 - \beta_1} ds + 2^{-m\mu_1} \int_{\Delta_{j_2r}} s_2^{-\beta_2}(s_1 - r2^{-m})^{-\beta_1} ds \left. \right)$$
$$\leq C(1 - \beta_2)^{-1}(2^{n(\beta_1 - \mu_1)} + 2^{m(\beta_1 - \mu_1)}) \to 0, \quad m, n \to \infty.$$

Further, from Hölder properties of f and g, it follows that for $u \leq (j_1 2^{-n}, j_2 2^{-n})$ we have the estimate $|\Delta_u \widetilde{f}_{mn}(s)| \leq 2(s_2 - u_2 + 2^{-n})^{\lambda_2} 2^{-n\mu_1} + C(s_2 - u_2 + 2^{-n})^{\mu_2}(s_1 - u_1)^{-n\mu_1}$, for $u \in (j_1 2^{-n}, r2^{-m}) \times (0, j_2 2^{-n})$ the estimate is $|\Delta_u \widetilde{f}_{mn}(s)| \leq 2(s_2 - u_2 + 2^{-n})^{\lambda_2}(s_1 - u_1 + 2^{-m})^{\mu_1} + C2^{-m\mu_1}(s_2 - u_2 + 2^{-n})^{\mu_2}$, and $\Delta_u \widetilde{f}_{mn}(s) = 0$ otherwise. Hence,

$$|\varphi_{mn2}(s)| \leq C2^{-n\mu_1}(s_1 - j_2 2^{-n})^{-\beta_1}(s_2 - (j_1 - 1)2^{-n})^{\lambda_2 \wedge \mu_2 - \beta_2}$$
$$+ C(s_1 + j_2 2^{-n} + 2^{-m})^{\mu_1 - \beta_1}(s_2 - j_2 2^{-m} + 2^{-n})^{\mu_2 \wedge \mu_2 - \beta_2},$$

and $\int_{\mathcal{P}} |\varphi_{mn2}(s)| ds \leq C2^{n(\beta_1+\beta_2-\mu_1-\mu_2\wedge\lambda_2)} \to 0, m, n \to \infty$. So, $|\Delta_{mn}^1| \to 0, m, n \to \infty$. Now we want to prove that $|\Delta_{mn}^2| \to 0, m, n \to \infty$. We can present Δ_{mn}^2 as

$$\Delta_{mn}^2 = \sum_{j_2=0}^{2^n-1} \Delta_{mn}^{2,j_2},$$

where

$$\Delta_{mn}^{2,j_2} = \sum_{j_1=0}^{2^n-1} \sum_{r\in A_{j_1}} \Delta_{jr}^1 f \Delta_{j_2 r}^1 g \Delta_{j_2 r}^2 g.$$

Moreover, Δ_{mn}^{2,j_2} can be presented as one-parameter generalized Lebesgue–Stieltjes integral $\int_0^1 \psi_{j_2}(u) d_1 g(u, j_2 2^{-n})$, where $\psi_{j_2}(u) = \Delta_{jr}^1 f \Delta_{j_2 r}^2 g \mathbf{1}_{\{r2^{-m}\leq u<(r+1)2^{-m}\}}$, $\psi(0) = 0$. Then $\int_0^1 \psi_{j_2}(u) d_1 g(u, j_2 2^{-n}) = \int_0^1 (D_{0+}^\beta \psi_{j_2})(u)(D_{1-}^{1-\beta} g_{1-})(u, j_2 2^{-n}) du$, where $1 - \mu_1 < \beta < 1/2$. Evidently, $|(D_{1-}^{1-\beta} g_{1-})(u, j_2 2^{-n})| \leq C$, therefore, it is sufficient to prove that

$$\sum_{j_2=0}^{2^n-1} \int_0^1 |(D_{0+}^\beta \psi_{j_2})(u)| du \to 0, m, n \to \infty.$$

Note that

$$(D_{0+}^\beta \psi_{j_2})(u) = (\Gamma(1-\beta))^{-1} \Big(\psi_{j_2}(u) u^{-\beta} + \beta \int_0^u (\psi_{j_2}(u) - \psi_{j_2}(z))(u-z)^{-1-\beta} dz\Big),$$

and $|\psi_{j_2}(u)| \leq C 2^{-n(\lambda_1+\mu_2)}$, whence

$$\sum_{j_2=0}^{2^n-1} \int_0^1 |\psi_{j_2}(u)| u^{-\beta} du \leq C \int_0^1 u^{-\beta} du \cdot 2^{n(1-\lambda_1-\mu_2)} \to 0, n \to \infty.$$

Further, for $j_1 2^{-n} \leq r 2^{-m} \leq u < (r+1) 2^{-m} \leq (j_1+1) 2^{-n}$,

$$\int_0^u (\psi_{j_2}(u) - \psi_{j_2}(z))(u-z)^{-1-\beta} dz = \int_0^{j_1 2^{-n}} + \int_{j_1 2^{-n}}^{r 2^{-m}},$$

and

$$|\psi_{j_2}(u) - \psi_{j_2}(z)| \leq |\psi_{j_2}(u)| + |\psi_{j_2}(z)| \leq C 2^{-n(\lambda_1+\mu_2)}.$$

From here,

$$\sum_{j_2=0}^{2^n-1} \int_0^1 |\int_0^{j_1 2^{-n}} (\psi_{j_2}(u) - \psi_{j_2}(z))(u-z)^{-1-\beta} dz| du$$
$$\leq C 2^{-n(\lambda_1+\mu_2)} \sum_{j_1,j_2=0}^{2^n-1} \sum_{r\in A_{j_1}} \int_{r2^{-m}}^{(r+1)2^{-m}} |\int_0^{j_1 2^{-n}} (u-z)^{-1-\beta} dz| du$$
$$\leq C 2^{n(1+\beta-\lambda_1-\mu_2)} \to 0, n \to \infty,$$

since under assumption $\lambda_1 + \mu_1 + \mu_2 > 2$ we can choose $\frac{1}{2} > \beta > 1 - \mu_1$ in such a way that $1 + \beta - \lambda_1 - \mu_2 < 0$. Finally, for $j_1 2^{-n} \leq z \leq u \leq (r+1)2^{-m}$

$$|\psi_{j_2}(u) - \psi_{j_2}(z)| \leq 2^{-n\mu_2}(u - z + 2^{-m})^{\lambda_1},$$

and

$$\sum_{j_2=0}^{2^n-1} \int_0^1 |\int_{j_1 r^{-n}}^{r 2^{-m}} (\psi_{j_2}(u) - \psi_{j_2}(z))(u-z)^{-1-\beta} dz| du$$
$$\leq C 2^{m(1+\beta_1-\lambda_1-\mu_2)} \to 0, m \to \infty.$$

□

Remark 2.2.18. For $f(s) = C$ $\Delta^2_{mn} = 0$, and it is easy to see from the bounds of Δ^1_{mn} that the theorem will hold under the assumption $\lambda_i, \mu_i > \frac{1}{2}, i = 1, 2$.

Remark 2.2.19. Multiple stochastic fractional integral with Hurst parameter less than $1/2$ was considered in (BJ06).

2.3 Wick Integration with Respect to fBm with $H \in [1/2, 1)$ as S^*-integration

2.3.1 Wick Products and S^*-integration

Recall (see Sections 1.4–1.5), that the random variable F on the probability space $S'(R)$ belongs to S^* if F admits the formal expansion (1.5.1) with finite negative norm

$$\|F\|^2_{-q} = \sum_{\alpha \in \mathcal{I}} \alpha! c_\alpha^2 (2\mathbb{N})^{-q\alpha} < \infty$$

for at least one $q \in \mathbb{N}$. Introduce the following notations:

(i) Let the function $Z : \mathbb{R} \to S^*$, and for any $F \in S$ we have that $\langle\!\langle Z(t), F \rangle\!\rangle \in L_1(\mathbb{R})$ as a function of $t \in \mathbb{R}$.
(ii) In this case, define $\int_{\mathbb{R}} Z(t) dt$ as the unique element of S^* such that

$$\left\langle\!\!\left\langle \int_{\mathbb{R}} Z(t) dt, F \right\rangle\!\!\right\rangle = \int_{\mathbb{R}} \langle\!\langle Z(t), F \rangle\!\rangle \, dt,$$

and say that Z is integrable in S^*.
(iii) Define the Wick products: for $F(\omega) = \sum_\alpha c_\alpha \mathcal{H}_\alpha(\omega)$, and $G(\omega) = \sum_\beta d_\beta \mathcal{H}_\beta(\omega)$, put $(F \Diamond G)(\omega) = \sum_{\alpha,\beta} c_\alpha d_\beta \mathcal{H}_{\alpha+\beta}(\omega)$.
According to the (HOUZ96), for $F, G, H \in S$ it holds that
(iv) $F \Diamond G = G \Diamond F$;
(v) $(F \Diamond G) \Diamond H = F \Diamond (G \Diamond H)$;
(vi) $H \Diamond (F + G) = H \Diamond F + H \Diamond G$;
(vii) $F \Diamond G \in S$ if $F, G \in S$; $F \Diamond G \in S^*$ if $F, G \in S^*$.

In this section we consider only the case $H \in [1/2, 1)$.

Theorem 2.3.1. *Let the process $Y(t) \in S^*$ and admit an expansion $Y(t) = \sum_\alpha c_\alpha(t)\mathcal{H}_\alpha(\omega)$, $t \in \mathbb{R}$, with the coefficients, satisfying the inequality*

$$K := \sup_\alpha \{\alpha! \, \|c_\alpha\|^2_{L_1(\mathbb{R})} (2\mathbb{N})^{-q\alpha}\} < \infty$$

for some $q > 0$.

Then the Wick product $Y(t) \diamond \dot{B}_t^M$ is S^*-integrable, and, moreover,

$$\int_\mathbb{R} Y(t) \diamond \dot{B}_t^M \, dt = \sum_{\alpha,k} \int_\mathbb{R} c_\alpha(t) M_+ \widetilde{h}_k(t) dt \cdot \mathcal{H}_{\alpha+\varepsilon_k}(\omega). \quad (2.3.1)$$

Proof. Consider only $\dot{B}_t^H$, and for arbitrary $\dot{B}_t^M$ the proof is the same. Since $\langle \widetilde{h}_k, \omega \rangle = \mathcal{H}_{\varepsilon_k}(\omega)$, we have that the Wick product $Y(t) \diamond \dot{B}_t^H \in S^*$ and equals $\sum_{\alpha,k} c_\alpha(t) M_+^H \widetilde{h}_k(t) \mathcal{H}_{\alpha+\varepsilon_k}(\omega)$. According to (HOUZ96, Lemmas 2.5.6 and 2.5.7), the S^*-integrability of $Y(t) \diamond \dot{B}_t^H$ follows from the inequality

$$\sum_{\beta \in \mathcal{I}} \beta! \left\| \sum_{\alpha,k:\alpha+\varepsilon_k=\beta} c_\alpha(t) M_+^H \widetilde{h}_k(t) \right\|^2_{L_1(\mathbb{R})} (2\mathbb{N})^{-p\beta} < \infty$$

for some $p > 0$. According to estimate (1.5.3), $\left| M_+^H \widetilde{h}_k(t) \right| \leq C k^{2/3 - H/2} < C k^{5/12}$ for any $k \geq 1$ and some $C > 0$.

Therefore,

$$\int_\mathbb{R} \left| c_\alpha(t) M_+^H \widetilde{h}_k(t) \right| dt \leq C k^{5/12} \|c_\alpha\|_{L_1(\mathbb{R})},$$

and

$$\left\| \sum_{\alpha,k:\alpha+\varepsilon_k=\beta} c_\alpha(t) M_+^H \widetilde{h}_k(t) \right\|^2_{L_1(\mathbb{R})} \leq \left(\sum_{\alpha,k:\alpha+\varepsilon_k=\beta} \left\| c_\alpha(t) M_+^H \widetilde{h}_k(t) \right\|_{L_1(\mathbb{R})} \right)^2$$

$$\leq C \left(\sum_{\alpha,k:\alpha+\varepsilon_k=\beta} k^{5/12} \|c_\alpha\|_{L_1(\mathbb{R})} \right)^2.$$

Consider the sum

$$S := \sum_{\beta \in \mathcal{I}} \beta! \left(\sum_{\alpha,k:\alpha+\varepsilon_k=\beta} k^{5/12} \|c_\alpha\|_{L_1(\mathbb{R})} \right)^2 (2\mathbb{N})^{-p\beta}$$

$$\leq \sum_{\beta \in \mathcal{I}} \beta! (l(\beta))^{5/6} \left(\sum_{\alpha,k:\alpha+\varepsilon_k=\beta} \|c_\alpha\|_{L_1(\mathbb{R})} \right)^2 (2\mathbb{N})^{-p\beta},$$

2.3 Wick Integration with Respect to fBm

where $l(\beta)$ equals the number of the last nonzero element in the index β (the length of the index β). Further, for any α, β there exists no more than one k, such that $\alpha + \varepsilon_k = \beta$. Therefore,

$$\left(\sum_{\alpha,k:\alpha+\varepsilon_k=\beta} \|c_\alpha\|_{L_1(\mathbb{R})}\right)^2 \leq l^2(\beta) \sum_{\alpha,k:\alpha+\varepsilon_k=\beta} \|c_\alpha\|^2_{L_1(\mathbb{R})}.$$

It means that

$$S \leq \sum_{\alpha,k}(\alpha+\varepsilon_k)!(l(\alpha+\varepsilon_k))^{17/6}\|c_\alpha\|^2_{L_1(\mathbb{R})}(2\mathbb{N})^{-p\alpha-p\varepsilon_k}$$

$$\leq K \sum_{\alpha,k} \frac{(\alpha+\varepsilon_k)!}{\alpha!}(l(\alpha+\varepsilon_k))^3 (2\mathbb{N})^{-(p-q)\alpha-p\varepsilon_k}$$

$$\leq K \sum_{\alpha,k}(|\alpha|+1)^4 2^{-|\alpha|(p-q)} k^{-p} < \infty,$$

for $p > q+1$, and we have established the S^*-integrability of $Y(t) \diamond \dot{B}^H_t$. Now, for any $F = \sum_{\beta,k} d_{\beta,k} \mathcal{H}_{\beta+\varepsilon_k}(\omega) \in S$, we have from the definition of the S^*-integral and of Wick product, that

$$\left\langle\!\!\left\langle \int_\mathbb{R} Y(t) \diamond \dot{B}^H_t \, dt, F \right\rangle\!\!\right\rangle = \int_\mathbb{R} \left\langle\!\!\left\langle \sum_{\alpha,k} c_\alpha(t) M^H_+ \widetilde{h}_k(t) \mathcal{H}_{\alpha+\varepsilon_k}(\omega), F \right\rangle\!\!\right\rangle dt \qquad (2.3.2)$$

$$= \int_\mathbb{R} \sum_{\alpha,k}(\alpha+\varepsilon_k)! c_\alpha(t) d_{\alpha,k} M^H_+ \widetilde{h}_k(t)(\omega) dt.$$

Note that

$$\sum_{\alpha,k}(\alpha+\varepsilon_k)! |d_{\alpha,k}|^2 (2\mathbb{N})^{2q(\alpha+\varepsilon_k)} =: C_q < \infty$$

for any $q \in \mathbb{N}$. Therefore

$$\sum_{\alpha,k} \int_\mathbb{R} (\alpha+\varepsilon_k)! |c_\alpha(t)| |d_{\alpha,k}| \left|M^H_+ \widetilde{h}_k(t)\right| dt \leq \sum_{\alpha,k}(\alpha+\varepsilon_k)! |d_{\alpha,k}| k^{5/12} \|c_\alpha\|_{L_1(\mathbb{R})}$$

$$\leq \left(\sum_{\alpha,k} \beta_k! |d_{\alpha,k}|^2 (2\mathbb{N})^{2q\beta_k} \sum_{\alpha,k} k^{5/6} \|c_\alpha\|^2_{L_1(\mathbb{R})} \beta_k! (2\mathbb{N})^{-2q(\alpha+\varepsilon_k)}\right)^{1/2}$$

$$\leq \left(C_q K \sum_{\alpha,k} k^{5/6} \frac{\beta_k!}{\alpha!} (2\mathbb{N})^{-q|\alpha|} k^{-2q}\right)^{1/2} < \infty$$

for $q > 11/12$, $\beta_k = \alpha + \varepsilon_k$, because $\sum_\alpha \frac{\beta_k!}{\alpha!}(2\mathbb{N})^{-q|\alpha|} \leq \sum_\alpha (|\alpha|+1) 2^{-q|\alpha|} < \infty$. So, we can change the signs of sum and integral in (2.3.2) and obtain

$$\left\langle\!\!\left\langle \int_\mathbb{R} Y(t) \diamond \dot{B}_t^H dt, F \right\rangle\!\!\right\rangle = \sum_{\alpha,k} (\alpha + \varepsilon_k)! d_{\alpha,k} \int_\mathbb{R} c_\alpha(t) M_+^H \widetilde{h}_k(t)(\omega) dt$$
$$= \left\langle\!\!\left\langle \sum_{\alpha,k} \int_\mathbb{R} c_\alpha(t) M_+^H \widetilde{h}_k(t)(\omega) dt, F \right\rangle\!\!\right\rangle,$$

whence (2.3.1) follows. □

Corollary 2.3.2. *Let* $Y(t) = \sum_\alpha c_\alpha(t) \mathcal{H}_\alpha(\omega) \in S^*$ *be a process such that* $\int_0^T EY^2(t) dt < \infty$ *for some* $T > 0$. *Then* $\sum_\alpha \alpha! \int_0^T c_\alpha^2(t) dt = \int_0^T EY^2(t) dt < \infty$, *whence* $K := \sup_\alpha \{\alpha! \|\bar{c}_\alpha\|_{L_1(\mathbb{R})}^2 (2\mathbb{N})^{-q\alpha}\} < \infty$ *for any* $q > 0$ *(hereafter we put* $\bar{c}_\alpha(t) := c_\alpha(t) \mathbf{1}_{[0,T]}(t))$.

So, we can use Theorem 2.3.1 and conclude that $Y(t) \diamond \dot{B}_t^M$ is S^*-integrable, and, moreover, equality (2.3.1) holds.

Corollary 2.3.3. *Let* $Y(t) \equiv 1$. *Then the previous corollary holds with* $c_0(t) = 1, c_\alpha(t) = 0$ *for* $\alpha \neq 0$, *whence*

$$\int_0^T \dot{B}_t^M dt = \sum_k \int_0^T M_+ \widetilde{h}_k(t) dt \cdot \mathcal{H}_{\varepsilon_k}(\omega) = B_T^M.$$

In this connection, we can say that the fractional noise is the S^*-*derivative of fBm.*

As a consequence, we can define $\int_\mathbb{R} Y_t \diamond dB_t^M := \int_\mathbb{R} Y_t \diamond \dot{B}_t^M dt$ for the process Y_t, satisfying the conditions of Theorem 2.3.1.

Now, let $Y \in L_2[0,T]$ be some nonrandom function, $H \in (1/2, 1)$.

Then $c_\alpha(t) = Y(t) = \bar{c}_\alpha(t)$, for $\alpha = 0$ and $c_\alpha \equiv 0$ for other α, so, by using Theorem 2.3.1, we obtain that

$$\int_0^T Y(t) \diamond \dot{B}_t^H dt = \sum_k \int_0^T Y(t) M_+^H \widetilde{h}_k(t) dt \cdot \langle \widetilde{h}_k, \omega \rangle.$$

Further, even for $Y \in L_1[0,T]$ we can replace the operator M_+^H and obtain $\int_0^T Y(t) M_+^H \widetilde{h}_k(t) dt = \int_0^T M_-^H Y(t) \widetilde{h}_k(t) dt$, whence

$$\int_0^T Y(t) \diamond \dot{B}_t^H dt = \sum_k \int_\mathbb{R} M_-^H \overline{Y}(t) \widetilde{h}_k(t) dt \cdot \langle \widetilde{h}_k, \omega \rangle$$
$$= \sum_k \int_\mathbb{R} M_-^H \overline{Y}(t) \widetilde{h}_k(t) dt \cdot \mathcal{H}_{\varepsilon_k}(\omega), \qquad (2.3.3)$$

where $\overline{Y}(t) = Y(t)\mathbf{1}_{[0,T]}(t)$. The right-hand side of (2.3.3) corresponds to (HOUZ96, representation (2.5.22)) of the integral $\int_0^T M_-^H Y(t) \diamond \dot{B}_t dt$, where $\dot{B}_t = \dot{B}_t^{1/2}$ is a white noise:

$$\int_0^T M_-^H Y(t) \diamond \dot{B}_t dt = \sum_{\alpha,k} \int_0^T c_\alpha(t)\widetilde{h}_k(t)dt \cdot \mathcal{H}_{\alpha+\varepsilon_k}(\omega).$$

Therefore, for $Y \in L_2^H[0,T]$

$$\int_0^T Y(t) \diamond \dot{B}_t^M dt = \int_\mathbb{R} M_- \overline{Y}(t) \diamond \dot{B}_t dt = \int_\mathbb{R} M_- \overline{Y}(t) \cdot \dot{B}_t dt. \qquad (2.3.4)$$

2.3.2 Comparison of Wick and Pathwise Integrals for "Markov" Integrands

In this subsection we can, without losing generality, consider instead of $S'(\mathbb{R})$ the probability space $\Omega = C_0(\mathbb{R}_+, \mathbb{R})$ of real-valued continuous functions on $\mathbb{R}_+$ with the initial value zero and the topology of local uniform convergence. There exists a probability measure P on $(\Omega, \mathcal{F})$, where $\mathcal{F}$ is the Borel σ-field, such that on the probability space $(\Omega, \mathcal{F}, P)$ the coordinate process $B : \Omega \to \mathbb{R}$ defined as,

$$B_t(\omega) = \omega(t), \ \omega \in \Omega$$

is the Wiener process.

(i) Recall the notion of a stochastic derivative. Let F be a square-integrable random variable, and suppose that the limit

$$\lim_{\beta \to 0} \frac{1}{\beta}\left(F(\omega. + \beta \int_0^\cdot h(s)ds) - F(\omega.)\right) \quad \text{exists in} \quad L_2(P)$$

for any $h \in L_2(\mathbb{R})$. Then this limit is called the directional derivative $D_h F$.

(ii) If the directional derivative $D_h F$, $h \in L_2(\mathbb{R})$, is absolutely continuous w.r.t. the measure $h(x)dx$, i.e.

$$D_h(F) = \int_\mathbb{R} \frac{dD_h(F)}{dh}(x) \cdot h(x)dx,$$

and $(dD_h(F))/(dh)$ does not depend on h, then the Radon–Nikodym derivative $(dD_h(F))/(dh)$ is called the stochastic derivative of F and is denoted by $D_x F$.

(iii) We have a chain rule for the stochastic derivative: if $D_x F$ exists and $\varphi \in C^1(\mathbb{R})$, then $D_x \varphi(F)$ has the stochastic derivative

$$D_x \varphi(F) = \varphi'(F) D_x F.$$

(iv) Let $u \in L_2(\mathbb{R})$ be a nonrandom function. Then it follows from (NP95, Proposition 5.5), that

$$D_x \int_{\mathbb{R}} u_s dB_s = u_x \quad \text{a.e.}$$

(v) Recall the notion of the class $\mathbb{D}_{1,2}$. This is the Banach space, obtained as a completion of the set $\mathcal{P}_0$ of smooth functionals $F = f(B_{t_1}, \ldots, B_{t_i})$, w.r.t. the norm $\|F\|_{1,2} := \|F\|_{L_2(P)} + \big\| \|D_xF\|_{HS} \big\|_{L_1(P)}$, where $F \in \mathcal{P}_0$, and $\|\cdot\|_{HS}$ denotes the Hilbert–Schmidt norm.

Denote $L_2^M(\mathbb{R}) = \{f : \mathbb{R} \to \mathbb{R} : \int_{\mathbb{R}} |M_-f(x)|^2 \, dx < \infty\}$.

Lemma 2.3.4. *Let $F \in \mathbb{D}_{1,2}$, $f \in L_2^M(\mathbb{R})$. Suppose that the integrals*

$$\int_{\mathbb{R}} (M_-f)(s) \cdot D_s F ds \text{ and } F \cdot \int_{\mathbb{R}} (M_-f)(s) dB_s = F \cdot \int_{\mathbb{R}} f(s) dB_s^M$$

belong to $L_2(P)$. Then $F \diamond \int_{\mathbb{R}} f(s) dB_s^M$ exists and

$$F \diamond \int_{\mathbb{R}} f(s) dB_s^M = \int_{\mathbb{R}} (F \cdot M_-f)(s) \delta B_s$$

$$= F \cdot \int_{\mathbb{R}} f(s) dB_s^M - \int_{\mathbb{R}} (M_-f)(s) \cdot D_s F ds. \quad (2.3.5)$$

Proof. By using (HOUZ96, Corollary 2.5.12) and (NP95, Theorem 3.2), we obtain for nonrandom f that

$$F \diamond \int_{\mathbb{R}} f(s) dB_s^M = F \diamond \int_{\mathbb{R}} (M_-f)(s) dB_s$$
$$= \int_{\mathbb{R}} (F \diamond M_-f)(s) \delta B_s = \int_{\mathbb{R}} (F \cdot M_-f)(s) \delta B_s$$
$$= F \cdot \int_{\mathbb{R}} (M_-f)(s) dB_s - \int_{\mathbb{R}} (M_-f)(s) \cdot D_s F ds$$
$$= F \cdot \int_{\mathbb{R}} f(s) dB_s^M - \int_{\mathbb{R}} (M_-f)(s) \cdot D_s F ds.$$

(Note that according to (NP95, Theorem 3.2), the Skorohod integral $\int_{\mathbb{R}} F \cdot (M_-f)(s) \delta B_s$ exists if and only if the difference $F \cdot \int_{\mathbb{R}} (M_-f)(s) dB_s - \int_{\mathbb{R}} (M_-f)(s) \cdot D_s F ds$ belongs to $L_2(P)$). □

Using this result, we can compare the Wick integral and the pathwise integral w.r.t. fBm B_t^H, $H \in (1/2, 1)$ (the latter integral coincides with Stratonovich integral). Therefore, now $M_\pm = M_\pm^H$.

Lemma 2.3.5. Let $\varphi \in C^1(\mathbb{R})$, $F_t = \varphi(B_t^H)$, $f(s) = \mathbf{1}_{[t,t+h]}(s)$, $t, h > 0$. If $\varphi'(B_t^H)$ and $F_t \cdot (B_{t+h}^H - B_t^H)$ belong to $L_2(P)$, then

$$F_t \Diamond (B_{t+h}^H - B_t^H) = F \cdot (B_{t+h}^H - B_t^H) \\ - H\varphi'(B_t^H)t^{2\alpha}h + c(\omega)(t^{2\alpha-1}h^2 + h^{2H}),$$

where $c(\omega)$ is a.s. finite and independent of t and h.

Proof. According to equation (2.3.5), we can rewrite formally the left-hand side of the previous equality:

$$F_t \Diamond (B_{t+h}^H - B_t^H) = F_t \cdot (B_{t+h}^H - B_t^H) \\ - \int_{\mathbb{R}} \left(M_-^H \mathbf{1}_{[t,t+h]} \right)(s) D_s \varphi(B_t^H) ds. \quad (2.3.6)$$

Further, according to the chain rule (iii), it holds that

$$D_s \varphi(B_t^H) = \varphi'(B_t^H) D_s B_t^H,$$

and

$$D_s B_t^H = D_s \int_{\mathbb{R}} \left(M_-^H \mathbf{1}_{[0,t]} \right)(u) dB_u = (M_-^H \mathbf{1}_{[0,t]})(s).$$

Therefore,

$$F_t \Diamond (B_{t+h}^H - B_t^H) = F_t \cdot (B_{t+h}^H - B_t^H) \\ - \varphi'(B_t^H) \int_{\mathbb{R}} \left(M_-^H \mathbf{1}_{[t,t+h]} \right)(s) \left(M_-^H \mathbf{1}_{[0,t]} \right)(s) ds,$$

and under the conditions of the lemma the right-hand side of equation (2.3.6) is well-defined. Finally,

$$\int_{\mathbb{R}} (M_-^H \mathbf{1}_{[t,t+h]})(s)(M_-^H \mathbf{1}_{[0,t]})(s) ds = E(B_{t+h}^H - B_t^H) B_t^H \\ = \frac{1}{2}((t+h)^{2H} - t^{2H} - h^{2H}) = Ht^{2\alpha}h + 2H\alpha\theta^{2\alpha-1}h^2 - h^{2H},$$

where $\theta \in (t, t+h)$. The lemma is proved. $\square$

Remark 2.3.6. Evidently, the assumption $E(\varphi(B_t^H))^{2+\varepsilon} < \infty$ for some $\varepsilon > 0$ is sufficient for $F_t(B_{t+h}^H - B_t^H)$ to belong to $L_2(P)$.

Now, fix some $T > 0$ and consider the sequence $\pi_n = \{0 = t_0^n < \cdots < t_n^n = T\}$ of partitions of $[0, T]$, such that $\pi_n \subset \pi_{n+1}$ and $|\pi_n| \to 0$ as $n \to \infty$. Suppose that

$$\varphi'(B_t^H) \in L_2(P), \quad \varphi(B_t^H) \in L_{2+\varepsilon}(P), \quad t \in [0,T] \quad (2.3.7)$$

for some $\varepsilon > 0$.

According to Lemma 2.3.5, we can write

$$\sum_{i=1}^{n} \varphi(B_{t_{i-1}^n}^H) \diamond \Delta B_{i,n}^H = \sum_{i=1}^{n} \varphi(B_{t_{i-1}^n}^H) \Delta B_{i,n}^H$$

$$- H \sum_{i=1}^{n} \varphi'(B_{t_{i-1}^n}^H)(t_{i-1}^n)^{2\alpha} \Delta t_{i,n} + R_n(T),$$

where $\Delta t_{i,n} = t_i^n - t_{i-1}^n$, $\Delta B_{i,n}^H = B_{t_i^n}^H - B_{t_{i-1}^n}^H$. Here $R_n(T)$ is a remainder term and $R_n(T) \to 0$ a.s. as $n \to \infty$. Furthermore, the process $C_t := \varphi(B_t^H)$ is Hölder continuous up to order H. Also, by Theorem 2.1.7, part 2), the sum $\sum_{i=1}^{n} \varphi(B_{t_{i-1}^n}^H) \Delta B_{i,n}^H$ converges a.s. as $n \to \infty$ to the pathwise integral $\int_0^T \varphi(B_s^H) dB_s^H$. Clearly,

$$\sum_{i=1}^{n} \varphi'(B_{t_{i-1}^n}^H)(t_{i-1}^n)^{2\alpha} \Delta t_{i,n} \to \int_0^T \varphi'(B_s^H) s^{2\alpha} ds \quad \text{a.s.}$$

Therefore,

$$\lim_{n \to \infty} \sum_{i=1}^{n} \varphi(B_{t_{i-1}^n}^H) \diamond \Delta B_{i,n}^H = \int_0^T \varphi(B_s^H) dB_s^H - H \int_0^T \varphi'(B_s^H) s^{2\alpha} ds \quad \text{a.s.}$$

Moreover, under assumption (2.3.7) and

$$E \int_0^T \left(\varphi(B_s^H) \right)^2 ds < \infty, \tag{2.3.8}$$

there exists the Wick integral $\int_0^T \varphi(B_s^H) \diamond dB_s^H$. Now we are in a position to prove that

$$\int_0^T \varphi(B_s^H) \diamond dB_s^H = \lim_{n \to \infty} \sum_{i=1}^{n} \varphi(B_{t_{i-1}^n}^H) \diamond \Delta B_{i,n}^H. \tag{2.3.9}$$

Theorem 2.3.7. *Under conditions (2.3.7) and*

$$E \sup_{s \leq T} \left(\varphi(B_s^H) \right)^2 + E \sup_{s \leq T} (\varphi'(B_s^H))^2 < \infty \tag{2.3.10}$$

equality (2.3.8) and (2.3.9), consequently, the equality

$$\int_0^T \varphi(B_s^H) \diamond dB_s^H = \int_0^T \varphi(B_s^H) dB_s^H - H \int_0^T \varphi'(B_s^H) s^{2\alpha} ds$$

holds a.s.

Proof. Let the random variables $F, G \in \mathbb{D}_{1,2}$. According to equality (2.3.5) and (NP95, Theorem 3.2), for $i \leq k$

$$E\left[F \diamond \Delta B_{i,n}^H \cdot G \diamond \Delta B_{k,n}^H\right]$$
$$= E\left[\int_\mathbb{R} F M_-^H \mathbf{1}_{[t_{i-1}^n, t_i^n]}(s) \delta B_s \cdot \int_\mathbb{R} G M_-^H \mathbf{1}_{[t_{k-1}^n, t_k^n]}(s) \delta B_s\right]$$
$$= E\left[\int_\mathbb{R} FG M_-^H \mathbf{1}_{[t_{i-1}^n, t_i^n]}(s) M_-^H \mathbf{1}_{[t_{k-1}^n, t_k^n]}(s) ds\right]$$
$$+ E\left[\int_{\mathbb{R}\times\mathbb{R}} D_t F D_s G M_-^H \mathbf{1}_{[t_{i-1}^n, t_i^n]}(t) M_-^H \mathbf{1}_{[t_{k-1}^n, t_k^n]}(s) ds\, dt\right] \quad (2.3.11)$$
$$= \frac{1}{2} E[FG r_{ik}]$$
$$+ E\left[\int_\mathbb{R} D_t F M_-^H \mathbf{1}_{[t_{i-1}^n, t_i^n]}(t) dt \cdot \int_\mathbb{R} D_s G M_-^H \mathbf{1}_{[t_{k-1}^n, t_k^n]}(s) ds\right],$$

where

$$r_{ik} = |t_{k-1}^n - t_i^n|^{2H} + (t_k^n - t_{i-1}^n)^{2H} - (t_k^n - t_i^n)^{2H} - (t_{k-1}^n - t_{i-1}^n)^{2H}.$$

Put in (2.3.11) $F = \varphi(B_{t_{i-1}^n}^H)$, $G = \varphi(B_{t_{k-1}^n}^H)$ and take the sum over $1 \leq i \leq k \leq n$. We obtain that

$$E\left(\sum_{i=1}^n \varphi(B_{t_{i-1}^n}^H) \diamond \Delta B_{i,n}^H\right)^2 = S_1^n + S_2^n,$$

where

$$S_1^n = \sum_{1 \leq i \leq k \leq n} E\varphi(B_{t_{i-1}^n}^H) \varphi(B_{t_{k-1}^n}^H) r_{ik},$$

and

$$S_2^n = \sum_{1 \leq i \leq k \leq n} E \int_\mathbb{R} \varphi'(B_{t_{i-1}^n}^H) M_-^H \mathbf{1}_{[t_{i-1}^n, t_i^n]}(t) M_-^H \mathbf{1}_{[0, t_{i-1}^n]}(t) dt$$
$$\times \int_\mathbb{R} \varphi'(B_{t_{k-1}^n}^H) M_-^H \mathbf{1}_{[t_{k-1}^n, t_k^n]}(s) M_-^H \mathbf{1}_{[0, t_{k-1}^n]}(s) ds$$
$$= \frac{1}{4} \sum_{1 \leq i \leq k \leq n} E\varphi'(B_{t_{i-1}^n}^H) \varphi'(B_{t_{k-1}^n}^H) \left((t_k^n)^{2H} - (t_{k-1}^n)^{2H} - (\Delta t_k^n)^{2H}\right)$$
$$\times \left((t_i^n)^{2H} - (t_{i-1}^n)^{2H} - (\Delta t_i^n)^{2H}\right).$$

Evidently,

$$|S_2^n| \leq H^2 E\left(\sum_{i=1}^n \left|\varphi'(B_{t_{i-1}^n}^H)\right| t_i^{2\alpha} \cdot \Delta t_i^n\right)^2. \quad (2.3.12)$$

If the partition π_n is uniform, i.e. $t_i^n = \frac{iT}{n}$, then for some $C_H > 0$

$$S_1^n \leq 2 \sum_{1 \leq i \leq n} E\left|\varphi(B_{t_{i-1}^n}^H)\right|^2 \left(\frac{iT}{n}\right)^{2H}$$
$$+ \left(\frac{T}{n}\right)^{2H} C_H \sum_{1 \leq i \leq k \leq n} \left|\varphi(B_{t_{i-1}^n}^H)\varphi(B_{t_{k-1}^n}^H)\right| \cdot \int_{i-1}^{i}\int_{k-1}^{k} (u-v)^{2\alpha-1} du\, dv.$$
(2.3.13)

Now it is very easy to conclude from (2.3.10)–(2.3.13), that the sums

$$S_n := \sum_{k=1}^{n} \varphi(B_{t_k^n}^H) \diamond \Delta B_{k,n}^H$$

form a Cauchy sequence in $L_2(P)$, at least, for uniform π_n. From the estimate

$$|\langle\!\langle F, g\rangle\!\rangle| \leq \|F\|_{L_2(P)} \|g\|_{L_2(P)}, \quad F \in L_2(P),\ g \in S,$$

we obtain that $\langle\!\langle S_n - S_m, g\rangle\!\rangle \to 0$, $n, m \to \infty$ for any $g \in S$. This means that $\{S_n\}$ is a Cauchy sequence in the weak sense. If we establish the weak convergence $S_n \to \widetilde{S} := \int_0^T \varphi(B_s^H) \diamond dB_s^H$, then the theorem will be proved, since the convergence will be in $L_2(P)$, as well. According to (2.3.1) and Corollary 2.3.2, we have that

$$\widetilde{S} = \int_0^T \varphi(B_t^H) \diamond \dot{B}_t^H\, dt = \sum_{\alpha, k} \int_0^T c_\alpha(t) M_+^H \widetilde{h}_k(t)\, dt \cdot \mathcal{H}_{\alpha+\varepsilon_k}(\omega),$$

$$S_n = \int_0^T \varphi_n(t) \diamond \dot{B}_t^H\, dt = \sum_{\alpha, k} \int_0^T c_\alpha^n(t) M_+^H \widetilde{h}_k(t)\, dt \cdot \mathcal{H}_{\alpha+\varepsilon_k}(\omega),$$

where

$$\varphi_n(t) = \sum_{i=1}^{n} \varphi(B_{t_{i-1}^n}^H) \mathbf{1}_{[t_{i-1}^n, t_i^n)}(t),$$

$$\varphi(B_t^H) = \sum_\alpha c_\alpha(t) \mathcal{H}_\alpha(\omega),\quad c_\alpha^n(t) = \sum_{i=1}^{n} c_\alpha(t_{i-1}^n) \mathbf{1}_{[t_{i-1}^n, t_i^n)}(t).$$

Denote $d_\alpha^n := c_\alpha - c_\alpha^n$. Then

$$S - S_n = \sum_\beta \sum_{\alpha, k:\, \alpha+\varepsilon_k=\beta} \int_0^T d_\alpha^n(t) M_+^H \widetilde{h}_k(t)\, dt \cdot \mathcal{H}_\beta(\omega).$$

Furthermore, for any $g = \sum_\beta g_\beta \mathcal{H}_\beta(\omega) \in S$ and any $q > 0$

2.3 Wick Integration with Respect to fBm

$$\left|\langle\langle \widetilde{S} - S_n, g \rangle\rangle\right| \leq \sum_{\beta} \beta! \left| g_{\beta} \sum_{\alpha, k: \alpha+\varepsilon_k=\beta} \int_0^T d_\alpha^n(t) M_+^H \widetilde{h}_k(t) dt \right|$$

$$\leq \left(\sum_{\beta} \beta! (g_\beta)^2 (2\mathbb{N})^{\beta q} \right)^{1/2}$$

$$\times \left(\sum_{\beta} \beta! \left\| \sum_{\alpha, k: \alpha+\varepsilon_k=\beta} \left| d_\alpha^n M_+^H \widetilde{h}_k \right| \right\|_{L_1[0,T]}^2 (2\mathbb{N})^{-\beta q} \right)^{1/2}.$$

We estimate only the second multiplicand. According to (1.5.3), for $H \in (1/2, 1)$ $\left| M_+^H \widetilde{h}_k(t) \right| \leq C k^{5/12}$ with constant C independent of t, k. So,

$$\left\| \sum_{\alpha, k: \alpha+\varepsilon_k=\beta} \left| d_\alpha^n M_+^H \widetilde{h}_k \right| \right\|_{L_1[0,T]}^2 \leq C \left(\sum_{\alpha, k: \alpha+\varepsilon_k=\beta} k^{5/12} \|d_\alpha^n\|_{L_1[0,T]} \right)^2$$

$$\leq C (l(\beta))^{5/6} \left(\sum_{\alpha, k: \alpha+\varepsilon_k=\beta} \|d_\alpha^n\|_{L_1[0,T]} \right)^2,$$

where $l(\beta)$ equals the number of nonzero entries in β. Further,

$$\sum_\beta \beta! (2\mathbb{N})^{-\beta q} \left\| \sum_{\alpha, k: \alpha+\varepsilon_k=\beta} \left| d_\alpha^n M_+^H \widetilde{h}_k \right| \right\|_{L_1[0,T]}^2$$

$$\leq \sum_\beta \beta! (2\mathbb{N})^{-\beta q} l(\beta)^{5/6} \left(\sum_{\alpha, k: \alpha+\varepsilon_k=\beta} \|d_\alpha^n\|_{L_1[0,T]} \right)^2$$

$$\leq \sum_\beta \beta! \, l(\beta)^{17/6} \sum_{\alpha: \exists k, \alpha+\varepsilon_k=\beta} \|d_\alpha^n\|_{L_1[0,T]} (2\mathbb{N})^{-\beta q}$$

$$\leq \sum_{\alpha, k} (\alpha + \varepsilon_k)! (l(\alpha + \varepsilon_k))^{17/6} \|d_\alpha^n\|_{L_1[0,T]}^2 (2\mathbb{N})^{-q(\alpha+\varepsilon_k)}$$

$$\leq \sup_\alpha \left\{ \alpha! \, \|d_\alpha^n\|_{L_1[0,T]}^2 \right\} \sum_{\alpha, k} \frac{(\alpha+\varepsilon_k)!}{\alpha!} (l(\alpha+\varepsilon_k))^{17/6} (2\mathbb{N})^{-q\alpha} (2\mathbb{N})^{-q\varepsilon_k}$$

$$\leq \sup_\alpha \left\{ \alpha! \, \|d_\alpha^n\|_{L_1[0,T]}^2 \right\} \sum_{\alpha, k} (|\alpha|+1)^{23/6} 2^{-|\alpha|q} k^{-q}.$$

The last series converges for $q > 1$, and it follows from the continuity of φ and condition (2.3.10), that

$$\sup_\alpha \left\{ \alpha! \, \|d_\alpha^n\|_{L_1[0,T]}^2 \right\} \leq \sum_\alpha \alpha! \, \|d_\alpha^n\|_{L_2[0,T]} \cdot T$$
$$= T \, \|\varphi(B^H_\cdot) - \varphi_n(\cdot)\|_{L_2[0,T]} \to 0, \; n \to \infty.$$
□

Theorem 2.3.7 can be generalized to the processes of the form
$$B_t^M := \sum_{k=1}^m \sigma_k B_t^{H_k}.$$

Suppose that $H_1 = \frac{1}{2}$ and $H_k \in (1/2, 1)$, $2 \leq k \leq m$.

Theorem 2.3.8. *Assume that conditions (2.3.7), (2.3.8) and (2.3.10) hold with B_t^H replaced by B_t^M. Then*

$$\int_0^T \varphi(B_t^M) \Diamond \, dB_t^M = \int_0^T \varphi(B_t^M) dB_t^M$$
$$- \sum_{i,k=1}^n \sigma_i \sigma_k \widetilde{C}_{H_i H_k}(II_i + II_k) \int_0^T \varphi'(B_s^M) s^{H_i + H_k - 1} ds + \frac{1}{2} \sigma_1^2 \int_0^T \varphi'(B_s^M) ds,$$

where

$$\widetilde{C}_{H_i H_k} = \begin{cases} \dfrac{C_{H_i}^{(3)} C_{H_k}^{(3)} B(H_i - 1/2, 2 - H_i - H_k)}{(H_i + H_k)(H_i + H_k - 1)\Gamma(H_i - 1/2)\Gamma(H_k - 1/2)}, \\ \quad H_i, H_k \in (1/2, 1), \\ \dfrac{C_{H_k}^{(3)}}{\Gamma(H_k + 3/2)}, \; H_i = 1/2, H_k \in (1/2, 1), \\ 0, H_i \in (1/2, 1), H_k = 1/2, \\ \dfrac{1}{2}, H_i = H_k = 1/2. \end{cases}$$

Proof. We start with (2.3.5) and conclude that
$$\varphi(B_t^M) \Diamond (B_{t+h}^M - B_t^M) = \varphi(B_t^M) \cdot (B_{t+h}^M - B_t^M)$$
$$- \varphi'(B_t^M) \sum_{i,k=1}^m \sigma_i \sigma_k \int_\mathbb{R} M_-^{H_i} \mathbf{1}_{[t,t+h]}(s) M_-^{H_k} \mathbf{1}_{[0,t]}(s) ds.$$

Further, for $f \in L_2^{H_i}(\mathbb{R}), g \in L_2^{H_k}(\mathbb{R}), H_i, H_k \in (1/2, 1)$
$$\int_\mathbb{R} M_-^{H_i} f(s) M_-^{H_k} g(s) ds = C_{i,k,H}^{(1)} \int_\mathbb{R} \int_s^\infty (x - s)^{H_i - 3/2} f(x) dx$$
$$\times \int_s^\infty (y - s)^{H_k - 3/2} g(y) dy \, ds = C_{i,k,H}^{(1)} \int_{\mathbb{R}^2} f(x) g(y) dx \, dy$$
$$\times \int_{-\infty}^{x \wedge y} (x - s)^{H_i - 3/2} (y - s)^{H_k - 3/2} ds,$$

2.3 Wick Integration with Respect to fBm

where $C^{(1)}_{i,k,H} = \frac{C^{(3)}_{H_i} C^{(3)}_{H_k}}{\Gamma(H_i-1/2)\Gamma(H_k-1/2)}$. Evidently,

$$\int_{-\infty}^{x \wedge y} (x-s)^{H_i-3/2}(y-s)^{H_k-3/2} ds$$
$$= |y-x|^{H_i+H_k-2} \left(C^{(2)}_{i,k,H} \mathbf{1}\{y > x\} + C^{(2)}_{k,i,H} \mathbf{1}\{y \le x\} \right).$$

with $C^{(2)}_{i,k,H} = \int_0^\infty z^{H_i-3/2}(1+z)^{H_k-3/2} dz = B(H_i-1/2, 2-H_i-H_k)$. Therefore,

$$\int_\mathbb{R} M_-^{H_i} f(s) M_-^{H_k} g(s) ds = C^{(1)}_{i,k,H} \int_\mathbb{R} f(x) |y-x|^{H_i+H_k-2}$$
$$\cdot \left(C^{(2)}_{i,k,H} \mathbf{1}\{x < y\} + C^{(2)}_{k,i,H} \mathbf{1}\{y < x\} \right) dx\, dy.$$

Let $f(x) = \mathbf{1}_{[t,t+h]}(x)$, $g(y) = \mathbf{1}_{[0,t]}(y)$. Then

$$\int_\mathbb{R} M_-^{H_i} \mathbf{1}_{[t,t+h]}(s) M_-^{H_k} \mathbf{1}_{[0,t]}(s) ds$$
$$= \prod_{j=1,2} C^{(j)}_{k,i,H} \int_0^t \int_t^{t+h} (y-x)^{H_i+H_k-2} dy\, dx$$
$$= \prod_{j=1,2} C^{(j)}_{k,i,H} ((H_i+H_k)(H_i+H_k-1))^{-1}$$
$$\times \left[(t+h)^{H_i+H_k} - t^{H_i+H_k} - h^{H_i+H_k} \right]$$
$$=: \widetilde{C}_{H_i H_k} \left[(t+h)^{H_i+H_k} - t^{H_i+H_k} - h^{H_i+H_k} \right]$$
$$= \widetilde{C}_{H_i H_k} \left[(H_i+H_k) t^{H_i+H_k-1} h + (H_i+H_k)(H_i+H_k-1) \theta^{H_i+H_k-1} h^2 \right.$$
$$\left. - h^{H_i+H_k} \right], \theta \in (t, t+h). \tag{2.3.14}$$

For $H_i = 1/2$ and $H_k \in (1/2, 1)$ we have that $M_-^{1/2} = I$ is identity operator, and

$$\int_\mathbb{R} M_-^{1/2} f(s) M_-^{H_k} g(s) ds = \frac{C^{(3)}_{H_k}}{\Gamma(H_k-1/2)} \int_\mathbb{R} f(s) \int_s^\infty g(y)(y-s)^{H_k-3/2} dy\, ds.$$

For f and g as above, the last integral equals

$$\frac{C^{(3)}_{H_k}}{\Gamma(H_k-1/2)} \int_0^t \int_t^{t+h} (y-s)^{H_k-3/2} dy\, ds$$
$$= \frac{C^{(3)}_{H_k}}{\Gamma(H_k+3/2)} \left[(t+h)^{H_k+1/2} - t^{H_k+1/2} - h^{H_k+1/2} \right]$$
$$=: \widetilde{C}_{\frac{1}{2} H_k} \left[(H_k+1/2) t^{H_k-1/2} h \right.$$
$$\left. + (H_k+1/2)(H_k-1/2) t^{H_k-2} h^2 - h^{H_k+1/2} \right]. \tag{2.3.15}$$

At last, for $H_i = H_k = 1/2$

$$\int_{\mathbb{R}} M_-^{1/2} \mathbf{1}_{[0,t]}(s) M_-^{1/2} \mathbf{1}_{[t,t+h]}(s) ds = 0. \qquad (2.3.16)$$

Now we can proceed as in Lemma 2.3.5 and Theorem 2.3.7, put $\widetilde{C}_{\frac{1}{2},\frac{1}{2}} := \frac{1}{2}$, take into account (2.3.14)–(2.3.16) and obtain the proof. □

2.3.3 Comparison of Wick and Stratonovich Integrals for "General" Integrands

Now we consider the general process F_t instead of $\varphi(B_t^M)$. Suppose that fBm $\{B_t^H, t \geq 0\}$ is "one-sided", $H \in (\frac{1}{2}, 1)$.

Theorem 2.3.9. *Let $\{F_t, \mathcal{F}_t, t \in [0,T]\}$ be the stochastic process satisfying the conditions*

(i) $F_t \in \mathbb{D}_{1,2}$ *for any* $t \in [0,T]$, $E|F_t|^{2+\varepsilon} < \infty$ *for any* $t \in [0,T]$ *and some* $\varepsilon > 0$, $\sup_{s,t \in [0,T]} |D_s F_t|$ *is bounded in probability;*

(ii) $\lim_{h \downarrow 0} \sup_{t \in [0,T]} |D_t F_s - D_t F_{s+h}| = 0$ *in probability;*

(iii) F_t *is a.s. Hölder continuous of order* $\alpha > 1 - H$ *(this condition implies the existence of the Stratonovich integral $\int_0^T F_t dB_t^H$, $H \in (1/2, 1)$);*

(iv) $E \int_0^T F_t^2 dt < \infty$ *(this condition implies the existence of the Wick integral $\int_0^T F_t \diamond dB_t^H$, according to Corollary 2.3.2);*

(v) *there exists a sequence of partitions $\{\pi_n, n \geq 1\}$ with $|\pi_n| \to 0$ as $n \to \infty$ such that the integral sums $\sum_{k=1}^n F_{t_{k-1}^n} \diamond \Delta B_{k,n}^H$ converge to $\int_0^T F_t \diamond dB_t^H$ in probability.*

Then

$$\int_0^T F_s \diamond dB_s^H = \int_0^T F_s dB_s^H - C_H^{(3)} \int_0^T \left(\int_0^s (s-t)^{\alpha-1} D_s F_t dt \right) ds.$$

Proof. Consider for any $0 \leq t < t+h \leq T$ the function $f(u) = \mathbf{1}_{[t,t+h]}(u)$. Then we take into account that $D_s F_t = 0$ for $s > t$ and $s < 0$ (since F_t is $\mathcal{F}_t$-adapted) and obtain that $\int_{\mathbb{R}} M_-^H f D_s F_t ds = C_H^{(3)} \int_0^t \int_t^{t+h} (u-s)^{\alpha-1} du D_s F_t ds$, where $\int_t^{t+h} (u-s)^{\alpha-1} du \leq \frac{h^\alpha}{\alpha}$. Hence,

$$E \left(\int_{\mathbb{R}} M_-^H f D_s F_t ds \right)^2 \leq \frac{\left(C_H^{(3)}\right)^2}{\alpha^2} h^{2\alpha} t E \int_0^t |D_s F_t|^2 ds < \infty.$$

Further, $F_t \cdot \int_{\mathbb{R}} M_-^H f dB_s = F_t \cdot (B_{t+h}^H - B_t^H)$, and, according to (i),

2.3 Wick Integration with Respect to fBm 155

$$E\left|F_t \cdot (B_{t+h}^H - B_t^H)\right|^2 \le \left(E|F_t|^{2+\varepsilon}\right)^{\frac{2}{2+\varepsilon}} \left(E\left|B_{t+h}^H - B_t^H\right|^{\frac{2(2+\varepsilon)}{\varepsilon}}\right)^{\frac{\varepsilon}{2+\varepsilon}} < \infty.$$

Therefore, $\int_{\mathbb{R}} M_-^H f \cdot D_s F_t ds$ and $F_t \cdot \int_{\mathbb{R}} M_-^H f dB_s$ belong to $L_2(P)$ and it follows from Lemma 2.3.4 that the integral sums $\sum_{k=1}^n F_{t_{k-1}^n} \Diamond \Delta B_{k,n}^H$ exist. Moreover,

$$\begin{aligned}
F_{t_{k-1}^n} \Diamond \Delta B_{k,n}^H &= F_{t_{k-1}^n} \cdot \Delta B_{k,n}^H - \int_{\mathbb{R}} M_-^H \mathbf{1}_{[t_{k-1}^n, t_k^n]}(s) D_s F_{t_{k-1}^n} ds \\
&= F_{t_{k-1}^n} \cdot \Delta B_{k,n}^H - \int_{\mathbb{R}} \mathbf{1}_{[t_{k-1}^n, t_k^n]}(s) \left(M_+^H (D. F_{t_{k-1}^n})\right)(s) ds \\
&= F_{t_{k-1}^n} \cdot \Delta B_{k,n}^H - C_H^{(3)} \int_{t_{k-1}^n}^{t_k^n} \int_0^{t_{k-1}^n} (s-u)^{\alpha-1} D_u F_{t_{k-1}^n} du\, ds.
\end{aligned}$$
(2.3.17)

Consider the difference,

$$\left| \sum_{k=1}^n \int_{t_{k-1}^n}^{t_k^n} \int_0^{t_{k-1}^n} (s-u)^{\alpha-1} D_u F_{t_{k-1}^n} du\, ds \right.$$

$$\left. - \int_0^T \int_0^s (s-u)^{\alpha-1} D_u F_{t_{k-1}^n} \mathbf{1}_{[t_{k-1}^n, t_k^n)}(s) du\, ds \right|$$

$$\le C \cdot \sup_{0 \le u \le t \le T} |D_u F_t| \cdot |\pi_n|^\alpha \cdot T \to 0, \quad (2.3.18)$$

as $n \to \infty$ in probability, according to (i). Further, according to (i) and (ii),

$$\left| \int_0^T \int_0^s (s-u)^{\alpha-1} D_u F_{t_{k-1}^n} \mathbf{1}_{[t_{k-1}^n, t_k^n)}(s) du\, ds \right.$$

$$\left. - \int_0^T \int_0^s (s-u)^{\alpha-1} D_u F_s du\, ds \right| \to 0 \quad (2.3.19)$$

in probability. Now, the proof follows from (v) and (2.3.17)–(2.3.19). □

Now consider one sufficient condition for (v) (condition (v) seems to be the most artificial among other conditions (i)–(iv)). To this end, consider the middle part of (2.3.11), from which we obtain that for any step processes $F_n(t) = \sum_{k=1}^n F_{k,n} \mathbf{1}_{[t_{k-1}^n, t_{k-1}^n)}(t)$ and $G_n(t) = \sum_{k=1}^n G_{k,n} \mathbf{1}_{[t_{k-1}^n, t_{k-1}^n)}(t)$

$$E\left[\sum_{k=1}^n F_n(t) \Diamond dB_t^H \cdot \sum_{k=1}^n G_n(t) \Diamond dB_t^H\right]$$

$$= E \int_{\mathbb{R}} M_-^H F_n(t) M_-^H G_n(t) dt + E \int_{\mathbb{R}^2} M_-^H D_s F_n(t) M_-^H D_t G_n(s) ds\, dt.$$
(2.3.20)

The next result was motivated by (Ben03a, Theorem 2.2.8).

Theorem 2.3.10. *Let the stochastic process $\{F_t, \mathcal{F}_t, t \in [0,T]\}$ satisfy the assumptions (i)–(iv) and*

(vi) $E \int_0^T F_t^2 dt < \infty;$
(vii) the operator $F_t : [0,T] \to \mathbb{D}_{1,2}$ is continuous in $L_2([0,T] \times P)$.

Then the integral sums $\sum_{k=1}^n F_{t_{k-1}^n} \diamond \Delta B_{k,n}^H$ exist, the integral $\int_0^T F_s \diamond dB_s^H$ exists and

$$\int_0^T F_s \diamond dB_s^H = \lim_{n \to \infty} \sum_{k=1}^n F_{t_{k-1}^n} \diamond \Delta B_{k,n}^H \quad in \quad L_2(P)$$

for any sequence of increasing partitions π_n with $|\pi_n| \to 0$ as $n \to \infty$.

Proof. Under condition (vi), the existence of sums $\sum_{k=1}^n F_{t_{k-1}^n} \diamond \Delta B_{k,n}^H$ and the integral $\int_0^T F_s \diamond dB_s^H$ was established in Theorem 2.3.9. Further, using (2.3.20) and (vii), we obtain that

$$E \left| \int_0^T F_t \diamond dB_t^H - \sum_{k=1}^n F_{t_{k-1}^n} \diamond \Delta B_{k,n}^H \right|^2$$
$$= E \int_{\mathbb{R}} \left[M_-^H (F_\cdot - F_\cdot^n)(t) \right]^2 dt$$
$$+ \int_{\mathbb{R}^2} E \left[M_-^H (D_t F_\cdot - D_t F_\cdot^n)(s) \right]^2 ds\, dt =: E_n < \infty$$

From the Hardy–Littlewood theorem (Theorem 1.1.1) with $q = 2$, $\alpha = H - 1/2$ and $p = \frac{1}{H}$

$$\int_{\mathbb{R}} \left[M_-^H (F_\cdot - F_\cdot^n)(t) \right]^2 dt \leq C_H \| F_\cdot - F_\cdot^n \|_{L_{\frac{1}{H}}[0,T]}^2$$

and from condition (vii) it follows that

$$\int_{\mathbb{R}} \left[M_-^H (D_t F_\cdot - D_t F_\cdot^n)(s) \right]^2 ds \leq C_H \| D_t F_\cdot - D_t F_\cdot^n \|_{L_{\frac{1}{H}}[0,T]}^2$$

whence from (vii) and (iv) we obtain that

$$E_n \leq C_H E \left(\| F_\cdot - F_\cdot^n \|_{L_{\frac{1}{H}}[0,T]}^2 + \int_0^T E \| D_t F_\cdot - D_t F_\cdot^n \|_{L_{\frac{1}{H}}[0,T]}^2 dt \right)$$
$$\leq C_H T^{2\alpha} E \left(\| F_\cdot - F_\cdot^n \|_{L_2[0,T]}^2 + \int_0^T \| D_t F_\cdot - D_t F_\cdot^n \|_{L_2[0,T]}^2 dt \right)$$
$$\leq C_H T^{2\alpha} \int_0^T E \| F_\cdot - F_\cdot^n \|_{1,2}^2 dt$$
$$\leq C_{H,1} T^{2\alpha} \| F - F^{(n)} \|_{L_2([0,T] \times P)} \to 0, \quad n \to \infty.$$

□

2.3.4 Reduction of Wick Integration w.r.t. Fractional Noise to the Integration w.r.t. White Noise

Recall that for nonrandom integrands $f \in L_2^H(\mathbb{R})$

$$\int_{\mathbb{R}} f(t) dB_t^H := \int_{\mathbb{R}} (M_-^H f)(t) dB_t.$$

In this subsection we reduce $\int_{\mathbb{R}} X_t \diamond \dot{B}_t^H dt$ to the corresponding integral $\int_{\mathbb{R}} (M_-^H X)(t) \diamond \dot{B}_t dt$ w.r.t. white noise.

Theorem 2.3.11. *Let the following conditions hold:*

$$E \int_{\mathbb{R}} |X_t|^2 dt < \infty \quad \text{and} \quad E \int_{\mathbb{R}} ((M_-^H |X_t|)(t))^2 dt < \infty.$$

Then

$$\int_{\mathbb{R}} X_t \diamond \dot{B}_t^H dt = \int_{\mathbb{R}} (M_-^H X)(t) \diamond \dot{B}_t dt \quad a.s.$$

Proof. According to Theorem 2.3.1 and Corollary 2.3.2, the condition $E \int_{\mathbb{R}} |X_t|^2 dt < \infty$ supplies the equality

$$\int_{\mathbb{R}} X_t \diamond \dot{B}_t^H dt = \sum_{\alpha,k} \int_{\mathbb{R}} c_\alpha(t) M_+^H \widetilde{h}_k(t) dt \cdot \mathcal{H}_{\alpha+\varepsilon_k}(\omega). \qquad (2.3.21)$$

First, replace the operator M_+^H in the last equality. Evidently,

$$\int_{\mathbb{R}} f(t) M_+^H g(t) dt = \int_{\mathbb{R}} M_-^H f(t) g(t) dt \qquad (2.3.22)$$

for $f \in L_p(\mathbb{R})$, $g \in L_q(\mathbb{R})$ with $p > 1, q > 1$ and $\frac{1}{p} + \frac{1}{q} = 1 + \alpha = H + 1/2$. Moreover, $\widetilde{h}_k \in L_q(\mathbb{R})$ for any $q > 1$. Since $E \int_{\mathbb{R}} |X_t|^2 dt = \sum_\alpha \alpha! \int_{\mathbb{R}} c_\alpha^2(t) dt < \infty$, we can take $p = 2$, $q = \frac{1}{H}$ and obtain from (2.3.22) that

$$\int_{\mathbb{R}} c_\alpha(t) M_+^H \widetilde{h}_k(t) dt = \int_{\mathbb{R}} (M_-^H c_\alpha)(t) \widetilde{h}_k(t) dt. \qquad (2.3.23)$$

Further, consider the formal expansion $Y_t := \sum_\alpha (M_-^H c_\alpha)(t) \mathcal{H}_\alpha(\omega)$. Again, from Corollary 2.3.2, the condition

$$E \int_{\mathbb{R}} Y_t^2 dt = \sum_\alpha \alpha! \int_{\mathbb{R}} |(M_-^H c_\alpha)(t)|^2 dt < \infty \qquad (2.3.24)$$

ensures the equality

$$\int_{\mathbb{R}} Y_t \diamond \dot{B}_t dt = \sum_{\alpha,k} \int_{\mathbb{R}} (M_-^H c_\alpha)(t) \widetilde{h}_k(t) dt\, \mathcal{H}_{\alpha+\varepsilon_k}(\omega). \qquad (2.3.25)$$

So, we want to know when (2.3.24) holds and we need the equality $Y_t = (M_-^H X)(t)$. This follows from the equalities

$$((M_-^H X)(t), \mathcal{H}_\alpha(\omega))_{L_2(P)} = (M_-^H c_\alpha)(t) = M_-^H (X_t, \mathcal{H}_\alpha(\omega))_{L_2(P)}, \qquad (2.3.26)$$

if they hold for any $\alpha \in \mathcal{I}$. Equalities (2.3.26) can be reduced to

$$\int_\Omega \left(\int_t^\infty (x-t)^{\alpha-1} X_x(\omega) dx \right) \mathcal{H}_\alpha(\omega) dP$$
$$= \int_t^\infty (x-t)^{\alpha-1} \left(\int_\Omega X_x(\omega) \mathcal{H}_\alpha(\omega) dP \right) dx \qquad (2.3.27)$$

for a.a. $t \in \mathbb{R}$. In turn, the Fubini theorem can be applied to (2.3.27) in the case when

$$E\left(\int_t^\infty (x-t)^{\alpha-1} |X_x(\omega)| dx \right)^2 < \infty \quad \text{for a.a. } t \in \mathbb{R} \qquad (2.3.28)$$

because $E\mathcal{H}_\alpha^2(\omega) = \alpha! < \infty$. Evidently, the condition $E \int_{\mathbb{R}} ((M_-^H |X|)(t))^2 dt < \infty$ ensures both (2.3.24) and (2.3.28). The proof now follows from (2.3.21), (2.3.23), (2.3.25) and (2.3.26). $\square$

2.4 Skorohod, Forward, Backward and Symmetric Integration w.r.t. fBm. Two Approaches to Skorohod Integration

Taking into account the definition of the integral for nonrandom function w.r.t. fBm: $\int_{\mathbb{R}} f(t) dB_t^H := \int_{\mathbb{R}} (M_-^H f)(t) dB_t$, and Theorem 2.3.11, it is desirable to define the integral $\int_{\mathbb{R}} f(t) dB_t^H$ for stochastic integrands in a similar way. Evidently, in this case, even for very simple and natural integrands, such as $f(t) = B_t^H$, we have that $(M_-^H B^H)(t) = C_H^{(3)} \int_t^\infty (x-t)^{\alpha-1} B_x^H dx$ is not adapted. So, we must in this case address the theory of integration of non-adapted processes. To this end, recall the definition of the Skorohod integral (see also the pioneer paper (Sko75)).

Let the stochastic process $X_t = X_t(\omega)$ be such that

$$EX_t^2 < \infty \quad \text{for all } t \in \mathbb{R}.$$

Then X_t admits a Wiener–Itô chaos expansion

$$X_t = \sum_{n=0}^\infty \int_{\mathbb{R}^n} f_n(s_1, \ldots, s_n, t) dB^{\otimes n}(s_1, \ldots, s_n),$$

2.4 Skorohod, Forward, Backward and Symmetric Integration

where the functions $f_n(\cdot) \in L_2(\mathbb{R}^n)$ and are symmetric in variables $(s_1, \ldots, s_n)$, for $n = 0, 1, 2, \ldots$ and for each $t \in \mathbb{R}$. See, for example, (HOUZ96, Theorem 2.2.5). Let $\widehat{f}_n(s_1, \ldots, s_n, s_{n+1})$ be the symmetrization of $f_n(s_1, \ldots, s_n, s_{n+1})$ with respect to $(n+1)$ variables $s_1, \ldots, s_n, s_{n+1}$.

Definition 2.4.1. Assume that

$$\sum_{n=0}^{\infty} (n+1)! \left\| \widehat{f}_n \right\|_{L_2(\mathbb{R}^{n+1})} < \infty.$$

Then we say that the process X is Skorohod integrable, write $X \in Dom(\delta)$, denote the Skorohod integral as $\int_{\mathbb{R}} X_t \delta B_t$, and define it as $\int_{\mathbb{R}} X_t \delta B_t := \sum_{n=0}^{\infty} \int_{\mathbb{R}^{n+1}} \widehat{f}_n(s_1, \ldots, s_{n+1}) dB^{\otimes(n+1)}(s_1, \ldots, s_{n+1})$. The Skorohod integral belongs to $L_2(P)$,

$$E \int_{\mathbb{R}} X_t \delta B_t = 0, \text{ and } E \left| \int_{\mathbb{R}} X_t \delta B_t \right|^2 = \sum_{n=0}^{\infty} (n+1)! \left\| \widehat{f}_n \right\|_{L_2(\mathbb{R}^{n+1})}.$$

Remark 2.4.2 ((NP95)). Define by $\mathbb{L}_{1,2}$ the class of stochastic processes $X \in L_2(\mathbb{R} \times \Omega)$ such that $X \in \mathbb{D}_{1,2}$ for almost all t, and there exists a measurable version of two-parameter process $D_s X_t$ satisfying the relation $E \int_{\mathbb{R}^2} (D_s X_t)^2 ds\, dt < \infty$. Then $\mathbb{L}_{1,2} \subset Dom(\delta)$.

Definition 2.4.3 ((Ben03a)). Let the stochastic process $X_t = X_t(\omega)$ be such that $(M_-^H X)(t)$ exists and belongs to $Dom(\delta)$. Then we define the Skorohod integral with respect to fBm B^H as

$$\int_{\mathbb{R}} X_t \delta B_t^H := \int_{\mathbb{R}} (M_-^H X)(t) \delta B_t$$

for the underlying Wiener process B.

Evidently, $E \int_{\mathbb{R}} X_t \delta B_t^H = 0$. Of course, we can define in the usual way the Skorohod integral with finite limits and indefinite integral $\int_0^t X_t \delta B_t^H$, $t \in [0, T]$. It is easy to compare now the Skorohod and Wick integral w.r.t. fBm.

Theorem 2.4.4. *Let* $M_-^H X \in Dom(\delta)$, $E \int_{\mathbb{R}} |X_t|^2 dt < \infty$ *and* $E \int_{\mathbb{R}} ((M_-^H |X|)(t))^2 dt < \infty$. *Then*

$$\int_{\mathbb{R}} X_t \delta B_t^H = \int_{\mathbb{R}} X_t \diamond \dot{B}_t^H dt.$$

Proof. According to (HOUZ96, Theorem 2.5.9), the condition $M_-^H X \in Dom(\delta)$ ensures the existence of $\int_{\mathbb{R}} (M_-^H X)(t) \diamond \dot{B}_t dt$ and the equalities:

$$\int_{\mathbb{R}} (M_-^H X)(t) \diamond \dot{B}_t dt = \int_{\mathbb{R}} (M_-^H X)(t) \delta B_t = \int_{\mathbb{R}} X_t \delta B_t^H.$$

Further, according to Theorem 2.3.11, in our case

$$\int_{\mathbb{R}} (M_-^H X)(t) \diamond \dot{B}_t dt = \int_{\mathbb{R}} X_t \diamond \dot{B}_t^H dt,$$

whence the proof follows. □

Remark 2.4.5. Let $Y \in L_2^H[0,T]$. Then Y is a Skorohod integrable adapted stochastic process. Indeed, it is nonrandom thus adapted. From (2.3.4) and (HOUZ96, Theorem 2.5.9), $Y(t) \diamond \dot{B}_t^M$ is S^*-integrable, and

$$\int_0^T Y(t) \diamond \dot{B}_t^M dt = \int_{\mathbb{R}} M_-\overline{Y}(t) \cdot \dot{B}_t dt$$

$$= \int_0^T M_-\overline{Y}(t) \delta B_t = \int_0^T M_-\overline{Y}(t) dB_t,$$

where δ means Skorohod integration, and the last integral is the Itô, and even the Wiener, integral. Note that, according to Corollary 1.9.4 (for $H > 1/2$, or $1/H < 2$) $L_2[0,T] \subset L_2^H[0,T]$. We obtain that the S^*-integral for nonrandom functions from $L_2[0,T]$ coincides with the Wiener integral $\int_0^T Y(t) dB_t^H$ from Definition 1.6.1.

Another approach to Skorohod integration w.r.t. fBm was developed in the papers (AN02), (Nua03), (Nua06). The main idea is to use the basic tools of a stochastic calculus of variations (Malliavin calculus) with respect to B^H. Recall some of these notions for $H \in (1/2, 1)$. (For $H \in (0, 1/2)$ see, for example, (AMN00).)

Let $\mathcal{S}$ be a family of smooth random variables of the form

$$F = f(B_{t_1}^H, \ldots, B_{t_n}^H)$$

with $f \in C_b^\infty(\mathbb{R}^n)$ and $t_i \in [0,T], 1 \leq i \leq n$. Let $\mathcal{H}$ be a closure of the linear space of step functions defined on $[0,T]$ with respect to the scalar product

$$\langle 1_{[0,t]}, 1_{[0,s]} \rangle_{\mathcal{H}} := 2\alpha H \int_0^t \int_0^s |r-u|^{2\alpha-1} du\, dr.$$

Then the derivative operator $D : \mathcal{S} \to L_p(\Omega, \mathcal{H})$ for $p \geq 1$ is defined as

$$D_H F = \sum_{i=1}^n \frac{\partial f}{\partial x_i}(B_{t_1}^H, B_{t_2}^H, \ldots, B_{t_n}^H) 1_{[0,t_i]}.$$

Let $D_{k,p}(\mathcal{H})$ be the Sobolev space, the closure of $\mathcal{S}$ with respect to the norm

$$\|F\|_{k,p}^p = E(|F|^p) + \sum_{j=1}^k E(\|D^\nu F\|_{\mathcal{H}^{\otimes j}}^p),$$

where D^j is the jth iteration of D. The Skorohod integral (divergence operator) δ_H is defined as the adjoint of $D_H : \mathbb{D}_{1,2}(\mathcal{H}) \subset L_2(\Omega) \to L_2(\Omega, \mathcal{H})$, defined by the means of the duality relationship

$$E(G\delta_H(u)) = E\langle D_H G, u\rangle_\mathcal{H}, u \in L_2(\Omega, \mathcal{H}), G \in S.$$

Its domain is denoted by $Dom(\delta_H)$.

Introduce the Banach space $|\mathcal{H}| \otimes |\mathcal{H}|$ as the class of all the measurable functions $\varphi : [0, T]^2 \to \mathbb{R}$ such that

$$\|\varphi\|^2_{|\mathcal{H}|\otimes|\mathcal{H}|}$$
$$:= (2\alpha H)^2 \int_{[0,T]^4} |\varphi_{u,v}||\varphi_{s,t}||s - u|^{2\alpha-1}|t - v|^{2\alpha-1} du\, dv\, ds\, dt < \infty,$$

and denote $|\mathcal{H}| := |R_H|$ with the norm $\|\cdot\|_{|R_H|,2}$ (see (1.6.7)). Denote also $\mathcal{S}_{|\mathcal{H}|}$ the family of $|\mathcal{H}|$-valued random variables of the form $F = \sum_{i=1}^n F_i h_i$, where $F_i \in S$ and $h_i \in |\mathcal{H}|$. Put $D^k F := \sum_{i=1}^n D^k F_i \otimes h_i$, and define the space $\mathbb{D}_{k,p}(|\mathcal{H}|)$ as the completion of $\mathcal{S}_{|\mathcal{H}|}$ with respect to the norm

$$\|F\|^p_{k,p,|\mathcal{H}|} = E(\|F\|^p_{|\mathcal{H}|}) + \sum_{i=1}^k E(\|D^i F\|^p_{\mathcal{H}^{\otimes i}\otimes|\mathcal{H}|}).$$

Then $\mathbb{D}_{1,2}(|\mathcal{H}|) \subset Dom(\delta_H)$. The basic property of the divergence operator is that for every $u \in \mathbb{D}_{1,2}(|\mathcal{H}|)$ we have

$$E(|\delta(u)|^2) \leq \|u\|^2_{\mathbb{D}_{1,2}(|\mathcal{H}|)}.$$

Consider the forward integral w.r.t. fBm ((AN02), (LT02)). It is defined as

$$\int_0^t u_s dB_s^{H,-} := P - \lim_{\varepsilon \to 0} \varepsilon^{-1} \int_0^t u_s(B_{(s+\varepsilon)\wedge t}^H - B_s^H) ds. \quad (2.4.1)$$

(Note that in a similar way the symmetric Stratonovich integral can be defined: $\int_0^t u_s dB_s^{H,-} := P - \lim_{\varepsilon \to 0} (2\varepsilon)^{-1} \int_0^t u_s(B_{(s+\varepsilon)\wedge t}^H - B_{(s-\varepsilon)\wedge t}^H) ds$, and also backward integral can be defined.) In (LT02) the ucp-limit is considered instead of the P-limit, where ucp-convergence is uniform convergence in probability on $[0, T]$. Moreover, it is mentioned in (AN02) that forward, backward and symmetric integrals with integrand u and w.r.t. fBm coincide with each other under the following suppositions: $u \in \mathbb{D}_{1,2}(|\mathcal{H}|)$ with $\int_0^t \int_0^t |D_s u_r||r - s|^{2\alpha-1} ds\, dr < \infty$ a.s.). Also, it was proved that for processes $u \in \mathbb{D}_{1,2}(|\mathcal{H}|)$ with $\int_0^t \int_0^t |D_s u_r||r - s|^{2\alpha-1} ds\, dr < \infty$ a.s. we have the equality

$$\int_0^t u_s dB_s^{H,s} = \delta_H(u) + 2\alpha H \int_0^t \int_0^t |D_s u_r||r - s|^{2\alpha-1} dr\, ds. \quad (2.4.2)$$

Evidently, for $u \in C^\beta[0,T]$ with $\beta + H > 1$ all the integrals, symmetric, forward, backward, and pathwise, coincide. We use this fact in order to establish the conditions of coincidence of Skorohod integrals introduced in (Ben03a) and in (AN02).

Theorem 2.4.6. *Fix a time interval $[0,T]$. Let $\phi \in C^1(\mathbb{R})$ and satisfy, together with its derivative ϕ', the growth condition $|\phi(x)| \leq C \exp(\lambda x^b)$ for some $\lambda > 0$ and $0 < b < 2$. Then the integrals $\delta_H(\phi(B^H))$ and $\int_0^t \phi(B_s^H)\delta B_s^H$ coincide on $[0,T]$ a.s.*

Proof. According to Proposition 3.3 (Nua06), under the condition of the theorem (even under the less restrictive condition $|\phi(x)| \leq C\exp(\lambda x^2)$ for $\lambda < (4T^{2H})^{-1}$), the divergence operator $\delta_H(\phi(B^H))$ exists on $[0,T]$ and satisfies the relation

$$\delta_H(\phi(B^H)) = \int_0^T \phi(B_s^H)dB_s^H - H\int_0^T \phi'(B_s^H)s^{2\alpha}ds \text{ a.s.,}$$

where $\int_0^T \phi(B_s^H)dB_s^H$ is the pathwise integral. According to Theorem 2.3.7, under conditions (2.3.10), which evidently hold now, the same equality is valid for the integral $\int_0^T \phi(B_s^H)\Diamond dB_s^H$. Therefore, $\delta_H(\phi(B^H))$ and $\int_0^T \phi(B_s^H) \Diamond dB_s^H$ coincide a.s. on $[0,T]$. Further, the conditions of Theorem 2.4.4 also hold now. Indeed, for example, $E\int_{\mathbb{R}}((M_-^H|X|)(t))^2 dt$ can be bounded in our case by $C\int_0^T |\phi(B_s^H)|^2 ds$. Therefore, $\int_0^T \phi(B_t^H)\delta B_t^H$ exists and equals $\int_0^T \phi(B_t^H) \Diamond \dot{B}_t^H dt$. Finally, we use Theorem 2.3.1 and Corollary 2.3.2 and obtain the proof.

□

Remark 2.4.7. A general S-transform approach to the stochastic fractional integration is presented in (Ben03b); see also (CC00) and (Cou07).

2.5 Isometric Approach to Stochastic Integration with Respect to fBm

2.5.1 The Basic Idea

Some special approach to stochastic integration w.r.t. fBm was considered in (MV00). We will work with a continuous stochastic process $\{X_t, 0 \leq t \leq T\}$ defined on a complete probability space $(\Omega, \mathcal{F}, P)$. Let $\mathcal{F}_t := \mathcal{F}_t^X$ be the sigma-field generated by X on $[0,t]$. We assume that $X_0 = 0$. Given a partition $\pi_n := \{t_i : 0 = t_0 < t_1 < \cdots < t_n = T\}$ and X a stochastic process, define ΔX_i by $\Delta X_i := X_{t_i} - X_{t_{i-1}}$ for $1 \leq i \leq n$. Assume first that the integrand f is a simple predictable process: $f_t = \sum_i f_i \mathbf{1}_{[t_{i-1}, t_i)}(t)$, where the random variables f_i are assumed to be $\mathcal{F}_{t_{i-1}}$ measurable and $t_i \in \pi_n$; denote the

2.5 Isometric Approach to Stochastic Integration with Respect to fBm

class of simple predictable processes by L^s. With such an $f \in L^s$ and any (continuous) process X, define the stochastic integral of f with respect to X by

$$(f, X) := \sum_i f_i \Delta X_i.$$

Assume now that $|\pi_n| \to 0$ as $n \to \infty$. If the process X is the standard Brownian motion B, $f := L_2(P \otimes \lambda)\text{-}\lim f^n$, where λ is the Lebesgue measure on $[0, T]$, one can define the integral (f, B) as the L_2-limit of the simple stochastic integrals $(f^{(n)}, B)$ using the classical Itô isometry

$$E(f^{(n)}, B)^2 = E \int_0^T (f_s^{(n)})^2 ds. \tag{2.5.1}$$

Assume now that the process X is any continuous stochastic process and f is a simple predictable process. Define now a semi-norm for (f, X) using (2.5.1). Note that such a semi-norm does not depend on the process X. It is the main feature of this approach. If the process X is the standard Brownian motion, then the semi-norm is a norm and the integrals of simple function converge to the classical stochastic integral defined by Itô. For an arbitrary integrator X, even if the semi-norm is a norm, it may happen that the integrals of simple functions of processes have no limit. However, they have a limit in the completion of the space integral sums with respect to this norm. In this sense we generalize the Itô construction of stochastic integrals.

In particular, we show that if X is a fractional Brownian motion B^H, then we can define a norm by putting

$$\|(f, B^H)\|_G := \left(E \int_0^T f_s^2 ds \right)^{1/2}$$

in the space G of random variables of the form $\{g \in G : G = (f, B^H), f \in L^s\}$.

Even more turns out to be true: for any $k \geq 2$ define random variables $(f, X^{(k)})$ by the formula

$$(f, X^{(k)}) := \sum_i f_i (\Delta X_i)^k$$

and define again a semi-norm for such random variables by putting

$$\|(f, X^{(k)})\|_{G^k} := \left(E \int_0^T f_s^2 ds \right)^{1/2}.$$

Again, if the process X is a fractional Brownian motion B^H, then $\|(f, (B^H)^{(k)})\|_{G^k}$ is a norm. Denote by $L_2^{pr}(P \otimes \lambda)$ the space of predictable process f with the property $E \int_0^T f_s^2 ds < \infty$. Now, let $f \in L_2^{pr}(P \otimes \lambda)$ be a predictable process and $f^{(n)}$ a sequence of simple predictable processes such that

$$\left\|f^{(n)} - f\right\|_{L^2(P \otimes \lambda)} \to 0$$

as $n \to \infty$. Define the higher-order generalized integral $(f, (B^H)^{(k)})$ as a limit in the Banach space $(\mathcal{J}^k, \|\cdot\|_{G^k})$, which is the space of some kind of extended random variables g, which are limits of the sequences of the form $(f, (B^H)^{(k)})$ with respect the norm $\|\cdot\|_{G^k}$.

2.5.2 First- and Higher-order Integrals with Respect to X

Wiener Integrals

Further, if $(Y, \|\cdot\|_Y)$ is a complete metric space, then the Y-lim stands for the limit on the space Y with respect to the norm $\|\cdot\|_Y$. Assume that f is a simple deterministic process, $f_t = \sum_{i=1}^m f_i \mathbf{1}_{[t_{i-1}, t_i)}(t)$. Then $\|\cdot\|_G$ is a norm if and only if

$$(f, X) = \sum_{i=1}^m f_i \Delta X_i = 0 \iff f_i = 0, 1 \leq i \leq m. \qquad (2.5.2)$$

Let $X = (X_t)_{t \in [0,T]}$ be a square integrable process with $EX_t = 0, X_0 = 0$, and write $R(t,s)$ for the covariance function, $R(t,s) = E X_t X_s$. Consider the quadratic forms

$$B_m = E((f, X))^2$$

where $f \in L^s$ has deterministic coefficients $f_i, 1 \leq i \leq m$. Then condition (2.5.2) is equivalent to the following:

The quadratic form B_m is positive definite for each $m \geq 1$. \qquad (2.5.3)

We can write B_m in terms of the correlation function R:

$$B_m = \sum_{i=1}^m [f_i^2 (R(t_i, t_i) - 2R(t_{i-1}, t_i) + R(t_{i-1}, t_{i-1}))]$$
$$+ 2 \sum_{i \neq j, i, j \leq m} f_i f_j [R(t_i, t_j) - R(t_{i-1}, t_j) - R(t_i, t_{j-1}) + R(t_{i-1}, t_{j-1})].$$
$$(2.5.4)$$

Put
$$\delta_{ii} := R(t_i, t_i) - 2R(t_{i-1}, t_i) + R(t_{i-1}, t_{i-1})$$

and
$$\delta_{ij} := R(t_i, t_j) - R(t_{i-1}, t_j) - R(t_i, t_{j-1}) + R(t_{i-1}, t_{j-1}).$$

Then condition (2.5.3) is equivalent to the property that the matrix $(\delta_{ij})_{i,j \leq m}$ is positive definite for each $m \geq 1$. Assume that condition (2.5.2) is valid for the process X and assume that $f \in L_2[0, T]$. Then there exists $f^n \in L^s$ such that $\|f^n - f\|_{L_2[0,T]} \to 0$ as $n \to \infty$. Moreover, the sequence (f^n, X)

2.5 Isometric Approach to Stochastic Integration with Respect to fBm

is a Cauchy sequence in the space $(E^s, \|\cdot\|_{E^s})$, where E^s is the subspace of L^s consisting of deterministic simple functions f. Complete E^s with respect the norm $\|\cdot\|_{E^s}$ and denote this Banach space by $\overline{E}$. Define now the integral $\int_0^T f_s dX_s$ as the limit of (f^n, X) in the space $\overline{E}$. We say that $\int_0^T f_s dX_s$ is the generalized Wiener integral with respect to process X. Note that L^s in dense in $L_2[0,T]$ and hence also E^s is dense in $\overline{E}$, by using the isometry.

We clarify the connection between random variables and Wiener integrals defined above. Let ζ^n be a sequence of random variables of the form

$$\zeta^n := (f^n, X)$$

with some $f^n \in L^s$. Assume now that $\zeta = P\text{-}\lim_n \zeta^n$ and $\|f - f^n\|_{L_2[0,T]} \to 0$, $n \to \infty$. We show later that it may happen that $P\{|\zeta| < \infty\} < 1$ or even $P\{|\zeta| < \infty\} = 0$. But even in the above situation the limit

$$\int_0^T f_s dX_s = \overline{E}\text{-}\lim_n (f^n, X)$$

defines the generalized Wiener integral. In this kind of situation we say that the random variable ζ is one of the representatives of $\int_0^T f_s dX_s$ in the space of random variables and $\int_0^T f_s dX_s$ is one of the representatives of the random variable ζ in the space $\overline{E}$: write this as $\zeta \leftrightarrow \int_0^T f_s dX_s$. It is easy to check that if X is a process with non-correlated increments and with the property

$$E X_t^2 > E X_s^2 \tag{2.5.5}$$

where $s < t$, then condition (2.5.2) is satisfied. Note first that condition (2.5.5) is equivalent to the condition $E(X_t - X_s)^2 > 0$ for $s < t$. Since the process X has non-correlated increments, we have that

$$E\Big(\sum_{i=1}^m f_i \Delta X_i\Big)^2 = \sum_{i=1}^m f_i^2 E(\Delta X_i)^2 = 0$$

if and only if $f_i = 0, i \leq m$. Note that if X is a square integrable martingale and $EX_t^2 > EX_s^2, s < t$, then (2.5.2) is satisfied.

Similarly, if X is a stationary process with so-called orthogonal vector measure $\varphi(d\lambda)$ such that the spectral measure $F(d\lambda) := E|\varphi(d\lambda)|^2$ is equivalent to the Lebesgue measure, then condition (2.5.2) is satisfied.

If the process X is the standard Brownian motion B, then

$$\|(f, B)\|_{E^s} = E(f, B)^2 = \|f\|_{L_2[0,T]}$$

and then the limits of simple integrals $(f^{(n)}, B)$ in the space $\overline{E}$ and in $L_2(P)$ are the same. Similarly, if the process X is a continuous square integrable martingale M with the angle bracket $\langle M \rangle_t = \int_0^t a_s ds$, where $1/K \leq Ea_s \leq K$, the limits in the space $\overline{E}$ and $L_2(P)$ are the same.

First-order Stochastic Integrals with Respect to X

Let $\mathcal{F} := \{\mathcal{F}_t, t \in [0,T]\}$ be a filtration on $(\Omega, \mathcal{F}, P)$ satisfying the usual conditions of right continuity and completeness.

The notation $X \in \mathcal{F}$ means that X_t is $\mathcal{F}_t$ measurable. So, let $X \in \mathcal{F}$ be a process and introduce the space G^s of random variables ξ:

$$\xi = \sum_{i=1}^{m} f_i \Delta X_i$$

where $f_i \in \mathcal{F}_{t_{i-1}}$ and $f_i \in L_2(P)$, $1 \leq i \leq m$, $m \geq 1$. Let f be as above, i.e., $f \in L^s$ and the coefficients f_i, $1 \leq i \leq m$ satisfy $f_i \in \mathcal{F}_{t_{i-1}}$ and $f_i \in L_2(P)$. Then we can define a surjection $\mathcal{I}$ from $L^s \to G^s$ by

$$\mathcal{I}(f) := (f, X) = \sum_{i=1}^{m} f_i \Delta X_i.$$

Introduce the following semi-norm on G^s:

$$\|(f, X)\|_{G^s} := \left(E \sum_{i=1}^{m} f_i^2 (t_i - t_{i-1}) \right)^{1/2}. \tag{2.5.6}$$

It is easy to check that the condition

$$(f, X) = 0 \text{ P-a.s. if and only if } f_i = 0 \text{ P-a.s. for } 1 \leq i \leq m \tag{2.5.7}$$

is a necessary and a sufficient condition for $\mathcal{I}$ to be a bijection and $\|\cdot\|_{G^s}$ to be a norm.

Let X be a square integrable process, which satisfies (2.5.7). Now let f be a predictable process with $E \int_0^T f_s^2 ds < \infty$. Then there exist processes $f^n \in L^s$ such that

$$E \int_0^T (f_s - f_s^n)^2 ds \to 0$$

as $n \to \infty$. Now L^s is the space of elementary "predictable" processes g, where $g_t := \sum_{i=1}^{m} f_i \mathbf{1}_{[t_{i-1}, t_i)}(t)$, and $f_i \in \mathcal{F}_{t_{i-1}}$, $1 \leq i \leq m$. Complete again the space G^s with respect to the norm $\|\cdot\|_{G^s}$. The integral $\int_0^T f_s dX_s =: \mathcal{I}(f)$ is defined using the extension of the isometry $\mathcal{I}$ on the completed Banach space $\overline{G}$. The sequence f^n is a Cauchy sequence with respect the norm $\|\cdot\|_{\overline{G}}$ and the integral $\int_0^T f_s dX_s$ is the limit of the elementary integrals (f^n, X) in the space $(\overline{G}, \|\cdot\|_{\overline{G}})$. We say that the integral $\int_0^T f_s dX_s$ defined for predictable $f \in L_2^{pr}(P \otimes \lambda)$ is the first order generalized stochastic integral with respect to the process X. Later we will use the notation $\int_0^T f_s dX_s^{(1)}$ for this integral. If ζ^n be a sequence of random variables of the form

$$\zeta^n := (f^n, X)$$

2.5 Isometric Approach to Stochastic Integration with Respect to fBm   167

with some $f^n \in L^s$ and assume that $\zeta = P\text{-lim}_n \zeta^n$ and $\|f - f^n\|_{L_2^{pr}(P \otimes \lambda)} \to 0$, $n \to \infty$. Hence also

$$\int_0^T f_s dX_s = \overline{G}\text{-}\lim_n (f^n, X).$$

It may happen that $P\{|\zeta| < \infty\} < 1$ or even $P\{|\zeta| < \infty\} = 0$. Again the random variable ζ is one of the representatives of the integral $\int_0^T f_s dX_s^{(1)}$ in the space of random variables and $\int_0^T f_s dX_s^{(1)}$ is one of the representatives of the random variable ζ in the space $\overline{G}$: write this again as $\zeta \leftrightarrow \int_0^T f_s dX_s^{(1)}$. The first-order integral is linear: $(af + bg, X) = a(f, X) + b(g, X)$.

Higher-order Stochastic Integrals with Respect to X

Let $(X, \mathcal{F})$ be again a stochastic process defined on $(\Omega, \mathcal{F}, P)$. Introduce the space $G^{s,k}$ of the random variables ξ:

$$\xi := \sum_{i=1}^m f_i (\Delta X_i)^k$$

where $k > 1$, $f_i \in \mathcal{F}_{t_{i-1}}$, $f_i \in L_2(P)$, $1 \leq i \leq m$. If $f \in L^s$ is a predictable step function, define a surjection $\mathcal{I}^k$ from L^s to $G^{s,k}$ by putting

$$\mathcal{I}^k(f) := (f, X^{(k)}) := \sum_{i=1}^m f_i (\Delta X_i)^k.$$

We suppose that any simple function has different values on the adjoining segments of the partition. With this assumption only one partition corresponds to a simple function, we have only one zero function and $\mathcal{I}^k$ is a surjection.

Introduce the following semi-norm on $G^{s,k}$:

$$\left\|(f, X^{(k)})\right\|_{G^{s,k}} := \left(E \sum_{i=1}^m f_i^2 (t_i - t_{i-1}) \right)^{1/2} = \|f\|_{L_2(P \otimes \lambda)}.$$

Let f and g be simple predictable processes, defined with respect to different partitions π_f and π_g. Consider $f + g$ on the partition $\pi := \pi_f \cup \pi_g$, put $(f, X^{(k)}) + (g, X^{(k)}) := (f + g, X^{(k)})$ and see that

$$\left\|(f, X^{(k)}) + (g, X^{(k)})\right\|_{G^{s,k}} \leq \left\|(f, X^{(k)})\right\|_{G^{s,k}} + \left\|(g, X^{(k)})\right\|_{G^{s,k}}. \quad (2.5.8)$$

Again it is easy to check that the condition

$$(f, X^{(k)}) = 0 \; P\text{-a.s. if and only if } f_i = 0 \text{ for } 1 \leq i \leq m,$$

$$\text{when } f \in L^s, \; f = \sum_{i=1}^m f_i \mathbf{1}_{[t_{i-1}, t_i)}(\cdot) \quad (2.5.9)$$

is a necessary and sufficient condition for $\mathcal{I}^k$ to be a bijection, for $G^{s,k}$ to be a linear space and for $\|\cdot\|_{G^{s,k}}$ to be a norm.

If f is a predictable process from $L_2^{pr}(P \otimes \lambda)$, take $f^n \in L^s$ such that $\|f - f^n\|_{L_2(P \otimes \lambda)} \to 0$. Assume that property (2.5.9) holds for the process X with some $k > 1$. Define the integral $\int_0^T f_s dX_s^{(k)} := \mathcal{I}^k(f)$ as the limit of $(f^n, X^{(k)})$ in the completed Banach space $(\overline{G}^k, \|\cdot\|_{\overline{G}^k})$, where $\overline{G}^k$ is the completion of $G^{s,k}$ with respect to norm $\|\cdot\|_{G^{s,k}}$. We say that such an integral $\int_0^T f_s dX_s^{(k)}$ is the kth order generalized stochastic integral of f with respect to the process X.

Assume now that property (2.5.9) holds for all $k \leq N$. Define the Banach space G^N by

$$G^N := \overline{G}^1 \times \overline{G}^2 \times \cdots \times \overline{G}^N$$

and define the norm in G^N by

$$\|\cdot\|_{G^N} := \sum_{k=1}^N \|\cdot\|_{\overline{G}^k}.$$

In view of (2.5.8), $\|\cdot\|_{G^N}$ satisfies the triangle inequality and hence it is really a norm.

The elements $g \in \overline{G}^N$ have the form

$$g = \sum_{k=1}^N \int_0^T f_k(s) dX_s^{(k)}$$

where f_k is a predictable process from $L_2(P \otimes \lambda)$. Note also that there is a bijection between such a g from $\overline{G}^N$ and $(f_1, \ldots, f_N) \in \otimes_{k=1}^N L_2^{pr}(P \otimes \lambda)$ equipped with the norm $\sum_{k=1}^N \|f_k\|_{L_2(P \otimes \lambda)}$.

The following examples clarify the definition of the generalized integrals of higher order. We assume that the process X satisfies property (2.5.9) for each $1 \leq k \leq N$ below.

Processes with bounded variation. Assume that the process X is a continuous process with bounded variation and consider the random variables X_T^m, where

$$X_T^m := \sum_{l=1}^N \sum_{k=1}^m (\Delta X_k)^l.$$

When $|\pi| \to 0$ we have that $X_T^m \xrightarrow{P} X_T$ and the right-hand side converges in the space $\overline{G}^N$ towards the element

$$\sum_{l=1}^N \int_0^T dX_s^{(l)}.$$

2.5 Isometric Approach to Stochastic Integration with Respect to fBm

Here the random variable X_T is a representative of the integral $\int_0^T dX_s^{(1)}$ and zero is a representative of the sum $\sum_{l=2}^{N} \int_0^T dX_s^{(l)}$.

Standard Brownian Motion. Assume that X is a standard Brownian motion, $X = B$. Define again the random variable X_T^m by

$$X_T^m := \sum_{l=1}^{N} \sum_{k=1}^{m} (\Delta B_k)^l.$$

Now, when $|\pi| \to 0$, $X_T^m \xrightarrow{P} B_T + T$, so the constant T is a representative of the integral $\int_0^T dB_s^{(2)}$ and zero is a representative of the sum $\sum_{l=3}^{N} \int_0^T dB_s^{(l)}$.

2.5.3 Generalized Integrals with Respect to fBm

Fractional Brownian Motion and Property (2.5.7)

Theorem 2.5.1. *Property* (2.5.7) *holds for fBm* $B^H, H \in (0,1)$.

Proof. Assume that $\sum_{i \leq m} f_i \Delta B_i^H = 0$ almost surely. Assume that m_0 is the largest index for which $P\{f_{m_0} \neq 0\} > 0$. Then from presentations (1.8.17)–(1.8.18) we have

$$\Delta B_{m_0}^H = \int_{t_{m_0-1}}^{t_{m_0}} m_H(t_{m_0}, s) dW_s + \int_0^{t_{m_0-1}} (m_H(t_{m_0}, s) - m_H(t_{m_0-1}, s)) dW_s$$

$$= A_{m_0} + B_{m_0},$$

For the term B_{m_0} we have $B_{m_0} \in \mathcal{F}_{t_{m_0-1}}$. Put $\Omega_c := \{\omega : |f_i| \leq c, i \leq m_0\}$. Then $\Omega_c \in \mathcal{F}_{t_{m_0-1}}$ and

$$\sum_{i=1}^{m_0} \mathbf{1}_{\Omega_c} f_i \Delta B_i^H = \sum_{i=1}^{m} \mathbf{1}_{\Omega_c} f_i \Delta B_i^H = 0.$$

Hence we can conclude the following:

$$0 = E\Big(\sum_{i=1}^{m_0} \mathbf{1}_{\Omega_c} f_i \Delta B_i^H\Big)^2$$

$$= E\Big(\Big(\sum_{i \leq m_0-1} \mathbf{1}_{\Omega_c} f_i \Delta B_i^H\Big) + f_{m_0} \mathbf{1}_{\Omega_c} B_{m_0-1} + f_{m_0} A_{m_0}\Big)^2. \quad (2.5.10)$$

The right-hand side of (2.5.10) is equal to

$$E\Big(\sum_{i\le m_0-1}(f_i\Delta B_i^H \mathbf{1}_{\Omega_c}) + f_{m_0}\mathbf{1}_{\Omega_c}B_{m_0-1}\Big)^2$$
$$+ E\Big(f_{m_0}^2 \mathbf{1}_{\Omega_c}\int_{t_{m_0-1}}^{t_{m_0}}(B^H(t_{m_0},s))^2 ds\Big).$$

Hence, from (2.5.10), since

$$\int_{t_{m_0-1}}^{t_{m_0}}(B^H(t_{m_0},s))^2 ds > 0$$

we have that $f_{m_0}\mathbf{1}_{\Omega_c} = 0$ almost surely for any $c > 0$ and so $f_{m_0} = 0$ P-a.s. This shows that condition (2.5.7) is fulfilled. Hence $f_i = 0$ for all $i \le m$. □

Fractional Brownian Motions and Property (2.5.9)

Theorem 2.5.2. *Property (2.5.9) holds for fBm B^H, $H \in (0,1)$.*

Proof. We know from Theorem 2.5.1 that the claim holds for $k = 1$. Assume now that $k > 1$ and let m_0, A_{m_0}, B_{m_0} and W be as in the proof of Theorem 2.5.1. Put $f_i^c := \mathbf{1}_{\Omega_c} f_i$. Note that $f_i^c \in \mathcal{F}_{t_{m_0-1}}$ for $i \le m_0$. Denote by χ the random variable

$$\chi := \sum_{i=1}^{m_0-1} f_i^c (\Delta B_i^H)^k.$$

For the random variable χ we have that $\chi \in \mathcal{F}_{t_{m_0-1}}$, and this fact is used below. Assume that $\sum_{i\le m}f_i(\Delta B_i^H)^k = 0$. With the above notation we have from this assumption that also

$$\chi + f_{m_0}^c \sum_{r=0}^{k}\binom{k}{r}(B_{m_0})^{k-r}(A_{m_0})^r = 0. \qquad (2.5.11)$$

Write the expression in (2.5.11) as

$$\Big(\chi + f_{m_0}^c \sum_{0\le r\le k,\ r\text{ even}}\binom{k}{r}(B_{m_0})^{k-r}(A_{m_0})^r\Big)$$
$$+ \Big(f_{m_0}^c \sum_{0\le r\le k,\ r\text{ odd}}\binom{k}{r}(B_{m_0})^{k-r}(A_{m_0})^r\Big) =: \chi_1 + \chi_2. \qquad (2.5.12)$$

The random variable A_{m_0} is a Gaussian random variable with zero expectation and hence for odd r $E(A_{m_0})^r = 0$ and by conditioning on $\mathcal{F}_{t_{m_0-1}}$ in (2.5.12) it is easy to see that $E(\chi_1\chi_2) = 0$. So from this we can conclude that $E\chi_2^2 = 0$, using also (2.5.11) and (2.5.12). But

2.5 Isometric Approach to Stochastic Integration with Respect to fBm

$$\chi_2^2 = f_{m_0}^2(\gamma_1 + \gamma_2)$$

with

$$\gamma_1 := \sum_{0 \le r \le k,\ r\ \text{odd}} \left(\binom{k}{r}(B_{m_0})^{k-r}(A_{m_0})^r\right)^2 \quad (2.5.13)$$

and

$$\gamma_2 := \sum_{r \ne q,\ r,q\ \text{odd}} \binom{k}{r}\binom{k}{q}(B_{m_0})^{2k-r-q}(A_{m_0})^{r+q}. \quad (2.5.14)$$

All the terms in (2.5.13) are nonnegative and since $r+q$ is even, the same holds for the expression (2.5.14), too. Note also that if $r=1$, then

$$k^2(B_{m_0})^{2k-2}(A_{m_0})^2 > 0$$

almost surely. But at the same time $E(f_{m_0}^2(\gamma_1 + \gamma_2)) = 0$. Hence $f_{m_0} = 0$ almost surely. From this follows that $f_i = 0$ almost surely for all $i \le m$. We have shown that fBm B^H satisfies property (2.5.9) for all $k \ge 1$. □

Some Properties of the Generalized Integrals

In this subsection we discuss some of the properties of the generalized integrals. At this stage we have results mostly on Wiener integrals.

Assume that B^H is again an fBm with index H. Take

$$f_s^n := n^\gamma \mathbf{1}_{(T/2-1/2n,\, T/2+1/2n]}(s).$$

Then $\|f^n\|_{L_2[0,T]}^2 = n^{2\gamma-1}$. If $H \in (1/2, 1)$, $1/2 < \gamma < H$, then $\|f^n\|_{L_2[0,T]} \to \infty$ and the generalized integral does not exist, but $E((f^n, B^H))^2 = n^{2\gamma - 2H} \to 0$, and the limit exists in $L_2(P)$. If $H < \gamma < 1/2$, then $E((f^n, B^H))^2 \to \infty$, but $\|f^n\|_{L_2[0,T]} \to 0$. Hence the integral exists in $\overline{G}$ and it is $= 0$, but the limit does not exist in $L_2(P)$. Note also that here we have that $|(f^n, B^H)| \xrightarrow{P} \infty$.

L_2-integrals and Wiener integrals, $H \in (1/2, 1)$. If B^H is an fBm with Hurst index $H \in (1/2, 1)$, then according to (1.9.2) we have the following estimate for L_2-integral, valid for any $p > 0$:

$$E \left| \int_0^T f_s dB_s^H \right|^p \le c_{H,p} \|f\|_{L_{\frac{1}{H}}[0,T]}^p. \quad (2.5.15)$$

Hence, if $(f^{(n)}, B^H)$ converges in $\overline{G}$, it also converges in $L_2(P)$.

L_2-integrals and Wiener integrals, $H \in (0, 1/2)$. Before the continuation, we prove the following theorem, which is the opposite to (2.5.15).

Theorem 2.5.3. *Let $f \in L^s$ and B^H is an fBm with Hurst index $H \in (0, 1/2)$. Then*

$$E \left| \int_0^T f_s dB_s^H \right|^2 \ge C \|f\|_{L_2[0,T]}^2. \quad (2.5.16)$$

172   2 Stochastic Integration with Respect to fBm and Related Topics

Proof. If $f \in L^s$ and $(f, B^H) = \sum_i f_i \Delta B_i^H$, then

$$E(f, B^H)^2 = \sum_i (f_i^2 E \Delta B_i^H)^2 + \sum_{i \neq k} f_i f_k E(\Delta B_i^H \Delta B_k^H). \qquad (2.5.17)$$

But $E(\Delta B_i^H \Delta B_k^H) < 0$ and hence

$$f_i f_k E(\Delta B_i^H \Delta B_k^H) \geq |f_i||f_k| E(\Delta B_i^H \Delta B_k^H).$$

Use this in (2.5.17) to obtain the inequality

$$E(f, B^H)^2 \geq E \left(\sum_i |f_i| \Delta B_i^H \right)^2.$$

Hence we can assume that $f_i \geq 0$ for all $i \leq n$ in proving (2.5.16).

Denote by $\mathcal{D}(\mathbb{R})$ the space of functions f with the two properties: $f \in C^\infty(\mathbb{R})$ and f has compact support.

Let $\phi \in \mathcal{D}(\mathbb{R})$. Then the Fourier transform $\hat{\phi}$ of ϕ belongs to $S(\mathbb{R}) \subset \mathcal{F}_H \subset L_2^H(\mathbb{R})$ (see Lemma 1.6.8), and moreover,

$$E \left| \int_\mathbb{R} \phi_t dB_t^H \right|^2 = E \left| \int_\mathbb{R} \phi'(t) B_t^H dt \right|^2 = c_H \int_\mathbb{R} |\hat{\phi}(\lambda)| |\lambda|^{-2\alpha} d\lambda, \qquad (2.5.18)$$

where c_H is some constant.

We want to prove that there exists a sequence $(\phi^n)_{n \geq 1}, \phi^n \in \mathcal{D}(\mathbb{R})$ such that

$$\int_\mathbb{R} (\phi^n)'(t) B_t^H dt \xrightarrow{L_2(P)} (f, B^H). \qquad (2.5.19)$$

To prove (2.5.19) it is sufficient to prove it for $f \in L^s$, $f_u = a\mathbf{1}_{[s,t)}(u)$, $s < t \leq T$ and $a > 0$. Take $\phi^n \in \mathcal{D}(\mathbb{R})$ such that $\operatorname{supp}(\phi^n) \subset [s - 1/n, t + 1/n]$ and $\phi^n = a$ on $[s + 1/n, t - 1/n]$. Then

$$\int_\mathbb{R} (\phi^n)'(u) B_u^H du = \int_{t-1/n}^{t+1/n} (\phi^n)'(u) B_u^H du + \int_{s-1/n}^{s+1/n} (\phi^n)'(u) B_u^H du$$

and, for example,

$$\left| aB_{t+1/n}^H - \int_{t-1/n}^{t+1/n} (\phi^n)'(u) B_u^H du \right| \leq \left| \int_{t-1/n}^{t+1/n} (\phi^n)'(u)(B_{t+1/n}^H - B_u^H) du \right|$$

$$\leq a \sup_{u \in [t-1/n, t+1/n]} |B_{t+1/n}^H - B_u^H|.$$

From self-similarity of B^H and Remark 1.10.7 with $f = 1, T = 2/n$

$$\sup_{u \in [t-1/n, t+1/n]} |B_{t+1/n}^H - B_u^H| \xrightarrow{L_2(P)} 0$$

2.5 Isometric Approach to Stochastic Integration with Respect to fBm   173

and so
$$E(f, B^H)^2 = \lim_n \int_{\mathbb{R}} |\widehat{\phi}^n(\lambda)| |\lambda|^{-2\alpha} d\lambda.$$

Since for any $\lambda \in \mathbb{R}$ $\widehat{f}(\lambda) = \lim_{n \to \infty} \widehat{\phi}^n(\lambda)$, we have, using the Fatou lemma and relation (2.5.18),

$$\int_{-\infty}^{\infty} |\widehat{f}(\lambda)|^2 |\lambda|^{-2\alpha} d\lambda \le \liminf_{n \to \infty} \int_{-\infty}^{\infty} |\widehat{\phi}^n(\lambda)|^2 |\lambda|^{-2\alpha} d\lambda = E \left| \sum_i f_i \Delta B_i^H \right|^2.$$

We have that

$$\int_{-\infty}^{\infty} |\widehat{f}(\lambda)|^2 |\lambda|^{-2\alpha} d\lambda \ge \varepsilon^{-2\alpha} \int_{|\lambda| > \varepsilon} |\widehat{f}(\lambda)|^2 d\lambda + \int_{|\lambda| \le \varepsilon} |\widehat{f}(\lambda)|^2 |\lambda|^{-2\alpha} d\lambda. \quad (2.5.20)$$

Put $\rho(\lambda) := |\lambda|^{-\alpha} \mathbf{1}_{[-\varepsilon, \varepsilon]}(\lambda)$. Since $H \in (0, 1/2)$, we have that $\rho \in L_1(\mathbb{R})$. Also,

$$\widehat{\rho}(t) := \int_{-\infty}^{\infty} e^{it\lambda} \rho(\lambda) d\lambda = \int_{-\varepsilon}^{\varepsilon} \cos(t\lambda) |\lambda|^{-\alpha} d\lambda.$$

This integral is finite and hence $\rho(\cdot)$ is the Fourier transform of $\widehat{\rho}(\cdot)$. Use the Parceval identity to obtain

$$\int_{|\lambda| < \varepsilon} |\widehat{f}(\lambda)|^2 |\lambda|^{-2\alpha} d\lambda$$
$$= \int_{\mathbb{R}} \left| \int_{\mathbb{R}} f(s) \left(\int_{-\varepsilon}^{\varepsilon} \cos((t-s)\lambda) |\lambda|^{-\alpha} d\lambda \right) ds \right|^2 dt. \quad (2.5.21)$$

Estimate the right-hand side of (2.5.21) from below by

$$\int_{-1}^{1} \left| \int_{0}^{T} f(s) \left(\int_{-\varepsilon}^{\varepsilon} \cos((t-s)\lambda) |\lambda|^{-\alpha} d\lambda \right) ds \right|^2 dt. \quad (2.5.22)$$

Take in (2.5.22) such an ε that $\varepsilon(T+1) \le \pi/3$. Then $\cos((t-s)\lambda) \ge 1/2$ and the left-hand side of inequality (2.5.21) can be estimated from below, using the estimate (2.5.22) and the chosen ε by the expression

$$\frac{1}{2} \left| \int_{0}^{T} f(s) ds \right|^2 \left(\int_{-\varepsilon}^{\varepsilon} |\lambda|^{-\alpha} d\lambda \right)^2 = \frac{2\varepsilon^{2-2\alpha}}{(1-\alpha)^2} |\widehat{f}(0)|^2,$$

but since f is nonnegative, we also have the estimate $|\widehat{f}(0)| \ge |\widehat{f}(\lambda)|$. Therefore, from the above estimates we obtain

$$\int_{|\lambda| \le \varepsilon} |\widehat{f}(\lambda)|^2 |\lambda|^{-2\alpha} d\lambda \ge \frac{\varepsilon^{1-2\alpha}}{(1-\alpha)^2} \int_{-\varepsilon}^{\varepsilon} |\widehat{f}(\lambda)|^2 d\lambda. \quad (2.5.23)$$

Take $C = \min\{\varepsilon^{-2\alpha}, \varepsilon^{1-2\alpha}/(1-\alpha)^2\}$ and use (2.5.23) in (2.5.20) to obtain

$$\int_{\mathbb{R}} |\widehat{f}(\lambda)|^2 |\lambda|^{1-2H} d\lambda \geq C \int_{\mathbb{R}} |\widehat{f}(\lambda)|^2 d\lambda = C_1 \|f\|_{L_2[0,T]}^2.$$

□

Random variables and the corresponding integrals. Assume first that $H \in (1/2, 1)$. Let $f^n \in L^s$ be such that $f = L_2(P)\text{-}\lim f^n$. Put $\zeta^n := (f^n, B^H)$ and assume that $\zeta := L_2(P)\text{-}\lim \zeta^n$. Let $g^n \in L^s$ be another sequence such that $\zeta = L_2(P)\text{-}\lim(g^n, B^H)$. Use the beginning of this subsection to conclude that the corresponding integral may not exist, and hence the representative of the random variable ζ need not to be unique in the space $\overline{E}$. On the other hand, it follows from inequality (2.5.15) that the integral $\int_0^T f_s dB_s^H$ has only one random variable as a representative.

If $H \in (0, 1/2)$ then the picture is the opposite. Namely, a random variable ζ can represent only one Wiener integral; this follows from Theorem 2.5.3. On the other hand, the zero Wiener integral has at least two representatives as extended random variables, namely $\zeta = 0$ and $\zeta = \infty$; this follows again from the beginning of this subsection.

2.6 Stochastic Fubini Theorem for Stochastic Integrals w.r.t. Fractional Brownian Motion

In this section we prove the generalization of stochastic Fubini theorem for the Wiener integrals with respect to fBm (Theorem 1.13.1). First, we consider pathwise integrals and the result is for the most part based on Hölder properties of fBm and of corresponding integrals. Then, the extension to Wick and Skorohod integration is more or less evident, due to comparison results of Sections 2.3 and 2.4.

Definition 2.6.1. *The nonrandom function $f : \mathbb{R} \to \mathbb{R}$ is called piecewise Hölder of order α on the interval $[T_1, T_2] \subset \mathbb{R}$ ($f \in C_{pw}^\alpha[T_1, T_2]$), if there exists a finite set of disjoint subintervals $\{[a_i, b_i), 1 \leq i \leq N \mid \bigcup_{i=1}^N [a_i, b_i] \cup T_2 = [T_1, T_2]\}$ and the function $f \in C^\alpha[a_i, b_i)$ for $1 \leq i \leq N$.*

As before, we denote

$$\|f\|_{C^\alpha[a_i, b_i)} := \sup_{a_i \leq t < b_i} |f(t)| + \sup_{a_i \leq s < t < b_i} \frac{|f(t) - f(s)|}{|t-s|^\alpha}.$$

Definition 2.6.2. *For $f \in C_{pw}^\alpha[T_1, T_2]$, let*

$$\|f\|_{C_{pw}^\alpha[T_1, T_2]} = \max_{1 \leq i \leq N} \|f\|_{C^\alpha[a_i, b_i)}.$$

2.6 Stochastic Fubini Theorem for Stochastic Integrals w.r.t. fBm

Let $f \in C^\alpha[a,b]$, $g \in C^\beta[a,b]$ with $\alpha + \beta > 1$. Then we know that the Riemann–Stieltjes integral exists,

$$\int_a^b f(t)dg(t) := \lim_{|\pi_n| \to 0} \sum_{k=0}^{k_n - 1} f(t_k^n) \Delta g(t_k^n), \qquad (2.6.1)$$

where $\pi_n = \{a = t_k^0 < t_k^1 < \cdots < t_k^{k_n} = b\}$, $\Delta g(t_k^n) = g(t_{k+1}^n) - g(t_k^n)$, $\pi_n \subset \pi_{n+1}$.

Moreover, according to (FdP01, Theorem 2.1), there exist the sequences $\{f_n, g_n\} \subset C^{(1)}[a,b]$ such that $\|f_n - f\|_{C^\alpha[a,b]} \to 0$, $n \to \infty$, $\|g_n - g\|_{C^\beta[a,b]} \to 0$, $n \to \infty$.

We shall use some bounds for integrals involving Hölder functions. They are proved in the next lemma.

Lemma 2.6.3. *Let $f \in C^\alpha[a,b]$, $g \in C^\beta[a,b]$, $\alpha + \beta > 1$, $f_m, g_m \in C^1[a,b]$, $m \geq 1$ and $\|f_m - f\|_{C^\alpha[a,b]} \to 0$, $\|g_m - g\|_{C^\beta[a,b]} \to 0$, as $m \to \infty$.*
Then 1) $\int_a^b f(t)dg(t) = \lim_{m \to \infty} \int_a^b f_m(t) g'_m(t) dt$;
2) the following estimate holds:

$$\left| \int_a^b f(t)dg(t) \right| \leq C \|f\|_{C^\alpha[a,b]} \cdot \|g\|_{C^\beta[a,b]} \cdot ((b-a)^{1+\varepsilon} \vee (b-a)^\beta);$$

3) if $f(a) = 0$, then

$$\left| \int_a^b f(t)dg(t) \right| \leq C \|f\|_{C^\alpha[a,b]} \cdot \|g\|_{C^\beta[a,b]} \cdot (b-a)^{1+\varepsilon}, \qquad (2.6.2)$$

where $0 < \varepsilon < \alpha + \beta - 1$, $C > 0$ is a constant not depending on α and β.

Proof. 1) Evidently,

$$\left| \int_a^b f(t)dg(t) - \int_a^b f_m(t) g'_m(t) dt \right| \leq \left| \int_a^b f(t)dg(t) - \sum_{k=1}^{k_n} f(t_k^n) \Delta g(t_k^n) \right|$$

$$+ \left| \int_a^b f_m(t) g'_m(t) dt - \sum_{k=1}^{k_n} f_m(t_k^n) \Delta g_m(t_k^n) \right|$$

$$+ \left| \sum_{k=1}^{k_n} f(t_k^n) \Delta g(t_k^n) - \sum_{k=1}^{k_n} f_m(t_k^n) \Delta g_m(t_k^n) \right|.$$

According to (2.6.1), for any fixed $\delta > 0$ we can choose π_n in such a way that

$$\left| \int_a^b f(t)dg(t) - \sum_{k=1}^{k_n} f(t_k^n) \Delta g(t_k^n) \right| < \delta. \qquad (2.6.3)$$

Further, according to (FdP01, Corollary 20),

$$\left| \int_a^b f_m(t) g'_m(t) dt - \sum_{k=1}^{k_n} f_m(t_k^n) \Delta g_m(t_k^n) \right| \leq C |\pi_n|^\varepsilon \cdot \|f_m\|_{C^{\alpha'}[a,b]} \cdot \|g_m\|_{C^{\beta'}[a,b]},\quad (2.6.4)$$

where $0 < \alpha' < \alpha$, $0 < \beta' < \beta$, and $\alpha' + \beta' = 1 + \varepsilon$. If $\|f_n - f\|_{C^\alpha[a,b]} \to 0$, $m \to \infty$, then $\|f_m - f\|_{C^{\alpha'}[a,b]} \to 0$, $m \to \infty$ for $0 < \alpha' < \alpha$, and $\|f_m\|_{C^{\alpha'}[a,b]} \leq C_1$, where C_1 does not depend on $m \geq 1$. Similarly, $\|g_m\|_{C^{\beta'}[a,b]} \leq C_2$. From these bounds and from (2.6.4) we obtain that

$$\left| \int_a^b f_m(t) g'_m(t) dt - \sum_{k=1}^{k_n} f_m(t_k^n) \Delta g_m(t_k^n) \right| \leq C_3 |\pi_n|^\varepsilon. \quad (2.6.5)$$

Choose such n that (2.6.3) holds and also $C_3 |\pi_n|^\varepsilon < \delta$; then for such fixed n we can choose such m that

$$\left| \sum_{k=1}^{k_n} f(t_k^n) \Delta g(t_k^n) - \sum_{k=1}^{k_n} f_m(t_k^n) \Delta g_m(t_k^n) \right| < \delta. \quad (2.6.6)$$

It is possible since $\sup_{t \in [a,b]} |g_m(t) - g(t)| \leq \|g_m - g\|_{C^{\beta'}[a,b]} \to 0$, and the same is true for f_m.

The proof of the first statement follows now from (2.6.3)–(2.6.6).

The third statement follows from 1) and (FdP01, Lemma 19), which states that the bound (2.6.2) holds for any $f \in C_0^{(1)}[a,b]$ (it means that $f \in C^{(1)}[a,b]$ and $f(a) = 0$) and $g \in C^{(1)}[a,b]$.

The second statement follows from 1) and (FdP01, Theorem 22). Indeed, according to 3)

$$\left| \int_a^b \big(f(t) - f(0)\big) dg(t) \right| \leq C \|f\|_{C^\alpha[a,b]} \cdot \|g\|_{C^\beta[a,b]} \cdot (b-a)^{1+\varepsilon},$$

whence

$$\left| \int_a^b f(t) dg(t) \right| \leq C \|f\|_{C^\alpha[a,b]} \cdot \|g\|_{C^\beta[a,b]} \cdot ((b-a)^{1+\varepsilon} \vee (b-a)^\beta).$$

$\square$

Further we consider $H \in (\frac{1}{2}, 1)$. Let $f \in C_{pw}^\beta[a,b]$ with $\beta > 1 - H$. In this case the sum $\sum_{i=1}^N \int_{a_i}^{b_i} f(t) dB_t^H$ exists. The next result means that this sum can be represented as a unique integral.

Lemma 2.6.4. *Let f be piecewise Hölder of order $\beta > 1 - H$ on the interval $[a, b]$. Then there exists the Riemann–Stieltjes integral*

2.6 Stochastic Fubini Theorem for Stochastic Integrals w.r.t. fBm

$$\int_a^b f(u)dB_u^H = \sum_{i=1}^N \int_{a_i}^{b_i} f(u)dB_u^H$$

and for an arbitrary sequence π_n of partitions of $[a,b]$ it can be represented as a limit

$$\int_a^b f(u)dB_u^H = \lim_{|\pi_n| \to 0} \sum_{k=1}^{k_n} f(u_k^n)\Delta B_{u_k^n}^H.$$

(We suppose that $\bigcup_{i=1}^N [a_i, b_i) = [a,b)$, $[a_i, b_i)$ are disjoint and $f \in C^\alpha[a_i, b_i)$).

Proof. Put $\pi_n^i := [a_i, b_i) \cap \pi_n$. Evidently, $|\pi_n^i| \leq |\pi_n|$. It follows from boundedness of f and continuity of B^H that

$$\sum_{j: u_j^n \in \pi_n^i} f(u_j^n)\Delta B_{u_j^n}^H \to \int_{a_i}^{b_i} f(u)dB_u^H,$$

even in the case when π_n^i does not contain a_i or(and) b_i.

Therefore, $\sum_{k: u_k^n \in \pi_n} f(u_k^n)\Delta B_{u_k^n}^H = \sum_{i=1}^N \sum_{k: u_k^n \in \pi_n^i} f(u_k^n)\Delta B_{u_k^n}^H$
$\to \sum_{i=1}^N \int_{a_i}^{b_i} f(u)dB_u^H = \int_a^b f(u)dB_u^H$, as $|\pi_n| \to 0$. □

Let $0 < T_1 < T_2$, $\Phi = \Phi(t, u, \omega) : \mathcal{P}_T := [T_1, T_2]^2 \times \Omega \to \mathbb{R}$ be the random function measurable in all the variables.

Theorem 2.6.5. *Let there exist the set $\Omega' \subset \Omega$ such that $P(\Omega') = 1$ and let for any $\omega \in \Omega'$ the function $\Phi(s, u, \omega)$ satisfy the conditions:*

1) $\forall s \in (T_1, T_2)$ $\Phi(t, \cdot, \omega)$ is piecewise Hölder of order $\beta > 1 - H$ in $u \in [T_1, T_2]$, and there exists $C = C(\omega) > 0$ such that $\|\Phi(t, \cdot, \omega)\|_{C_{pw}^\beta[T_1, T_2]} \leq C$;

2) the function $\int_{T_1}^{T_2} \Phi(t, u, \omega)dB_u^H$ is Riemann integrable in the interval $[T_1, T_2]$.

Then there exist the repeated integrals

$$I_1 := \int_{T_1}^{T_2}\left(\int_{T_1}^{T_2} \Phi(t,u,\omega)dB_u^H\right)dt \quad \text{and} \quad I_2 := \int_{T_1}^{T_2}\left(\int_{T_1}^{T_2} \Phi(t,u,\omega)dt\right)dB_u^H,$$

and $I_1 = I_2$ P-a.s.

Proof. We fix $\omega \in \Omega'$ and omit ω throughout the proof. The integral $\int_{T_1}^{T_2} \Phi(t,u)dB_u^H$ exists according to Lemma 2.6.4 and condition 1); the repeated integral I_1 exists according to condition 2). Since $\Phi(t, \cdot)$ is piecewise Hölder, then from the evident bound $\int_{T_1}^{T_2} |\Phi(t, u_1) - \Phi(t, u_2)|\,ds \leq C(T_2 - T_1)|u_1 - u_2|^\alpha$ we obtain that $\int_{T_1}^{T_2} \Phi(t,u)ds$ is piecewise Hölder of order α in $u \in [T_1, T_2]$. Further, since B^H is Hölder up to order $H > \frac{1}{2}$ and $\alpha + H > 1$, the integral I_2 also exists. The integral I_1 can be presented as a limit of integral sums,

$$I_1 = \lim_{|\pi_n|\to 0} \sum_{k=0}^{k_n-1} \int_{T_1}^{T_2} \Phi(t_k^n, u) dB_u^H \Delta t_k^n. \qquad (2.6.7)$$

For any point $t_k^n \in \pi_n$, according to condition 1), there exists a finite number of points $\{u_{1,k} < u_{2,k} < \cdots < u_{l(k),k}\}$ such that $\Phi(\cdot, u)$ is Hölder between them. Denote

$$\{T_1 = u_0 < u_1 < u_2 < \cdots < u_{L(n)} = T_2\}$$
$$:= \bigcup_{k=1}^{k_n} \{u_{1,k} < u_{2,k} < \cdots < u_{l(k),k}\} \cup \{T_1, T_2\}.$$

For any interval $[u_i, u_{i+1}]$ we consider the sequence of partitions $\pi_{i,r}, r \geq 1$ of the form

$$\pi_{i,r} := \{u_i = u_{i,r}^{(0)} < u_{i,r}^{(1)} < \cdots < u_{i,r}^{(m_r)} = u_{i+1}\}, |\pi_{i,r}| \to 0, r \to \infty.$$

Then $\tilde{\pi}_r := \bigcup_{i=0}^{L(n)-1} \pi_{i,r} \cup \{T_1, T_2\} := \{T_1 = u_r^{(0)} < \cdots < u_r^{(N_r)} = T_2\}$ is a partition of interval $[T_1, T_2]$ w.r.t. argument u, its diameter $|\tilde{\pi}_r| = \max_{1 \leq i \leq L(n)-1} |\pi|_{i,r}$, and $|\tilde{\pi}_r| \to 0, r \to \infty$.

Estimate the difference $|I_1 - I_2|$:

$$|I_1 - I_2| \leq \left| I_1 - \sum_{k=0}^{k_n-1} \sum_{j=0}^{N_r-1} \Phi(t_k^n, u_r^{(j)}) \Delta B_{u_r^{(j)}}^H \Delta t_k^n \right|$$
$$+ \left| I_2 - \sum_{j=0}^{N_r-1} \sum_{k=0}^{k_n-1} \Phi(t_k^n, u_r^{(j)}) \Delta t_k^n \Delta B_{u_r^{(j)}}^H \right| =: \Delta_1^{n,r} + \Delta_2^{n,r}. \quad (2.6.8)$$

Further,

$$\Delta_1^{n,r} \leq \left| I_1 - \sum_{k=0}^{k_n-1} \int_{T_1}^{T_2} \Phi(t_k^n, u) dB_u^H \cdot \Delta t_k^n \right|$$
$$+ \sum_{k=0}^{k_n-1} \left| \int_{T_1}^{T_2} \Phi(t_k^n, u) dB_u^H - \sum_{j=0}^{N_r-1} \Phi(t_k^n, u_r^{(j)}) \Delta B_{u_r^{(j)}}^H \right| \Delta t_k^n.$$

Since Φ is piecewise Hölder, then, according to Lemma 2.6.4,

$$\left| \int_{T_1}^{T_2} \Phi(t_k^n, u) dB_u^H - \sum_{j=0}^{N_r-1} \Phi(t_k^n, u_r^{(j)}) \Delta B_{u_r^{(j)}}^H \right| \to 0, r \to \infty.$$

According to (2.6.7), $\left| I_1 - \sum_{k=0}^{k_n-1} \int_{T_1}^{T_2} \Phi(t_k^n, u) dB_u^H \cdot \Delta t_k^n \right| \to 0, n \to \infty$.

Therefore,

2.6 Stochastic Fubini Theorem for Stochastic Integrals w.r.t. fBm

$$\lim_{n\to\infty}\lim_{r\to\infty} \Delta_1^{n,r} = 0. \qquad (2.6.9)$$

Further,

$$\Delta_2^{n,r} \leq \left| I_2 - \sum_{j=0}^{N_r-1} \int_{T_1}^{T_2} \Phi(t, u_r^{(j)}) dt \cdot \Delta B_{u_r^{(j)}}^H \right|$$

$$+ \left| \sum_{j=0}^{N_r-1} \sum_{k=0}^{k_n-1} \int_{t_k^n}^{t_{k+1}^n} \left(\Phi(t, u_r^{(j)}) - \Phi(t_k^n, u_r^{(j)}) \right) dt \cdot \Delta B_{u_r^{(j)}}^H \right|. \qquad (2.6.10)$$

The second term can be expanded as

$$\left| \sum_{k=0}^{k_n-1} \int_{t_k^n}^{t_{k+1}^n} \sum_{j=0}^{N_r-1} \left(\Phi(t, u_r^{(j)}) - \Phi(t_k^n, u_r^{(j)}) \right) \Delta B_{u_r^{(j)}}^H dt \right| \qquad (2.6.11)$$

$$= \left| \sum_{k=0}^{k_n-1} \sum_{i=0}^{L(N)-1} \int_{t_k^n}^{t_{k+1}^n} \sum_{u_r^{(j)} \in \pi_{i,r}} \left(\Phi(t, u_r^{(j)}) - \Phi(t_k^n, u_r^{(j)}) \right) \Delta B_{u_r^{(j)}}^H dt \right|.$$

Since the function $\Phi(s, u) - \Phi(t_k^n, u)$ is Hölder on any interval $[u_i, u_{i+1})$, we have that

$$\lim_{|\pi_{i,r}|\to 0} \sum_{u_r^{(j)} \in \pi_{i,r}} \left(\Phi(t, u_r^{(j)}) - \Phi(t_k^n, u_r^{(j)}) \right) \Delta B_{u_r^{(j)}}^H$$

$$= \int_{u_i}^{u_{i+1}} \left(\Phi(t, u) - \Phi(t_k^n, u) \right) dB_u^H. \qquad (2.6.12)$$

Moreover, $\forall \, 0 \leq i \leq L(n) - 1$ the sequence $f_i^r(t, t_k^n) := \sum_{u_r^{(j)} \in \pi_{i,r}} \left(\Phi(t, u_r^{(j)}) - \Phi(t_k^n, u_r^{(j)}) \right) \Delta B_{u_r^{(j)}}^H$ has the integrable dominant. Indeed, we can use the bounds from (FdP01, Corollary 20), Lemma 2.6.3, and the boundedness of Hölder norms, and obtain that

$$|f_i^r(t, t_k^n)| \leq \left| f_i^r(t, t_k^n) - \int_{u_r^{(j)}}^{u_{r+1}^{(j)}} \left(\Phi(t, u) - \Phi(t_k^n, u) \right) dB_u^H \right|$$

$$+ \left| \int_{u_r^{(j)}}^{u_{r+1}^{(j)}} \left(\Phi(t, u) - \Phi(t_k^n, u) \right) dB_u^H \right|$$

$$\leq C |\pi_{i,r}|^\varepsilon \cdot \| \Phi(t, \cdot) - \Phi(t_k^n, \cdot) \|_{C[u_r^{(j)}, u_{r+1}^{(j)}]^{\beta'}} \cdot \| B^H \|_{C[u_r^{(j)}, u_{r+1}^{(j)}]^{H'}}$$

$$+ \left| \int_{u_r^{(j)}}^{u_{r+1}^{(j)}} \left(\Phi(t,u) - \Phi(t_k^n, u) \right) dB_u^H \right|$$

$$\leq C + \left| \int_{u_r^{(j)}}^{u_{r+1}^{(j)}} \left(\Phi(t,u) - \Phi(t_k^n, u) \right) dB_u^H \right|, \qquad (2.6.13)$$

where $\beta' < \beta$, $H' < H$ and $\beta' + H' > 1$.

Using the second statement of Lemma 2.6.3 and condition 1) of this theorem, we obtain the bound

$$\left| \int_{u_r^{(j)}}^{u_{r+1}^{(j)}} \left(\Phi(t,u) - \Phi(t_k^n, u) \right) dB_u^H \right|$$

$$\leq C \left\| \Phi(t, \cdot) - \Phi(t_k^n, \cdot) \right\|_{C_{pw}^{\alpha'}[T_1, T_2]} \cdot \left\| B^H \right\|_{C^{H'}[T_1, T_2]} \leq C. \qquad (2.6.14)$$

Estimates (2.6.13) and (2.6.14) mean that we can use the Lebesgue dominant convergence theorem and obtain that

$$\lim_{r \to \infty} \int_{t_k^n}^{t_{k+1}^n} f_i^r(t, t_k^n) dt = \int_{t_k^n}^{t_{k+1}^n} \int_{u_i}^{u_{i+1}} \left(\Phi(t,u) - \Phi(t_k^n, u) \right) dB_u^H dt,$$

where the integrand $\int_{u_i}^{u_{i+1}} \left(\Phi(t,u) - \Phi(t_k^n, u) \right) dB_u^H$ is measurable and bounded in t.

Therefore,

$$\lim_{r \to \infty} \sum_{k=0}^{k_n-1} \sum_{i=0}^{L(n)-1} \int_{t_k^n}^{t_{k+1}^n} \sum_{u_r^{(j)} \in \pi_{i,r}} \left(\Phi(t, u_r^{(j)}) - \Phi(t_k^n, u_r^{(j)}) \right) \Delta B_{u_r^{(j)}}^H dt$$

$$= \sum_{k=0}^{k_n-1} \int_{t_k^n}^{t_{k+1}^n} \int_{T_1}^{T_2} \left(\Phi(t,u) - \Phi(t_k^n, u) \right) dB_u^H dt$$

$$= \int_{T_1}^{T_2} \left(\int_{T_1}^{T_2} \Phi(t,u) dB_u^H \right) dt - \sum_{k=0}^{k_n-1} \int_{T_1}^{T_2} \Phi(t_k^n, u) dB_u^H \Delta t_k^n. \qquad (2.6.15)$$

According to condition 2) of this theorem, the integral $\int_{T_1}^{T_2} \Phi(t,u) dB_u^H$ is Riemann integrable in t, therefore

$$\lim_{n \to \infty} \sum_{k=0}^{k_n-1} \int_{T_1}^{T_2} \Phi(t_k^n, u) dB_u^H \Delta t_k^n = \int_{T_1}^{T_2} \left(\int_{T_1}^{T_2} \Phi(t,u) dB_u^H \right) dt. \qquad (2.6.16)$$

From Lemma 2.6.4,

$$\left| I_2 - \sum_{r=0}^{L(n)-1} \int_{T_1}^{T_2} \Phi(t, u_j^{(r)}) dt \cdot \Delta B_{u_j^{(r)}}^H \right| \to 0, \text{ as } n \to \infty. \qquad (2.6.17)$$

Now the proof follows from (2.6.8)–(2.6.17). □

2.6 Stochastic Fubini Theorem for Stochastic Integrals w.r.t. fBm

Let $I(t) = \int_0^t f(s)dB_s^H$ for some stochastic process f with trajectories from $C^\beta[0,T]$ with $\beta + H > 1$. Consider the integral $\left(H \in (\frac{1}{2},1)\right)$ $J_1(t) = \int_0^t l_H(t,s)I(s)ds$ that will appear in connection with the Girsanov theorem and stochastic differential equations in subsections 2.8.2 and 3.2.3, and also, let $J_2(t) = \int_0^t f(u)\left(\int_u^t l_H(t,s)ds\right)dB_u^H$.

Lemma 2.6.6. *Both the integrals, J_1 and J_2, exist and $J_1 = J_2$ P-a.s.*

Proof. It follows from (FdP01) that the trajectories of $I(t)$, $t \in [0,T]$ are Hölder of order $H - \varepsilon$ for any $0 < \varepsilon < H$, whence the existence of $J_1(t)$ follows. Further, elementary calculations

$$\int_{u_1}^{u_2}(t-s)^{-\alpha}s^{-\alpha}ds \leq \frac{1}{2}\left[\int_{u_1}^{u_2}(t-s)^{-2\alpha}ds + \int_{u_1}^{u_2}s^{-2\alpha}ds\right] \leq (u_2 - u_1)^{1-2\alpha}$$

demonstrate that the function $f(u) \cdot \int_u^t l_H(t,s)ds$ is Hölder up to order $\beta \wedge (1-2\alpha) > 1 - H$, and $J_2(t)$ exists. We can present these integrals in the following way:

$$J_1 = \int_0^t \left(\int_0^t \Phi(s,u)dB_u^H\right)ds, \quad J_2 = \int_0^t \left(\int_0^t \Phi(s,u)ds\right)dB_u^H,$$

where $\Phi(s,u) = l_H(t,s)f(u)\mathbf{1}_{\{0 \leq u \leq s\}}$.

The function Φ will satisfy both the conditions of Theorem 2.6.5, if we put $T_1 = \delta$ and $T_2 = t - \delta$ for any $0 < \delta < \frac{t}{2}$. In particular, $\Phi(s,\cdot)$ is piecewise Hölder of order β on $[\delta, t-\delta]$ with one point $u = s$ of Hölder discontinuity for any $s \in [\delta, t-\delta]$.

Therefore, the following equality holds a.s.:

$$\int_\delta^{t-\delta} l_H(t,s)\int_\delta^s f(u)dB_u^H ds = \int_\delta^{t-\delta} f(u)\int_u^{t-\delta} l_H(t,s)ds dB_u^H.$$

The last equality can be rewritten as

$$J_1 - R_1 = J_2 - R_2, \qquad (2.6.18)$$

where

$$R_1 = \int_0^\delta l_H(t,s)\left(\int_0^s f(u)dB_u^H\right)ds + \int_\delta^{t-\delta} l_H(t,s)\left(\int_0^\delta f(u)dB_u^H\right)ds$$
$$+ \int_{t-\delta}^t l_H(t,s)\left(\int_0^s f(u)dB_u^H\right)ds =: R_{11} + R_{12} + R_{13};$$

$$R_1 = \int_0^\delta f(u)\left(\int_u^t l_H(t,s)ds\right)dB_u^H + \int_\delta^{t-\delta} f(u)\left(\int_{t-\delta}^t l_H(t,s)ds\right)dB_u^H$$
$$+ \int_{t-\delta}^t f(u)\left(\int_u^t l_H(t,s)ds\right)dB_u^H =: R_{21} + R_{22} + R_{23}.$$

According to (FdP01, Theorem 22), there exists $C > 0$ such that $\left|\int_0^s f(u)dB_u^H\right| \leq Cs^{H-\varepsilon}$ for any fixed $0 < \varepsilon < \frac{1}{2}$. Therefore,

$$|R_{11}| \leq C \int_0^\delta s^{\frac{1}{2}-\varepsilon}(t-s)^{-\alpha}ds \leq Ct^{1-\alpha}(1-\alpha)^{-1}\delta^{\frac{1}{2}-\varepsilon} \to 0 \text{ as } \delta \to 0.$$

Similarly,
$$|R_{12}| \leq C_1 \delta^{H-\varepsilon} \cdot \delta^{-\alpha} \cdot \delta^{1-\alpha} \to 0 \text{ and } |R_{13}| \leq C_2 t^{\frac{1}{2}-\varepsilon} \delta^{1-\alpha} \to 0 \text{ as } \delta \to 0,$$
where C_1 and C_2 are some constants, possibly depending on ω.

As mentioned above, the process $f(u) \cdot \int_u^t l_H(t,s)ds$ is Hölder of order $\beta \wedge (1-2\alpha) > 1-H$. Therefore, by using again (FdP01, Theorem 22), we obtain the bounds $|R_{21}| \leq C\delta^{H-\varepsilon}$, $|R_{22}| \leq C_1(t-2\delta)^{H-\varepsilon}$, and $|R_{23}| \leq C\delta^{H-\varepsilon}$ with some constants C, C_1, depending on ω. Taking in (2.6.18) a limit as $\delta \to 0$, we obtain from all these estimates that $J_1 = J_2$ a.s. □

2.7 The Itô Formula for Fractional Brownian Motion

2.7.1 The Simplest Version

First, we present a very elegant proof of the Itô formula involving fBm from (Shi01).

Lemma 2.7.1. *Let B^H be an fBm with $H \in (1/2, 1)$, $F \in C^2(\mathbb{R})$. Then for any $t > 0$*

$$F(B_t^H) = F(0) + \int_0^t F'(B_u^H)dB_u^H.$$

Proof. The Taylor formula with the reminder term in the integral form gives us

$$F(x) = F(y) + F'(y)(x-y) + \int_y^x F''(u)(x-u)du.$$

Let the sequence of partitions $\pi_n = \{0 = t_0^n < t_1^n < \cdots < t_{k_n}^n = t\}$, $|\pi_n| \to 0$, $n \to \infty$. Then $F(B_t^H) - F(0) = \sum_{k=1}^{k_n} \left[F(t_k^n) - F(t_{k-1}^n)\right]$
$= \sum_{k=1}^{k_n} F'(B_{t_{k-1}^n}^H)(B_{t_k^n}^H - B_{t_{k-1}^n}^H) + R_t^n$, where $R_t^n = \sum_{k=1}^{k_n} \int_{B_{t_{k-1}^n}^H}^{B_{t_k^n}^H} F''(u)(B_{t_k^n}^H - u)du$.
Further, $\sup_{0 \leq u \leq t} |F''(B_u^H)| < \infty$ a.s. and for $H \in (1/2, 1)$, and

$$P\text{-}\lim_{n \to \infty} \sum_{k=1}^{k_n} \left|B_{t_k^n}^H - B_{t_{k-1}^n}^H\right|^2 = 0.$$

Therefore $|R_t^n| \leq \frac{1}{2} \sup_{0 \leq u \leq t} |F''(B_u^H)| \sum_{k=1}^{k_n} \left|B_{t_k^n}^H - B_{t_{k-1}^n}^H\right|^2 \xrightarrow{P} 0$. Even if we do not know that the limit of integral sums $\sum_{k=1}^{k_n} F'(B_{t_{k-1}^n}^H)(B_{t_k^n}^H - B_{t_{k-1}^n}^H)$ exists (but we know it from Theorem 2.1.7), we can obtain this existence now and, moreover,

$$F(B_t^H) - F(0) = \int_0^t F'(B_u^H) dB_u^H.$$

□

2.7.2 Itô Formula for Linear Combination of Fractional Brownian Motions with $H_i \in [1/2, 1)$ in Terms of Pathwise Integrals and Itô Integral

Denote $C^{\beta-}[a,b] = \bigcap_{0<\gamma<\beta} C^\gamma[a,b]$.

Theorem 2.7.2. *Let the process $X_t = \sum_{i=1}^m \sigma_i B_t^{H_i}$, where $H_1 = 1/2$ and $H_i \in (1/2, 1)$ for $2 \leq i \leq m$. Let the function $F \in C^2(\mathbb{R})$. Then for any $t > 0$*

$$F(X_t) = F(0) + \sigma_1 \int_0^t F'(X_s) dW_s + \sum_{i=2}^m \sigma_i \int_0^t F'(X_s) dB_s^{H_i} + \frac{\sigma_1^2}{2} \int_0^t F''(X_s) ds.$$

Proof. Note that $\int_0^t |F'(X_s)|^2 ds < \infty$ and $\int_0^t |F''(X_s)| ds < \infty$ a.s., so, the Itô integral $\int_0^t F'(X_s) dW_s$ exists and is a local square-integrable martingale, and the Lebesgue integral $\int_0^t F''(X_s) ds$ also exists. As to integrals $\int_0^t F'(X_s) dB_s^{H_i}$ for $2 \leq i \leq m$, they exist as pathwise integrals because $X \in C^{1/2-}[0,t]$, $B^{H_i} \in C^{H_i-}[0,t]$ and $H_i + 1/2 > 1$. Further calculations are obvious: we use the Taylor formula and pass to the limit, as usual, taking into account that for any $1 \leq i \leq m$ and $2 \leq j \leq m$ $\sum_{k=1}^{k_n} \left(B_{t_k^n}^{H_i} - B_{t_{k-1}^n}^{H_i} \right) \left(B_{t_k^n}^{H_j} - B_{t_{k-1}^n}^{H_j} \right) \xrightarrow{P} 0$ as $n \to \infty$. □

Now, consider the process $Y_t = \sum_{i=1}^m \sigma_i B_t^{H_i}$, where $H_i \in (1/2, 1)$ for any $1 \leq i \leq m$. We can forecast that in this case the class $C^1(\mathbb{R})$ of functions can be used.

Theorem 2.7.3. *Let $Y_t = \sum_{i=1}^m \sigma_i B_t^{H_i}$, where $H_i \in (1/2, 1)$ for any $1 \leq i \leq m$. Let $F \in C^1(\mathbb{R})$, and $F' \in C^\beta[0,t]$ with $(\beta+1) \min H_i > 1$ for any $t > 0$. Then for any $t > 0$*

$$F(Y_t) - F(0) = \sum_{i=1}^m \sigma_i \int_0^t F'(Y_s) dB_s^{H_i}. \qquad (2.7.1)$$

Proof. Clearly, condition $(\beta + 1) \min H_i > 1$ ensures the existence of $\int_0^t F'(Y_s) dB_s^{H_i}$ as the limit of Riemann sums for any $i > 1$. Consider convolutions $F_n = F * \varphi_n$ with φ_n from Lemma 2.1.8. Then $F_n \in C^\infty(\mathbb{R})$, formula (2.7.1) holds for any F_n and for any $1 - \min H_i < \gamma < \beta \cdot \min H_i$ we have that

$D^\gamma_{0+}F'_n \to D^\gamma_{0+}F'$ in $L_1[a,b]$ as $n \to \infty$ for any $a,b \in \mathbb{R}$, which can be proved similarly to (2.1.10). Therefore,

$$\left|\int_0^t (F'(Y_s) - F'_n(Y_s))dB^{H_i}_s\right|$$

$$\leq \sup_{0\leq s\leq t}\left|D^{1-\gamma}_{t-}B^{H_i}_{t-}(s)\right| \int_{-\sup_{0\leq s\leq t}|Y_s|}^{\sup_{0\leq s\leq t}|Y_s|} \left|D^\gamma_{0+}F'_n(s) - D^\gamma_{0+}F'(s)\right|ds \to 0,$$

whence the proof follows. □

Remark 2.7.4. Theorems 2.7.2 and 2.7.3 can be extended to the functions F of several variables, depending also on t. The Itô formula has the following form: let $Y^i_t = \int_0^t f_i(s)dB^{H_i}_s$, where $H_1 = 1/2$, $H_i \in (1/2, 1)$, $2 \leq i \leq m-1$, $Y^m_t = \int_0^t g(s)ds$, $\int_0^t f_i^2(s)ds < \infty$ a.s., $f_i \in C^{\beta_i}[0,t]$ a.s. for $\beta_i + H_i > 1$, $\int_0^t |g(s)|\,ds < \infty$ a.s., $F = F(t,x) : \mathbb{R}_+ \times \mathbb{R}^n \to \mathbb{R}$, $F \in C^1(\mathbb{R}^+) \times C^2(\mathbb{R})$ $\times C^1(\mathbb{R}^{n-1})$, the integrals $\int_0^t \left(\frac{\partial F}{\partial x_1}(Z_s)f_1(s)\right)^2 ds < \infty$, $\int_0^t \left|\frac{\partial F}{\partial t}(Z_s)\right|ds < \infty$, $\int_0^t \left|\frac{\partial^2 F}{\partial x_1^2}(Z_s)\right|f_1^2(s)ds < \infty$, and $\int_0^t \left|\frac{\partial F}{\partial x_i}(Z_s)\right||g(s)|\,ds < \infty$ a.s., $\frac{\partial F}{\partial x_i}(Z_s)f_i \in C^\gamma[0,t]$ a.s. for $\gamma + H_i > 1$ and any $t > 0$, where $Z_s = (s, Y^1_s, \ldots, Y^m_s)$. Then

$$F(t, Y^1_t, \ldots, Y^m_t) = F(0) + \int_0^t \frac{\partial F}{\partial t}(Z_s)ds + \sum_{i=1}^{m-1} \int_0^t \frac{\partial F}{\partial x_i}(Z_s)f_i(s)dB^{H_i}_s$$

$$+ \int_0^t \frac{\partial F}{\partial x_m}(Z_s)g(s)ds + \frac{1}{2}\int_0^t \frac{\partial^2 F}{\partial x_1^2}(Z_s)f_1^2(s)ds. \quad (2.7.2)$$

In particular, for the process $Y_t = \int_0^t a(s)dB^H_s + \int_0^t b(s)ds$ we have that

$$F(t, Y_t) = F(0, Y_0) + \int_0^t F'_t(s, Y_s)ds + \int_0^t F'_x(s, Y_s)b(s)ds$$

$$+ \int_0^t F'_x(s, Y_s)a(s)dB^H_s, \quad H \in (1/2, 1). \quad (2.7.3)$$

2.7.3 The Itô Formula in Terms of Wick Integrals

The next result is a direct consequence of Theorems 2.3.8 and 2.7.3.

Theorem 2.7.5. *Let the function $F = F(t,x) : \mathbb{R}_+ \times \mathbb{R} \to \mathbb{R}$ be continuously differentiable in t and twice continuously differentiable in x. Let Y_t be as in Theorem 2.7.2, $E\left|\frac{\partial F}{\partial x}(t, Y_t)\right|^{2+\varepsilon} < \infty$, $t > 0$ for some $\varepsilon > 0$, $E \sup_{0\leq s\leq t}\left[\left(\frac{\partial F}{\partial x}(s, Y_s)\right)^2 + \left(\frac{\partial^2 F}{\partial x^2}(s, Y_s)\right)^2\right] < \infty$, $t > 0$. Then*

$$F(t, Y_t) - F(0,0) = \int_0^t \frac{\partial F}{\partial t}(s, Y_s) ds + \int_0^t \frac{\partial F}{\partial x}(s, Y_s) \Diamond dY_s$$

$$+ \sum_{i,k=1}^m \sigma_i \sigma_k \tilde{C}_{H_i, H_k}(H_i + H_k) \int_0^t \frac{\partial^2 F}{\partial x^2}(s, Y_s) s^{H_i + H_k - 1} ds. \tag{2.7.4}$$

2.7.4 The Itô Formula for $H \in (0, 1/2)$

We use the integral representation of fBm via the underlying Wiener process B on the finite interval $[0, t]$:

$$B_t^H = \int_0^t m_H(t, s) dB_s$$
$$= C_H^{(6)} t^\alpha \int_0^t u^{-\alpha}(t-u)^\alpha dB_u - C_H^{(6)} \alpha \int_0^t s^{\alpha-1} \left(\int_0^s u^{-\alpha}(s - u^{-\alpha}) dB_u \right) ds.$$

Let the function $F \in C^3(\mathbb{R})$ and we want to expand $F(B_t^H)$. Note that $B_t^H = B_{t,t}^H$, where for $0 < z < t$ $B_{z,t}^H = C_H^{(6)} z^\alpha \int_0^z u^{-\alpha}(t-u)^\alpha dB_u - C_H^{(6)} \alpha \int_0^z s^{\alpha-1} \left(\int_0^s u^{-\alpha}(s-u)^{-\alpha} dB_u \right) ds$. Therefore

$$F(B_t^H) = F(0) + \int_0^t F'(B_{z,t}^H) d_z B_{z,t}^H + \frac{1}{2}(C_H^{(6)})^2 \int_0^t F''(B_{z,t}^H)(t-z)^{2\alpha} dz$$

$$= F(0) + \alpha C_H^{(6)} \int_0^t F'(B_{z,t}^H) z^{\alpha-1} \int_0^z u^{-\alpha}(t-u)^\alpha dB_u dz$$

$$+ C_H^{(6)} \int_0^t F'(B_{z,t}^H)(t-z)^\alpha dB_z$$

$$- \alpha C_H^{(6)} \int_0^t F'(B_{z,t}^H) z^{\alpha-1} \left(\int_0^z u^{-\alpha}(t-u^{-\alpha}) dB_u \right) dz$$

$$+ \frac{1}{2}(C_H^{(6)})^2 \int_0^t F''(B_{z,t}^H)(t-z)^{2\alpha} dz. \tag{2.7.5}$$

Further,

$$B_{z,t}^H = B_z^H + \alpha C_H^{(6)} z^\alpha \int_0^z u^{-\alpha} \int_z^t (v-u)^{\alpha-1} dv \, dB_u$$
$$= B_z^H + \alpha C_H^{(6)} z^\alpha \int_z^t \int_0^z u^{-\alpha}(v-u)^{\alpha-1} dB_u dv, \tag{2.7.6}$$

whence

$$F'(B_{z,t}^H) = F'(B_z^H) + \int_z^t F'' \left(B_z^H + \alpha C_H^{(6)} z^\alpha \int_z^r \int_0^z u^{-\alpha}(v-u)^{\alpha-1} dB_u dv \right)$$
$$\times \alpha C_H^{(6)} z^\alpha \int_0^z u^{-\alpha}(r-u)^{\alpha-1} dB_u dr =: F'(B_z^H) + \phi(F'', z, t), \tag{2.7.7}$$

and similar relation holds for $F'''(B_{z,t}^H)$. But

$$\int_z^r \int_0^z u^{-\alpha}(v-u)^{\alpha-1} dB_u dv = \frac{1}{\alpha} \int_0^z u^{-\alpha}\left[(r-u)^\alpha - (z-u)^\alpha\right] dB_u. \quad (2.7.8)$$

Substituting (2.7.6)–(2.7.8) into (2.7.5), we obtain the following result.

Theorem 2.7.6. *Let $H \in (0, 1/2)$, B^H be an fBm with Hurst index H, represented as $B_t^H = \int_0^t m_H(t,s) dB_s$. Denote $Y_{r,z} := C_H^{(6)} \int_0^z u^{-\alpha}(r-u)^\alpha dB_u$, $0 \leq z \leq r$, $Y_z := Y_{z,z}$. Then*

$$F(B_t^H) = F(0) + \int_0^t F'(B_z^H) \alpha z^{\alpha-1} Y_{t,z} dz + C_H^{(6)} \int_0^t F'(B_z^H)(t-z)^\alpha dB_z$$

$$- \alpha \int_0^t F'(B_z^H) z^{\alpha-1} Y_{t,z} dz + \frac{1}{2}(C_H^{(6)})^2 \int_0^t F''(B_z^H)(t-z)^{2\alpha} dz + R_t,$$

where

$$R_t = \alpha \int_0^t \phi(F'', z, t) \alpha z^{\alpha-1} Y_{t,z} dz + C_H^{(6)} \int_0^t \phi(F'', z, t)(t-z)^\alpha dB_z$$

$$- \alpha \int_0^t \phi(F'', z, t) z^{\alpha-1} Y_{t,z} dz + \frac{1}{2}(C_H^{(6)})^2 \int_0^t \phi(F''', z, t)(t-z)^{2\alpha} dz.$$

Remark 2.7.7. The different approaches to the Itô formula for fBm with $H \in (1/2, 1)$ are contained in the papers (Lin95), (DH96), (DU99), (AN02), (DHP00), (BO04), (CCM03), (FdP01). An elegant version of the Itô formula for $F(B_t^H)$ for any $H \in (0,1)$ was obtained by C. Bender in (Ben03a) and (Ben03c), but in terms of distributions. If the distribution F is of function type, continuous at 0 and of polynomial growth, the form of such an Itô formula coincides with (2.7.4) for $m = 1$. For the other forms of the Itô formula for fBm with $H \in (0, 1/2)$ see also (Nua03), (GRV03), (ALN01), (AMN00), (CN05).

2.7.5 Itô Formula for Fractional Brownian Fields

First, we prove one auxiliary result for Hölder two-parameter functions. Let the function

$$F: \mathbb{R} \to \mathbb{R}, \ F \in C^3(\mathbb{R}), F''' \text{ is the Lipschitz function, } f(t) := F(g(t)),$$
$$g \in C^{\mu_1 \mu_2}(\mathbb{R}_+^2) \text{ with } \mu_i > 1/2, \ i = 1, 2. \quad (2.7.9)$$

Let the rectangle $\mathcal{P}_t = [0,t] \subset \mathbb{R}_+^2$ be fixed, $\pi_n^i := \left\{0 = t_0^{i,n} < \cdots t_{2^n}^{i,n} = t_i\right\}$, where $t_k^{i,n} = \frac{kt_i}{2^n}$, $f_{ik} = f(\frac{it_1}{2^n}, \frac{kt_2}{2^n})$,

$$\Delta_{ik}^1 f = f_{i+1k} - f_{ik}, \Delta_{ik}^2 f = f_{ik+1} - f_{ik}, \Delta_{ik} f = \Delta_{ik+1}^1 f - \Delta_{ik}^1 f.$$

Lemma 2.7.8. *Under assumption* (2.7.9) $\lim_{n\to\infty} I_j^n = 0$, $1 \leqslant j \leqslant 7$, *where*

$$I_1^n = \sum_{i,k=0}^{2^n-1} \Delta_{ik}^1 f \Delta_{ik} g, \quad I_2^n = \sum_{i,k=0}^{2^n-1} \Delta_{ik}^2 f \Delta_{ik} g, \quad I_3^n = \sum_{i,k=0}^{2^n-1} f_{ik} \Delta_{ik} g \Delta_{ik}^1 g,$$

$$I_4^n = \sum_{i,k=0}^{2^n-1} f_{ik} \Delta_{ik} g \Delta_{ik}^2 g, \quad I_5^n = \sum_{i,k=0}^{2^n-1} \Delta_{ik}^1 f (\Delta_{ik}^2 g)^2, \quad I_6^n = \sum_{i,k=0}^{2^n-1} (\Delta_{ik}^1 f)^2 \Delta_{ik}^2 g,$$

$$I_7^n = \sum_{i,k=0}^{2^n-1} F'''(g_{i,k}) \Delta_{ik}^1 g (\Delta_{ik}^2 g)^2.$$

Proof. Consider I_1^n (I_2^n is similar). We can rewrite $I_1^n = \int_{\mathcal{P}_t} \tilde{f}_n dg$, where $\tilde{f}_n = \Delta_{ik}^1 f$ for $s \in \Delta_{ik}^n := \left[\frac{it_1}{2^n}, \frac{(i+1)t_1}{2^n}\right) \times \left[\frac{kt_2}{2^n}, \frac{(k+1)t_2}{2^n}\right)$. Further,

$$\int_{\mathcal{P}_t} \tilde{f}_n dg = \int_{\mathcal{P}_t} (D_{0+}^{\alpha_1 \alpha_2} \tilde{f}_n)(s)(D_{1-}^{1-\alpha_1, 1-\alpha_2} g_{1-})(s) ds,$$

where $1 - \mu_1 < \alpha_i < \mu_i$, $i = 1, 2$. Since $\left|(D_{1-}^{1-\alpha_1, 1-\alpha_2} g_{1-})(s)\right| \leqslant C$ for some $C > 0$, it is sufficient to prove that $\lim_{n\to\infty} \int_{\mathcal{P}_t} \left|(D_{0+}^{\alpha_1 \alpha_2} \tilde{f}_n)(s)\right| ds = 0$, and in turn, for this purpose it is sufficient to prove that $\int_{\mathcal{P}_t} |\phi_{n,i}(s)| ds \to 0$, $1 \leqslant i \leqslant 4$, where $\phi_{n,1}(s) = s_1^{-\alpha} s_2^{-\alpha} \tilde{f}_n(s)$, $\phi_{n,2}(s) = s_2^{-\alpha_2} \int_0^{s_1} (\tilde{f}_n(s) - \tilde{f}_n(u, s_2))(s_1 - u)^{-1-\alpha_1} du$, $\phi_{n,3}(s) = s_1^{-\alpha_1} \int_0^{s_2} (\tilde{f}_n(s) - \tilde{f}_n(s_1, v))(s_2 - v)^{-1-\alpha_2} dv$, $\phi_{n,4}(s) = \int_{[0,s]} \Delta_{u,v} \tilde{f}_n(s)(s_1 - u)^{-1-\alpha_1}(s_2 - v)^{-1-\alpha_2} du dv$. The relation $\int_{\mathcal{P}_t} |\phi_{n,1}(s)| ds \to 0$ is evident. Further, if $\frac{it_1}{2^n} \leqslant s < \frac{(i+1)t_1}{2^n}$, then $|\phi_{n,2}(s)| \leqslant C s_2^{-\alpha_2} \int_0^{i2^{-n}} (s_1 - u_1)^{-1-\alpha_1} du \cdot 2^{-n\mu_1}$, whence $\int_{\mathcal{P}_t} |\phi_{n,2}(s)| ds \leqslant C \int_0^{t_2} s_2^{-\alpha_2} ds_2 \cdot 2^{n(\alpha_1 - \mu_1)} \to 0$, $n \to \infty$. Similarly, $\int_{\mathcal{P}_t} |\phi_{n,3}(s)| ds \to 0$, $n \to \infty$. Finally, $\int_{\mathcal{P}_t} |\phi_{n,4}(s)| ds \leqslant C 2^{-n\mu_1}$

$$\times \sum_{i,k=0}^{2^n-1} \int_{\Delta_{ik}^n} \int_{[0, t_k^{i,n}]} (s_1 - u)^{-1-\alpha_1} (s_2 - v + 2^{-n})^{\mu_2 - \alpha_2 - 1} du\, dv\, ds_1\, ds_2$$

$= C' 2^{n(\alpha_1 + \alpha_2 - \mu_1 - \mu_2)} \to 0$, $n \to \infty$. Of course, similar estimates hold for I_3^n and I_4^n. As to I_5^n, I_6^n and I_7^n, their estimates resemble each other, so, we consider only I_5^n. Note that

$$\lim_{n\to\infty} S_n := \lim_{n\to\infty} \sum_{i=0}^{2^n-1} f(t_{i2^n}^n)(\Delta_{i2^n}^1 g_{i+12^n})^2 \leqslant \lim_{n\to\infty} C \cdot 2^n \cdot 2^{-2n\mu_1} = 0.$$

Now, present the sum S_n as

$$S_n = \sum_{i,k=0}^{2^n-1} (f_{ik}(\Delta_{ik} g)^2 + 2 f_{ik} \Delta_{ik} g \Delta_{ik}^1 g + \Delta_{ik}^2 f (\Delta_{ik}^1 g)^2 + \Delta_{ik}^2 f (\Delta_{ik} g)^2$$

$$+ 2\Delta_{ik}^2 f \Delta_{ik}^1 g \Delta_{ik} g) =: \sum_{1 \leqslant i \leqslant 5} S_{n,i},$$

where $S_{n,1} \leq C \cdot 2^{-2n(\mu_1+\mu_2-1)} \to 0$, $n \to \infty$, similarly, $S_{n,4} \to 0$, $S_{n,5} \to 0$, $n \to \infty$. According to previous estimates $\lim_{n\to\infty} S_{n,2} = \lim_{n\to\infty} I_3^n = 0$. Therefore, $\lim_{n\to\infty} I_5^n = \lim_{n\to\infty} S_{n,3} = 0$. □

Remark 2.7.9. Let $F : \mathbb{R} \to \mathbb{R}$, $F \in C^3(\mathbb{R})$ and F''' is the Lipschitz function, the field $g(t)$ is a linear combination of the fractional Brownian fields,

$$g(t) = \sum_{i=1}^{m} \sigma_i B_t^{H_1^i H_2^i} \text{ with } H_j^i > \frac{1}{2}, \; j=1,2, \; 1 \leq i \leq m.$$

Clearly, the previous lemma holds for such $g(t)$ and $f(t) = F(g(t))$.

Theorem 2.7.10. *For any $t \in \mathbb{R}_+^2$*

$$F(g(t)) = F(0) + \int_{\mathcal{P}_t} F'(g) dg + \int_{\mathcal{P}_t} F''(g) d_1 g \, d_2 g.$$

Proof. According to the one-parameter Itô formula (Theorem 2.7.3)

$$F(g(t)) = F(0) + \int_0^{t_1} F'(g(s_1, t_2)) d_1 g(s_1, t_2)$$

$$= F(0) + \lim_{n\to\infty} \sum_{i=0}^{2^n} f(t_{i,2^n}^n) \Delta_{i,2^n}^1 g_{i+1,2^n} \text{ a.s.}$$

The prelimit sum can be presented as

$$\sum_{i,k=0}^{2^n-1} F'(g(t_{ik}^n)) \Delta_{ik} g + \sum_{i,k=0}^{2^n-1} F''(g(t_{ik}^n)) \Delta_{ik}^1 g \Delta_{ik}^2 g + \sum_{i,k=0}^{2^n-1} F''(g(s_n^{ik})) \Delta_{ik} g \Delta_{ik}^2 g$$

$$+ \frac{1}{2} \sum_{i,k=0}^{2^n-1} F'''(g(\theta_{ik}^n))(\Delta_{ik}^2 g)^2 \Delta_{ik}^1 g + \frac{1}{2} \sum_{i,k=0}^{2^n-1} F'''(g(\theta_{ik}^n))(\Delta_{ik}^2 g)^2 \Delta_{ik} g,$$

(2.7.10)

where $\theta_{ik}^n \in \Delta_{ik}^n$. According to Theorem 2.2.9, $\sum_{i,k=0}^{2^n-1} F'(g(t_{ik}^n)) \Delta_{ik} g \to \int_{\mathcal{P}_t} F'(g) dg$ a.s. Furthermore, according to Theorem 2.2.17 and Lemma 2.7.8, $\sum_{i,k=0}^{2^n-1} F''(g(t_{ik}^n)) \Delta_{ik}^1 g \Delta_{ik}^2 g \to \int_{\mathcal{P}_t} F''(g) d_1 g \, d_2 g$, $\sum_{i,k=0}^{2^n-1} F''(g(s_n^{ik})) \Delta_{ik} g \Delta_{ik}^2 g \to 0$, $\frac{1}{2} \sum_{i,k=0}^{2^n-1} F'''(g(t_{ik}^n))(\Delta_{ik}^2 g)^2 \Delta_{ik}^1 g \to 0$, $\frac{1}{2} \sum_{i,k=0}^{2^n-1} F'''(g(t_{ik}^n))(\Delta_{ik}^2 g)^2 \Delta_{ik} g \to 0$, and due to the Lipschitz properties of F''', $\frac{1}{2} \sum_{i,k=0}^{2^n-1} F'''(g(\theta_{ik}^n))(\Delta_{ik}^2 g)^2 \Delta_{ik}^1 g \to 0$,

$\frac{1}{2} \sum_{i,k=0}^{2^n-1} F'''(g(\theta_{ik}^n))(\Delta_{ik}^2 g)^2 \Delta_{ik} g \to 0$, $n \to \infty$, a.s., and the assertion of the theorem is proved. □

Remark 2.7.11. The theorem holds even for $F \in C^2(\mathbb{R})$, such that F'' is the Lipschitz function. To prove this, we must rewrite the sum of second and fourth term on the right-hand side of (2.7.10) as $\sum_{i,k=0}^{2^n-1} F''(g(\theta_{ik}^n))\Delta_{ik}^1 g \Delta_{ik}^2 g$. Then we can prove that this sum has a limit $\int_{\mathcal{P}_t} F''(g) d_1 g\, d_2 g$, similarly to Theorem 2.2.17. Also, the sum of third and fifth terms can be rewritten as $\sum_{i,k=0}^{2^n-1} F''(g(\theta_{ik}^n))\Delta_{ik} g \Delta_{ik}^2 g$, and we can prove that its limit is zero.

2.7.6 The Itô Formula for $H \in (0,1)$ in Terms of Isometric Integrals, and Its Applications

Definitions

If $f \in L^2(P \otimes \lambda)$, f is predictable, π is a partition, then f_π is the step function $f_\pi = \sum_i f(t_{i-1})\mathbf{1}_{[t_{i-1}, t_i)}(t)$.

Define the class of functions Φ as follows: $\vec{f} \in \Phi$ if the following conditions are satisfied:

(i) $\vec{f} := (f^i : i \geq 1)$, where $f^i \in L^2(P \otimes \lambda)$, f^i is predictable and $\sum_i \|f^i\|_{L^2(P \otimes \lambda)} < \infty$.

(ii) $\vec{f}$ is uniformly tight: $P\{\sup_{t \leq T} \sup_i |f^i(t)| > C\} \to 0$ as $C \to \infty$.

(iii) The random variable u defined by $u := \sum_i (f_\pi^i, (B^H)^{(i)})$ (for the notations see Section 2.5.2) does not depend on the partition π, and the series converges absolutely with probability one, when $\vec{f} \in \Phi$.

Write $(\vec{f}, \overrightarrow{B^H})$ for the sum $\sum_i (f_\pi^i, (B^H)^{(i)})$, and put $\mathcal{U} := \{u : u = (\vec{f}, \overrightarrow{B^H}), \vec{f} \in \Phi\}$. Let Φ_p be the projection of Φ to the first p coordinates.

The following example shows that $\mathcal{U}$ is nonempty.

Example 2.7.12. Assume that $f \in C_b^\infty(\mathbb{R})$: then

$$f(B_T^H) - f(0) = \sum_{i=1}^n \Delta f(B_{t_i}^H)$$

and if $f^k := (1/k!)f^{(k)}, k \geq 1$, then

$$f(B_T^H) - f(0) = (\vec{f}, \overrightarrow{B^H}),$$

$f(B_T^H) - f(0) \in \mathcal{U}$ and $\vec{f} \in \Phi, (f^1, \ldots, f_p) \in \Phi_p$ for any $p \geq 1$.

Lemma 2.7.13. *If $u \in \mathcal{U}, u = (\vec{f}, \overrightarrow{B^H})$ with $u = 0$, then $f^i = 0, i \geq 1$.*

Proof. Since u does not depend on the partition, take first the partition $\{0, T\}$. The random variable u has a representation

$$u = \sum_i f_0^i (B_T^H)^i, \qquad (2.7.11)$$

where f_0^i are real numbers, since $\mathcal{F}_0$ is the trivial σ-algebra. But since $u = 0$ from (2.7.11) it follows that for almost all $y \in \mathbb{R}$ we have that $\sum_i f_0^i y^i = 0$ and hence $f_0^i = 0$ for all $i \geq 1$.

Next, consider the partition $\{0, t, T\}$. We have that

$$u = \sum_i f_0^i (B_t^H)^i + \sum_i f_t^i (B_T^H - B_t^H)^i = 0.$$

From the above we get that $f_0^i = 0$ for all $i \geq 1$ and hence also $f_t^i = 0$ for all $i \geq 1$. □

The Itô Formula for Isometric Integrals

The following is an analogue of the Itô formula in this context.

Theorem 2.7.14. *Assume that the Hurst index H satisfies $H \in (0, 1/2)$. There exists one-to-one correspondence between $\mathcal{U}$ and the set*

$$\mathcal{V} := \left\{ v : v := \sum_{i=1}^{[1/H]} (f^i, (B^H)^{(i)}) \right\}.$$

Proof. We must show that there exists one-to-one correspondence between $\mathcal{U}$ and $\Phi_{[1/H]}$. Assume that $f \in \Phi_{[1/H]}$. Then there exists a vector $\vec{g} \in \Phi$ such that $f^i = g^i$ for $i \leq [1/H]$. Assume that $\vec{h}$ is another element from Φ such that $f^i = h^i$ for $i \leq [1/H]$. Put $u := (\vec{g}, \overrightarrow{B^H})$ and $v := (\vec{h}, \overrightarrow{B^H})$. Then

$$u - v = \sum_{i=\lfloor 1/H \rfloor + 1}^{\infty} (g^i - h^i, (B^H)^{(i)}).$$

On one hand, since u and v are independent of the partition π, we can take a partition π such that $|\pi| < 1$. Then for any $\varepsilon > 0$ we have that

$$P\{|u - v| > \varepsilon\} \leq P(D) + P\{|u - v| > \varepsilon, \Omega \setminus D\} \qquad (2.7.12)$$

and D is the set $D := \{\sup_{t \leq T} \sup_i |f_t^i - g_t^i| \geq C\}$. But

$$P\{|u - v| > \varepsilon, \Omega \setminus D\} \leq \frac{C}{\varepsilon} \sum_{i > 1/H} E \sum_k |\Delta B_k^H|^i$$

and since
$$E\sum_k |\Delta B_k^H|^i \le CT(|\pi|)^{Hi-1}$$
we have that
$$P\{|u-v| > \varepsilon, \Omega \setminus D\} \to 0$$
as $|\pi| \to 0$. By property (iii) of Φ we can choose C such that $P(D) < \delta$ for any $\delta > 0$. Use these estimates in (2.7.12) to conclude that $u = v$. On the other hand, if $u = (\overrightarrow{f}, \overrightarrow{B^H}) = (\overrightarrow{h}, \overrightarrow{B^H})$ we have from Lemma 2.7.13 that $\overrightarrow{f} = \overrightarrow{h}$. To finish, note that from Example 2.7.12 it follows that the random variable $f(B_T^H) - f(0)$ is a representative of $\sum_{i=1}^{l} 1/H](1/i!) \int_0^T f^{(i)}(x_s) dB_s^{H(i)}$. □

Example 2.7.15 (Fractional Doleans exponent). Assume that $[1/H] = 2p$, where $p \in \mathbb{N}$. Then the random variable $y_t = \exp(B_t^H - t/(2p)!) - 1$ is a representative of
$$\sum_{i=1}^{2p-1} \frac{1}{i!} \int_0^t y_s d(B_s^H)^{(i)}.$$
We say that y is the Doleans exponent of B^H.

2.8 The Girsanov Theorem for fBm and Its Applications

2.8.1 The Girsanov Theorem for fBM

Consider the kernel $l_H(t,s) = C_H^{(5)} s^{-\alpha}(t-s)^{-\alpha}$, $0 < s < t$. Let $\mathcal{F}_t = \sigma\{B_s^H, 0 \le s \le t\} = \sigma\{B_s, 0 \le s \le t\}$, where B is underlying Wiener process in the representation
$$M_t^H = \int_0^t l_H(t,s) dB_s^H, \quad B_t = \widehat{\alpha} \int_0^t s^\alpha dM_s^H.$$
Assume that the random process $\{\phi_t, t \ge 0\}$ is adapted to filtration $\mathcal{F}_t$ and satisfies
$$\int_0^t l_H(t,s)|\phi_s| ds < \infty, \; t > 0, \; P\text{-a.s.} \tag{2.8.1}$$
Assume also that we have the representation
$$\int_0^t l_H(t,s)\phi_s ds = \widetilde{\alpha} \int_0^t \delta_s ds, \; t > 0, \tag{2.8.2}$$
with some $\mathcal{F}_t$-adapted process δ satisfying
$$\int_0^t |\delta_s| ds < \infty, \; P\text{-a.s.}, \; t > 0, \tag{2.8.3}$$

and
$$E\int_0^t s^{2\alpha}\delta_s^2 ds < \infty, \ t > 0. \tag{2.8.4}$$

Define a square-integrable martingale L by $L_t := \int_0^t s^\alpha \delta_s dB_s$.

Theorem 2.8.1. *Assume that we have (2.8.1)–(2.8.4) and the martingale L satisfies*
$$E\exp\{L_t - 1/2\langle L\rangle_t\} = 1, \ t > 0.$$

Then the process $\widetilde{B}_t^H := B_t^H - \int_0^t \phi_s ds$ is an fBm with respect to measure Q, where the measure Q is defined by
$$\left.\frac{dQ}{dP}\right|_{\mathcal{F}_t} = \exp\left\{L_t - \frac{1}{2}\langle L\rangle_t\right\}.$$

Proof. Note first that the integral
$$\widetilde{M}_t^H := \int_0^t l_H(t,s)d\widetilde{B}_s^H = \int_0^t l_H(t,s)dB_s^H - \int_0^t l_H(t,s)\phi_s ds \tag{2.8.5}$$

exists, since both integrals exist as pathwise integrals (the first integral was studied in Section 1.8 and (2.8.2) ensures the existence of the second integral). Moreover, from (2.8.2) it follows that
$$\widetilde{M}_t^H = M_t^H - \widetilde{\alpha}\int_0^t \delta_s ds = \widetilde{\alpha}\left(\int_0^t s^{-\alpha}dB_s - \int_0^t \delta_s ds\right).$$

Evidently, $\left[\widetilde{M}^H\right]_t := P\text{-lim}_{|\pi|\to 0}\sum_{t_i\in\pi}(\widetilde{M}_{t_i}^H - \widetilde{M}_{t_{i-1}}^H)^2$ exists and equals $\left[\widetilde{M}^H\right]_t = t^{1-2\alpha}$. Therefore, for any $\theta\in\mathbb{R}$ we have for $\widehat{M}_t^H := \widehat{\alpha}\widetilde{M}_t^H$ that

$$\theta\widehat{M}_t^H - \frac{\theta^2}{2}\left[\widehat{M}^H\right]_t + L_t - \frac{1}{2}\langle L\rangle_t = \theta\int_0^t s^{-\alpha}dB_s - \theta\int_0^t \delta_s ds - \frac{\theta^2}{2}\frac{t^{1-2\alpha}}{1-2\alpha}$$
$$+ \int_0^t s^\alpha \delta_s dB_s - \frac{1}{2}\int_0^t s^{2\alpha}\delta_s^2 ds = \int_0^t (\theta s^{-\alpha} + s^\alpha\delta_s)dB_s$$
$$- \frac{1}{2}\int_0^t (\theta^2 s^{-2\alpha} - 2\delta_s\theta + \delta_s^2 s^{2\alpha})ds =: R_t - \frac{1}{2}\langle R\rangle_t, \tag{2.8.6}$$

where R is a square-integrable martingale given by $R_t := \int_0^t (\theta s^{-\alpha} + s^\alpha \delta_s)dB_s$. But (2.8.6) means that the process
$$K_t := \exp\left\{\theta\widehat{M}_t^H - \frac{\theta^2}{2}\left[\widehat{M}^H\right]_t + L_t - \frac{1}{2}\langle L\rangle_t\right\}$$

is a local P-martingale. This implies, in turn, that the process $\exp\left\{\theta\widehat{M}_t^H - \frac{\theta^2}{2}\left[\widehat{M}^H\right]_t\right\}$ is a local Q-martingale. From (Ell82, Theorem

13.22), we can conclude that $\widehat{M}^H$ is a local Q-martingale with the angle bracket $\langle \widehat{M}^H \rangle_t = \int_0^t s^{-2\alpha} ds$ and so $\widetilde{M}_t = \widetilde{\alpha} \int_0^t s^{-\alpha} d\widetilde{B}_s$, where $\widetilde{B}$ is a standard Brownian motion with respect to Q (and is obtained from B by subtracting a drift). This means that

$$\int_0^t l_H(t,s) d\widetilde{B}_s^H = \widetilde{\alpha} \int_0^t s^{-\alpha} d\widetilde{B}_s. \qquad (2.8.7)$$

Now, using two representations for $\widetilde{B}^H$, (2.8.5) and (2.8.7), we can obtain (1.8.17) for $\widetilde{B}^H$ and then conclude from Remark 1.8.2 that it is the fBm with respect to the measure Q. $\square$

2.8.2 When the Conditions of the Girsanov Theorem Are Fulfilled? Differentiability of the Fractional Integrals

If we analyze the conditions of the Girsanov theorem, we see that condition (2.8.2) is a principal concern. Now we shall establish that in one particular but important case this condition holds. Let the process $I(t) := \int_0^t l_H(t,s)\phi(s) ds$ with $\phi(t) = \int_0^t a(s,\omega) dB_s^H$, where the integrand $a = a(s,\omega) : \mathbb{R} \times \Omega \to \mathbb{R}$ is measurable in its variables and for a.a. $\omega \in \Omega$ is Hölder in s with some index $\beta \in (1/2, 1)$. According to Theorem 2.1.7, the integral $\phi(t)$ exists as a pathwise integral for $\omega \in \Omega'$, $P(\Omega') = 1$. Moreover, according to Lemma 2.6.6, there exists a repeated integral $J(t) := \int_0^t a(u,\omega) \int_u^t l_H(t,s) ds \, dB_u^H$ and the equality $I(t) = J(t)$ holds for $\omega \in \Omega'$.

Lemma 2.8.2. *Let $a \in C^\rho[0,t]$ for any $t > 0$ and for any $\omega \in \Omega'$, $P(\Omega') = 1$, $\rho \in (1/2, 1)$. Then for any $t > 0$ $I(t)$ admits the representation*

$$I(t) = C_H^{(5)} t^{1-2\alpha} \int_0^t \delta_s ds,$$

where $\delta_s = s^{2\alpha-2} \int_0^s u^{1-\alpha}(s-u)^{-\alpha} a(u,\omega) dB_u^H$, and $\delta \in L_1[0,t]$ for any $t > 0$, $\omega \in \Omega'$.

Proof. Further we suppose everywhere that $\omega \in \Omega'$ and argument ω will be omitted. We rewrite $J(t)$ as

$$J(t) = t^{1-2\alpha} \int_0^t \int_{u/t}^1 a(u) l_H(1,s) ds \, dB_u^H$$

$$= C_H^{(5)} t^{1-2\alpha} \int_0^t \int_u^t s^{2\alpha-2}(s-u)^{-\alpha} u^{1-\alpha} a(u) ds \, dB_u^H =: C_H^{(5)} t^{1-2\alpha} M(t).$$

Consider now the function

$$N(t) := \int_0^t s^{2\alpha-2} \int_0^s (s-u)^{-\alpha} u^{1-\alpha} a(u) dB_u^H \, ds.$$

The following results ensure its existence:
(i) According to (NVV99, Lemma 2.1), for the function $g \in C^\beta[0,T]$ with $0 < \gamma + \beta < 1$, $f(0) = 0$ the integral $\int_0^t (t-u)^\gamma dg(u)$ exists and equals

$$\int_0^t (t-u)^\gamma dg(u) = \lim_{\varepsilon \to 0} (\varepsilon^\gamma (g(t-\varepsilon) - g(t))$$
$$+ t^\gamma g(t) + \gamma \int_0^{t-\varepsilon} (g(u) - g(t))(t-u)^{\gamma-1} du). \quad (2.8.8)$$

(ii) According to Lemma 2.6.3, for $f \in C^\gamma[a,b]$, $g \in C^\beta[a,b]$, $\gamma + \beta > 1$, $0 < \varepsilon' < \gamma + \beta - 1$

$$\left| \int_a^b f(t) dg(t) \right| \leq C \|f\|_{C^\gamma[a,b]} \|g\|_{C^\beta[a,b]} ((b-a)^{1+\varepsilon'} \vee (b-a)^\beta), \quad (2.8.9)$$

where C does not depend of f and g. Using (2.8.8)–(2.8.9), we obtain the following estimates for $0 < s_1 < s_2 < t$:

$$\left| \int_{s_1}^{s_2} a(z)(s_2-z)^{-\alpha} dB_z^H \right| = \left| \lim_{\varepsilon \to 0} \left(-\varepsilon \int_{s_2-\varepsilon}^{s_2} a(v) dB_v^H \right. \right.$$
$$\left. \left. + (s_2-s_1)^{-\alpha} \int_{s_1}^{s_2} a(z) dB_z^H + \alpha \int_{s_1}^{s_2-\varepsilon} (s_2-z)^{-1-\alpha} \int_z^{s_2} a(v) dB_v^H dz \right) \right|$$
$$\leq \lim_{\varepsilon \to 0} \left(C \|a\|_{C^\rho[0,t]} \|B^H\|_{C^{H'}[0,t]} \left((s_2-s_1)^{1-\alpha+\varepsilon'} \vee (s_2-s_1)^{-\alpha+H'} \right) \right.$$
$$\left. + \alpha \int_{s_1}^{s_2-\varepsilon} (s_2-z)^{-1-\alpha} \left((s_2-z)^{1+\varepsilon'} \vee (s_2-z)^{H'} \right) dz \right), \quad (2.8.10)$$

where H' is any constant not exceeding H and $0 < \varepsilon < \rho + H - 1$. Evidently, the right-hand side of (2.8.10) can be estimated by $CK_1(t)(s_2-s_1)^{-\alpha+H'}$, where $K_1(t) \leq \|a\|_{C^\rho[0,t]} \|B^H\|_{C^{H'}[0,t]} (t \vee 1)^{1+\varepsilon-H'}$, C does not depend on ρ, B^H, t. Further,

$$\int_{s_1}^{s_2} (s_2-u)^{-\alpha} u^{1-\alpha} a(u) dB_u^H = \int_{s_1}^{s_2} u^{1-\alpha} d \left(\int_{s_1}^u (s_2-z)^{-\alpha} a(z) dB_z^H \right)$$
$$= s_2^{1-\alpha} \int_{s_1}^{s_2} (s_2-z)^{-\alpha} a(z) dB_z^H - (1-\alpha) \int_{s_1}^{s_2} u^{-\alpha} \int_{s_1}^u (s_2-z)^{-\alpha} a(z) dB_z^H du$$
$$=: L(s_1, s_2).$$

The estimate

$$|L(s_1, s_2)| \leq C s_2^{1-\alpha} K_1(t)(s_2-s_1)^{-\alpha+H'}$$
$$+ C(1-\alpha) K_1(t) \int_{s_1}^{s_2} u^{-\alpha} (u-s_1)^{-\alpha+H'} du$$
$$\leq CK_1(t) \left(s_2^{1-\alpha} (s_2-s_1)^{-\alpha+H'} + (s_2-s_1)^{1-2\alpha+H'} \right) \quad (2.8.11)$$

means that $|L(0,s)| \leqslant CK_1(t)s^{1-2\alpha+H'}$.

Now it is clear that

$$|N_t| \leqslant CK_1(t)\int_0^t s^{2\alpha-2}s^{1-2\alpha+H'}ds \leqslant CK_1(t)t^{H'} < \infty.$$

Consider the function

$$N_\varepsilon(t) := \int_0^t s^{2\alpha-2}\mathbf{1}_{\{s\in[\varepsilon,t]\}}\int_0^{s-\varepsilon} u^{1-\alpha}(s-u)^{-\alpha}a(u)dB_u^H ds.$$

Evidently, for any $\varepsilon > 0$ the function

$$\phi_\varepsilon(s,u) := \mathbf{1}_{\{s\in[\varepsilon,t],0\leqslant u\leqslant s-\varepsilon\}}s^{2\alpha-2}u^{1-\alpha}(s-u)^{-\alpha}a(u)$$

is piecewise-Hölder in u with index $\rho \wedge (1-\alpha) > 1/2$ ($u = s - \varepsilon$ is the point of Hölder discontinuity), and the function

$$\psi_\varepsilon(s) := \int_0^t \phi_\varepsilon(s,u)dB_u^H = s^{2\alpha-2}\mathbf{1}_{\{s\in[\varepsilon,t]\}}\int_0^{s-\varepsilon}(s-u)^{-\alpha}u^{1-\alpha}a(u)dB_u^H$$

is Riemann integrable on $[0,t]$. Therefore, $\phi_\varepsilon(s,u)$ satisfies the conditions of the stochastic Fubini Theorem 2.6.5, whence $N_\varepsilon(t)$ exists and equals

$$M_\varepsilon(t) := \int_0^{t-\varepsilon} u^{1-\alpha}a(u)\int_{u+\varepsilon}^t s^{2\alpha-2}(s-u)^{-\alpha}ds\, dB_u^H.$$

Further,

$$|N(t) - N_\varepsilon(t)| \leqslant \left|\int_\varepsilon^t s^{2\alpha-2}\int_{s-\varepsilon}^s u^{1-\alpha}(s-u)^{-\alpha}a(u)dB_u^H ds\right|$$
$$+ \left|\int_0^\varepsilon s^{2\alpha-2}\int_0^s u^{1-\alpha}(s-u)^{-\alpha}a(u)dB_u^H ds\right|$$
$$\leqslant \int_\varepsilon^t s^{2\alpha-2}CK_1(t)(s^{1-\alpha}\varepsilon^{-\alpha+H'} + \varepsilon^{1-2\alpha+H'})ds$$
$$+ \int_0^\varepsilon s^{2\alpha-2}CK_1(t)s^{1-2\alpha+H'}ds$$
$$\leq CK_1(t)(\varepsilon^{-\alpha+H'} + \varepsilon^{H'}) \to 0, \ \varepsilon \to 0.$$

For $M(t) - M^\varepsilon(t)$ we use one of the integral transformations from (NVV99, Lemma 2.2): for $\mu \in \mathbb{R}$, $\nu > -1$, $c > 1$ the integral $\int_1^c t^\mu(t-1)^\nu dt$
$= \int_0^{1-1/c} s^\nu(1-s)^{-\mu-\nu-2}ds$, and as a result obtain the bound

$$|M(t) - M_\varepsilon(t)| \leqslant C \left| \int_0^{t-\varepsilon} a(u) u^{1-\alpha} \int_u^{u+\varepsilon} s^{2\alpha-2}(s-u)^{-\alpha} ds \, dB_u^H \right|$$

$$+ C \left| \int_{t-\varepsilon}^t a(u) u^{1-\alpha} \int_u^t s^{2\alpha-2}(s-u)^{-\alpha} ds \, dB_u^H \right|$$

$$= C \left| \int_0^{t-\varepsilon} a(u) \int_0^{\frac{\varepsilon}{u+\varepsilon}} s^{-\alpha}(1-s)^{-\alpha} ds \, dB_u^H \right|$$

$$+ C \left| \int_{t-\varepsilon}^t a(u) \int_0^{1-\frac{u}{t}} s^{-\alpha}(1-s)^{-\alpha} ds \, dB_u^H \right| =: A_1(\varepsilon) + A_2(\varepsilon).$$

According to the stochastic Fubini theorem 2.6.5,

$$A_1(\varepsilon) = C \int_0^{\varepsilon/t} s^{-\alpha}(1-s)^{-\alpha} \int_0^{t-\varepsilon} a(u) dB_u^H \, ds$$

$$+ C \int_{\varepsilon/t}^t s^{-\alpha}(1-s)^{-\alpha} \int_0^{\frac{\varepsilon(1-s)}{s}} a(u) dB_u^H \, ds$$

and

$$A_2(\varepsilon) = C \int_0^{\varepsilon/t} s^{-\alpha}(1-s)^{-\alpha} \int_{t-\varepsilon}^{t(1-s)} a(u) dB_u^H \, ds.$$

Therefore,

$$|A_1(\varepsilon)| \leqslant C \left| \int_0^{t-\varepsilon} a(u) s B_u^H \right| \left(1 - \frac{\varepsilon}{t}\right)^{-\alpha} \left(\frac{\varepsilon}{t}\right)^{1-\alpha}$$

$$+ C K_1(t) \int_{\varepsilon/t}^1 s^{-\alpha}(1-s)^\alpha \left(\frac{\varepsilon(1-s)}{s}\right)^{H'} ds \to 0, \ \varepsilon \to 0,$$

and

$$|A_2(\varepsilon)| \leqslant C K_1(t) \int_0^{\varepsilon/t} s^{-\alpha}(1-s)^{-\alpha} (\varepsilon - ts)^{H'} ds \to 0, \ \varepsilon \to 0.$$

Therefore, $N(t) = M(t)$, and our lemma is proved. □

3

Stochastic Differential Equations Involving Fractional Brownian Motion

3.1 Stochastic Differential Equations Driven by Fractional Brownian Motion with Pathwise Integrals

3.1.1 Existence and Uniqueness of Solutions: the Results of Nualart and Răşcanu

Consider the function $\sigma = \sigma(t,x) : [0,T] \times \mathbb{R} \to \mathbb{R}$ satisfying the assumptions: σ is differentiable in x, there exist $M > 0$, $0 < \gamma, \kappa \leq 1$ and for any $R > 0$ there exists $M_R > 0$ such that

(i) σ is Lipschitz continuous in x:

$$|\sigma(t,x) - \sigma(t,y)| \leq M|x-y|, \quad \forall t \in [0,T], x, y \in \mathbb{R};$$

(ii) x-derivative of σ is local Hölder continuous in x:

$$|\sigma_x(t,x) - \sigma_x(t,y)| \leq M_R|x-y|^\kappa, \quad \forall |x|, |y| \leq R, t \in [0,T];$$

(iii) σ is Hölder continuous in time:

$$|\sigma(t,x) - \sigma(s,x)| + |\sigma_x(t,x) - \sigma_x(s,x)| \leq M|t-s|^\gamma, \quad \forall x \in \mathbb{R}, t,s \in [0,T].$$

Let $0 < \beta < 1/2$, $f \in W_0^\beta[0,T]$, $g \in W_1^{1-\beta}[0,T]$. We need some preliminary estimates, in addition to Lemmas 2.1.9 and 2.1.10.

Consider on $W_0^\beta[0,T]$ the norm, equivalent to $\|\cdot\|_{0,\beta}$:

$$\|f\|_{0,\beta,\lambda} := \sup_{t \in [0,T]} e^{-\lambda t} \varphi_f^\beta(t).$$

Lemma 3.1.1 ((NR00)). *Let assumptions (i)–(iii) hold with $\gamma > \beta$. Then the following statements hold.*
1. There exists the integral $G^{(\sigma)}(f)(t) := \int_0^t \sigma(\cdot, f(\cdot))dg, t \in [0,T]$.
2. $G^{(\sigma)}(f) \in C^{1-\beta}[0,T] \subset W_0^\beta[0,T]$.

3. $\|G^{(\sigma)}(f)\|_{1-\beta} \leq C_1 \Lambda_{1-\beta}(g)(1+\|f\|_{0,\beta})$.
4. $\|G^{(\sigma)}(f)\|_{0,\beta,\lambda} \leq C_2 \Lambda_{1-\beta}(g) \lambda^{2\beta-1}(1+\|f\|_{0,\beta,\lambda}), \lambda \geq 1$,
where C_1 and C_2 depend only on M, β, γ, T and $|\sigma(0,0)|$.
5. For any $f, h \in W_0^\beta[0,T]$ such that $f_T^* \vee h_T^* \leq R$

$$\|G^{(\sigma)}(f) - G^{(\sigma)}(h)\|_{0,\beta,\lambda} \leq C_3 \lambda^{2\beta-1} \Lambda_{1-\beta}(g)(1+C_f+C_h)\|f-h\|_{0,\beta,\lambda},$$

where $C_f := \sup_{r \in [0,T]} \int_0^r \frac{|f_r - f_s|^\kappa}{(r-s)^{\beta+1}} ds$, C_3 depends only on $M, \beta, \gamma, R, M_R, T$.

Proof. We prove only statement 5; the others can be proved in a similar, but more simple way. It is easy to check via the Taylor formula in the integral form that the function σ satisfying (i)–(iii) admits the following bound: for any $R > 0$, $t_i \in [0,T], i = 1, 2$, and $|x_i| \leq R, 1 \leq i \leq 4$

$$\begin{aligned}&|\sigma(t_1,x_1) - \sigma(t_2,x_2) - \sigma(t_1,x_3) + \sigma(t_2,x_4)| \\ &\leq M|x_1 - x_2 - x_3 + x_4| + M|x_1 - x_3||t_2 - t_1|^\gamma \\ &+ M_R|x_1 - x_3|(|x_1 - x_2|^\kappa + |x_3 - x_4|^\kappa).\end{aligned} \quad (3.1.1)$$

Therefore, from Lemma 2.1.9, part 1,

$$\begin{aligned}&\|G^{(\sigma)}(f) - G^{(\sigma)}(h)\|_{0,\beta,\lambda} \\ &\leq C_{\beta,T}^1 \Lambda_{1-\beta}(g) \sup_{t \in [0,T]} e^{-\lambda t} \int_0^t ((t-r)^{-2\beta} + r^{-\beta}) \varphi_{\sigma(\cdot,f(\cdot))-\sigma(\cdot,h(\cdot))}(r) dr \\ &\leq C_{\beta,T}^1 \Lambda_{1-\beta}(g) \sup_{r \in [0,T]} (e^{-\lambda r} \varphi_{\sigma(\cdot,f(\cdot))-\sigma(\cdot,h(\cdot))}(r)) \\ &\quad \times \int_0^t e^{-\lambda(t-r)}((t-r)^{-2\beta} + r^{-\beta}) dr.\end{aligned} \quad (3.1.2)$$

The last integral in (3.1.2) can be estimated by

$$\begin{aligned}&\int_0^\infty e^{-\lambda u} u^{-2\beta} du + \int_0^t e^{-\lambda u}(t-u)^{-\beta} du \\ &= \lambda^{2\beta-1} \int_0^\infty e^{-u} u^{-2\beta} du + \lambda^{\beta-1} \int_0^{\lambda t} e^{-u}(\lambda t - u)^{-\beta} du \\ &\leq \lambda^{2\beta-1} C_{1,\beta} + \lambda^{\beta-1} C_{2,\beta}\end{aligned} \quad (3.1.3)$$

with $C_{1,\beta} = \int_0^\infty e^{-u} u^{-2\beta} du$, $C_{2,\beta} = \sup_{z \geq 0} \int_0^z e^{-u}(z-u)^{-\beta} du$.
Evidently, for $\lambda \geq 1$

$$\int_0^t e^{-\lambda(t-r)}((t-r)^{-2\beta} + r^{-\beta}) dr \leq \lambda^{2\beta-1}(C_{1,\beta} + C_{2,\beta}). \quad (3.1.4)$$

Further, from the Lipschitz property (i) and (3.1.1), it follows that

$$\begin{aligned}&\varphi_{\sigma(\cdot,f(\cdot))-\sigma(\cdot,h(\cdot))}(r) \leq M|f(r)-h(r)| + M\int_0^r |f(r)-f(s)-h(r)\\ &+h(s)|(r-s)^{-\beta-1} ds + \frac{M}{\gamma-\beta}|f(r)-h(r)|r^{\gamma-\beta} \\ &+ M_R|f(r)-h(r)|\left(\int_0^r \frac{|f(r)-f(s)|^\kappa}{|r-s|^{\beta+1}} ds + \int_0^r \frac{|h(r)-h(s)|^\kappa}{|r-s|^{\beta+1}} ds\right).\end{aligned} \quad (3.1.5)$$

The proof follows now directly from (3.1.2)–(3.1.5), with $C_3 = (C_{1,\beta} + C_{2,\beta})(M + M_R)\left(1 + \frac{T^{\gamma-\beta}}{\gamma-\beta}\right)$. □

3.1 SDEs Driven by fBm with Pathwise Integrals   199

The next lemma describes the situation with the Lebesgue integrals. Let the function $b = b(t,x) : [0,T] \times \mathbb{R} \to \mathbb{R}$ satisfy the assumptions

(iv) for any $R \geq 0$ there exists $L_R > 0$ such that

$$|b(t,x) - b(t,y)| \leq L_R |x-y|, \quad \forall |x|, |y| \leq R, \forall t \in [0,T];$$

(v) there exists the function $b_0 \in L_p[0,T]$ and $L > 0$ such that

$$|b(t,x)| \leq L|x| + b_0(t), \quad \forall (t,x) \in [0,T] \times \mathbb{R}.$$

Lemma 3.1.2. *Let $0 < \beta < 1/2$, assumptions (iv) and (v) hold with $p = \beta^{-1}$, $f \in W_0^\beta[0,T]$. Then the following statements hold.*

1. *There exists the Lebesgue integral $F^{(b)}(f)(t) := \int_0^t b(s, f(s)) ds$, $t \in [0,T]$.*
2. $F^{(b)}(f) \in C^{1-\beta}[0,T]$.
3. $\|F^{(b)}(f)\|_{1-\beta} \leq C_4(1 + f_T^*) \leq C_4(1 + \|f\|_{0,\beta})$.
4. $\|F^{(b)}(f)\|_{0,\beta,\lambda} \leq C_5 \lambda^{2\beta-1}(1 + \|f\|_{0,\beta,\lambda})$,

where $\lambda \geq 1$, C_4 and C_5 depend only on β, T, L and $\|b_0\|_{L_p[0,T]}$.

5. *Let $f, h \in W_0^\beta[0,T]$ with $f_T^* \vee h_T^* \leq R$. Then*

$$\|F^{(b)}(f) - F^{(b)}(h)\|_{0,\beta,\lambda} \leq C_6 \lambda^{\beta-1} \|f - h\|_{0,\beta,\lambda}, \quad \lambda \geq 1,$$

where C_6 depends on β, R, T, L_R.

Proof. We prove only statement 4. Indeed, from Lemma 2.1.10,

$$\varphi^\beta_{F^{(b)}(f)}(t) \leq C^3_{\beta,T} \int_0^t \frac{|b(s,f(s))|}{(t-s)^\beta} ds \leq C^3_{\beta,T} \int_0^t \frac{(L|f(s)| + b_0(s))}{(t-s)^\beta} ds$$
$$\leq C^3_{\beta,T} \left(L \int_0^t \frac{|f(s)|}{(t-s)^\beta} ds + \left(\int_0^t (t-s)^{-\frac{\beta}{1-\beta}} ds \right)^{1-\beta} \|b_0\|_{L_{1/\beta}[0,T]} \right) \quad (3.1.6)$$
$$\leq C^3_{\beta,T} \left(L \int_0^t \frac{|f(s)|}{(t-s)^\beta} ds + c_\beta t^{1-2\beta} B_{0,\beta} \right),$$

where $c_\beta = \left(\frac{1-\beta}{1-2\beta} \right)^{1-\beta}$, $B_{0,\beta} = \|b_0\|_{L_p[0,T]}$, $C^3_{\beta,T} = T^\beta + 1/\beta$.

Hence

$$\|F^{(b)}(f)\|_{0,\beta,\lambda} \leq C^3_{\beta,T} \cdot L \cdot \sup_{t \in [0,T]} e^{-\lambda t} \int_0^t \frac{|f(s)|}{(t-s)^\beta} ds$$
$$+ C^3_{\beta,T} c_\beta B_{0,\beta} \sup_{t \in [0,T]} e^{-\lambda t} t^{1-2\beta}$$
$$\leq C^3_{\beta,T} \cdot L \cdot \sup_{s \in [0,T]} e^{-\lambda s} |f(s)| \int_0^t e^{-\lambda u} u^{-\beta} du$$
$$+ C^3_{\beta,T} c_\beta B_{0,\beta} \lambda^{2\beta-1} \sup_{z \geq 0} e^{-z} z^{1-2\beta} \leq C_5 \lambda^{2\beta-1}(1 + \|f\|_{0,\beta,\lambda}),$$

where $C_5 = C^3_{\beta,T}(L \cdot \Gamma(1-\beta) + c_\beta B_{0,\beta} \sup_{z \geq 0} e^{-z} z^{1-2\beta})$. □

Now, let $0 < \beta < 1$ be fixed, $g \in W_1^{1-\beta}[0,T]$. Consider the (deterministic) differential equation

$$X_t = X_0 + \int_0^t b(s, X_s) ds + \int_0^t \sigma(s, X_s) dg_s, \quad t \in [0,T], \quad (3.1.7)$$

where $X_0 \in \mathbb{R}$, and the coefficients $\sigma, b : [0,T] \times \mathbb{R} \to \mathbb{R}$ are measurable functions satisfying (i)–(v) with $p = 1/\beta, 0 < \gamma, \kappa \leq 1$ and $0 < \beta < \beta_0 = \frac{1}{2} \wedge \gamma \wedge \frac{\kappa}{1+\kappa}$.

Theorem 3.1.3. *Equation (3.1.7) has the unique solution $X \in W_0^\beta[0,T]$. This solution belongs also to the space $C^{1-\beta}[0,T]$.*

Proof. Let the function $f \in W_0^\beta[0,T]$. Then, according to statements 3 of Lemmas 3.1.1 and 3.1.2, $G^{(\sigma)}(f) \in C^{1-\beta}[0,T]$ and $F^{(b)}(f) \in C^{1-\beta}[0,T]$. So, if X is the solution of (3.1.7) and $X \in W_0^\beta[0,T]$, then $X = X_0 + F^{(b)}(X)(t) + G^{(\sigma)}(X)(t) \in C^{1-\beta}[0,T]$.

Now we prove the uniqueness. Let X and Y be two solutions from $C^{1-\beta}[0,T]$ and $\|X\|_{C^{1-\beta}[0,T]} \vee \|Y\|_{C^{1-\beta}[0,T]} \leq R$. Then from statements 5 of Lemmas 3.1.1 and 3.1.2, for $\beta < \gamma$

$$\|X - Y\|_{0,\beta,\lambda} \leq \|F^{(b)}(X) - F^{(b)}(Y)\|_{0,\beta,\lambda} + \|G^{(\sigma)}(X) - G^{(\sigma)}(Y)\|_{0,\beta,\lambda}$$
$$\leq (C_3 \Lambda_{1-\beta}(g) \lambda^{2\beta-1}(1 + C_X + C_Y) + C_6 \lambda^{\beta-1})\|X - Y\|_{0,\beta,\lambda}, \quad \lambda \geq 1.$$

Note, that for $\beta < \frac{\kappa}{1+\kappa}$ and for $(1-\beta)$-Hölder X and Y

$$C_X + C_Y \leq 2R \sup_{r \in [0,T]} \int_0^r (r-s)^{(1-\beta)\kappa - \beta - 1} ds \leq C_7,$$

where C_7 depends on R, T and β. Take λ sufficiently large such that for $\beta < 1/2$

$$C_3 \Lambda_{1-\beta}(g) \lambda^{2\beta-1} C_7 + C_6 \lambda^{\beta-1} \leq 1/2$$

and obtain

$$\|X - Y\|_{0,\beta,\lambda} \leq 1/2 \|X - Y\|_{0,\beta,\lambda}$$

whence $X = Y$ on $[0,T]$.

Now prove the existence by a fixed-point theorem. Consider the operator $A : W_0^\beta[0,T] \to C^{1-\beta}[0,T] \subset W_0^\beta[0,T]$ of the form $AX = X_0 + \int_0^t b(s, X_s) ds + \int_0^t \sigma(s, X_s) ds$. Then for all $\lambda \geq 1$ from Lemmas 3.1.1 and 3.1.2 for any $u \in W_0^\beta[0,T]$ it follows that

$$\|AX\|_{0,\beta,\lambda} \leq |X_0| + \|F^{(b)}(X)\|_{0,\beta,\lambda} + \|G^{(\sigma)}(X)\|_{0,\beta,\lambda}$$
$$\leq |X_0| + C_5 \lambda^{2\beta-1}(1 + \|X\|_{0,\beta,\lambda}) + C_2 \Lambda_{1-\beta}(g) \lambda^{2\beta-1}(1 + \|X\|_{0,\beta,\lambda})$$
$$\leq \lambda^{2\beta-1}(C_5 + C_2 \Lambda_{1-\beta}(g))(1 + \|X\|_{0,\beta,\lambda}) + |X_0|.$$

If $\lambda_0^{2\beta-1}(C_5 + C_2 \Lambda_{1-\beta}(g)) < 1/2$ and $\|X\|_{0,\beta,\lambda_0} \leq 2(1 + |X_0|)$, then $\|AX\|_{0,\beta,\lambda} \leq 2(1 + |X_0|)$. So $A(B_0) \subset B_0$, where

$$B_0 = \left\{ X \in W_0^\beta[0,T] : \|X\|_{0,\beta,\lambda_0} \leq 2(1 + |X_0|) \right\}.$$

For all $X \in B_0$ $\|X\|_{0,\beta} \leq 2(1 + |X_0|)e^{\lambda_0 T}$. Further, for any $X, Y \in B_0$ and $\lambda \geq 1$ from the same lemmas

$$\|AX - AY\|_{0,\beta,\lambda} \leq C_8 \lambda^{2\beta-1}(1 + C_X + C_Y)\|X - Y\|_{0,\beta,\lambda}, \qquad (3.1.8)$$

where $C_8 = C_3 \Lambda_{1-\beta}(g) + C_6$.

If $X \in A(B_0) \subset B_0$ then there exists $\overline{X} \in B_0$ such that $X = A(\overline{X}) \in C^{1-\beta}[0,T]$, and from statements 3 of Lemmas 3.1.1 and 3.1.2

$$\|X\|_{C^{1-\beta}[0,T]} \leq |X_0| + \|F^{(b)}(X)\|_{C^{1-\beta}[0,T]} + \|G^{(\sigma)}(X)\|_{C^{1-\beta}[0,T]}$$
$$\leq (C_1 \Lambda_{1-\beta}(g) + C_4)(1 + \|X\|_{0,\beta}) \leq C_9,$$

where $C_9 = (C_1 \Lambda_{1-\beta}(g) + C_4)(1 + 2(1 + |X_0|)e^{\lambda_0 T})$.

Therefore, for such X

$$C_X \leq C_{10} := \frac{C_9}{\kappa - \beta(1+\kappa)} T^{\kappa - \beta(1+\kappa)}. \qquad (3.1.9)$$

From (3.1.8)–(3.1.9), for any $X, Y \in A(B_0)$

$$\|AX - AY\|_{0,\beta,\lambda_1} \leq \frac{1}{2}\|X - Y\|_{0,\beta,\lambda_1}, \qquad (3.1.10)$$

for such λ_1 that $C_8 \lambda_1^{2\beta-1}(1 + 2C_{10}) \leq \frac{1}{2}$.

Denote by $\rho_i(\cdot,\cdot)$, $i = 0,1$ the equivalent metrics generated by norms $\|\cdot\|_{0,\beta,\lambda_0}$ and $\|\cdot\|_{0,\beta,\lambda_1}$, correspondingly.

Let $X_{n+1} = AX_n$, $n \geq 0$. Then $X_n \in A(B_0)$, $n \geq 1$, and $\rho_1(X_n, X_m) \leq 2^{-n}\rho_1(X_2, X_1) \to 0$ for $m \geq n \to \infty$. Since the metric space $(W_0^\beta[0,T], \rho_1)$ is complete, there exists $X^* \in W_0^\beta[0,T]$ such that $X_n \xrightarrow{\rho_1} X^*$, $n \to \infty$. Evidently, $\rho_0(X_n, X^*) \to 0$, whence $\|X^*\|_{0,\beta,\lambda_0} \leq 2(1 + |X_0|)$, and $X^* \in B_0$. Moreover, $C_{X_n} \leq C_{10}$ and it follows from convergence in ρ_0 that X_n uniformly converges to X^* on $[0,T]$, whence $C_X \leq C_{10}$. Therefore, from (3.1.10),

$$\rho_1(AX_n, AX^*) = \|AX_n - AX^*\|_{0,\beta,\lambda_1}$$
$$\leq \frac{1}{2}\|X_n - X^*\|_{0,\beta,\lambda_1} = \frac{1}{2}\rho_1(X_n, X^*) \to 0, n \to \infty,$$

and it means that $X^* = AX^*$. □

Now, consider the SDE with fBm B_t^H, $H \in (1/2, 1)$ on a complete probability space $(\Omega, \mathcal{F}, P)$:

$$X_t = X_0 + \int_0^t b(s, X_s)ds + \int_0^t \sigma(s, X_s)dB_s^H, \quad t \in [0,T]. \qquad (3.1.11)$$

In this case we can reformulate Theorem 3.1.3 in such a way:

Theorem 3.1.4. *Let the coefficients b and σ satisfy (i)–(v) with $p = (1 - H + \varepsilon)^{-1}$ with some $0 < \varepsilon < H - 1/2, \gamma > 1 - H, \kappa > H^{-1} - 1$ (the constants M, M_R, R, L_R and the function b_0 can depend on ω).*

Then there exists the unique solution $\{X_t, t \in [0,T]\}$ of equation (3.1.11), $X \in L_0(\Omega, \mathcal{F}, P, W_0^{1-H+\varepsilon}[0,T])$ with a.a. trajectories from $C^{H-\varepsilon}[0,T]$.

Remark 3.1.5. Theorem 3.1.4 admits evident generalization to the multidimensional case. Consider the equation on $\mathbb{R}^d$

$$X_t^i = X_0^i + \int_0^t b_i(s, X_s)ds + \sum_{j=1}^m \int_0^t \sigma_{ji}(s, X_s)dB_s^{H_j}, \quad 1 \leq i \leq d, t \in [0,T],$$
(3.1.12)

where the processes B^{H_j} are fBms with Hurst index $H_j \in (1/2, 1), 1 \leq j \leq m$. Denote by $\sigma = (\sigma_{ji})_{i,j=1}^{d,m}$ the matrix of "diffusions" and $b = (b_i)_{i=1}^d$ the "drift" vector, $|\sigma| := (\sum_{i,j} |\sigma_{ji}|^2)^{1/2}$, $|b| := (\sum_i (b_i)^2)^{1/2}$, and suppose that assumptions (i)–(v) hold with these notations, $H = \min_{1 \leq j \leq m} H_j$, $p = (1 - H + \varepsilon)^{-1}$, $\gamma > 1 - H$, $\kappa > H^{-1} - 1$.

Then there exists the unique vector solution X_t of equation (3.1.12) on $[0,T]$ in $L_0(\Omega, \mathcal{F}, P, W_0^{1-H+\varepsilon}[0,T])$ with a.a. trajectories from $C^{H-\varepsilon}[0,T]$.

3.1.2 Norm and Moment Estimates of Solution

We consider equation (3.1.7), suppose that the assumptions of Theorem 3.1.3 hold and, in addition, the coefficient σ satisfies the following growth condition:
(v') $|\sigma(t,x)| \leq M(1 + |x|^\mu)$ for some $0 \leq \mu \leq 1$.

Lemma 3.1.6. *The solution of (3.1.7) satisfies the estimate*

$$\|X\|_{0,\beta} \leq C_0 \exp(C_1(\Lambda_{1-\beta}(g))^{\tilde{\kappa}}),$$

where $0 < \beta < \beta_0 = 1/2 \wedge \gamma \wedge \frac{\kappa}{1+\kappa}$,

$$\tilde{\kappa} = \begin{cases} (1-2\beta)^{-1}, & \text{if } \mu = 1, \\ (1-\beta)^{-1}, & \text{if } 0 \leq \mu < \frac{1-2\beta}{1-\beta}, \\ > \frac{\mu}{1-2\beta}, & \text{if } \frac{1-2\beta}{1-\beta} \leq \mu < 1, \end{cases}$$
(3.1.13)

and the constants C_0 and C_1 depend on T, β, μ and on the constants from conditions (i)–(v).

Proof. Evidently,

$$\varphi_X^\beta(t) \leq |X_0| + \varphi_{F^{(b)}(X)}^\beta(t) + \varphi_{G^{(\sigma)}(X)}^\beta(t).$$
(3.1.14)

From (3.1.6)

$$\begin{aligned}\varphi_{F^{(b)}(X)}^\beta(t) &\leq C_{\beta,T}^3 (L \int_0^t \frac{|X_u|}{(t-u)^\beta} du + c_\beta t^{1-2\beta} B_{0,\beta}) \\ &\leq L C_{\beta,T}^3 \int_0^t \frac{|X_u|}{(t-u)^\beta} du + C_{\beta,T}^4,\end{aligned}$$
(3.1.15)

$$|G_t^{(\sigma)}(X)| \le \Lambda_{1-\beta}(g)\left(\int_0^t \frac{|\sigma(s,X_s)|}{s^\beta}ds + \beta\int_0^t\int_0^r \frac{|\sigma(r,X_r)-\sigma(u,X_u)|}{(r-u)^{\beta+1}}du\,dr\right)$$
$$\le \Lambda_{1-\beta}(g)\left(M\int_0^t \frac{|X_s|^\mu}{s^\beta}ds + M\int_0^t\int_0^r \frac{|X_r-X_u|}{(r-u)^{\beta+1}}du\,dr\right.$$
$$\left. + M\frac{t^{1-\beta}}{1-\beta} + M\frac{t^{\gamma-\beta+1}}{(\gamma-\beta)(\gamma-\beta+1)}\right)$$
$$\le C_{\beta,\gamma,T}\Lambda_{1-\beta}(g) + M\Lambda_{1-\beta}(g)\left(\int_0^t \frac{|X_s|^\mu}{s^\beta}ds + \int_0^t\int_0^r \frac{|X_r-X_u|}{(r-u)^{\beta+1}}du\,dr\right),$$
(3.1.16)

and, similarly to (2.1.15)–(2.1.16)

$$\int_0^t \frac{|G_t^{(\sigma)}(X) - G_s^{(\sigma)}(X)|}{(t-s)^{\beta+1}}ds \le M\Lambda_{1-\beta}(g)(C_{\beta,\gamma,T} + \int_0^t |X_u|^\mu(t-u)^{-2\beta}du$$
$$+ \int_0^t (t-u)^{-\beta}\int_0^u |X_u - X_v|(u-v)^{-\beta-1}dv\,du).$$
(3.1.17)

Let us estimate the "worst" integral $\int_0^t |X_u|^\mu (t-u)^{-2\beta}du$:

$$\int_0^t |X_u|^\mu (t-u)^{-2\beta}du \le \left(\int_0^t \left(\frac{|X_u|^\mu}{(t-u)^\rho}\right)^p du\right)^{1/p} \left(\int_0^t \frac{ds}{(t-s)^{(2\beta-\rho)q}}\right)^{1/q},$$
(3.1.18)

where we must choose $\mu p = 1, (2\beta - \rho)q < 1$, whence $\rho > 2\beta + \mu - 1$, and estimate (3.1.18) takes the form

$$\int_0^t |X_u|^\mu(t-u)^{-2\beta}du \le C_{\beta,\mu,T}\left(\int_0^t \frac{|X_u|}{(t-u)^{\rho/\mu}}du\right)^\mu$$
$$\le C_{\beta,\mu,T}\left(1 + \int_0^t \frac{|X_u|}{(t-u)^\nu}du\right),$$
(3.1.19)

where $\nu = \frac{\rho}{\mu} > \frac{2\beta+\mu-1}{\mu}$ (for $\mu = 1$ we put $\nu = 2\beta$).

From (3.1.14)–(3.1.19) we obtain that $\varphi_X^\beta(t)$ admits an estimate

$$\varphi_X^\beta(t) \le K_1(1 + \Lambda_{1-\beta}(g)) + K_2(1 + \Lambda_{1-\beta}(g)) \cdot \int_0^t \varphi_X^\beta(u)((t-u)^{-\nu} + u^{-\beta})du$$

with constants K_1 and K_2 depending on on T, β, μ and on the constants from conditions (i)–(v). Evidently,

$$(t-u)^{-\nu} + u^{-\beta} = \frac{u^\beta + (t-u)^\nu}{u^\beta(t-u)^\nu} \le (t^\beta + t^\nu)u^{-\beta}(t-u)^{-\nu}.$$

For $\mu > \frac{1-2\beta}{1-\beta}$ we have that $\nu > \beta$; for $0 < \mu \le \frac{1-2\beta}{1-\beta}$ we can put $\nu = \beta > \frac{2\beta+\mu-1}{\mu}$. In any case

$$\varphi_X^\beta(t) \le K_1(1 + \Lambda_{1-\beta}(g)) + K_2(1 + \Lambda_{1-\beta}(g))t^\nu \int_0^t \varphi_X^\beta(u)u^{-\nu}(t-u)^{-\nu}du.$$
(3.1.20)

In (NR00) the following version of the Gronwall lemma was proved: if $0 \le c < 1$, $a, b \ge 0$, $x: \mathbb{R}_+ \to \mathbb{R}_+$ is a continuous function such that for each $t \in [0, T]$

$$x_t \leq a + bt^c \int_0^t (t-s)^{-c} s^{-c} ds, \tag{3.1.21}$$

then

$$x_t \leq C_3 \exp\{C_4 t b^{\frac{1}{1-c}}\}, \tag{3.1.22}$$

where C_3 and C_4 depend only on a, b, c. The proof follows from (3.1.20)–(3.1.22). □

In the case of equation (3.1.11) $g(t) = B^H(t, \omega)$ and instead of $\Lambda_{1-\beta}(g)$ we have the random variable $G := \frac{1}{\Gamma(1-\beta)} \sup_{0 \leq s < t \leq T} |(D_{t-}^{1-\beta} B_{t-}^H)(s)|$. It was considered and estimated in Lemma 1.17.1 and Remark 1.17.2.

Corollary 3.1.7. *It follows from Lemmas 3.1.6 and 1.17.1 that the moments of solution of SDE (3.1.7) admit the following estimate: if the coefficients M, M_R, L, L_R do not depend on ω, $p \geq \frac{1}{\beta}$, $1 - H < \beta < \frac{1}{2} \wedge \gamma \wedge \frac{\kappa}{1+\kappa}$ and we can take the value $\widetilde{\kappa}$ from (3.1.13) not exceeding 2 (it means that $\beta < \frac{1}{4}$ for $\mu = 1$, therefore, $H > 3/4$ for $\mu = 1$ and $\beta < \frac{1}{2} - \frac{\mu}{4}$ if $\frac{1-2\beta}{1-\beta} \leq \mu < 1$), then $E\|X\|_{0,\beta}^q < \infty$ for any $q > 0$.*

3.1.3 Some Other Results on Existence and Uniqueness of Solution of SDE Involving Processes Related to fBm with ($H \in (1/2, 1)$)

It follows from the results of Subsection 3.1.1, that it is possible to consider an SDE involving fBm with $H \in (1/2, 1)$ as an ordinary differential equation for any $\omega \in \Omega', P(\Omega') = 1$. Therefore, the results for the ordinary differential equations with the Hölder continuous forcing can be applied. One of these results belongs to Ruzmaikina (Ruz00). Another approach was developed in the papers (CQ00), (GA98), (GA99a), (GA99b), (Jum93), (IT99), (KKA98a), (Kli98), (KZ99), (MN00), (Mis03), (Zah99) (Zah01) and (Zah05). For example, in the papers (Zah99) and (Zah05) the author considers SDEs of the form

$$dX_t^i = \sum_{j=0}^m \sigma_{ji}(t, X_t) dZ_t^{j,-} + b_i(t, X_t) dt, \tag{3.1.23}$$
$$t \in [0, T], X_{t_0} = X_0, 0 \leq t_0 < T,$$

under the following assumptions:

(vi) $\sigma_{ji} \in C^1(\mathbb{R}^d \times [0, T], \mathbb{R}^d)$ and all partial derivatives are locally Lipschitz in $x \in \mathbb{R}^d$;

(vii) $b_i \in C(\mathbb{R}^d \times [0, T], \mathbb{R}^d)$ is locally Lipschitz in $x \in \mathbb{R}^d$ (with probability 1 in the random case). Here $1 \leq i \leq d$.

Also, X_0 is an arbitrary vector random variable. The integrals w.r.t. the processes Z_t^j are the generalized stochastic forward integrals. What are they and what processes can we consider here? (Recall that the forward (not generalized) integrals were introduced in the Section 2.4.)

Suppose that Y is a stochastic càglàd (left continuous with right limits) process and Z a stochastic càdlàg (right continuous with left limits) process on $[0,T]$.

Then the generalized stochastic forward integral is defined as

$$\int_0^t Y dZ^- := \lim_{\varepsilon \to 0} \varepsilon \int_0^1 u^{\varepsilon-1} \int_0^t Y_s \frac{Z_{t-}(s+u) - Z_{t-}(s)}{u} ds\, du, \quad (3.1.24)$$

whenever the right-hand side is determined, where lim stands for uniform on $[0,T]$ convergence in probability (ucp-convergence), and $\int_0^t = \lim_{\delta \downarrow 0} \int_\delta^1$ a.s. We use the same notation as for the forward integral in the Section 2.4.

Similarly, the generalized quadratic variation process (bracket) is defined as

$$[Z]_t := \lim_{\varepsilon \to 0} \varepsilon \int_0^1 u^{\varepsilon-1}, \int_0^t \frac{1}{u}(Z_{t-}(s+u) - Z_{t-}(s))^2 ds\, du + (Z_t - Z_{t-})^2 \quad (3.1.25)$$

whenever the convergence holds uniformly in probability. If Z is a semimartingale and Y an adapted càglàd process then integral (3.1.24) agrees with the usual Itô integral $\int_{0+}^{t-} Y dZ$, and notion of the generalized bracket coincides with the classical one. If Z is a continuous process with the generalized bracket $[Z]$, and the function $F = F(t,x) : \mathbb{R} \times [0,T] \to \mathbb{R}$, $F \in C^1([0,T]) \times C^1(\mathbb{R})$, then the simple Itô formula holds for $0 \le s < t \le T$:

$$F(t, Z_t) = F(s, Z_s) + \int_s^t \frac{\partial F}{\partial x}(u, Z_u) dZ_u^- \\ + \int_s^t \frac{\partial F}{\partial t}(u, Z_u) du + \frac{1}{2} \int_s^t \frac{\partial^2 F}{\partial x^2}(u, Z_u) d[Z]_u.$$

Now we suppose that the paths of $Z^j, 1 \le j \le m$ from equation (3.1.23) belong to the Sobolev–Slobodeckij space $W_3^{H-} := \bigcap_{\beta < H} W_3^\alpha$, $H \in (1/2, 1)$, where the norm in W_3^β is given by

$$\|f\|_{W_3^\beta} := \|f\|_{L_2[0,T]} + \left(\int_0^T \int_0^T \frac{|f(s) - f(t)|^2}{(t-s)^{2\beta+1}} ds\, dt \right)^{1/2}.$$

We suppose also, that Z^0 is a continuous process with the generalized bracket. Then the sample paths of Z^0 belong to the Sobolev–Slobodeckij space $W_3^{1/2-}$ (for the details see (Zah05)).

Definition 3.1.8. A local solution $X = (X^1, \ldots, X^d)$ of SDE (3.1.23) is a process with the generalized bracket admitting the integral representation

$$X_t^i = X_0^i + \sum_{j=0}^m \int_{t_0}^t \sigma_{ji}(s, X_s) dZ_s^{j,-} + \int_{t_0}^t b_i(s, X_s) ds,$$

in some neighborhood of t_0.

To formulate the main results, it is necessary to consider an auxiliary partial differential equation on $\mathbb{R}^d \times \mathbb{R} \times [0,T]$,

$$\frac{\partial h}{\partial z}(y,z,t) = \sigma_0(t, h(y,z,t)), \quad h(Y_0, Z_0, t_0) = X_0, \quad (3.1.26)$$

where $Z_0 = Z^0(t_0)$, $\sigma_0 = (\sigma_{01}, \ldots, \sigma_{0d})$ and Y_0 is an arbitrary random vector in $\mathbb{R}^d$. Now, the main result of the paper (Zah05) is stated as below.

Theorem 3.1.9 ((Zah05)). *Under suppositions (vi) and (vii), any representation $X(t) = h(Y_t, Z_t^0, t)$ with the function h satisfying equation (3.1.26) and $Y \in W_3^{H-}$ locally determined in some neighborhood of the point $t_0 \in [0,T]$ by the following matrix representation:*

$$\begin{aligned}
dY_t = &\left(\frac{\partial h}{\partial y}(t, Y_t, Z_t^0)\right)^{-1} \Big(\sum_{j=1}^m \sigma_j(t, h(t, Y_t, Z_t^0)) dZ_t^{j,-} \\
&+ (b(t, h(t, Y_t, Z_t^0)) - \frac{\partial h}{\partial t}(t, Y_t, Z_t^0)) dt \\
&- \frac{1}{2} \frac{\partial \sigma_0}{\partial x}(t, h(t, Y_t, Z_t^0)) \sigma_0(t, h(t, Y_t, Z_t^0)) d[Z^0]_t \Big), \\
Y_{t_0} = &Y_0,
\end{aligned} \quad (3.1.27)$$

provides a pathwise local solution of the SDE (3.1.23). (Here we omit index i everywhere.) If X is an arbitrary solution of (3.1.23), then it agrees with any of the above representations on the common interval of definition.

3.1.4 Some Properties of the Stochastic Differential Equations with Stationary Coefficients

Now we consider the multidimensional stochastic differential equation driven by the vector fBm $B_t^H = (B_t^{1,H}, \ldots, B_t^{m,H})$ with the same Hurst index $H \in (1/2, 1)$ and with coefficients, stationary in time:

$$X_t = X_0 + \int_0^t b(X_s)ds + \int_0^t \sigma(X_s)dB_s^H, \quad t \geq 0, \quad (3.1.28)$$

or

$$X_t^i = X_0^i + \int_0^t b_i(X_s)ds + \sum_{j=1}^m \int_0^t \sigma_{ji}(X_s)dB_s^{j,H}, \quad i = 1, \ldots, d,$$

where the processes $B^{j,H}$, $j = 1, \ldots, m$ are fBms with Hurst parameter H defined on the complete probability space $(\Omega, \mathcal{F}, P)$, X_0 is a d-dimensional random variable, and the coefficients $\sigma_{ji}, b_i : \mathbb{R}^d \to \mathbb{R}$ are measurable functions.

The conditions of existence and uniqueness of solution of the equation (3.1.28) on any interval $[0,T]$, consequently on $\mathbb{R}_+$, according to Theorem 3.1.4 can be reduced to the following ones:

(i') Lipschitz continuity of b and σ:
$$|\sigma(x) - \sigma(y)| + |b(x) - b(y)| \leq M|x - y|, \quad x, y \in \mathbb{R}^n;$$

(ii') growth conditions:
$$|b(x)| \leq C(1 + |x|), \quad |\sigma(x)| \leq (1 + |x|^\mu), \quad x \in \mathbb{R}^n$$

for some $\mu \in [0, 1)$ (this condition previously was used only for the estimate of the norm of the solution of SDE (3.1.7));

(iii') local Hölder continuity of $\partial_{x_i} \sigma$:
$$|\partial_{x_i}\sigma(x) - \partial_{x_i}\sigma(y)| \leq M_R |x - y|^\kappa$$

for $1 \leq i \leq d$, $|x|, |y| \leq R$ and some $\kappa > \frac{1}{H} - 1$.

Existence of Pathwise Solution for Bounded Coefficients

Now we relax the conditions on coefficients to obtain the existence (not uniqueness) of the solution of equation (3.1.28).

Theorem 3.1.10. *Let the coefficient b be bounded and continuous, coefficient σ be bounded and Hölder of order $1 > \rho > 1/H - 1$. Then equation (3.1.28) has a pathwise solution.*

Proof. We consider the sequence $\{\psi_n(x), x \in \mathbb{R}^d, n \geq 1\}$ of smooth kernels, such that $\psi_n \geq 0$; $\psi_n = 0$, $|x| \geq \frac{1}{n}$; $\psi_n \in C^\infty(\mathbb{R}^d)$; $\int_{\mathbb{R}^d} \psi_n(x)dx = 1$.
Introduce the functions
$$b_n(x) = \int_{\mathbb{R}^d} b(y)\psi_n(x - y)dy, \quad \sigma_n(x) = \int_{\mathbb{R}^d} \sigma(y)\psi_n(x - y)dy.$$

Then for any $x \in \mathbb{R}^d$ and any $1 \leq i \leq d$
$$|\partial_{x_i} b_n(x)| = \left|\int_{\mathbb{R}^d} b(y)\partial_{x_i}\psi_n(x - y)dy\right|$$
$$\leq \|b\|_\infty \int_{\mathbb{R}^d} |\partial_{x_i}\psi_n(x - y)|\, dy \leq C_n \|b\|_\infty,$$

where $\|b\|_\infty := \sup_{x \in \mathbb{R}^d} |b(x)|$.

The same estimate is true for σ_n, and it means that b_n and σ_n are Lipschitz continuous, with the constants possibly depending on n. Further, for any $x \in \mathbb{R}^d$, $|b_n(x)| = \left|\int_{\mathbb{R}^d} b(y)\psi_n(x - y)dy\right| \leq \|b\|_\infty$ i.e. b_n are bounded functions.

The same is true for σ_n. Finally, for any $N > 0$, $x, y \in \mathbb{R}^d$, $|x| \leq N$, $|y| \leq N$ and $1 \leq i \leq d$

$$|\partial_{x_i}\sigma_n(x) - \partial_{x_i}\sigma_n(y)| \leq \|\sigma\|_\infty \int_{\mathbb{R}^d} |\partial_{x_i}\psi_n(x-z) - \partial_{x_i}\psi_n(y-z)|\,dz$$

$$\leq \|\sigma\|_\infty \sup_{|z|\leq\frac{1}{n}, 1\leq i,j\leq d} \left|\partial^2_{x_i,x_j}\psi_n(z)\right|(N+1)^d|x-y|,$$

i.e. $\partial_{x_i}\sigma_n$ satisfy the local Lipschitz conditions. These estimates demonstrate that b_n and σ_n satisfy conditions (i')–(iii') that in turn ensure the pathwise existence (and uniqueness) of the solution of the equation

$$X^n_t = X_0 + \int_0^t b_n(X^n_s)ds + \int_0^t \sigma_n(X^n_s)dB^H_s. \tag{3.1.29}$$

We fix some $\omega \in \Omega$ and denote by C different constants even if they depend on Ω. According to Theorem 3.1.4 and Remark 3.1.5, the solution X^n_t is Hölder continuous of order $H-\delta$ for any $\delta > 0$; from Hölder continuity of σ we obtain that

$$|\sigma_n(x) - \sigma_n(y)| \leq \int_{\mathbb{R}^d} |\sigma(x-z) - \sigma(y-z)||\psi_n(z)|\,dz \leq C|x-y|^\rho, \tag{3.1.30}$$

therefore, $\sigma_n(X^n_s)$ belongs to the space $W^\beta_2[0,T]$ for any $\beta < H\rho$. By using estimate (2.1.14) and the boundedness of σ_n for any $0 \leq s \leq t$, we obtain for each $1 - H < \beta < \rho H$ (this is possible since $\rho > 1/H - 1$) the estimate

$$\left|\int_s^t \sigma_n(X^n_u)dB^H_u\right|$$

$$\leq G\left(\int_s^t \frac{|\sigma_n(X^n_u)|}{(u-s)^\beta}du + \int_s^t \int_s^u \frac{|\sigma_n(X^n_u) - \sigma_n(X^n_y)|}{(u-y)^{\beta+1}}dy\,du\right)$$

$$\leq G\|\sigma\|_\infty \frac{(t-s)^{1-\beta}}{1-\beta} + CG\int_s^t \int_s^u \frac{|X^n_u - X^n_y|^\rho}{(u-y)^{\beta+1}}dy\,du.$$

Here $G_T := \Lambda_{1-\beta}(B^H)$ (see Section 1.17) and $EG^p_T < \infty$ for any $p > 0$ (Lemma 1.17.1 and Remark 1.17.2). Finally, we can estimate

$$|X^n_t - X^n_s| \leq \left|\int_s^t b_n(X^n_u)du\right| + \left|\int_s^t \sigma_n(X^n_u)dB^H_u\right| \leq \|b\|_\infty(t-s)$$

$$+ G_T \frac{\|\sigma\|_\infty}{1-\alpha}(t-s)^{1-\alpha} + CG_T \int_s^t \int_s^u \frac{|X^n_u - X^n_y|^\rho}{(u-y)^{\beta+1}}dy\,du. \tag{3.1.31}$$

Consider any fixed interval $[0,T]$ and denote

$$\|X\|_{1-\beta,T} := \sup_{0\leq s<t\leq T} \frac{|X_t - X_s|}{(t-s)^{1-\beta}}.$$

Check, first, that the inequality $\|X^n\|_{1-\beta,T} < \infty$ holds for any $T > 0$. Indeed, X^n are Hölder of order $H - \varepsilon$ for any $\varepsilon > 0$ with constant, possibly depending on n (Theorem 3.1.4 and Remark 3.1.5), therefore $\|X^n\|_{1-\beta,T} < \infty$ a.s. Now, from (3.1.31),

$$\|X^n\|_{1-\beta,T} \leq \|b\|_\infty T^\beta + G_T \frac{\|\sigma\|_\infty}{1-\beta}$$
$$+ CG_T \sup_{0 \leq s < t \leq T} |t-s|^{\beta-1} \int_s^t \int_s^u \left(\frac{|X_u^n - X_y^n|}{(u-y)^{1-\beta}}\right)^\rho |u-y|^{\rho-\rho\beta-1-\beta} dy\, du$$
$$\leq C + C(\|X^n\|_{1-\beta,T})^\rho \cdot \sup_{0 \leq s < t \leq T} (t-s)^{\beta-1+(1-\beta+\rho-\rho\beta)}$$
$$\leq C + C(\|X^n\|_{1-\beta,T})^\rho,$$

under the condition $\rho - \rho\beta - 1 - \beta > -1$, i.e. $\rho > \frac{\beta}{1-\beta}$, which is possible for some $\beta > 1 - H$, since $\rho > 1/H - 1$. Note that for $0 < \rho < 1$ the equality $P = C(1+P^\rho)$ has the unique root $P_0 > 0$ and the inequality $P \leq C(1+P^\rho)$ holds for $P \leq P_0$. Therefore, $\|X^n\|_{1-\beta,T} \leq P_0(T,\rho)$, where $P_0(T,\rho)$ depends only on T and ρ, not on n. This means that

$$|X_t^n - X_s^n| \leq P_0(T,\rho)(t-s)^{1-\beta}, \tag{3.1.32}$$

which according to the Arcela criterion means that the sequence $\{X_t^n, t \in [0,T]\}$, $n \geq 1$ is tight for any $\omega \in \Omega$ in the space $C[0,T]$. Evidently, we can conclude that there exists $\{X_t^{n_k}, t \in [0,T]\}$, $n_k \geq 1$, such that $X_t^{n_k} \to X_t$ in the space $C[0,T]$. We can suppose that $X_t^n \to X_t$ in $C[0,T]$. Now it is sufficient to prove that $X(t)$ is a solution of (3.1.28). Let consider some auxiliary estimates. First,

$$\left|\int_0^t (b_n(X_s^n) - b(X_s)) ds\right| \leq \int_0^t |b_n(X_s^n) - b_n(X_s)|\, ds + \int_0^t |b_n(X_s) - b(X_s)|\, ds. \tag{3.1.33}$$

Further, for any $x, y \in \mathbb{R}^d$

$$|b_n(x) - b_n(y)| = \left|\int_{\mathbb{R}^d} (b(x-u) - b(y-u))\psi_n(u)du\right|$$
$$\leq \int_{|u| \leq \frac{1}{n}} |b(x-u) - b(y-u)|\psi_n(u) du \leq \sup_{|u| \leq \frac{1}{n}} |b(x-u) - b(y-u)|. \tag{3.1.34}$$

The process $\{X_t, t \in [0,T]\}$ is continuous on $[0,T]$, so bounded for any $\omega \in \Omega$. Let $C(T,\omega) = \sup_{0 \leq s \leq T} |X_s|$. For any $\varepsilon > 0$ there exists $\eta > 0$ such that

$$\sup_{|s-z|<\eta,\, |s| \leq C(T,\omega)+1} |b(s) - b(z)| < \varepsilon. \tag{3.1.35}$$

For any $\eta > 0$ and any $\omega \in \Omega$ there exists such $n_0 \in N$ that $|X_s^n - X_s| < \eta$, $n \geq n_0$, $s \in [0,T]$. From these estimates,

$$|b_n(X_s^n) - b_n(X_s)| \leq \sup_{|u| \leq \frac{1}{n}} |b_n(X_s^n - u) - b_n(X_s - u)|$$
$$\leq \sup_{|s| \leq C(T,\omega) + \frac{1}{n}} \sup_{|s-z| < \eta} |b(s) - b(z)| < \beta, \quad n \geq n_0. \quad (3.1.36)$$

Since $\beta > 0$ is arbitrary, we obtain that a.s.

$$\int_0^t |b_n(X_s^n) - b_n(X_s)|\, ds \to 0, \quad n \to \infty. \quad (3.1.37)$$

The second term on the right-hand side of (3.1.33) can be estimated in such a way:

$$|b_n(X_s) - b(X_s)| \leq \int_{\mathbb{R}^d} |b(X_s - u) - b(X_s)| \psi_n(u)\, du$$
$$\leq \sup_{|z| \leq C(T,\omega)} \sup_{|u| \leq \frac{1}{n}} |b(z-u) - b(z)| \to 0, \quad n \to \infty,$$

since b is a continuous function. Moreover, b_n are bounded, which implies the convergence $\int_0^t |b_n(X_s) - b(X_s)|\, ds \to 0$, $n \to \infty$ a.s. We obtain that $\int_0^t b_n(X_s^n)\, ds \to \int_0^t b(X_s)\, ds$ a.s., $t > 0$. Furthermore,

$$\left|\int_0^t (\sigma_n(X_s^n) - \sigma(X_s))\, dB_s^H\right| \leq \left|\int_0^t (\sigma_n(X_s^n) - \sigma_n(X_s))\, dB_s^H\right|$$
$$+ \left|\int_0^t (\sigma_n(X_s) - \sigma(X_s))\, dB_s^H\right|. \quad (3.1.38)$$

Now we can estimate the first term of (3.1.38) for any $1 - H < \beta < \frac{1}{2}$:

$$\left|\int_0^t (\sigma_n(X_s^n) - \sigma_n(X_s))\, dB_s^H\right| \leq G \int_0^t \frac{|\sigma_n(X_s^n) - \sigma_n(X_s)|}{s^\beta}\, ds$$
$$+ G \int_0^t \int_0^u \frac{|\sigma_n(X_u^n) - \sigma_n(X_r^n) + \sigma_n(X_r) - \sigma_n(X)|}{(u-r)^{1+\beta}}\, dr\, du. \quad (3.1.39)$$

Similarly to estimates (3.1.34)–(3.1.37), we obtain that a.s. $\sup_{s \leq T} |\sigma_n(X_s^n) - \sigma_n(X_s)| \to 0$ and $\int_0^t |\sigma_n(X_s^n) - \sigma_n(X_s)|\, s^{-\beta}\, ds \to 0$, while $n \to \infty$. Further, recall the estimate (3.1.30). For any sufficiently small $\varepsilon > 0$, present $\int_0^t \int_0^u$ on the right-hand side of (3.1.39) as $\int_0^t \int_0^u$
$= \int_\varepsilon^t \int_0^{u-\varepsilon} + \int_0^\varepsilon \int_0^u + \int_\varepsilon^t \int_{u-\varepsilon}^u$, and the integrals on the right-hand side can be estimated as

$$\int_\varepsilon^t \int_0^{u-\varepsilon} \leq 2 \sup_{s\leq T} |\sigma_n(X_s^n) - \sigma_n(X_s)| \int_\varepsilon^t \int_0^{u-\varepsilon} \frac{dr\,du}{(u-r)^{1+\beta}}$$

$$\leq C\varepsilon^{-\alpha} \cdot \sup_{s\leq T} |\sigma_n(X_s^n) - \sigma_n(X_s)| \to 0$$

a.s. for any fixed $\varepsilon > 0$. Further, from (3.1.32),

$$\int_\varepsilon^t \int_{u-\varepsilon}^u \leq C \int_\varepsilon^t \int_{u-\varepsilon}^u \frac{|X_u^n - X_r^n|^\rho + |X_r - X_u|^\rho}{(u-r)^{1+\beta}} dr\,du$$

$$\leq C \int_0^t \int_{u-\varepsilon}^u (u-r)^{\rho(1-\beta)-1-\beta} dr\,du = C\varepsilon^{\rho(1-\beta)-\beta} \qquad (3.1.40)$$

is small for small $\varepsilon > 0$, and moreover, C does not depend on n. The integral

$$\int_0^\varepsilon \int_0^u \leq C\varepsilon^{\rho(1-\beta)-\beta+1}. \qquad (3.1.41)$$

Therefore, since $\varepsilon > 0$ is arbitrary, $\int_0^t \int_0^u \to 0$ a.s. while $n \to \infty$.

The second term of (3.1.38) can be estimated as

$$\left| \int_0^t (\sigma_n(X_s) - \sigma(X_s)) dB_s^H \right| \leq G \int_0^t \frac{|\sigma_n(X_s) - \sigma(X_s)|}{s^\beta} ds$$

$$+ G \int_0^t \int_0^u \frac{|\sigma_n(X_u) - \sigma_n(X_r) - \sigma(X_u) + \sigma(X_r)|}{(u-r)^{1+\beta}} dr\,du.$$

Evidently,

$$|\sigma_n(X_s) - \sigma(X_s)| \leq \int_{\mathbb{R}^d} |\sigma(X_s - u) - \sigma(X_s)| \psi_n(u) du$$

$$\leq \sup_{|u|\leq \frac{1}{n}} |u|^\rho \leq \frac{1}{n^\rho} \to 0, \quad n \to \infty$$

and $\int_0^t |\sigma_n(X_s) - \sigma(X_s)| s^{-\beta} ds \to 0$ a.s., $n \to \infty$. Now, as before, $\int_0^t \int_0^u = \int_\varepsilon^t \int_0^{u-\varepsilon} + \int_0^\varepsilon \int_0^u + \int_\varepsilon^t \int_{u-\varepsilon}^u$ and $\int_\varepsilon^t \int_0^{u-\varepsilon} \leq 2\frac{1}{n^\rho} \cdot \varepsilon^{-\beta} \to 0$ for any fixed $\varepsilon > 0$ and other integrals can be estimated as in (3.1.40)–(3.1.41). So, $\int_0^t \sigma_n(X_s^n) dB_s^H \to \int_0^t \sigma(X_s) dB_s^H$ a.s. while $n \to \infty$ and $X_t = \int_0^t \sigma(X_s) dB_s^H + \int_0^t b(X_s) ds$. The theorem is proved. □

Remark 3.1.11. By similar, but even more simple arguments we can prove the existence of the solution of the equation

$$X_t = X_0 + \int_0^t b(X_s) ds + \int_0^t f(s) dB_s^H,$$

where b is bounded and continuous, $f \in C^{1-H}[0,T]$, X_0 is a real-valued random variable.

Differentiability and Local Differentiability of the Solution

Here we shall use the elements of Malliavin calculus with respect to fBm B^H, contained in Section 2.4.

Suppose that we consider some subspace $\Omega_1 \subset \Omega$ and restrict $\mathcal{F}$ and P to Ω_1. Denote the mathematical expectation w.r.t. restricted measure P_1 as E_1.

Definition 3.1.12. Random variable F belongs locally to the space $D_{1,p}(\mathcal{H})$ on $[0, T]$ if there exists a sequence $\Omega_1 \subset \Omega_2 \subset \ldots \subset \Omega$ such that $\bigcup_{n=1}^{\infty} \Omega_n = \Omega$ and $\|F\|_{1,p,n}^p := E_n(|F|^p) + E_n(\|DF\|_{\mathcal{H}}^p) < \infty$.

In this case we say that F is *locally differentiable*, $F \in D_{1,p,loc}$. According to Lemma 1.5.4 (Nua95), see also (Nua98), we can formulate the sufficient conditions of local differentiability. Let $\{F_r, r \geq 1\}$ be a sequence of r.v. from $D_{1,p,loc}$ satisfying the conditions

(viii) $F_r \to F$ in any $L^p(\Omega_n)$, $n \geq 1$,
(ix) $\sup_r \|F_r\|_{1,p,n} < \infty$ for any $n \geq 1$.

Then F belongs to $D_{1,p,loc}$.

Remark 3.1.13. Suppose that there exists a localizing sequence $(\Omega_n, n \geq 1)$, such that $F \in D_{1,p,loc}$ for any $p > 1$. Then we say that $F \in D_{1,\infty,loc}$.

Consider equation (3.1.28) and suppose that its coefficients X_0, b and σ satisfy conditions (i')–(ii') and

(x) $b \in C^1(\mathbb{R}^d)$;
(xi) $|\partial_{x_i}\sigma(x) - \partial_{x_i}\sigma(y)| \leq M|x - y|$, $\forall x, y \in \mathbb{R}^d$;
(xii) $X_0 \in D_{1,\infty} := \bigcap_{p \geq 1} D_{1,p}(\mathcal{H})$, X_0 is a bounded $\mathcal{F}_0$-adapted random variable.

Theorem 3.1.14. *1. Let conditions (i')–(ii') and (x)–(xii) hold. Then the unique solution X_t of equation (3.1.28) is locally differentiable in the sense that $X_t^i \in D_{1,\infty,loc}$ for any $1 \leq i \leq d$ with the same localizing sequence.*
2. Let equation (3.1.28) be semilinear, i.e. $\sigma(x) = \sigma x$, conditions (i'), (ii'), (x), (xii) hold for b and X_0 and $H > 3/4$. Then $X_t^i \in D_{1,\infty}$ for any $1 \leq i \leq d$.

Proof. 1. Let $T > 0$ be fixed. According to Theorem 3.1.4 and Remark 3.1.5, under conditions (i')–(i'') and (x) equation (3.1.28) has the unique solution X_t on the interval $[0, T]$. Moreover, it can be obtained by successive approximations, $\{X_t^{(n)}, n \geq 0\}$, $t \in [0, T]$ where $X_t^{(0)} = X_0 \in D_{1,\infty}$. Further we consider the case $d = 1$, for technical simplicity; in the general case they are similar. We use induction. Suppose that $X_t^{(k)} \in D_{1,\infty}$, $1 \leq k \leq n$, and the derivatives $D_s X_t^{(k)}$, $0 \leq s \leq t \leq T$, $1 \leq k \leq n$ are Hölder continuous of order $1 - \beta$ for some $1 - H < \beta < 1/2$. Since the approximations $X_t^{(n)}$ are Hölder continuous of any order not exceeding H, and from conditions (i') and (x) σ' and b' are bounded, $\sigma'(X_r^{(n)}) D_s X_r^{(n)}$ is Hölder continuous in r of order $1 - \beta$.

3.1 SDEs Driven by fBm with Pathwise Integrals

Therefore, the integrals $\int_s^t \sigma'(X_r^{(n)}) D_s X_r^{(n)} dB_r^H$ and $\int_s^t b'(X_r^{(n)}) D_s X_r^{(n)} dr$ exist, $0 \leq s \leq t \leq t \leq T$ and

$$D_s X_t^{(n+1)} = \sigma(X_s^{(n)}) + \int_s^t \sigma'(X_r^{(n)}) D_s X_r^{(n)} dB_r^H + \int_s^t b'(X_r^{(n)}) D_s X_r^{(n)} dr.$$

Hence for any $1 - H < \beta < \frac{1}{2}$, from (2.1.14),

$$|D_s X_t^{(n+1)}| \leq |\sigma(X_s^{(n)})| + M \int_s^t |D_s X_r^{(n)}| dr + MG \int_s^t |D_s X_r^{(n)}|(r-s)^{-\beta} dr$$

$$+ G \int_s^t \int_s^r |\sigma'(X_r^{(n)}) D_s X_r^{(n)} - \sigma'(X_u^{(n)}) D_s X_u^{(n)}|(r-u)^{-1-\beta} du \, dr, \quad (3.1.42)$$

where $G = 1/\Gamma(\beta) \sup_{0<s<t<T} \left|(D_{t-}^{1-\beta} B_{t-}^H)(s)\right|$, $E \exp\{pG^\delta\} < \infty$ for any $p > 0$ and $0 < \delta < 2$.

Further we denote by C different constants not depending on ω. Note that

$$|\sigma(X_s^n)| \leq C \left(1 + |X_s^{(n)}|^\mu\right).$$

Further, it follows from condition (ii') and Lemma 3.1.6, that $\sup_{0 \leq s \leq T} \left|X_s^{(n)}\right| \leq C \exp\{CG^{\widetilde{\kappa}}\}$, where

$$\widetilde{\kappa} = \begin{cases} \frac{1}{1-2\beta}, & \text{if } \mu = 1; \\ > \frac{\mu}{1-2\beta}, & \text{if } \frac{1-2\beta}{1-\beta} \leq \mu < 1; \\ \frac{1}{1-\beta}, & \text{if } 0 \leq \mu < \frac{1-2\beta}{1-\beta}. \end{cases}$$

Finally, $|\sigma(X_s^n)| \leq C \exp\{CG^{\widetilde{\kappa}}\}$, and from (3.1.42) it follows that

$$\left|D_s X_t^{(n+1)}\right| \leq C \exp\{CG^{\widetilde{\kappa}}\} + M \int_s^t \left|D_s X_r^{(n)}\right| dr + MG \int_s^t \frac{\left|D_s X_r^{(n)}\right|}{(r-s)^\beta} dr$$

$$+ MG \int_s^t \int_s^r \frac{\left|X_r^{(n)} - X_u^{(n)}\right| \left|D_s X_r^{(n)}\right|}{(r-u)^{1+\beta}} du \, dr$$

$$+ CG \int_s^t \int_s^r \frac{\left|D_s X_r^{(n)} - D_s X_u^{(n)}\right|}{(r-u)^{1+\beta}} du \, dr. \quad (3.1.43)$$

It follows from Lemmas 3.1.1, 3.1.2 and 3.1.6 that

$$|X_s^n - X_r^n| \leq \exp\{CG^{\widetilde{\kappa}}\} |s - r|^{1-\beta} \quad (3.1.44)$$

for any $1 - H < \beta < \frac{1}{2}$. In this case $1 - 2\beta > 0$ and from (3.1.43)

$$\left|D_s X_t^{(n+1)}\right| \leq C \exp\{CG^{\widetilde{\kappa}}\} + M \int_s^t \left|D_s X_r^{(n)}\right| dr$$

$$+ MG \int_s^t \frac{\left|D_s X_r^{(n)}\right|}{(r-s)^\beta} dr + MG \exp\{CG^{\widetilde{\kappa}}\} \int_s^t \left|D_s X_r^{(n)}\right| (r-s)^{1-2\beta} dr$$

$$+ CG \int_s^t \int_s^r \frac{\left|D_s X_r^{(n)} - D_s X_u^{(n)}\right|}{(r-u)^{1+\beta}} du\, dr,$$

or, briefly,

$$\left|D_s X_t^{(n+1)}\right| \leq C \exp\{CG^{\widetilde{\kappa}}\}$$

$$+ CG \exp\{CG^{\widetilde{\kappa}}\} \int_s^t \frac{\left|D_s X_r^{(n)}\right|}{(r-s)^\beta} dr + CG \int_s^t \int_s^r \frac{\left|D_s X_r^{(n)} - D_s X_u^{(n)}\right|}{(r-u)^{1+\beta}} du\, dr.$$

Further, for any $0 \leq u < r \leq T$

$$D_s X_r^{(n+1)} - D_s X_u^{(n+1)} = \int_u^r \sigma'(X_v^{(n)}) D_s X_v^{(n)} dB_v^H + \int_u^r b'(X_v^{(n)}) D_s X_v^{(n)} dv,$$

and we obtain by similar estimates that

$$\left|D_s X_r^{(n+1)} - D_s X_u^{(n+1)}\right| \leq CG \exp\{CG^{\widetilde{\kappa}}\} \int_u^r \frac{\left|D_s X_u^{(n)}\right|}{(v-u)^\beta} dv$$

$$+ CG \int_u^r \int_u^v \frac{\left|D_s X_v^{(n)} - D_s X_z^{(n)}\right|}{(v-z)^{1+\beta}} dz\, dv, \quad (3.1.45)$$

whence

$$\int_s^r \frac{\left|D_s X_r^{(n+1)} - D_s X_u^{(n+1)}\right|}{(r-u)^{1+\beta}} du$$

$$\leq CG \exp\{CG^{\widetilde{\kappa}}\} \int_s^r \frac{1}{(r-u)^{1+\beta}} \int_u^r \frac{\left|D_s X_v^{(n)}\right|}{(v-u)^\beta} dv\, du$$

$$+ CG \int_s^r \frac{1}{(r-u)^{1+\beta}} \int_u^r \int_u^v \frac{\left|D_s X_v^{(n)} - D_s X_z^{(n)}\right|}{(v-z)^{1+\beta}} dz\, dv.$$

Denote $\varphi_n^1(t) := \left|D_s X_t^{(n)}\right|$, $\varphi_n^2(t) := \int_s^t \frac{\left|D_s X_t^{(n)} - D_s X_u^{(n)}\right|}{(t-u)^{1+\beta}} du$, $\varphi_0^1(t) := D_s X_0$, $\varphi_0^2(t) := 0$, $\widetilde{C}_1(\omega) := C \exp\{CG^{\widetilde{\kappa}}\}$, $\widetilde{C}_2(\omega) := CG \exp\{CG^{\widetilde{\kappa}}\}$, $\widetilde{C}_3(\omega) := CG$. Then for $\varphi_n(t) := \varphi_n^1(t) + \varphi_n^2(t)$

3.1 SDEs Driven by fBm with Pathwise Integrals

$$\varphi_{n+1}^1(t) \leq \tilde{C}_1(\omega) + \tilde{C}_2(\omega) \int_s^t \varphi_n^1(r)(r-s)^{-\beta} dr + \tilde{C}_3(\omega) \int_s^t \varphi_n^2(r) dr$$

$$\leq \tilde{C}_1(\omega) + \left(\tilde{C}_2(\omega) + \tilde{C}_3(\omega) T^\beta\right) \int_s^t \varphi_n(r)(r-s)^{-\beta} dr;$$

$$\varphi_{n+1}^2(t) \leq \tilde{C}_2(\omega) \int_s^t \frac{1}{(t-u)^{1+\beta}} \int_u^t \frac{\varphi_n^1(v)}{(v-u)^\beta} dv\, du$$

$$+ \tilde{C}_3(\omega) \int_s^t \frac{1}{(t-u)^{1+\beta}} \int_u^t \varphi_n^2(v) dv\, du$$

$$= \tilde{C}_2(\omega) \int_s^t \varphi_n^1(v) \int_s^v \frac{du}{(v-u)^\beta (t-u)^{1+\beta}} dv$$

$$+ \tilde{C}_3(\omega) \int_s^t \varphi_n^2(v) \int_s^v \frac{du}{(t-u)^{1+\beta}} dv.$$

Since $\int_s^v (v-u)^{-\beta}(t-u)^{-1-\beta} du \leq C(t-v)^{-2\beta}$, with $C = \int_0^\infty u^{-\beta}(1+u)^{-1-\beta} du$, we have that

$$\varphi_{n+1}^2(t) \leq \tilde{C}_2(\omega) C \int_s^t \varphi_n^1(v)(t-v)^{-2\beta} dv + \tilde{C}_3(\omega) C \int_s^t \varphi_n^2(v)(t-v)^{-\beta} dv$$

$$\leq C \left(\tilde{C}_2(\omega) + \tilde{C}_3(\omega) T^\beta\right) \int_s^t \varphi_n(v)(t-v)^{-2\beta} dv.$$

Finally,

$$\varphi_{n+1}(t) \leq C(\omega) \left(1 + \int_0^t \varphi_n(v) \left((t-v)^{-2\beta} + v^{-2\beta}\right) dv\right)$$

$$\leq C(\omega) \left(1 + t^{2\beta} \int_0^t \varphi_n(v)(t-v)^{-2\beta} v^{-2\beta} dv\right),$$

where $C(\omega) = C\tilde{C}_1(\omega) \vee \left(\tilde{C}_2(\omega) + \tilde{C}_3(\omega) T^\beta \vee 1\right)$ for some $C > 0$.

It is very easy to check by induction, similarly to (3.1.20)–(3.1.21), that

$$\varphi_n(t) \leq C(\omega) C_1 \exp\{C_2 t \left(C(\omega)\right)^{\frac{1}{1-2\beta}}\} =: \psi(t),$$

where C_1 and C_2 depend only on β. In particular, $\varphi_n^1 \leq \psi(t)$ and $\varphi_n^2 \leq \psi(t)$. Evidently, $\sup_{s \leq t \leq T} \psi(t) =: \tilde{C}(\omega) < \infty$ a.s., and from (3.1.45) it follows that

$$\left| D_s X_r^{(n+1)} - D_s X_u^{(n+1)} \right| \leq C(\omega) \tilde{C}(\omega) \left(\frac{(r-u)^{1-\beta}}{1-\beta} + (r-u)\right),$$

i.e. $D_s X_r^{(n+1)}$ is Hölder continuous of index $1-\beta$ (it is necessary for induction). Denote $\Omega_k := \{\omega : \tilde{C}(\omega) \leq k\}$. Then

$$E_k \sup_{0 \le s \le t \le T} \left| D_s X_t^{(n)} \right|^p \le k^p < \infty. \tag{3.1.46}$$

Moreover,

$$\left| X_t^{(n+1)} - X_t \right| \le \int_0^t \left| b(X_s^{(n)}) - b(X_s) \right| ds + \left| \int_0^t (\sigma(X_s^{(n)}) - \sigma(X_s)) dB_s^H \right|$$

$$\le M \int_0^t \left| X_s^{(n)} - X_s \right| ds + C(\omega) \int_0^t \frac{\left| X_s^{(n)} - X_s \right|}{s^\beta} ds$$

$$+ C(\omega) \int_0^t \int_0^r \frac{\left| \sigma(X_r) - \sigma(X_r^{(n)}) - \sigma(X_n) + \sigma(X_u^{(n)}) \right|}{(r-u)^{1+\beta}} du \, dr.$$

From Lemma 7.1 (NR00), conditions (i'), (x) and from (3.1.44), it follows that

$$\left| \sigma(X_r) - \sigma(X_r^{(n)}) - \sigma(X_u) + \sigma(X_u^{(n)}) \right|$$

$$\le C \left| X_r - X_u - X_r^{(n)} + X_u^{(n)} \right| + C \left| X_r^{(n)} - X_r \right| \left(|X_r - X_u| + \left| X_r^{(n)} - X_u^{(n)} \right| \right)$$

$$\le C \left| X_r^{(n)} - X_r \right| \left(C \exp\{CG^{\tilde{\kappa}}\} |u - r|^{1-\beta} \right) + C \left| X_r - X_u - X_r^{(n)} + X_u^{(n)} \right|.$$

Then

$$\left| X_t^{(n+1)} - X_t \right| \le C(\omega) \Big(\int_0^t \int_0^r \frac{\left| X_r - X_u - X_r^{(n)} + X_u^{(n)} \right|}{(r-u)^{1+\beta}} du \, dr$$

$$+ \int_0^t \left| X_s^{(n)} - X_s \right| s^{-\beta} ds \Big).$$

By similar estimates we obtain

$$\int_0^t \frac{\left| X_t^{(n+1)} - X_u^{(n+1)} - X_t - X_u \right|}{(t-u)^{1+\beta}} du$$

$$\le C(\omega) \int_0^t \frac{du}{(t-u)^{1+\beta}} \Big(\int_u^t \left| X_s^{(n)} - X_s \right| s^{-\beta} ds$$

$$+ \int_u^t \int_u^r \frac{\left| X_r - X_v - X_r^{(n)} + X_v^{(n)} \right|}{(r-v)^{1+\beta}} du \, dr \Big).$$

Denote $\xi_n^1(t) := \left| X_t^{(n)} - X_t \right|$, $\xi_n^2(t) := \int_0^t \left| X_t^{(n)} - X_u^{(n)} - X_t + X_u \right|$
$\times (t-u)^{-1-\beta} du$, then

$$\xi_{n+1}^1(t) \le C(\omega) \Big(\int_0^t \xi_n^1(s) s^{-\beta} ds + \int_0^t \xi_n^2(s) ds \Big),$$

$$\xi_{n+1}^2(t) \leq C(\omega)\Big(\int_0^t \xi_n^1(s)s^{-\beta}\int_0^s \frac{du}{(t-u)^{1+\beta}}ds$$
$$+\int_0^t \xi_n^2(s)\int_0^s \frac{du}{(t-u)^{1+\beta}}ds\Big) \leq C(\omega)\int_0^t s^{-\beta}(t-s)^{-\beta}(\xi_n^1(s)+\xi_n^2(s))ds.$$

Let $\xi_n(t) = \xi_n^1(t) + \xi_n^2(t)$, then $\xi_{n+1}(t) \leq C(\omega)t^{2\beta}\int_0^t s^{-2\beta}(t-s)^{-2\beta}\xi_n(s)ds$. Denote $C_4(\omega) := \sup_{0\leq s\leq T} |\xi_0(s)|$. Then it is easy to obtain that

$$\xi_n(t) \leq (C(\omega))^n C_4(\omega) \frac{\Gamma(1-2\beta)^{n+1}}{\Gamma(n(1-2\beta))} t^{n(1-2\beta)}. \tag{3.1.47}$$

Hence, $E_k \sup_{0\leq t\leq T} |\xi_n(t)|^p \leq Ck^{(n+1)p} < \infty$ for some $C > 0$ and any $p > 0$, where $\Omega_k = \{\omega : \tilde{C}(\omega) \leq k, C_4(\omega) \leq k\}$. Finally, we obtain from (3.1.47) that $E_k \sup_{0\leq t\leq T} |X_t^{(n)} - X_t|^p \to 0$, $n \to \infty$. Together with (3.1.46) it means that $X(t) \in D_{1,\infty,loc}$.

2. Let equation (3.1.28) be semilinear, i.e. it has a form

$$X_t = X_0 + \int_0^t b(X_s)ds + \sigma \int_0^t X_s dB_s^H,$$

where b satisfies conditions (i'), (ii'), (x), X_0 satisfies condition (xii). Then

$$\left|D_s X_t^{(n+1)}\right| \leq \tilde{C}_1(\omega) + C(\omega)\int_s^t \frac{|D_s X_r^n|}{(r-s)^\beta}dr$$
$$+ C(\omega)|\sigma|\int_s^t \int_s^r \frac{|D_s X_r^n - D_s X_u^n|}{(r-u)^{1+\beta}}du\,dr,$$

$$\left|D_s X_r^{n+1} - D_s X_u^{n+1}\right|$$
$$\leq C(\omega)\int_u^r \frac{|D_s X_z^n|}{(z-u)^\beta}dz + C(\omega)|\sigma|\int_u^r \int_u^z \frac{|D_s X_z^n - D_s X_v^n|}{(z-v)^{1+\beta}}dv\,dz,$$

or in terms of $\varphi_n^1(t)$ and $\varphi_n^2(t)$ from Part 1 of the proof,

$$\varphi_n^1(t) \leq \tilde{C}_1(\omega) + C(\omega)\int_s^t \varphi_n^1(r)(r-s)^{-\beta}dr + C(\omega)|\sigma|\int_s^t \varphi_n^2(r)dr,$$
$$\varphi_n^2(t) \leq C(\omega)\int_s^t \varphi_n^1(v)(t-v)^{-2\beta}dv + C(\omega)\int_s^t \varphi_n^2(v)(t-v)^{-\beta}dv.$$

Repeating the same estimates as in Part 1 but with other constants, we obtain

$$\varphi_n^1(t) \leq \tilde{C}_1(\omega)\exp\{CG^{\tilde{\kappa}}(t-s)\}.$$

Evidently, $E|\varphi_n^1(t)|^p \leq C_p < \infty$ since now, of course, $\mu = 1$, $\widetilde{\kappa} = \frac{1}{1-2\beta}$, and for $H > \frac{3}{4}$ the coefficient $\beta > 1 - H$ can be chosen in such a way that $\frac{1}{1-2\beta} < 2$. Moreover, in this, semilinear, case,

$$\left|X_t^{(n+1)} - X_t\right|$$
$$\leq C(\omega) \int_0^t \frac{|X_s^n - X_s|}{s^\beta} ds + C(\omega) \int_0^t \int_0^r \frac{\left|X_r - X_r^{(n)} - X_u + X_u^{(n)}\right|}{(r-u)^{1+\beta}} du\, dr,$$

$$\int_0^t \frac{\left|X_t^{(n+1)} - X_t - X_u^{(n+1)} + X_u\right|}{(t-u)^{1+\beta}} du$$

$$\leq \int_0^t \frac{du}{(t-u)^{1+\beta}} \left(C(\omega) \int_u^t \left|X_s^{(n)} - X_s\right| s^{-\beta} ds \right.$$
$$\left. + C(\omega) \int_u^t \int_u^r \frac{\left|X_r - X_v - X_r^{(n)} + X_v^{(n)}\right|}{(r-v)^{1+\beta}} dv\, dr \right)$$

$$\leq C(\omega) \int_0^t \left|X_s^{(n)} - X_s\right| s^{-\beta} (t-s)^{-\beta} ds$$
$$+ C(\omega) \int_0^t (t-s)^{-\beta} \int_0^s \frac{\left|X_s - X_v - X_s^{(n)} + X_v^{(n)}\right|}{(s-v)^{1+\beta}} ds,$$

or

$$\xi_{n+1}^1(t) \leq C(\omega) \int_0^t \xi_n^1(s) s^{-\beta} ds + C(\omega) \int_0^t \xi_n^2(s) ds,$$
$$\xi_{n+1}^2(t) \leq C(\omega) \int_0^t \xi_n^1(s) s^{-\beta}(t-s)^{-\beta} ds + C(\omega) \int_0^t \xi_n^2(s)(t-s)^{-\beta} ds,$$

where $\xi_n^1(t) = \left|X_t^{(n)} - X_t\right|$, $\xi_n^2(t) = \int_0^t \frac{|X_t^{(n)} + X_u^{(n)} - X_t + X_u|}{(t-u)^{1+\beta}} du$.

Repeating the same estimates as in Part 1, but with other constants, we obtain:

$$\sup_{0 \leq t \leq T} \xi_n(t) \leq \frac{C^{n+1} G^{n+1} \widetilde{C}_4(\omega)}{\Gamma(n(1-2\beta))},$$

where $\xi_n(t) = \xi_n^1(t) + \xi_n^2(t)$,

$$\widetilde{C}_4(\omega) = \sup_{0 \leq t \leq T} |\xi_0(t)| = \sup_{0 \leq t \leq T} \left(|X_0| + |X_t| + \int_0^t \frac{|X_t - X_r|}{(t-r)^{1+\beta}} dr \right).$$

According to Corollary 3.1.7, $E\widetilde{C}_4^p(\omega) < \infty$ for any $p \geq 1$ if $H > \frac{3}{4}$. Clearly, $EG^p < \infty$ for any $p \geq 1$. Therefore, $E \sup_{0 \leq t \leq T} \left|X_t^{(n)} - X_t\right|^p \leq C_p < \infty$ and we obtain the proof. □

Remark 3.1.15. It is easy to see that under conditions (i')–(ii') and (x)–(xii) the derivative $D_s X_t$ satisfies the equation

$$D_s X_t = \sigma(X_s) + \int_s^t \sigma'(X_r) D_s X_r dB_r + \int_s^t b'(X_r) D_s X_r dr. \quad (3.1.48)$$

Remark 3.1.16. For differentiability and local differentiability of the solutions of SDE involving fBm see also (NS05) and (MS07b).

Smoothness of the Functionals of the Solution

We consider equation (3.1.28) and suppose that the coefficients b and σ satisfy the conditions of Theorem 3.1.10 and the condition

(xiii) $b, \sigma \in C^1(\mathbb{R})$.

Note that under these conditions equation (3.1.28) has a pathwise solution. Let X_t be any solution of (3.1.28) and the function $F \in C^2(\mathbb{R})$. Then for any fixed $T > 0$ $\int_0^T |F(X_s)b(X_s)|\, ds < \infty$ a.s. Suppose that the process $F'(X_s)\sigma(X_s) \in D_{1,2}(|\mathcal{H}|)$ and a.s.

$$\int_0^T \int_s^T |D_s(F'(X_u)\sigma(X_u))|\, |u-s|^{2\alpha-1}\, du\, ds < \infty.$$

According to the Itô formula (2.7.3) and equality (2.4.2), it holds that

$$F(X_t) = F(X_0) + \int_0^t F'(X_s)b(X_s)ds + \int_0^t F'(X_s)\sigma(X_s)dB_s^H$$

$$= F(X_0) + \int_0^t F'(X_s)b(X_s)ds + \int_0^t F'(X_s)\sigma(X_s)\delta B_s^H$$

$$+ C_H \int_0^t \int_s^t D_s(F'(X_u)\sigma(X_u))\, |u-s|^{2\alpha-1}\, du\, ds. \quad (3.1.49)$$

By using this equality, we can prove the following result. Denote

$$\varepsilon_s^t := \exp\left\{ \int_s^t b'(X_u)du + \int_s^t \sigma'(X_u)dB_u^H \right\}, \quad 0 \le s < t \le T.$$

Theorem 3.1.17. *Let the conditions of Theorem 3.1.10, condition (xiii) and the following conditions hold:*

(xiv) $E \int_0^T |F'(X_t)b(X_t)|\, dt < \infty$, the function $f(s) := E F'(X_s) b(X_s)$ is continuous on $[0, T]$;

(xv) $F'(X_s)\sigma(X_s) \in D_{1,2}(|\mathcal{H}|)$ and

$$E \int_0^T \int_s^T |D_s(F'(X_u)\sigma(X_u))|\, |u-s|^{2\alpha-1}\, du\, ds < \infty.$$

Then the function $\varphi(t) := EF(X_t)$ is differentiable in t and

$$\varphi'(t) = EF'(X_t)b(X_t)$$
$$+ 2\alpha HE\left(\int_0^t (F'(X_s)\sigma'(X_s))'\sigma(X_s)|s-t|^{2\alpha-1}(\varepsilon_0^s)^{-1}ds \cdot \varepsilon_0^t\right).$$

Proof. From the Itô formula (3.1.49) and conditions (xiv)–(xv) it follows that

$$\varphi(t) = EF(X_0) + \int_0^t EF'(X_s)b(X_s)ds$$
$$+ C_H \int_0^t \int_s^t ED_s(F'(X_u)\sigma(X_u))|u-s|^{2\alpha-1}\,du\,ds, \quad C_H = 2\alpha H.$$

Note that the mathematical expectation of the divergence operator $E\int_0^t F'(X_s)\sigma(X_s)\delta B_s^H = 0$. Therefore, under (xiv) and (xv) we can differentiate φ and obtain that

$$\varphi'(t) = EF'(X_t)b(X_t) + 2\alpha H \int_0^t ED_s(F'(X_t)\sigma(X_t))|t-s|^{2\alpha-1}\,ds.$$

Further, from the chain rule, Theorem 3.1.14 and Remark 3.1.15 $D_s(F'(X_t)\sigma(X_t)) = (F'(X_s)\sigma(X_s))'D_sX_t$, the derivative D_sX_t exists and satisfies linear differential equation (3.1.48), whence

$$D_sX_t = \sigma(X_s)\varepsilon_s^t \mathbf{1}\{s \le t\}.$$

Therefore

$$\varphi'(t) = EF'(X_t)b(X_t)$$
$$+ 2\alpha HE\left(\int_0^t (F'(X_s)\sigma'(X_s))'\sigma(X_s)|s-t|^{2\alpha-1}(\varepsilon_0^s)^{-1}ds \cdot \varepsilon_0^t\right).$$

$\square$

3.1.5 Semilinear Stochastic Differential Equations Involving Forward Integral w.r.t. fBm

León and Tudor in their paper (LT02) established the existence of a global solution of a semilinear stochastic differential equation with forward integrals (for the definition and properties of forward integral see Section 2.4). Let $p > 1$ and $\gamma \in (0,1)$. A process $u \in \mathbb{D}_{1,p}(|\mathcal{H}|)$ belongs to $\mathbb{L}_\gamma^{1,p}$ if

$$\|u\|_{\mathbb{L}_\gamma^{1,p}}^p := E(\|u\|_{L_{\frac{1}{\gamma}}[0,T]}^p) + E(\|Du\|_{L_{\frac{1}{\gamma}}[0,T]^2}^p) < \infty. \qquad (3.1.50)$$

It follows from (AN02) that $\mathbb{L}_\gamma^{1,p} \subset Dom(\delta_H)$ for any $0 \le \gamma \le H$.

The next statement from (LT02) establishes the relationship between the forward integral (understood in the sense of ucp-convergence) and the divergence operator that we denote here as $\int_0^t u_s \delta B_s^H$ (for P-convergence such a statement was proved by (AN02), see (2.4.1)). Here something like condition (3.1.50) is needed.

Theorem 3.1.18 ((LT02)). *1) Let $\{u_t, t \in [0,T]\}$ be a stochastic process, $u \in \mathbb{L}_\gamma^{1,2}$ for some $1/2 < \gamma < H$ and the trace condition holds,*

$$\int_0^T \int_0^t |D_s u_r| |r-s|^{2\alpha-1} ds\, dr < \infty \ a.s.$$

Then both the integrals, $\int_0^t u_s dB_s^{H,-}$ and $\int_0^t u_s \delta B_s^H$, exist for any $t \in [0,T]$ and

$$\int_0^t u_s dB_s^{H,-} = \int_0^t u_s \delta B_s^H + 2\alpha H \int_0^t \int_0^T D_s u_r |r-s|^{2\alpha-1} ds\, dr.$$

2) Now consider the semilinear stochastic differential equation

$$X_t = X_0 + \int_0^t b(s, X_s) ds + \int_0^t \sigma_s X_s dB_s^{H,-}, \quad t \in [0,T], \qquad (3.1.51)$$

where the coefficients $b : \Omega \times [0,T] \times \mathbb{R} \to \mathbb{R}$ and $\sigma : \Omega \times [0,T] \to \mathbb{R}$ are measurable, X_0 is random variable, b and σ satisfy the following assumptions:

(xvi) For all $\omega \in \Omega, t \in [0,T]$ and $x, y \in \mathbb{R}$

$$|b(\omega, t, x) - b(\omega, t, y)| \le \kappa(\omega)|x - y|,$$
$$|b(\omega, t, 0)| \le \kappa(\omega)$$

for some random variable $\kappa(\omega)$.
(xvii) σ is forward integrable and there is $\varepsilon_0 > 0$ such that

$$\lim_{c \to \infty} \sup_{0 < \varepsilon < \varepsilon_0} P\left\{\int_0^T |\int_0^r \sigma_s \varepsilon^{-1}(B_{(s+\varepsilon) \wedge T}^H - B_s^H) ds - \int_0^r \sigma_s dB_s^{H,-}|\right.$$
$$\left.\times |\sigma_r \varepsilon^{-1}(B_{(r+\varepsilon) \wedge T}^H - B_r^H)| dr > c\right\} = 0.$$

(xviii) for all $c > 0$

$$\lim_{\varepsilon \to 0} P\left\{\sup_{0 \le t \le T} \left|\int_0^t (\int_0^r \sigma_s \varepsilon^{-1}(B_{(s+\varepsilon) \wedge T}^H - B_s^H) ds - \int_0^r \sigma_s dB_s^{H,-})\right.\right.$$
$$\left.\left.\times \sigma_r \varepsilon^{-1}(B_{(r+\varepsilon) \wedge T}^H - B_r^H) dr\right| > c\right\} = 0.$$

Also, denote by $\mathcal{A}$ the class of all processes X such that (σX) is forward integrable and for any $c > 0$ and $t \in [0, T]$

$$\lim_{\varepsilon \to 0} \lim_{\eta \to 0} P\left\{\left|\int_0^s \sigma_s X_s \exp\{-\int_0^s \sigma_r \varepsilon^{-1}(B_{(r+\varepsilon) \wedge T}^H - B_r^H) dr\}\right.\right.$$
$$\left.\left.\times (\eta^{-1}(B_{(s+\eta) \wedge T}^H - B_s^H) - \varepsilon^{-1}(B_{(s+\varepsilon) \wedge T}^H - B_s^H)) ds\right| > c\right\} = 0.$$

Theorem 3.1.19 ((LT02)). *Under assumptions (xvi)–(xviii) equation (3.1.51) has a unique solution in the class $\mathcal{A}$ that is given by the unique solution of the equation*

$$X_t = \exp\left\{\int_0^t \sigma_s dB_s^{H,-}\right\} X_0 + \int_0^t \exp\left\{\int_u^t \sigma_s dB_s^{H,-}\right\} b(u, X_u) du, t \in [0,T].$$

Here are some classes of coefficients satisfying assumptions (xvii) and (xviii).

Example 3.1.20 ((LT02)). Assume that the stochastic process $\{\sigma_t, t \in [0,T]\}$ satisfies the following conditions:

(xix) $\sigma \in \mathbb{L}_\gamma^{1,2}$ for some $1/2 < \gamma < H$, and for some $t_0 \in [0,T]$

$$E\left(\int_0^T |D_s \sigma_{t_0}|^{\frac{1}{\gamma}} ds\right)^{2\gamma} < \infty;$$

(xx) there exists β such that for all $s,t \in [0,T]$

$$E|\sigma_t - \sigma_s| \leq c|s-t|^{\beta/2}$$

and

$$E\left(\int_0^T |D_u(\sigma_s - \sigma_t)|^{\frac{1}{\gamma}} du\right)^{2\gamma} \leq c|s-t|^\beta;$$

(xxi) $E|\sigma_t|^2 < \infty$ and $E(\int_0^T |D_r \sigma_t||t-r|^{2\alpha-1} dr)^2 < \infty$, $t \in [0,T]$;
(xxii) there exists $\mu \in (0, H)$ such that
  (a) $\lim_{c \to \infty} \sup_{0 < \varepsilon < \varepsilon_0} P\{\theta_\varepsilon > c\} = 0$, where $\varepsilon_0 > 0$ and

$$\theta_\varepsilon = \varepsilon^{-1-\mu+H}\left(\int_0^T \left(\int_0^r \int_{(s-\varepsilon^\mu)\vee 0}^{(s+\varepsilon^\mu)\wedge T} |D_u \sigma_s||s-u|^{2\alpha-1} du\, ds\right)^2 dr\right)^{1/2},$$

  (b) $\theta_\varepsilon \xrightarrow{P} 0$ as $\varepsilon \to 0$,
  (c) $\beta > 2(1-H+\mu)$ and $H - \mu > 1/2 \vee \mu$.

Then σ satisfies assumptions (xvi) and (xvii).

Example 3.1.21. Let $\{\sigma_t, t \in [0,T]\}$ be an absolutely continuous process of the form

$$\sigma_t = \sigma_0 = \int_0^t \dot{\sigma}_s ds, \quad t \in [0,T],$$

with $\sigma_0, \dot{\sigma} \in \mathbb{L}_\gamma^{1,2}$ for some $1/2 < \gamma < H$, and σ satisfies conditions (xxi) and (xxii). Then σ satisfies assumptions (xvii) and (xviii).

3.1.6 Existence and Uniqueness of Solutions of SDE with Two-Parameter Fractional Brownian Field

In this subsection we use the notations introduced in Subsection 2.2.4. We continue with the estimates of the two-parameter generalized Lebesgue–Stieltjes integrals (the first result in this direction was formulated in Lemma 2.2.16), and use these estimates to obtain the conditions of existence and uniqueness of solution of SDE involving the two-parameter fBm. The estimates of the norms of integrals on the whole duplicate the corresponding estimates from Lemmas 3.1.1–3.1.2 and Theorem 3.1.3, but are much more technical. Therefore we omit the proofs. For details, see (MisIl03).

Another approach to SDEs involving a two-parameter fractional Brownian field was developed in (TT03).

Denote $\mathcal{P}_T = [0, T_1] \times [0, T_2] \subset \mathbb{R}_+^2$ and introduce the following norm on the space $W_0^{\beta_1, \beta_2}(\mathcal{P}_T)$:

$$\|f\|_{\beta_1, \beta_2, \lambda_1, \lambda_2} := \sup_{t \in \mathcal{P}_T} e^{-\lambda_1 t_1 - \lambda_2 t_2} \varphi_f^{\beta_1, \beta_2}(t).$$

Also, recall that $\|f\| = \sup_{t \in \mathcal{P}_T} |f(t)|$.

Lemma 3.1.22. *Let the function $\sigma : \mathcal{P}_T \times \mathbb{R} \to \mathbb{R}$ and satisfy the following conditions:*

(xxiii) 1) $\sigma \in C^3(\mathcal{P}_T \times \mathbb{R})$;
2) $\exists\, C > 0$ such that $|D\sigma(t, x)| \leq C$, where the symbol D stands for any differentiation that is possible according to item 1) and $(t, x) \in \mathcal{P}_T \times \mathbb{R}$;
3) $|\sigma(r, 0)| \leq C$;

Also, let $f \in W_0^{\beta_1, \beta_2}(\mathcal{P}_T)$, $g \in W_1^{1-\beta_1, 1-\beta_2}(\mathcal{P}_T)$, for some $0 < \beta_i < \frac{1}{2}$, $i = 1, 2$.
Then the following statements hold:

1) $\|\sigma(\cdot, f(\cdot))\|_{0, \beta_1, \beta_2} \leq C_{\beta_1, \beta_2, T}(1 + \|f\|)(1 + \|f\|_{0, \beta_1, \beta_2})^2$;
2) *The generalized Lebesgue–Stieltjes integral*

$$G_t^{(\sigma)}(f) := \int_{\mathcal{P}_T} \sigma(s, f_s) dg_s$$

exists, belongs to the spaces $C^{1-\beta_1, 1-\beta_2}$ and $W_0^{\beta_1, \beta_2}(\mathcal{P}_T)$ and admits in these spaces the following estimates:
(a) $\|G^{(\sigma)}(f)\|_{1-\beta_1, 1-\beta_2} \leq C_{\beta_1, \beta_2, T} \Lambda_{1-\beta_1, 1-\beta_2}(g)(1 + \|f\|)$
$\times (1 + \|f\|_{0, \beta_1, \beta_2})^2$;
(b) $\|G^{(\sigma)}(f)\|_{\beta_1, \beta_2, \lambda_1, \lambda_2} \leq C_{\beta_1, \beta_2, T} \Lambda_{1-\beta_1, 1-\beta_2}(g) \lambda_1^{-1+2\beta_1} \lambda_2^{-1+2\beta_2}$

$$\times (1 + \|f\|^2)\left(1 + \|f\|_{\beta_1, \beta_2, \lambda_1, \lambda_2} + \|f\|_{\beta_1, \beta_2, \frac{\lambda_1}{2}, \frac{\lambda_2}{2}}^2\right). \tag{3.1.52}$$

Here $C_{\beta_1,\beta_2,T}$ depends only on β_1, β_2 and T.

Remark 3.1.23. All the estimates hold for $f_s = g_s = B_s^{H_1 H_2}$ with $H_i \in (\frac{1}{2}, 1)$, $i = 1, 2$.

Remark 3.1.24. Let the function σ be bounded, $f_1(x) = f(x) + C_0$, where $C_0 \in \mathbb{R}$ be some constant. Then $\|G^{(\sigma)}(f_1)\|_{\beta_1,\beta_2,\lambda_1,\lambda_2}$ can be estimated by the right-hand side of (3.1.52), i.e. this estimate does not depend on C_0.

Lemma 3.1.25. *Let the function σ satisfy the condition*
  (xxv) $\sigma \in C^3(\mathcal{P}_T) \times C^5(\mathbb{R})$, *and conditions (xxiii), 2) and 3) hold. Also, let $f, h \in W_0^{\beta_1,\beta_2}$ and $g \in W_1^{1-\beta_1,1-\beta_2}$ for some $0 < \beta_i < \frac{1}{2}$ and $i = 1, 2$. Then*

$$\|G^{(\sigma)}(f) - G^{(\sigma)}(h)\|_{\beta_1,\beta_2,\lambda_1,\lambda_2} \leq \frac{C_{\beta_1,\beta_2,T} \Lambda_{1-\beta_1,1-\beta_2}(g)}{\lambda_1^{1-2\beta_1} \lambda_2^{1-2\beta_2}} (1 + \|f\| + \|h\|)^2$$

$$\times (1 + \|f\|_{0,\beta_1,\beta_2} + \|h\|_{0,\beta_1,\beta_2})^2 \left(\|f - h\|_{\beta_1,\beta_2,\lambda_1,\lambda_2} + \|f - h\|^2_{\beta_1,\beta_2,\frac{\lambda_1}{2},\frac{\lambda_2}{2}} \right),$$

$\lambda_i \geq 1$, $i = 1, 2$.

Lemma 3.1.26. *1) Let the function $b = b(t,x) : \mathcal{P}_T \times \mathbb{R} \to \mathbb{R}$ be of linear growth: $|b(t,x)| \leq C(1 + |x|)$. Also, let $f \in W_0^{\beta_1,\beta_2}(\mathcal{P}_T)$. Then the integral $F_t^{(b)}(f) := \int_{\mathcal{P}_t} b(s, f(s)) ds \in C^1(\mathcal{P}_T)$ for $t \in \mathcal{P}_T$ and*

$$\|F^{(b)}(f)\|_{\beta_1,\beta_2,\lambda_1,\lambda_2} \leq \frac{C_{\beta_1,\beta_2,T}}{\lambda_1^{1-\beta_1} \lambda_2^{1-\beta_2}} (1 + \|f\|_{\beta_1,\beta_2,\lambda_1,\lambda_2}).$$

2) If the function b is bounded, we have the same situation as described in Remark 3.1.24.

3) If $f, h \in W_0^{\beta_1,\beta_2}(\mathcal{P}_T)$ and $\|f\| \leq N$, $\|h\| \leq N$, then

$$\|F^{(b)}(f) - F^{(b)}(h)\|_{\beta_1,\beta_2,\lambda_1,\lambda_2} \leq \frac{C_{\beta_1,\beta_2,T,N}}{\lambda_1^{1-\beta_1} \lambda_2^{1-\beta_2}} \|f - h\|_{\beta_1,\beta_2,\lambda_1,\lambda_2},$$

$\lambda_i \geq 1$, $i = 1, 2$, *where $C_{\beta_1,\beta_2,T,N}$ depends on β_1, β_2, T and N.*

Consider now a stochastic differential equation on the plane,

$$X_t = X_0 + \int_{\mathcal{P}_t} b(s, X_s) ds + \int_{\mathcal{P}_t} \sigma(s, X_s) dB_s^{H_1,H_2} = X_0 + F_t^{(b)}(X) + G_t^{(\sigma)}(X),$$

(3.1.53)

where $t \in \mathcal{P}_T \subset \mathbb{R}_+^2$, B^{H_1,H_2} is the fractional Brownian field with the Hurst indices $H_i \in (\frac{1}{2}, 1)$, $\sigma, b : \mathcal{P}_T \times \mathbb{R} \to \mathbb{R}$ are measurable bounded functions, σ satisfies conditions (xxv), (xxiii), 2), and the function $b(s,x)$ is continuous in s and Lipschitz in x.

The two-parameter process $X_t : \mathcal{P}_T \times \Omega \to \mathbb{R}$ will be called a solution of (3.1.53) if it converts (3.1.53) into identity for a.a. $\omega \in \Omega$ and any $t \in \mathcal{P}_T$, and the integral $G_t^{(\sigma)}(X)$ exists for a.a. $\omega \in \Omega$ as the two-parameter generalized Lebesgue–Stieltjes integral. The proof of the main result, stated in the next theorem, relies, in particular, on the boundedness of the coefficients and on Remark 3.1.24.

Theorem 3.1.27. *Under the conditions mentioned above, SDE* (3.1.53) *has a unique solution in the class* $W_0^{\beta_1,\beta_2}(\mathcal{P}_T)$ *and for a.a.* $\omega \in \Omega$, $X \in C^{1-\beta_1, 1-\beta_2}$ *for any* $1 - H_i < \beta_i < \frac{1}{2}$, $i = 1, 2$.

3.2 The Mixed SDE Involving Both the Wiener Process and fBm

Real objects varying in time (climate and weather derivative, prices on the stock market etc.) can have a component with a long memory (that is modeled by fBm with $H \in (1/2, 1)$) and also a component without memory (that is modeled by a Wiener process). Therefore, it is natural to consider stochastic differential equations involving both Brownian and fractional Brownian motions. We refer to such equations as mixed stochastic differential equations(and, correspondingly, to such models as mixed models).

The conditions of existence of a local solution of the mixed SDE were formulated in Theorem 3.1.9. Of course, we would like to establish the conditions of the existence of a global solution. We start with the semilinear SDE.

3.2.1 The Existence and Uniqueness of the Solution of the Mixed Semilinear SDE

Consider an SDE of the form

$$X_t = X_0 + \int_0^t b(s, X_s)ds + \sigma_1 \int_0^t X_s dW_s + \sigma_2 \int_0^t X_s dB_s^H, t \in [0, T], \quad (3.2.1)$$

where X_0 is an $\mathcal{F}_0$-measurable random variable, σ_1 and σ_2 are real numbers, $\{W_t, \mathcal{F}_t, t \in [0, T]\}$ and $\{B_t^H, \mathcal{F}_t, t \in [0, T]\}$ are a Wiener process and fBm, correspondingly, on the same probability space $(\Omega, \mathcal{F}, \mathcal{F}_t, t \in [0, T])$, without any suppositions on their dependence.

Theorem 3.2.1. *Let the function b satisfy Lipschitz and linear growth conditions in x:*

$$|b(t, x) - b(t, y)| \leq L|x - y|, |b(t, x)| \leq L(1 + |x|), \quad L > 0, x, y \in \mathbb{R},$$

and is continuous in both variables, $b \in C([0, T] \times \mathbb{R})$.

Then there exists the unique solution $\{X_t, t \in [0, T]\}$ of equation (3.2.1), *and the trajectories of X a.s. belong to $C^{1/2-}[0, T]$.*

Proof. First, we use Theorem 3.1.9 and construct a local solution. In this order we consider an auxiliary system of partial differential equations (3.1.27) that now acquires the following form:

$$\begin{cases} \frac{\partial h}{\partial Z_j}(Y,(Z_1,Z_2)) = \sigma_j h(Y,(Z_1,Z_2)), \ j = 1, 2, \\ h(Y_0, 0, 0) = X_0. \end{cases}$$

The solution of this system has the form

$$h(Y,(Z_1,Z_2)) = (Y - Y_0 + X_0)\exp\{\sigma_1 Z_1 + \sigma_2 Z_2\}, \qquad (3.2.2)$$

where $Z_1(t) = W_t, Z_2(t) = B_t^H$.

Now we try to construct the local solution X_t of equation (3.2.1) in the form of $X_t = h(Y_t,(Z_1(t), Z_2(t)))$, where the trajectories of Y a.s. belong to $C^1[0,T], Y(0) = Y_0$ be some $\mathcal{F}_0$-measurable random variable. Applying the Itô formula (2.7.2) from Remark 2.7.4, we obtain that

$$dX_t = \sum_{i=1}^{2} \frac{\partial h}{\partial Z_i}(Y_t, Z_1(t), Z_2(t))dZ_i(t)$$

$$+ \frac{\partial h}{\partial Y}(Y_t, Z_1(t), Z_2(t))Y_t' dt + \frac{1}{2}\sigma_1^2 h(Y_t, Z_1(t), Z_2(t))dt. \qquad (3.2.3)$$

Comparing (3.2.1) and (3.2.3), we get the ordinary differential equation for the process Y:

$$\begin{cases} Y_t' = (c_1(t))^{-1} b(t,(Y_t - Y_0 + X_0)c_1(t)) - \frac{1}{2}\sigma_1^2(Y_t - Y_0 + X_0) =: f(t, Y), \\ Y(0) = Y_0, \end{cases}$$
$$(3.2.4)$$

where $c_1(t) = \exp\{\sigma_1 Z_1(t) + \sigma_2 Z_2(t)\}$.

Further we fix $\omega \in \Omega$ and put for this ω $L_1(T) := \max_{0 \le t \le T}(c_1(t))^{-1} > 0$, $L_2(T) := \max_{0 \le t \le T} c_1(t) > 0$, $D_1 = LL_1(T)$, $D_2 = L + \frac{1}{2}\sigma_1^2$. Then for $t \le a_0$ and $|Y_t - Y_0| \le b_0$ with some $a_0, b_0 > 0$ we have that $M := \max_{0 \le t \le T} |f(t, Y_t)| \le L(L_1(T) + b_0 + |X_0|) + \frac{1}{2}\sigma_1^2(b_0 + |X_0|) = D_1 + D_2(b_0 + |X_0|) =: M_0$, and by the Picard theorem, the solution of equation (3.2.4) exists and is unique on the interval $[0, l^{(0)}]$, where $l^{(0)} := \min(a_0, b_0/M) \ge \min(a_0, b_0/M_0) =: t_0$; consequently, the solution exists on $[0, t_0]$.

By using (3.2.2), the solution at the point t_0 can be bounded by $|h(Y_{t_0}, Z_1(t_0), Z_2(t_0))| \le |Y_{t_0} - Y_0 + X_0|L_2(T) \le (b_0 + |X_0|)L_2(T)$. Evidently, the trajectories of the solution belong to $C^{1/2-}[0, t_0]$, since Y is continuously differentiable (recall that $b \in C([0,T] \times \mathbb{R}))$ and $\exp\{\sigma_1 Z_1(t) + \sigma_2 Z_2(t)\} = \exp\{\sigma_1 W_t + \sigma_2 B_t^H\} \in C^{1/2-}[0, t_0]$.

Now we want to extend the solution for $[0, T]$. The value X_{t_0} will be the new initial value $X_0^{(1)}$, and

$$|X_0^{(1)}| \le (b_0 + |X_0|)L_2(T).$$

Now, for $|t - t_0| \le a_1, |Y_t - Y_{t_0}| \le b_1$, for some a_1 and $b_1 > 0$, the solution of (3.2.4) exists on the interval $[t_0, t_1]$, where $t_1 - t_0 = \min(a_1, b_1/M_1)$ with $M_1 = D_1 + D_2(b_1 + (b_0 + |X_0|)L_2(T))$. In the nth step of this procedure of the extension of the solution we obtain $t_n - t_{n-1} = \min(a_n, b_n/M_n)$ with $M_n = D_1 + D_2(b_n + \sum_{k=0}^{n-1} b_{n-1-k} L_2^{k+1}(T) + |X_0| L_2^{n+1}(T))$ and the solution exists on $[t_{n-1}, t_n]$.

Now we have two possibilities: if $|X_0| \le 1$ we can put $b_n = 1, 0 \le k \le n$ and $b_n/M_n = (D_1 + D_2(\sum_{k=0}^{n} L_2^k(T) + |X_0| L_2^{n+1}(T)))^{-1}$
$\ge (D_1 + D_2 \frac{L_2^{n+2}(T)-1}{L_2(T)-1})^{-1} =: K_n$. If $|X_0| > 1$ then we put $b_k = |X_0|$, $0 \le k \le n$ and in this case also $b_n/M_n \ge K_n$. For both the cases put $a_n = K_n$, $n \ge 0$ and $t_n - t_{n-1} = a_n, t_n = \sum_{k=0}^{n} a_k$.

(a) Let $L_2(T) \le 1$. Then the series $\sum_{n \ge 0} a_n$ diverges, so, there exists only a finite number of aforementioned steps and we obtain the existence of a solution on the whole interval $[0, T]$.

(b) Let $L_2(T) > 1$. Then the series $\sum_{n \ge 0} a_n$ converges, possibly, its sum $S \le T$ and we obtain the existence of a solution on $[0, S)$. Therefore, we have established the existence of a finite solution on $[0, \frac{S}{2}]$. By the same method we can extend it on $[\frac{S}{2}, S]$ with the same step $\frac{S}{2}$, since the size of step does not depend on the initial value X_0. So, we can extend the solution with the step $\frac{S}{2}$ on the whole $[0, T]$. The uniqueness of the solution follows from Theorem 3.1.9. It follows from its construction (see (3.2.2) and (3.2.4)), that the trajectories of solution belong to $C^{1/2-}[0, T]$. □

3.2.2 The Existence and Uniqueness of the Solution of the Mixed SDE for fBm with $H \in (3/4, 1)$

Now we consider a mixed SDE without any semilinear restrictions but only for $H \in (3/4, 1)$.

Existence and Uniqueness of Solution of Mixed SDE for fBm with $H \in (3/4, 1)$ and with Stabilizing Term

We follow here the approach of (MP07). Consider the following mixed SDE:

$$X_t = X_0 + \int_0^t a(s, X_s)ds + \int_0^t b(s, X_s)dW_s$$
$$+ \int_0^t c(s, X_s)dB_s^H + \varepsilon \int_0^t c(s, X_s)dV_s, \ t \in [0, T], \quad (3.2.5)$$

where $a, b, c : [0, T] \times \mathbb{R} \to \mathbb{R}$ are measurable functions, V, W are independent Wiener processes, $\varepsilon > 0$ and B^H is independent of W and V fractional

Brownian motion with $H \in (3/4,1)$, X_0 is independent of W, B^H and V. The integral $\varepsilon \int_0^t c(s, X_s) dV_s$ will play the role of the stabilizing term. It permits us to establish the existence and uniqueness of the solution of (3.2.5), adapted to the filtration

$$\mathcal{F}'_t, \ t \geq 0, \text{ where } \mathcal{F}'_t = \sigma\{X_0, W_s, (\varepsilon V_s + B_s^H) | s \in [0,t]\}. \qquad (3.2.6)$$

The results are valid also for the case when $b = 0$. If $\varepsilon = 0$ and $b = 0$, we obtain equation (3.1.6) with $g = B^H$, whose existence and uniqueness conditions were formulated in Theorem 3.1.4. As we shall see, the stabilizing term permits us to avoid the smoothness condition on c, for example, the existence and Hölder properties of $\partial_x c(s,x)$. The main result that we use in the proof was stated by Cheridito (Che01b). For the completeness of exposition we shall present it here. Its proof originated in the papers (HH76) and (Hit68).

Proposition 3.2.2. 1. Let $\{W_t, t \in [0,T]\}$ be a Wiener process, $\{B_t^H, t \in [0,T]\}$ be an independent fBm with $H \in (3/4,1)$, $\gamma \in \mathbb{R} \setminus \{0\}$,

$$M_t^{H,\gamma} := B_t + \gamma B_t^H, \quad t \in [0,T],$$

with its own filtration $\{F_t^{M^{H,\gamma}}, t \in [0,T]\}$.

Then $\{M_t^{H,\gamma}, F_t^{M^{H,\gamma}}, t \in [0,T]\}$ is equivalent to Brownian motion; consequently it is a semimartingale.

2. There exists a unique real-valued Volterra kernel $h = h_\gamma \in L_2[0,T]^2$ such that

$$B_t := M_t^{H,\gamma} - \int_0^t \int_0^s h(s,u) dM_u^{H,\gamma} ds, \quad t \in [0,T]$$

is a Brownian motion. Furthermore,

$$M_t^{H,\gamma} = B_t - \int_0^t \int_0^s r(s,u) dB_u ds, \quad t \in [0,T], \qquad (3.2.7)$$

where $r = r_\gamma \in L_2[0,T]^2$.

As a consequence, the process $N_t^{H,\varepsilon} := B_t^H + \varepsilon V_t = \varepsilon(V_t + \frac{1}{\varepsilon} B_t^H) = \varepsilon M_t^{H, \frac{1}{\varepsilon}}$ can be represented as

$$N_t^{H,\varepsilon} = \varepsilon V'_t + \int_0^t \int_0^s \varepsilon r_\varepsilon(s,u) dV'_u ds, \qquad (3.2.8)$$

where V' is some Wiener process with respect to filtration $\mathcal{F}_t := \sigma\{\varepsilon V_s + B_s^H, s \in [0,t]\}$ and, from the independence of V, W, B^H and X_0, it is a Wiener process w.r.t. $\{\mathcal{F}'_t, t \in [0,T]\}$. Using (3.2.26), we can rewrite the equation (3.2.4) in the form

3.2 The Mixed SDE Involving Both the Wiener Process and fBm

$$X_t = X_0 + \int_0^t a(s, X_s)ds + \int_0^t b(s, X_s)dW_s$$
$$+ \varepsilon \int_0^t c(s, X_s)dV'_s + \int_0^t c(s, X_s) \int_0^s \varepsilon r_\varepsilon(s, u)dV'_u ds. \quad (3.2.9)$$

The drift coefficient of equation (3.2.9) equals $a(s,x) + c(s,x,\omega)$, where $c(s, x, \omega) = c(s, x)\int_0^s \varepsilon r_\varepsilon(s, u)dV'_u$. Evidently, the random variable $\int_0^s \varepsilon r_\varepsilon(s, u)dV'_u$ is not bounded, but we can consider the sequence of stopping times $\tau^M = \inf\{t \in [0,T] : \int_0^t (\int_0^s \varepsilon r_\varepsilon(s, u)dV'_u)^2 ds > M\} \wedge T$, and consider the sequence of corresponding stopped equations. The existence and uniqueness of the solutions of these equations can be established by standard methods and then it is easy to pass to the limit when $M \to \infty$. Finally, we obtain the following result (note that in this section we begin with the new numeration of the conditions).

Theorem 3.2.3. *Let the following conditions hold:*

*(i) The functions $|a(s, 0)| + |b(s, 0)| + |c(s, 0)| \le L, s \in [0, T]$ and
$|a(s, x)| + |b(s, x)| + |c(s, x)| \le L(1 + |x|)$, for some constant $L > 0$;*
(ii) there exists an increasing function $l(s) : [0, T] \to \mathbb{R}$ such that $\forall x, y \in \mathbb{R}$

$$|a(s, x) - a(s, y)| + |b(s, x) - b(s, y)| + |c(s, x) - c(s, y)| \le l(s)|x - y|;$$

(iii) the initial value X_0 is square-integrable.

Then equation (3.2.9), and consequently equation (3.2.5), has on $[0,T]$ the unique $\mathcal{F}'_t$-adapted solution X_t.

The Existence and Uniqueness of the Solution of the Mixed SDE Involving fBm with $H \in (3/4, 1)$ as the Limit Result for the Equations with the Stabilizing Term

Now we want to pass to the limit as $\varepsilon \to 0$ in equation (3.2.5). Let $\varepsilon = 1/N, N \ge 1$, and consider the sequence of the equations with the stabilizing term

$$X_t^N = X_0 + \int_0^t a(s, X_s^N)dt + \int_0^t b(s, X_s^N)dW_t$$
$$+ \int_0^t c(s, X_s^N)dB_s^H + \frac{1}{N}\int_0^t c(s, X_s^N)dV_s, \quad t \in [0, T]. \quad (3.2.10)$$

Let the coefficients a, b, c and X_0 satisfy conditions (i), (ii) and (iii). Then, according to Theorem 3.2.3, equation (3.2.10) has a unique strong solution, say $\{X_t^N, t \in [0, T]\}$. Evidently, the solutions are adapted to different filtrations $\mathcal{F}_t^N = \sigma\{X_0, W_s, (N^{-1}V_s + B_s^H), s \in [0, t]\}$. The aim of this section is to establish the conditions of existence and uniqueness of the solution of the limit mixed equation

$$X_t = X_0 + \int_0^t a(s, X_s)ds + \int_0^t b(s, X_s)dW_s + \int_0^t c(s, X_s)dB_s^H, \quad t \in [0, T]. \tag{3.2.11}$$

Let the coefficients of equation (3.2.11) satisfy assumption (iii) and the following ones: there exist such constants $B, L, M > 0, \gamma \in (1 - H, 1)$ and $\kappa \in (3/2 - H, 1)$ that
(iv) all the coefficients are bounded:

$$|a(s, x)| + |b(s, x)| + |c(s, x)| \leq L, \forall s \in [0, T], \forall x \in \mathbb{R};$$

(v) all the coefficients are Lipschitz in x:

$$|a(t, x) - a(t, y)| + |(b(t, x) - b(t, y)| + |c(t, x) - c(t, y)| \leq L|x - y|,$$

$\forall t \in [0, T], \forall x, y \in \mathbb{R}$,
(vi) the x-derivative of the function c exists and is Hölder continuous in t:
$\forall s, t \in [0, T], \forall x \in \mathbb{R}$

$$|c(s, x) - c(t, x)| + |\partial_x c(s, x) - \partial_x c(t, x)| \leq L|s - t|^\gamma.$$

(vii) the x-derivative of the function c is Hölder continuous in x:

$$|\partial_x c(t, x) - \partial_x c(t, y)| \leq L|x - y|^\kappa,$$

for $\forall t \in [0, T], \forall x, y \in \mathbb{R}$.

Remark 3.2.4. Note that for $H \in [3/4, 1)$ $3/2 - H > 1/H - 1$, so condition (vii) is more restrictive than the corresponding condition (ii) used in Theorem 3.1.4. In general, this last group of conditions is evidently more strong than conditions (i)–(ii) of Theorem 3.2.3.

Now consider for $\beta < (1/2 \wedge \gamma \wedge \kappa/2 \wedge (\kappa - \frac{1}{2}))$ some "stochastic analog" of the functional space of Besov type:

$$W^\beta[0, T] := \{Y = Y_t(\omega)|(t, \omega) \in [0, T] \times \Omega, \|Y\|_\beta < \infty\}$$

with the norm

$$\|Y\|_\beta := \sup_{t \in [0,T]} \left(E(Y_t)^2 + E\left(\int_0^t \frac{|Y_t - Y_s|}{(t - s)^{1+\beta}} ds \right)^2 \right),$$

and prove that the solution of SDE (3.2.10) belongs to this space for any $N > 1$. We shall denote different constants as C if they do not depend on N and it is unimportant to the stated results. First of all we prove the Hölder continuity of the solution of equation (3.2.10), by using (1.17.1) and (1.17.2).

Theorem 3.2.5. *For any $\delta \in (0, 1/2)$ the solution of equation (3.2.10) is Hölder continuous with parameter $1/2 - \delta$.*

3.2 The Mixed SDE Involving Both the Wiener Process and fBm

Proof. Consider $|X_r^N - X_z^N|$ for $0 < z < r < T$:

$$|X_r^N - X_z^N| \leq \left|\int_z^r a(u, X_u)du\right| + \left|\int_z^r b(u, X_u)dW_u\right| + \frac{1}{N}\left|\int_z^r c(u, X_u)dV_u\right|$$

$$+ \left|\int_z^r c(u, X_u)dB_u^H\right| \leq L(r-z) + C\xi_{r,\delta}^b|r-z|^{1/2-\delta} + \frac{C}{N}\xi_{r,\delta}^c|r-z|^{1/2-\delta}$$

$$+ \Lambda_{1-\beta}(B^H)\int_z^r \frac{|c(u, X_u^N)|du}{u^\beta}$$

$$+ \Lambda_{1-\beta}(B^H)\int_z^r \int_z^u \frac{|c(u, X_u^N) - c(v, X_v^N)|}{(u-v)^{1+\beta}}dv\,du$$

$$\leq C_r'(\omega)(r-z)^{1/2-\delta} + C_r'(\omega)\int_z^r \int_z^u \frac{|X_u^N - X_v^N|}{(u-v)^{1+\beta}}dv\,du,$$

where

$$C_t'(\omega) := C(\Lambda_{1-\beta}(B^H) \vee \xi_{t,\delta}^b \vee \xi_{t,\delta}^c \vee 1), \quad (3.2.12)$$

$\xi_{t,\delta}^b$ and $\xi_{t,\delta}^c$ are defined by (1.17.2), $C_t'(\omega) \leq C_T'(\omega)$ and $C_T'(\omega)$ has the moments of any order.

Therefore, for $\delta < 1/2 - \beta$ we have that

$$\phi_{r,s} := \int_s^r \frac{|X_r^N - X_z^N|}{(r-z)^{1+\beta}}dz \leq C_r'(\omega)\bigg(\int_s^r (r-z)^{-1/2-\delta-\beta}dz$$

$$+ \int_s^r \frac{1}{(r-z)^{1+\beta}} \int_z^r \int_z^u \frac{|X_u^N - X_v^N|}{(u-v)^{1+\beta}}dv\,du\,dz\bigg)$$

$$\leq C_r'(\omega)\bigg((r-u)^{1/2-\beta-\delta} + \int_s^r (r-u)^{-\beta}\phi_{u,s}du\bigg).$$

From the modified Gronwall inequality (Lemma 7.6 (NR00)) it follows that

$$\phi_{r,s} \leq C_r'(\omega)(r-s)^{1/2-\beta-\delta}\exp\{C_r'(\omega)^{\frac{1}{1-\beta}}\}.$$

Return to $|X_r^N - X_z^N|$:

$$|X_r^N - X_z^N| \leq C_r'(\omega)(r-z)^{1/2-\delta}$$

$$+ C_r'(\omega)\exp\{C_r'(\omega)^{\frac{1}{1-\beta}}\}\int_z^r (v-z)^{1/2-\beta-\delta}dv \leq \widetilde{C}_r(\omega)(r-z)^{1/2-\delta},$$

where $\widetilde{C}_r(\omega) = C_r'(\omega)\exp\{C_r'(\omega)^{\frac{1}{1-\beta}}\}$, and the theorem is proved for $0 < \delta < 1/2 - \beta$, and consequently for $0 < \delta < 1/2$. □

Introduce the random variable $\widetilde{C}(\omega) := \sup_{0 \leq t \leq T} \widetilde{C}_t(\omega)$. It also has moments of any order. Now we want to prove that the solution of (3.2.10) belongs to the space $\{W^\beta[0, T], \|\cdot\|_\beta\}$ for all $N > 1$.

Theorem 3.2.6. *Under assumptions (iii)–(vi) the solution of equation (3.2.10) belongs to the space $W^\beta[0,T]$ of Besov type with norm $\|\cdot\|_\beta$ for all $N > 1$ and any $\beta < (1/2 \wedge \gamma \wedge \kappa/2 \wedge \kappa - \frac{1}{2})$.*

Proof. In order to prove the statement of this theorem, we want to estimate

$$A_1^N(t) + A_2^N(t) := E(X_t^N)^2 + E\left(\int_0^t \frac{|X_t^N - X_s^N|}{(t-s)^{1+\beta}} ds\right)^2.$$

First, for $A_1^N(t)$ we have that

$$E(X_t^N)^2 \leq 5E(X_0)^2 + 5E\left(\int_0^t a(s,X_s^N)ds\right)^2 + 5E\left(\int_0^t b(s,X_s^N)dW_s\right)^2$$

$$+ 5E\left(\int_0^t c(s,X_s^N)dB_s^H\right)^2 + 5E(\frac{1}{N}\int_0^t c(s,X_s^N)dV_s)^2. \quad (3.2.13)$$

Evidently, $E\left(\int_0^t a(s,X_s^N)ds\right)^2 \leq L^2 T^2$,
$E\left(\int_0^t b(s,X_s^N)dW_s\right)^2 \leq L^2 T$, $E\left(\frac{1}{N}\int_0^t c(s,X_s^N)dV_s\right)^2 \leq \frac{L^2 T}{N^2} \leq L^2 T$.
Further, for $\delta < 1/2 - \beta$ we have that

$$E\left(\int_0^t c(s,X_s^N)dB_s^H\right)^2 \leq E\left(\overline{C}^2(\omega)\left(\int_0^t \frac{c(s,X_s^N)}{s^\beta}ds\right.\right.$$
$$+ \int_0^t \int_0^s \frac{|c(s,X_s^N) - c(u,X_u^N)|}{(s-u)^{1+\beta}} du\, ds\right)^2\right) \leq CE\left(\overline{C}^2(\omega)\left(t\int_0^t \frac{L^2}{s^{2\beta}}ds\right.\right.$$
$$+ \left(\int_0^t \int_0^s \frac{L(s-u)^\gamma + L\widetilde{C}(\omega)(s-u)^{1/2-\delta}}{(s-u)^{1+\beta}} du\, ds\right)^2\right)\right)$$

$$\leq C(E\overline{C}^2(\omega)(L^2 T^{2-2\beta} + L^2 T^{2(1-\beta+\gamma)}) + L^2 E(\widetilde{C}^2(\omega)\overline{C}^2(\omega))T^{3-2\beta-2\delta})$$

with $\overline{C}(\omega) = \Lambda_{1-\beta}(B^H)$. From all these estimates it follows that $A_1^N(t) < \infty$. Consider now $A_2^N(t)$. We have that

$$A_2^N(t) \leq 4E\left(\int_0^t \frac{|\int_s^t a(u,X_u^N)du|}{(t-s)^{1+\beta}} ds\right)^2$$

$$+ 4E\left(\int_0^t \frac{|\int_s^t b(u,X_u^N)dW_u|}{(t-s)^{1+\beta}} ds\right)^2 + 4N^{-2} E\left(\int_0^t \frac{|\int_s^t c(u,X_u^N)dV_u|}{(t-s)^{1+\beta}} ds\right)^2$$

$$+ 4E\left(\int_0^t \frac{|\int_s^t c(u,X_u^N)dB_u^H|}{(t-s)^{1+\beta}} ds\right)^2. \quad (3.2.14)$$

Evidently,

$$E\left(\int_0^t \frac{|\int_s^t a(u, X_u)du|}{(t-s)^{1+\beta}} ds\right)^2 \leq CL^2 t^{2-2\beta}.$$

Now, let $\rho \in (\beta, 1/2)$, then we have the estimate

$$E\left(\int_0^t \frac{|\int_s^t b(u, X_u)dW_u|}{(t-s)^{1+\beta}} ds\right)^2 \leq Ct^{1-2\rho} \int_0^t \frac{E|\int_s^t b(u, X_u)dW_u|^2}{(t-s)^{2+2\beta-2\rho}} ds$$

$$\leq Ct^{1-2\rho} \int_0^t \frac{\int_s^t b^2(u, X_u)du}{(t-s)^{2+2\beta-2\rho}} ds \leq CL^2 t^{1-2\beta}, \quad (3.2.15)$$

and similarly,

$$E\left(\int_0^t \frac{|\int_s^t c(u, X_u)dV_u|}{(t-s)^{1+\beta}} ds\right)^2 \leq CL^2 t^{1-2\beta}.$$

Now we estimate $E^N := E\left(\int_0^t |\int_s^t c(u, X_u)dB_u^H|(t-s)^{-1-\beta} ds\right)^2$. Since

$$\left|\int_s^t c(u, X_u)dB_u^H\right| \leq \overline{C}(\omega)\left(\int_s^t |c(u, X_u)|(u-s)^{-\beta} du\right.$$

$$\left. + \int_s^t \int_s^u |c(u, x_u^N) - c(r, X_r^N)|(u-r)^{-1-\beta} dr\, du\right) \leq \overline{C}(\omega)$$

$$\times \left(\int_s^t |c(u, X_u)|(u-s)^{-\beta} du + \int_s^t \int_s^u \frac{L(u-r)^\gamma + L\widetilde{C}(\omega)(u-r)^{1/2-\delta}}{(u-r)^{1+\beta}} dr\, du\right),$$

we have that for $\delta < 1/2 - \beta$ E^N can be bounded by

$$E\left(\overline{C}(\omega) \int_0^t \frac{L(t-s)^{1-\beta} + L(t-s)^{1+\gamma-\beta} + L\widetilde{C}(\omega)(t-s)^{3/2-\delta-\beta}}{(t-s)^{1+\beta}} ds\right)^2$$

$$\leq C(L^2 t^{2-4\beta} E\overline{C}^2(\omega) + L^2 t^{2+2\gamma-4\beta} E\overline{C}^2(\omega) + L^2 t^{3-2\delta-4\beta} E\overline{C}^2(\omega)\widetilde{C}^2(\omega)).$$

$$(3.2.16)$$

Therefore, $A_2^N(t)$ satisfies the inequality

$$A_2^N(t) \leq C(L^2 T^{2-2\beta} + L^2 T^{1-2\beta} + L^2 T^{2-4\beta} E\overline{C}^2(\omega)$$

$$+ L^2 T^{2+2\gamma-4\beta} E\overline{C}^2(\omega) + L^2 T^{3-2\delta-4\beta} E\overline{C}^2(\omega)\widetilde{C}^2(\omega)) < \infty. \quad (3.2.17)$$

Finally, the statement of our theorem follows from inequalities (3.2.13)–(3.2.17) with sufficiently small $\delta > 0$. □

Introduce for any $R > 1$ the stopping time τ_R by

$$\tau_R := \inf\{t : C'_t(\omega) \geq R\} \wedge T, \qquad (3.2.18)$$

where $C'_t(\omega)$ is defined by (3.2.12). Evidently, for any $\omega \in \Omega$ there exists $R(\omega)$ such that $\tau_R = T$ for all $R > R(\omega)$.

Define the processes $\{X^N_{\tau_R \wedge t}, N \geq 1, t \in [0,T]\}$ as the solutions of equation (3.2.10) stopped at the moment τ_R, and prove that they are fundamental in the norm $\|\cdot\|_\beta$ of the space $W^\beta[0,T]$.

Theorem 3.2.7. *Under assumptions (iii)–(vi) the sequence $\{X^N_{t \wedge \tau_R}, N \geq 1, t \in [0,T]\}$ of solutions of equations (3.2.10) is fundamental in the norm $\|\cdot\|_\beta$ for any $\beta < (1/2 \wedge \gamma \wedge \kappa/2 \wedge \kappa - \frac{1}{2})$.*

Proof. Consider

$$A_1^{N,M}(t) + A_2^{N,M}(t) := E(X^N_{t \wedge \tau_R} - X^M_{t \wedge \tau_R})^2$$
$$+ E\left(\int_0^t \frac{|X^N_{t \wedge \tau_R} - X^M_{t \wedge \tau_R} - X^N_{s \wedge \tau_R} + X^M_{s \wedge \tau_R}|}{(t-s)^{1+\beta}} ds\right)^2$$
$$= E(X^N_{t \wedge \tau_R} - X^M_{t \wedge \tau_R})^2 + E\left(\int_0^{\tau_R \wedge t} \frac{|X^N_{t \wedge \tau_R} - X^M_{t \wedge \tau_R} - X^N_s + X^M_s|}{(t-s)^{1+\beta}} ds\right)^2.$$

First, for $A_1^{N,M}(t)$ we have the estimate

$$A_1^{N,M}(t) \leq 4E\left(\int_0^{\tau_R \wedge t}(a(s, X^N_s) - a(s, X^M_s))ds\right)^2$$
$$+ 4E\left(\int_0^{\tau_R \wedge t}(b(s, X^N_s) - b(s, X^M_s))dW_s\right)^2$$
$$+ 4E\left(\int_0^{\tau_R \wedge t}(c(s, X^N_s) - c(s, X^M_s))dB^H_s\right)^2$$
$$+ 4E\left(\int_0^{\tau_R \wedge t}\left(\frac{c(s, X^N_s)}{N} - \frac{c(s, X^M_s)}{M}\right)dV_s\right)^2 =: 4(I_1 + I_2 + I_3 + I_4).$$

Then $I_1 \leq CTL^2 \int_0^t E(X^N_{s \wedge \tau_R} - X^M_{s \wedge \tau_R})^2 ds$, $I_2 \leq CL^2 \int_0^t E(X^N_{s \wedge \tau_R} - X^M_{s \wedge \tau_R})^2 ds$, $I_4 \leq CL^2 T(N^{-2} + M^{-2})$. Now we are in a position to estimate I_3:

$$I_3 \leq 2R^2\Bigg(E\Big(\int_0^{\tau_R \wedge t} |c(s, X^N_s) - c(s, X^M_s)| s^{-\beta} ds\Big)^2$$
$$+ E\Big(\int_0^{\tau_R \wedge t}\int_0^s |c(s, X^N_s) - c(s, X^M_s) - c(u, X^N_u) + c(u, X^M_u)|$$
$$\times (s-u)^{-1-\beta} du\, ds\Big)^2\Bigg) = 2R^2(I_4 + I_5).$$

3.2 The Mixed SDE Involving Both the Wiener Process and fBm

Further,

$$I_4 \leq CL^2 T^{1-2\beta} E \int_0^{\tau_R \wedge t} (X_s^N - X_s^M)^2 ds = CL^2 T^{1-2\beta} \int_0^t A_1^{N,M}(s) ds.$$

By using Lemma 7.1 (NR00), we estimate I_5 as

$$I_5 \leq 3E \left(\int_0^{\tau_R \wedge t} \int_0^s \frac{L|X_s^N - X_s^M - X_u^N + X_u^M|}{(s-u)^{1+\beta}} du \, ds \right)^2$$
$$+ 3E \left(\int_0^{\tau_R \wedge t} \int_0^s \frac{L^2 |X_s^N - X_s^M|(s-u)^\gamma}{(s-u)^{1+\beta}} du \, ds \right)^2$$
$$+ 3E \left(\int_0^{\tau_R \wedge t} \int_0^s \frac{L|X_s^N - X_s^M|(|X_u^N - X_u^M|^\kappa + |X_s^M - X_u^M|^\kappa)}{(s-u)^{1+\beta}} du \, ds \right)^2$$
$$= 3(I_6 + I_7 + I_8).$$

Here

$$I_6 \leq CTL^2 \int_0^t E \left(\int_0^{s \wedge \tau_R} \frac{|X_{s \wedge \tau_R}^N - X_{s \wedge \tau_R}^M - X_u^N + X_u^M|}{(s-u)^{1+\beta}} du \right)^2 ds,$$

$$I_7 \leq CTL^2 \int_0^t s^{2(\gamma-\beta)} E|X_{s \wedge \tau_R}^N - X_{s \wedge \tau_R}^M|^2 ds,$$

$$I_8 \leq E \left(\int_0^{\tau_R \wedge t} \int_0^s \frac{L|X_s^N - X_s^M| 2R(s-u)^{\kappa(1/2-\delta)}}{(s-u)^{1+\beta}} du \, ds \right)^2$$
$$\leq CTL^2 R^2 \int_0^t s^{\kappa - 2\kappa\delta - 2\beta} E|X_{s \wedge \tau_R}^N - X_{s \wedge \tau_R}^M|^2 ds,$$

where we choose δ in such a way that $\kappa - 2\kappa\delta - 2\beta > 0$. It is possible since $\beta < \kappa - 1/2$ so $\kappa - 2\beta > 1/2 - \beta > 0$. Finally,

$$I_5 \leq C \int_0^t \left(A_2^{N,M}(s) + (s^{2(\gamma-\beta)} + CR^2 s^{\kappa - 2\kappa\delta - 2\beta}) A_1^{N,M}(s) \right) ds,$$

and

$$A_1^{N,M}(t) \leq CR^2 \int_0^t A_1^{N,M}(s) ds + CR^2 \int_0^t A_2^{N,M}(s) ds$$
$$+ C(N^{-2} + M^{-2}). \quad (3.2.19)$$

Return to $A_2^{N,M}(t)$. It admits the following estimate:

3 Stochastic Differential Equations Involving Fractional Brownian Motion

$$A_2^{N,M}(t) \leq C \left(E \left(\int_0^{T_R \wedge t} \frac{\int_s^{T_R \wedge t}(a(u, X_u^N) - a(u, X_u^M))du}{(t-s)^{1+\beta}} ds \right)^2 \right.$$

$$+ E \left(\int_0^{T_R \wedge t} \frac{\int_s^{T_R \wedge t}(b(u, X_u^N) - b(u, X_u^M))dW_u}{(t-s)^{1+\beta}} ds \right)^2$$

$$+ E \left(\int_0^{T_R \wedge t} \frac{\int_s^{T_R \wedge t}(c(u, X_u^N) - c(u, X_u^M))dB_u^H}{(t-s)^{1+\beta}} ds \right)^2$$

$$+ E \left(\int_0^{T_R \wedge t} \frac{\int_s^{T_R \wedge t}\left(\frac{c(u, X_u^N)}{N} - \frac{c(u, X_u^M)}{M}\right)dV_u}{(t-s)^{1+\beta}} ds \right)^2 \right)$$

$$= C(I_9 + I_{10} + I_{11} + I_{12}).$$

Further, for $\beta < \rho < 1/2$

$$I_9 \leq CT^{1-2\rho} E \int_0^{T_R \wedge t} \frac{(t-s) \int_0^{T_R \wedge t} L^2 |X_u^N - X_u^M|^2 du}{(t-s)^{2+2\beta-2\rho}} ds$$

$$\leq CT^{1-2\beta} \int_0^t E\left(X_{s \wedge \tau_R}^N - X_{s \wedge \tau_R}^M\right)^2 ds \leq CT^{1-2\beta} \int_0^t A_1^{N,M}(s) ds,$$

$$I_{10} \leq CT^{1-2\rho} \int_0^t \frac{\int_s^t E|X_{u \wedge \tau_R}^N - X_{u \wedge \tau_R}^M|^2 du}{(t-s)^{2+2\beta-2\rho}} ds$$

$$\leq CT^{1-2\rho} \int_0^t \frac{A_1^{N,M}(s)}{(t-s)^{1+2\beta-2\rho}} ds.$$

For I_{12} we have $I_{12} \leq CT^{1-2\beta}(N^{-2} + M^{-2})$. Now consider I_{11}:

$$I_{11} \leq CR^2 T^{1-2\rho}(I_{13} + I_{14}),$$

where

$$I_{13} \leq CE \int_0^{T_R \wedge t} \frac{\int_s^{T_R \wedge t}\left(X_{u \wedge \tau_r}^N - X_{u \wedge \tau_r}^M\right)^2 du \int_s^t (u-s)^{-2\beta} du}{(t-s)^\nu} ds$$

$$\leq C \int_0^t A_1^{N,M}(s)(t-s)^{-1+2\rho-4\beta} ds,$$

3.2 The Mixed SDE Involving Both the Wiener Process and fBm

$$I_{14} \leq CE \int_0^{T_R \wedge t} \left(\left(\int_s^{T_R \wedge t} \int_s^u \frac{L|X_u^N - X_u^M - X_v^N + X_v^M|}{(u-v)^{1+\beta}} dv\, du \right)^2 \right.$$

$$+ \left(\int_s^{T_R \wedge t} \int_s^u L|X_u^N - X_u^M|(u-v)^{\rho-1-\beta} dv\, du \right)^2$$

$$+ \left(\int_s^{T_R \wedge t} \int_s^u L|X_u^N - X_u^M| \left(|X_u^N - X_v^N|^\kappa + |X_u^M - X_v^M|^\kappa \right) \right.$$

$$\left. \left. \times (u-v)^{-1-\beta} dv\, du \right)^2 \right) (t-s)^{-\nu} ds =: C(I_{15} + I_{16} + I_{17}),$$

where $\nu = 2 + 2\beta - 2\rho, \rho > \beta$. In turn,

$$I_{15} \leq CT^{2\rho-2\beta} \int_0^t E\left(\int_0^{s \wedge T_R} \frac{|X_{s \wedge T_R}^N - X_{s \wedge T_R}^M - X_u^N + X_u^M|}{(s-u)^{1+\beta}} du \right)^2 ds$$

$$= CT^{2\rho-2\beta} \int_0^t A_2^{N,M}(s) ds,$$

$$I_{16} \leq C \int_0^t \frac{E\left(\int_s^{T_R \wedge t} |X_u^N - X_u^M|(u-s)^{\gamma-\beta} du \right)^2}{(t-s)^\nu} ds$$

$$\leq CT^{2\rho+2\gamma-4\beta} \int_0^t A_1^{N,M}(s) ds,$$

where $\beta < \gamma, \beta < \rho$. Furthermore,

$$I_{17} \leq CR^2 E \int_0^{T_R \wedge t} \frac{\left(\int_s^{T_R \wedge t} \int_s^u |X_u^N - X_u^M|(u-v)^{\kappa(1/2-\delta)-1-\beta} dv\, du \right)^2}{(t-s)^\nu} ds,$$

where we chose $0 < \delta < 1/2 - \beta/\kappa$; note that $\beta < \kappa - 1/2$. Similarly to I_{16},

$$I_{17} \leq CR^2 T^{\kappa - 2\kappa\delta + 2\rho - 4\beta} \int_0^t A_1^{N,M}(s) ds,$$

where $\kappa - 2\kappa\delta + 2\rho - 4\beta > 0$ for sufficiently small δ since $\rho > \beta$ and $\kappa > 2\alpha$. Therefore we have

$$I_{14} \leq CR^2 \int_0^t \left(A_1^{N,M}(s) + A_2^{N,M}(s) \right) ds.$$

Hence

$$I_{11} \leq CR^4 \int_0^t \left(\frac{A_1^{N,M}(s)}{(t-s)^{1+2\beta-2\rho}} + A_2^{N,M}(s) \right) ds.$$

Finally,

$$A_2^{N,M}(t) \le CR^4 \left(\int_0^t A_1^{N,M}(s)(t-s)^{-1-2\beta+2\rho} ds + \int_0^t A_2^{N,M}(s) ds \right)$$
$$+ C(N^{-2} + M^{-2}). \quad (3.2.20)$$

From (3.2.19) and (3.2.20) we obtain that the sum $A_1^{N,M}(t) + A_2^{N,M}(t)$ admits the same estimate as $A_2^{N,M}(t)$, i.e.

$$A_1^{N,M}(t) + A_2^{N,M}(t) \le CR^4 \int_0^t \left(A_1^{N,M}(s)(t-s)^{-1-2\beta+2\rho} + A_2^{N,M}(s) \right) ds$$
$$+ C(N^{-2} + M^{-2});$$

taking into account that $\rho > \beta$ and using the modified Gronwall lemma (NR00), we obtain that

$$A_1^{N,M}(t) + A_2^{N,M}(t)$$
$$\le CR^4(N^{-2} + M^{-2}) \exp\{t(CR^4)^{1/(2\rho-2\beta)}\}, \quad (3.2.21)$$

and we can put, for example, $\rho := 1/4 + \beta/2$. When $N, M \to 0$, we obtain that the right-hand side of (3.2.21) tends to zero, whence the proof follows. □

Theorem 3.2.8. *The SDE (3.2.11) has a solution on the interval $[0, T]$, and this solution is unique.*

Proof. Since the space $\{W^\beta[0, T], \|\cdot\|_\beta\}$ is complete, from Theorem 3.2.6 we can define
$$X_{\tau_R \wedge t} := \lim_{N \to \infty} X_{\tau_R \wedge t}^N,$$
where the limit is taken in space $W_\beta[0, T]$ (in particular, we have that the limit exists in $L_2(\Omega \times [0, T])$). Using similar estimates and Theorem 3.2.6, we can prove that $X_{\tau_R \wedge t}$ is the unique solution of the original equation (3.2.11) on the interval $[0, \tau_R]$.

From the definition (3.2.18) of τ_R we have $\tau_{R_1} \le \tau_{R_2}$ for $R_1 \le R_2$. So $X_{\tau_{R_1}}$ and $X_{\tau_{R_2}}$ coincide a.s. on the interval $[0, \tau_{R_1}]$. Where $R \to \infty$ we obtain the existence and uniqueness of the solution of SDE (3.2.11) on the whole interval $[0, T]$. □

3.2.3 The Girsanov Theorem and the Measure Transformation for the Mixed Semilinear SDE

Consider equation (3.2.1) and suppose that W is underlying Wiener process for B^H and that the coefficient $b(t, x)$ satisfies the condition of Theorem 3.2.1 and can be presented as $b(t, x) = e(t, x)x$, where $e \in C_b(\mathbb{R}_+ \times \mathbb{R})$. Denote $\hat{e}(t, x) := e(t, x)t^{-\alpha}$, $\alpha = H - \frac{1}{2}$, $H \in (\frac{1}{2}, 1)$. Now we try to change the measure P for another probability measure Q such that $Q_T \ll P_T$, where $P_T := P|_{\mathcal{F}_T}$,

3.2 The Mixed SDE Involving Both the Wiener Process and fBm 239

$Q_T := Q|_{\mathcal{F}_T}$, and such that the drift $e(t, X_t)X_t dt$ will be annihilated under Q_T. First, let some probability measure $\widetilde{Q}$ satisfy the assumptions

$$\left.\frac{d\widetilde{Q}}{dP}\right|_{\mathcal{F}_T} = \exp\left\{\int_0^T \varphi_s dW_s - \frac{1}{2}\int_0^T \varphi_s^2 ds\right\}$$

and

$$E\exp\left\{\int_0^T \varphi_s dW_s - \frac{1}{2}\int_0^T \varphi_s^2 ds\right\} = 1 \qquad (3.2.22)$$

with $E\int_0^T \varphi_s^2 ds < \infty$.

Then from the Girsanov theorem the process $W_t - \int_0^t \varphi_s^2 ds =: \widehat{W}_t$ will be a Wiener process under the measure $\widetilde{Q}_T$. Also, let the measure $\bar{Q}$ be such that

$$\left.\frac{d\bar{Q}}{dP}\right|_{\mathcal{F}_T} = \exp\left\{L_T - \frac{1}{2}\langle L\rangle_T\right\},$$

and

$$E\exp\left\{L_T - \frac{1}{2}\langle L\rangle_T\right\} = 1, \qquad (3.2.23)$$

where $L_t = \int_0^t s^\alpha \delta_s dW_s$, $M_t^H = \int_0^t l_H(t,s) dB_s^H$, $W_t = \widehat{a}\int_0^t s^\alpha dM_s^H$, $\int_0^t l_H(t,s)\psi_s ds = \widetilde{\alpha}\int_0^t \delta_s ds$, $t > 0$ with $E\int_0^t s^{2\alpha}\delta_s^2 ds < \infty$, $\int_0^t |\delta_s| ds < \infty$, P-a.s., $t > 0$ (see Subsection 2.8.1). Then the process $\widehat{B}_t^H := B_t^H - \int_0^t \psi_s ds$ is an fBm w.r.t. measure $\bar{Q}|_{\mathcal{F}_T}$. Now we need in the equality $\widetilde{Q}|_{\mathcal{F}_T} = \bar{Q}|_{\mathcal{F}_T} = Q|_{\mathcal{F}_T}$. Hence, in particular, $L_t = \int_0^t \varphi_s dW_s$, whence $\varphi_s = s^\alpha \delta_s$. Therefore we want to find φ and ψ in such a way that common drift equals

$$\sigma_1 \varphi_t + \sigma_2 \psi_t = -e(t, X_t), \quad t \in [0, T]. \qquad (3.2.24)$$

Now we apply the Abel rearrangement to the relation

$$\int_0^t l_H(t,s)\psi_s ds = \widetilde{\alpha}\int_0^t \delta_s ds = \widetilde{\alpha}\int_0^t s^{-\alpha}\varphi_s ds :$$

$$C_H^{(5)}\int_0^t (t-u)^{\alpha-1}\int_0^u (u-s)^{-\alpha}s^{-\alpha}\psi_s ds\, du$$

$$= \widetilde{\alpha}\int_0^t (t-u)^{\alpha-1}\int_0^u s^{-\alpha}\varphi_s ds\, du,$$

or

$$B(\alpha, 1-\alpha)C_H^{(5)}\int_0^t s^{-\alpha}\psi_s ds = \widetilde{\alpha}\int_0^t \frac{(t-u)^\alpha}{\alpha} u^{-\alpha}\varphi_u du,$$

whence after differentiation

$$(\alpha C_H^{(6)})^{-1} t^{-\alpha} \psi_t = \tilde{\alpha} \int_0^t (t-u)^{\alpha-1} u^{-\alpha} \varphi_u du. \quad (3.2.25)$$

Substituting (3.2.25) into (3.2.24), we obtain that

$$\sigma_1 \varphi_t + \sigma_2 C_H^{(9)} t^\alpha \int_0^t (t-u)^{\alpha-1} u^{-\alpha} \varphi_u du = -e(t, X_t), C_H^{(9)} = \alpha C_H^{(6)} \tilde{\alpha}. \quad (3.2.26)$$

Denote $\theta_t := t^{-\alpha} \varphi_t$, then

$$\sigma_1 \theta_t + \sigma_2 C_H^{(9)} \int_0^t (t-u)^{\alpha-1} \theta_u du = -\hat{e}(t, X_t). \quad (3.2.27)$$

Equation (3.2.27) is a Volterra equation with weak singularity, and its unique solution has the form

$$\theta_t = -\frac{\hat{e}(t, X_t)}{\sigma_1} - \frac{1}{\sigma_1} \int_0^t \sum_{n=1}^\infty \rho^n \frac{(t-s)^{n\alpha-1}}{\Gamma(n\alpha)} \hat{e}(s, X_s) ds,$$

where $\rho = \sigma_2 C_H^{(9)} \Gamma(\alpha)$. Now we must check conditions (3.2.22) and (3.2.23). Evidently, it is sufficient to check Novikov's condition: $E \exp\left\{\frac{1}{2} \int_0^T \varphi_t^2 dt\right\} < \infty$ and $E \exp\left\{\frac{1}{2} \langle L \rangle_T\right\} < \infty$. But $\varphi_t = -\frac{e(t, X_t)}{\sigma_1} - \frac{1}{\sigma_1} t^\alpha \int_0^t \sum_{n=1}^\infty \rho^n \frac{(t-s)^{n\alpha-1}}{\Gamma(n\alpha)} \hat{e}(s, X_s) ds$ and is bounded since e is bounded. Further, $\delta_s = \tilde{\alpha} s^{-\alpha} \varphi_s$, and Novikov's condition evidently holds for the function L, too. So, we have proved the following result.

Theorem 3.2.9. *Under our suppositions equation (3.2.1) under measure Q obtains the differential form*

$$dX_t = \sigma_1 X_t d\widehat{W}_t + \sigma_2 X_t d\widehat{B}_t^H, \quad X(0) = X_0,$$

and its solution has a form

$$X_t = X_0 \exp\{\sigma_1 \widehat{W}_t + \sigma_2 \widehat{B}_t^H - 1/2 \sigma_1^2 t\}.$$

3.3 Stochastic Differential Equations with Fractional White Noise

3.3.1 The Lipschitz and the Growth Conditions on the Negative Norms of Coefficients

Now we return to Wick integration with respect to fBm (see Sections 1.5 and 2.3). Consider the SDE of the form

3.3 Stochastic Differential Equations with Fractional White Noise

$$X_t = X_0 + \int_0^t b(s, X_s) ds + \sum_{j=1}^m \int_0^t \sigma_j(s, X_s) \Diamond \dot{B}_s^{H_j} ds, \qquad (3.3.1)$$
$$t \in [0, T],$$

where all $H_j \in [1/2, 1)$ are different, $\dot{B}^{H_j}$ are the fractional noises. The equation, similar to (3.3.1), but with white noise was studied by Våge (Vage96). Note that the proof of the existence and uniqueness result in (Vage96) is in fact based not on the structure of white noise, but on its inclusion into S^*, and this fact holds for fractional noise also, see Lemma 1.5.3. According to Theorem 1 (Vage96), the negative norm of the Wick products admits the following estimate:
$\|F \Diamond G\|_{-r} \leq C_{r,q} \|F\|_{-r} \|G\|_{-q}$ for random variables $F \in S_{-r}$, $G \in S_{-q}$, $r < q - 1$. According to Lemma 1.5.3, $\dot{B}_t^{H_k} \in S_{-q}$ for any $q > 7/3$, in particular, $\dot{B}_t^{H_k} \in S_{-3}$ and, moreover, $\sup_{t \geq 0} \|\dot{B}_t^{H_k}\|_{-q} \leq C_q$ for $q > 7/3$ and some $C_q > 0$.

Therefore, for any $r > 0$ and $F \in S_{-r}$ $\|F \Diamond \dot{B}_t^{H_k}\|_{-r} \leq C \|F\|_{-r}$.

Suppose now that the coefficients b and σ and the initial value X_0 of equation (3.3.1) satisfy the conditions:

(i) for any $1 \leq j \leq m$ and some $r > 0$, $b, \sigma_j : [0, T] \times S_{-r} \to S_{-r}$, $X_0 \in S_{-r}$, the functions $b(t, X_t)$ and $\sigma_j(t, X_t)$, $1 \leq j \leq m$ are strongly measurable on $[0, T]$ for any $X \in C([0, T], S_{-r})$;

(ii) for some $r > 0$

$$\|b(t, x) - b(t, y)\|_{-r} + \sum_{j=1}^m \|\sigma_j(t, x) - \sigma_j(t, y)\|_{-r} \leq C \|x - y\|_{-r}, \quad 0 \leq t \leq T,$$

$$\|b(t, x)\|_{-r} + \sum_{j=1}^m \|\sigma_j(t, x)\|_{-r} \leq c(1 + \|x\|_{-r}), \quad 0 \leq t \leq T.$$

It follows from strong measurability of σ_j and Theorem 6 (Vage96) that $\sigma_j(t, x) \Diamond \dot{B}_t^{H_j}$ is also strongly measurable. Further, condition (ii) ensures the existence of $\int_0^t b(s, X_s) ds$ and $\int_0^t \sigma_j(s, x) \Diamond \dot{B}_s^{H_j} ds, 0 \leq t \leq T$, that can be considered as the Bochner integrals in S_{-r} for $X \in C([0, T], S_{-r})$.

The next result can be proved with the help of the standard method of successive approximations (similar proof for white noise is contained in (Vage96)).

Theorem 3.3.1. *Under conditions (i) and (ii) equation (3.3.1) has on $[0, T]$ the unique solution $X \in C([0, T], S_{-r})$.*

3.3.2 Quasilinear SDE with Fractional Noise

As mentioned in (Vage96), simultaneous fulfilment of the Lipschitz and growth conditions on the negative norms of coefficients is very restrictive. To avoid this, we consider the quasilinear equation of the form

$$X_t = X_0 + \int_0^t b(s, X_s, \omega) ds + \sum_{j=1}^m \int_0^t \sigma_j(s) X_s \Diamond \dot{B}_s^{H_j} ds, \qquad (3.3.2)$$

where $H_j \in [1/2, 1)$, the coefficients and the initial value X_0 satisfy the following conditions:

(iii) $\sigma_j(s), 1 \leq j \leq m$ are nonrandom functions, $\sigma_j \in L_{1/H_j}[0,T]$;
(iv) the function $b(s,x,\omega) : [0,T] \times \mathbb{R} \times S' \to \mathbb{R}$ is measurable in all the arguments,

$$|b(s,x,\omega)| \leq C(1+|x|), \ \omega \in S'(\mathbb{R}), \ s \in [0,T], \ x \in \mathbb{R};$$
$$|b(s,x,\omega) - b(s,y,\omega)| \leq C|x-y|, \ x,y \in \mathbb{R}, \ \omega \in S'(\mathbb{R}), \ s \in [0,T];$$

(v) $X_0 \in L_p(\Omega) = L_p(S'(\mathbb{R}))$ for some $p > 0$.

Theorem 3.3.2. *Under conditions (iii)–(v) the equation (3.3.2) has on $[0,T]$ the unique solution $X \in L_{p'}(\Omega)$ for any $p' < p$.*

Proof. Let, for simplicity, $m = 1$, $H_1 = H \in (1/2, 1)$. Consider the differential form of equation (3.3.2)

$$\frac{dX_t}{dt} = b(t, X_t) + \sigma(t) X_t \diamond \dot{B}_t^H, \quad X(0) = X_0. \tag{3.3.3}$$

Put $\sigma_t(s) := \sigma(s) 1_{[0,t]}(s)$, and suppose that $J_\sigma(t)$ is the Wick exponent of the form $J_\sigma(t) = \exp^\diamond(-\int_0^t \sigma(s) dB_s^H)$, where the Wick exponent is defined as $\exp^\diamond X := \sum_{n=0}^\infty \frac{X^{\diamond n}}{n!}$. Then, according to formula (1.6.1),

$$J_\sigma(t) = \exp^\diamond\left\{-\int_\mathbb{R} (M_-^H \sigma_t)(s) dW_s\right\}.$$

Denote also $Z_t := J_\sigma(t) \diamond X_t$.

By the rules of stochastic differentiation (see, for example, (HOUZ96)),

$$\frac{dZ_t}{dt} = J_\sigma(t) \diamond \frac{dX_t}{dt} - \sigma(t) \frac{dJ_\sigma(t)}{dt} \diamond X_t \diamond \dot{B}_t^H,$$

and we obtain from (3.3.3) that

$$\frac{dZ_t}{dt} = \frac{dJ_\sigma(t)}{dt} \diamond b(t, X_t). \tag{3.3.4}$$

Now we use the Gjessing lemma (Gje94), which states that

$$\frac{dJ_\sigma(t)}{dt} \diamond b(t, X_t, \omega) = \frac{dJ_\sigma(t)}{dt} \cdot b(t, T_{-(M_-^H \sigma_t)} X_t, \omega + M_-^H \sigma_t), \tag{3.3.5}$$

where T is the shift operator, $T_{\omega_0} F(\omega) = F(\omega + \omega_0)$ for any $\omega_0 \in S'(\mathbb{R})$.

Similarly, $Z_t = J_\sigma(t) \cdot T_{-(M_-^H \sigma_t)} X_t$, and from (3.3.4)–(3.3.5) we obtain that Z_t is the solution of the ordinary differential equation

$$\frac{dZ_t}{dt} = \frac{dJ_\sigma(t)}{dt} \cdot b(t, J_\sigma^{-1}(t) \cdot Z_t, \omega + M_-^H \sigma_t), \ Z_0 = X_0, \tag{3.3.6}$$

for any $\omega \in S'(\mathbb{R})$. Equation (3.3.6) differs from the corresponding equation (3.6.15) for the white noise (see the book (HOUZ96)) only with the function $M_-^H \sigma_t$ instead of σ_t. However, it has the same structure, which means that conditions (iii)–(v) ensure the existence and the uniqueness of the solution of equation (3.3.2) for any $\omega \in S'(\mathbb{R})$ on the interval $[0,T]$. Now we estimate the moments of the solution X_t.

First, from conditions (iii)–(iv)

$$|Z_t| \leq |X_0| + \int_0^t J_\sigma(s)|a(s, J_\sigma^{-1}(s)Z_s, \omega - (M_-^H \sigma_s))|ds$$
$$\leq |X_0| + C \int_0^t J_\sigma(s)(1 + J_\sigma^{-1}(s)|Z_s|)ds$$
$$\leq |X_0| + C \int_0^t J_\sigma(s)ds + C \int_0^t |Z_s|ds,$$

and from the Gronwall inequality it follows that

$$|Z_t| \leq (|X_0| + C \int_0^T J_\sigma(s)ds) \exp\{CT\},$$
$$E|Z_t|^p \leq \exp\{pCT\} 2^p (E|X_0|^p + CE \int_0^T |J_\sigma(s)|^p ds). \tag{3.3.7}$$

Since

$$E|J_\sigma(s)|^p + E\exp^\diamond\{-p \int_\mathbb{R} (M_-^H \sigma_t)(s)dW_s\}$$
$$= \exp\{p^2 \|M_-^H \sigma_t\|_{L_2(\mathbb{R})}^2\},$$

and condition (iii) and inequality (1.9.2) ensure that $(M_-^H \sigma_t) \in L_2(\mathbb{R})$, therefore we obtain from (3.3.7) that $E|Z_t|^p < \infty$ for any $p > 0$. Further, $T_{-(M_-^H \sigma_t)} X_t = Z_t J_\sigma^{-1}(t)$, and $E|J_\sigma^{-1}(t)|^q < \infty$ for any $q > 0$, therefore $T_{-(M_-^H \sigma_t)} X_t \in L_{p'}(\Omega)$ for any $p' < p$. Since $M_-^H \sigma_t \in L_2(\mathbb{R})$, we obtain from Corollary 2.10.5 (HOUZ96) that $X \in L_{p'}(\Omega)$ for any $p' < p$. □

3.4 The Rate of Convergence of Euler Approximations of Solutions of SDE Involving fBm

The numerical solution of stochastic differential equations driven by Wiener process is essentially based on the method of time discretization and has a long history. We refer to the monograph (KP92), which contains an almost complete theory of the numerical solution of such SDEs with regular coefficients. The paper (KP94) is devoted to the Euler approximations for SDEs driven by semimartingales. Concerning the numerical solution of SDEs driven by fBm, we mention first the paper (GA98), where the equations with the modified fBm (which is a special semimartingale) are studied. The papers (Nou05; NN06) study Euler approximations for the homogeneous one-dimensional SDEs involving fBm and having bounded coefficients with bounded derivatives up to third order. It is proved that the error of the approximation is a.s. equivalent to $\delta^{2\alpha} \xi_t$, and the process ξ_t is given explicitly. These papers also discuss the Crank–Nicholson and the Milstein schemes for SDEs driven by fBm. Here we present the results on the rate of convergence of Euler approximations of solutions for SDE with nonstationary coefficients. Of course, our approach differs from those proposed in (Nou05),(NN06).

3.4.1 Approximation of Pathwise Equations

Consider the multidimensional equation (3.1.12) with the coefficients satisfying the $\mathbb{R}^d$-version of assumptions (i)–(v) of Subsection 3.1.1 with $H_j = H, 1 \leq j \leq m, b_0(t) = L$ (see Remark 3.1.5 for additional notations). Under these assumptions, this equation has the unique solution $\{X_t, t \in [0,T]\}$ and for a.a. $\omega \in \Omega$ the trajectories of the solution belong to $C^{H-}[0,T]$.

Now, let $t \in [0,T], \delta = \frac{T}{N}, \tau_n = \frac{nT}{N} = n\delta, n = 0,\ldots,N$. Consider the discrete Euler approximations of the solution of equation (3.1.12),

$$\widetilde{Y}^{i,\delta}_{\tau_{n+1}} = \widetilde{Y}^{i,\delta}_{\tau_n} + b_i(\tau_n, \widetilde{Y}^{\delta}_{\tau_n})\delta + \sum_{j=1}^{m} \sigma_{ji}(\tau_n, \widetilde{Y}^{\delta}_{\tau_n})\Delta B^{j,H}_{\tau_n}, \quad \widetilde{Y}^{i,\delta}_0 = X^i_0,$$

and the corresponding continuous interpolations

$$Y^{i,\delta}_t = \widetilde{Y}^{i,\delta}_{\tau_n} + b_i(\tau_n, \widetilde{Y}^{\delta}_{\tau_n})(t-\tau_n) + \sum_{j=1}^{m} \sigma_{ji}(\tau_n, \widetilde{Y}^{\delta}_{\tau_n})(B^{j,H}_t - B^{j,H}_{\tau_n}), \quad t \in [\tau_n, \tau_{n+1}]. \tag{3.4.1}$$

The continuous interpolations satisfy the equation

$$Y^{i,\delta}_t = X^i_0 + \int_0^t b_i(t_u, Y^{\delta}_{t_u})du + \sum_{j=1}^{m} \int_0^t \sigma_{ji}(t_u, Y^{\delta}_{t_u})dB^{j,H}_u, \tag{3.4.2}$$

where $t_u = \tau_{n_u}, n_u = \max\{n : \tau_n \leq u\}$.

For simplicity we denote the vector of solutions as $X_t = (X^i_t)_{i=1,\ldots,d}$, the vector of the continuous approximations as $Y^{\delta}_t = (Y^{\delta,i}_t)_{i=1,\ldots,d}$.

Theorem 3.4.1. *1) Let the modification of conditions (i)–(v') from Section 3.1 hold for the vector case, with $\gamma > 1 - H$, $\kappa = \mu = 1$, $L_R = L$, $M_R = M$ and $b_0(t) = L$.*

Then for any $\varepsilon > 0$ and $0 < \rho < H$ there exist $\delta_0 > 0$ and $\Omega_{\varepsilon,\delta_0,\rho} \subset \Omega$ such that $P(\Omega_{\varepsilon,\delta_0,\rho}) > 1 - \varepsilon$ and for any $\omega \in \Omega_{\varepsilon,\delta_0,\rho}$, $\delta < \delta_0$ one has $|Y^{\delta}_t| \leq C(\omega)$, $|Y^{\delta}_s - Y^{\delta}_r| \leq C(\omega)(t_s - t_r)^{H-\rho}, 0 \leq r < s \leq T$.

2) If, instead of (v) and (v') we assume that b and σ are bounded functions, then $|Y^{\delta}_t| \leq C(\omega)$, $|Y^{\delta}_s - Y^{\delta}_r| \leq C(\omega)(s - r)^{H-\rho}, 0 \leq r < s \leq T$.

In both the cases $C(\omega)$ does not depend on δ.

Proof. 1) We can always assume that $\delta \leq 1$. It follows immediately from (i) and (iii), Section 3.1.1 and (3.4.2) that for any $\beta \in (1 - H, \gamma \wedge 1/2)$

3.4 The Rate of Convergence of Euler Approximations   245

$$|Y_t^{i,\delta}| \le |X_0^i| + \int_0^t |b_i(t_u, Y_{t_u}^\delta)|\, du + \sum_{j=1}^m \left|\int_0^t \sigma_{ji}(t_u, Y_{t_u}^\delta) dB_u^{j,H}\right|$$

$$\le |X_0^i| + L\int_0^t \left(1 + |Y_{t_u}^\delta|\right) du + G_T \sum_{j=1}^m \int_0^t |\sigma_{ji}(t_u, Y_{t_u}^\delta)|\, u^{-\beta} du$$

$$+ G_T \sum_{j=1}^m \int_0^t \int_0^r |\sigma_{ji}(t_r, Y_{t_r}^\delta) - \sigma_{ji}(t_u, Y_{t_u}^\delta)|(r-u)^{-\beta-1} du\, dr$$

$$\le |X_0^i| + \left(mMG_T \frac{T^{1-\beta}}{1-\beta} + LT\right) + (mMG_T + LT^\beta) \int_0^t |Y_{t_u}^\delta|\, u^{-\beta} du$$

$$+ MG_T \int_0^t \int_0^{t_r} \left((t_r - t_u)^\gamma + |Y_{t_r}^\delta - Y_u^\delta| + |Y_u^\delta - Y_{t_u}^\delta|\right)$$

$$\times (r-u)^{-\beta-1} du\, dr,$$
(3.4.3)

where $G_T := \Lambda_{1-\beta}(B^H)$. (We use here the equality $t_r = t_u$ for $t_r \le u < r$.)
Denote $C_1(\omega) := |X_0| + \left(mMG_T \frac{T^{1-\alpha}}{1-\alpha} + LT\right)$, $C_2(\omega) := (mMG_T + LT^\beta)$.
Further, note that $t_r - t_u \le r - u + \delta$. Also, it follows from representations (3.4.1) and (3.4.2) that for any $\rho \in (0, H)$

$$|Y_u^\delta - Y_{t_u}^\delta| \le L\left(1 + |Y_{t_u}^\delta|\right)(u - t_u) + M \cdot C(\omega, \rho)\left(1 + |Y_{t_u}^\delta|\right)(u - t_u)^{H-\rho}$$

$$\le C_3(\omega)\left(1 + |Y_{t_u}^\delta|\right)(u - t_u)^{H-\rho},$$
(3.4.4)

where the value $C(\omega, \rho)$ appears in the relation
$|B_t^H - B_s^H| \le C(\omega, \rho)|t-s|^{H-\rho}, s, t \in [0, T], C_3(\omega) = LT^{1-H+\rho} + M \cdot C(\omega, \rho)$.
Moreover, for $\gamma > \beta$

$$P_t := \int_0^t \int_0^{t_r} (t_r - t_u)^\gamma (r - u)^{-\beta-1} du\, dr$$

$$\le \int_0^t \int_0^{t_r} ((r-u)^\gamma + \delta^\gamma)(r-u)^{-\beta-1} du\, dr$$

$$\le (\gamma - \beta)^{-1} \int_0^t r^{\gamma-\beta} dr + \beta^{-1}\delta^\gamma \int_0^t (r - t_r)^{-\beta} dr,$$

and for any $k \ge 0$ and any power $\pi > -1$

$$\int_{\tau_k}^{\tau_{k+1}} (r - t_r)^\pi dr = \int_{\tau_k}^{\tau_{k+1}} (r - \tau_k)^\pi dr = C_1 \delta^{\pi+1} \text{ with } C_1 = (\pi + 1)^{-1},$$

whence

$$\int_0^t (r - t_r)^{-\beta} dr \le \int_0^T (r - t_r)^{-\beta} dr = C_1 N\delta^{1-\beta} = C_1 \delta^{-\beta}. \tag{3.4.5}$$

Therefore

3 Stochastic Differential Equations Involving Fractional Brownian Motion

$$P_t \leq C_1 T^{\gamma-\beta+1} + \beta^{-1} C_1 \delta^{\gamma-\beta} \leq C_1 T^{\gamma-\beta+1} + \beta^{-1} C_1 =: C_2. \qquad (3.4.6)$$

Estimate now

$$Q_t := \int_0^t \int_0^{t_r} |Y_u^\delta - Y_{t_u}^\delta|(r-u)^{-\beta-1} du\, dr,$$

using (3.4.4) and (3.4.5):

$$Q_t \leq C_3(\omega)(1 + Y_t^{\delta,*}) \int_0^t \int_0^{t_r} (u - t_u)^{H-\rho}(r-u)^{-\beta-1} du\, dr$$

$$\leq C_3(\omega)(1 + Y_t^{\delta,*}) \delta^{H-\rho} \beta^{-1} \int_0^t (r - t_r)^{-\beta} dr \leq C_4(\omega)(1 + Y_t^{\delta,*}) \delta^{H-\beta-\rho}$$

$$(3.4.7)$$

with $C_4(\omega) = C_3(\omega) \beta^{-1} \cdot C_1$. Note that $Y_t^{\delta,*} := \sup_{0 \leq s \leq t} |Y_s^\delta| < \infty$ for any $t \in [0, T]$ a.s. Substituting (3.4.6) and (3.4.7) into (3.4.3), we obtain that

$$|Y_t^\delta| \leq C_5(\omega) + mC_2(\omega) \int_0^t |Y_{t_u}^\delta| u^{-\beta} du + mC_4(\omega) Y_t^{\delta,*} \delta^{H-\beta-\rho}$$

$$+ C_6(\omega) \int_0^t \int_0^{t_r} \varphi_{r,u} du\, dr,$$

$$(3.4.8)$$

where $\varphi_{r,u} := |Y_{t_r}^\delta - Y_u^\delta|(r-u)^{-\beta-1}$, for $0 < v < t_u < T$, $0 < \beta < 1$ with $C_5(\omega) = mC_1(\omega) + mG_T C_1 + mC_1 G_T C_3(\omega) + C_4(\omega)$, $C_6(\omega) = MG_T$. To simplify the notations, in what follows we remove subscripts from $C(\omega)$ and C, writing $C(\omega)$ for all constants depending on ω and C for all nonrandom constants.

Summing up everything, we can write

$$Y_t^{\delta,*} \leq C(\omega)\left(1 + Y_t^{\delta,*} \delta^{H-\beta-\rho} + \int_0^t |Y_{t_u}^\delta| u^{-\beta} du + \int_0^t \int_0^{t_r} \varphi_{r,u} du\, dr\right).$$

$$(3.4.9)$$

In turn, we can estimate $\int_0^{t_s} \varphi_{s,u} du$. First, similarly to the previous estimates,

$$|Y_{t_s}^\delta - Y_u^\delta| \leq C(\omega)\left[\int_u^{t_s} \left(1 + |Y_{t_v}^\delta|\right) dv + \int_u^{t_s} \left(1 + |Y_{t_v}^\delta|\right)(v-u)^{-\beta} dv\right.$$

$$\left. + \int_u^{t_s} \int_u^{t_v} |\sigma(t_v, Y_{t_v}^\delta) - \sigma(t_z, Y_{t_z}^\delta)|(v-z)^{-\beta-1} dz\, dv\right]$$

$$\leq C(\omega)\left[(t_s - u)^{1-\beta} + \int_u^{t_s} |Y_{t_v}^\delta|(v-u)^{-\beta} dv\right.$$

$$+ \delta^\gamma \int_u^{t_s} (v - t_v)^{-\beta} dv + \int_u^{t_s} \int_u^{t_v} \varphi_{v,z} dz\, dv$$

$$\left. + \int_u^{t_s} \int_u^{t_v} |Y_z^\delta - Y_{t_z}^\delta|(v-z)^{-\beta-1} dz\, dv\right];$$

$$(3.4.10)$$

3.4 The Rate of Convergence of Euler Approximations 247

multiplying by $(s-u)^{-\beta-1}$ and integrating over $[0, t_s]$, we obtain that

$$\int_0^s \varphi_{s,u} du \leq C(\omega) \sum_{i=1}^5 Q_s^i, \qquad (3.4.11)$$

where

$$Q_s^1 := \int_0^s (t_s - u)^{1-\beta}(s-u)^{-\beta-1} du \leq \int_0^{t_s} (s-u)^{-2\beta} du \leq C; \qquad (3.4.12)$$

$$Q_s^2 := \int_0^{t_s} (s-u)^{-\beta-1} \int_u^{t_s} |Y_{t_v}^\delta|(v-u)^{-\beta} dv$$
$$= \int_0^{t_s} |Y_{t_v}^\delta| \int_0^v (v-u)^{-\beta}(s-u)^{-\beta-1} du\, dv \leq C \int_0^{t_s} |Y_{t_v}^\delta|(s-v)^{-2\beta} dv, \qquad (3.4.13)$$

where $C = \int_0^\infty (1+y)^{-\beta-1} y^{-\beta} dy$;

$$Q_s^3 := \delta^\gamma \int_0^{t_s} (s-u)^{-\beta-1} \int_u^{t_s} (v-t_v)^{-\beta} dv\, du$$
$$\leq \beta^{-1} \delta^\gamma \int_0^{t_s} (s-v)^{-\beta}(v-t_v)^{-\beta} dv. \qquad (3.4.14)$$

Let $t_s = n\delta$ for some $0 < n \leq N$. The last integral can be estimated as

$$I := \int_0^{t_s} (s-v)^{-\beta}(v-t_v)^{-\beta} dv = \sum_{k=0}^{n-2} \int_{k\delta}^{(k+1)\delta} + \int_{(n-1)\delta}^{(n-1/2)\delta} + \int_{(n-1/2)\delta}^{n\delta},$$

where

$$\int_{k\delta}^{(k+1)\delta} \leq (s-(k+1)\delta)^{-\beta} \int_{k\delta}^{(k+1)\delta} (v-\tau_v)^{-\beta} dv \leq C(s-(k+1)\delta)^{-\beta} \delta^{1-\beta},$$

and the last two integrals are bounded by $C\delta^{1-2\beta}$. Therefore, $I \leq C\delta^{-\beta}$. Further, using estimate (3.4.4), we can conclude that

$$Q_s^4 := \int_0^{t_s} (s-u)^{-\beta-1} \int_u^{t_s} \int_u^{t_v} \varphi_{v,z} dz\, dv\, du$$
$$\leq \int_0^{t_s} \int_0^{t_v} \int_0^{z \wedge v} \varphi_{v,z}(s-u)^{-\beta-1} du\, dz\, dv \qquad (3.4.15)$$
$$\leq C \int_0^{t_s} (s-v)^{-\beta} \int_0^{t_v} \varphi_{v,z} dz\, dv.$$

Finally, similarly to the previous estimates,

$$Q_s^5 := \int_0^{t_s}(s-u)^{-\beta-1}\int_u^{t_s}\int_u^{t_v}|Y_z^\delta - Y_{t_z}^\delta|(v-z)^{-\beta-1}dz\,dv\,du$$

$$\leq C(\omega)\int_0^{t_s}(s-u)^{-\beta-1}\int_u^{t_s}\int_u^{t_v}(v-z)^{-\beta-1}dz\,dv\,du \cdot \delta^{H-\rho}\left(1+\left|Y_{t_s}^{\delta,*}\right|\right)$$

$$\leq C(\omega)\left(1+\left|Y_{t_s}^{\delta,*}\right|\right)\delta^{H-\rho-\beta}.$$
(3.4.16)

Now, denote $\psi_s := Y_s^{\delta,*} + \int_0^{t_s}\varphi_{s,u}du$. Then it follows from (3.4.9) and (3.4.11)–(3.4.16) that for any $t \in [0,T]$ (including $t = k\delta$)

$$\psi(t) \leq C(\omega)\left(1 + Y_t^{\delta,*}\delta^{H-\beta-\rho} + \int_0^t\left((t-v)^{-2\beta} + v^{-\beta}\right)\psi_v dv\right).$$

Let $\varepsilon > 0$ be fixed. Note that all constants $C(\omega)$ are finite a.s. and independent of δ. Thus, we can choose $\delta_0 > 0, \rho$ small enough such that $H - \beta - \rho > 0$, and $\Omega_{\varepsilon,\delta_0,\rho}$ such that $C(\omega)\delta_0^{H-\beta-\rho} \leq 1/2$ on $\Omega_{\varepsilon,\delta_0,\rho}$ and $P(\Omega_{\varepsilon,\delta_0,\rho}) > 1 - \varepsilon$. Then for any $\omega \in \Omega_{\varepsilon,\delta_0,\rho}$

$$\psi_t \leq C(\omega) + \frac{1}{2}\psi_t + C(\omega)\int_0^t\left((t-v)^{-2\beta} + v^{-\beta}\right)\psi_v dv,$$

whence

$$\psi_t \leq C(\omega)\left(1 + t^{2\beta}\int_0^t(t-v)^{-2\beta}v^{-2\beta}\psi_v dv\right),$$

and it follows immediately from the last equation and (3.1.22)–(3.1.23) that $\psi_t \leq C(\omega)$ whence, in particular, $|Y_t^\delta| \leq C(\omega)$, $t \in [0,T]$. Moreover, from (3.4.10) with $u = t_r$, $r \leq s$, taking into account that $\int_{t_r}^{t_s}(v-t_v)^{-\beta}dv = (1-\beta)^{-1}\delta^{-\beta}(t_s - t_r)$, we obtain the bound

$$|Y_{t_s}^\delta - Y_{t_r}^\delta| \leq C(\omega)\left((t_s - t_r)^{1-\beta} + \delta^{\gamma-\beta}(t_s - t_r) + (t_s - t_r)\right.$$
$$\left. + \delta^{H-\rho}\int_{t_r}^{t_s}(v-t_v)^{-\beta}dv\right) \leq C(\omega)(t_s - t_r)^{1-\beta},$$

and statement 1) is proved.

2) Let $|b(t,x)| \leq b$, $|\sigma(t,x)| \leq \sigma$. Then it is very easy to see that estimate (3.4.8) will take the form

$$|Y_t^\delta| \leq C(\omega)\left(1 + \int_0^t\int_0^{t_r}\varphi_{r,v}du\,dr\right),$$

(3.4.10) will take the form

$$|Y_{t_s}^\delta - Y_u^\delta| \leq C(\omega)\left((t_s - u)^{1-\beta} + (\delta^\gamma + \delta^{H-\rho})\int_u^{t_s}(v-t_v)^{-\beta}dv\right.$$
$$\left. + \int_u^{t_s}\int_u^{t_v}\varphi_{v,z}dz\,dv\right),$$

and instead of (3.4.11)–(3.4.16) we obtain

$$\int_0^{t_s} \varphi_{s,u} du \leq C(\omega)\left(1 + \int_0^{t_s} (s-v)^{-\beta} \int_0^{t_v} \varphi_{v,z} dz\, dv\right),$$

whence the proof follows. □

Remark 3.4.2. It is easy to see that we proved a little more than Theorem 3.2.3 states. Namely, we proved that the norm in Besov space, $\sup_{0 \leq s \leq T} \psi_s$, is bounded by $C(\omega)$ on $\Omega_{\varepsilon,\delta_0,\rho}$, with $C(\omega)$ not depending on δ.

Now we establish the estimates of the rate of convergence of our approximations (3.4.2) for the solution of equation (3.1.12) with pathwise integral w.r.t. fBm. We establish even more, namely, the estimate of convergence rate for the norm of the difference $X_t - Y_t^\delta$ in some Besov space, similarly to the result of Theorem 3.4.1. Denote

$$\Delta_{u,s}(X, Y^\delta) := \left|X_s - Y_s^\delta - X_u + Y_u^\delta\right|.$$

Theorem 3.4.3. *Let the modification of conditions (i)–(v') from Section 3.1 hold for the vector case, with $\gamma > 1 - H$, $\kappa = \mu = 1$, $L_R = L$, $M_R = M$ and $b_0(t) = L$, and suppose also that:*

1) the coefficient b is Hölder continuous in time: $|b(t,x) - b(s,x)| \leq C|t-s|^\theta$, $C > 0$, $2\alpha < \theta \leq 1$, $\alpha = H - 1/2$;

2) the exponent γ from condition (iii)(Section 3.1) satisfies $\gamma > H$.

Then:

1. For any $\varepsilon > 0$, $\beta \in (1 - H, 1/2)$ and any sufficiently small $\rho > 0$ there exists $\delta_0 > 0$ and $\Omega_{\varepsilon,\delta_0,\rho}$ such that $P(\Omega_{\varepsilon,\delta_0,\rho}) > 1 - \varepsilon$ and for any $\omega \in \Omega_{\varepsilon,\delta_0,\rho}$, $\delta < \delta_0$

$$U_\delta := \sup_{0 \leq s \leq T}\left(\left|X_s - Y_s^\delta\right| + \int_0^{t_s} \left|\Delta_{u,s}(X, Y^\delta)\right|(s-u)^{-\beta-1} du\right)$$
$$\leq C(\omega) \cdot \delta^{2\alpha - \rho},$$

where $C(\omega)$ does not depend on δ and ε (but depends on ρ);

2. If, in addition, the coefficients b and σ are bounded, then for any $\rho \in (0, 2\alpha)$ there exists $C(\omega) < \infty$ a.s. such that $U_\delta \leq C(\omega)\delta^{2\alpha-\rho}$, $C(\omega)$ does not depend on δ.

Proof. 1. Denote $Z_t^\delta := \sup_{0 \leq s \leq t}\left|X_s - Y_s^\delta\right|$. Then

$$Z_t^\delta \leq \sup_{0\leq s\leq t} \int_0^s |b(u, X_u) - b(t_u, Y_{t_u}^\delta)| du$$

$$+ \sup_{0\leq s\leq t} \sum_{i,j=1}^m \left| \int_0^s (\sigma_{ji}(u, X_u) - \sigma_{ji}(t_u, Y_{t_u}^\delta)) dB_u^{j,H} \right|$$

$$\leq \int_0^t |b(u, X_u) - b(u, Y_u^\delta)| du + \int_0^t |b(u, Y_u^\delta) - b(t_u, Y_u^\delta)| du$$

$$+ \int_0^t |b(t_u, Y_u^\delta) - b(t_u, Y_{t_u}^\delta)| du \tag{3.4.17}$$

$$+ \sup_{0\leq s\leq t} \sum_{i,j=1}^m \left| \int_0^s (\sigma_{ji}(u, X_u) - \sigma_{ji}(u, Y_u^\delta)) dB_u^{j,H} \right|$$

$$+ \sup_{0\leq s\leq t} \sum_{i,j=1}^m \left| \int_0^s (\sigma_{ji}(u, Y_u^\delta) - \sigma_{ji}(t_u, Y_u^\delta)) dB_u^{j,H} \right|$$

$$+ \sup_{0\leq s\leq t} \sum_{i,j=1}^m \left| \int_0^s (\sigma_{ji}(t_u, Y_u^\delta) - \sigma_{ji}(t_u, Y_{t_u}^\delta)) dB_u^{j,H} \right| =: \sum_{k=1}^6 I_k.$$

Now we estimate separately all these terms. Evidently,

$$I_1 \leq L \int_0^t Z_u^\delta du. \tag{3.4.18}$$

Condition 1) implies that for $\delta < 1$

$$I_2 \leq C \int_0^t |u - t_u|^\theta du \leq C\delta^\theta \leq C\delta^{2\alpha}. \tag{3.4.19}$$

It follows from Theorem 3.4.1 that for any $\varepsilon > 0$ and any $\rho \in (0, H)$ there exists $\delta_0 > 0$ and $\Omega_{\varepsilon,\delta_0,\rho} \subset \Omega$ such that $P(\Omega_{\varepsilon,\delta_0,\rho}) > 1 - \varepsilon$ and $C(\omega)$ independent of ε and δ such that for for any $\omega \in \Omega_{\varepsilon,\delta_0,\rho}$ it holds that $|Y_t^\delta - Y_s^\delta| \leq C(\omega)|t - s|^{H-\rho}$. In what follows we assume that $\delta < \delta_0 < 1$. Therefore

$$I_3 \leq L \cdot C(\omega)\delta^{H-\rho} \cdot t \leq C(\omega)\delta^{H-\rho}, \quad \omega \in \Omega_{\varepsilon,\delta_0,\rho}. \tag{3.4.20}$$

Now we go on with I_4. For $1 - H < \beta < 1/2$

$$I_4 \leq C(\omega) \sum_{i,j=1}^m \left[\int_0^t |\sigma_{ji}(u, X_u) - \sigma_{ji}(u, Y_{t_u}^\delta)| u^{-\beta} du \right.$$

$$+ \int_0^t \int_0^r |\sigma_{ji}(r, X_r) - \sigma_{ji}(u, X_u) - \sigma_{ji}(r, Y_r^\delta) + \sigma_{ji}(u, Y_u^\delta)| \tag{3.4.21}$$

$$\left. \times (r - u)^{-\beta - 1} du\, dr \right] =: I_7 + I_8.$$

Evidently,
$$I_7 \leq C(\omega)\int_0^t Z_u^\delta u^{-\beta}du. \qquad (3.4.22)$$

According to (3.1.1), under conditions (i)–(iii)
$$|\sigma(t_1,x_1) - \sigma(t_2,x_2) - \sigma(t_1,x_3) + \sigma(t_2,x_4)| \leq M|x_1 - x_2 - x_3 + x_4|$$
$$+ M|x_1 - x_3|\left(|t_2 - t_1|^\gamma + |x_1 - x_2|^\kappa + |x_3 - x_4|^\kappa\right). \qquad (3.4.23)$$

Therefore, $I_8 \leq \sum_{k=9}^{12} I_k$, where
$$I_9 = C(\omega)\int_0^t\int_0^r |X_r - Y_r^\delta|(r-u)^{\gamma-\beta-1}du\,dr,$$
$$I_{10} = C(\omega)\int_0^t\int_0^r |X_r - Y_r^\delta||X_r - X_u|^\kappa(r-u)^{-\beta-1}du\,dr,$$
$$I_{11} = C(\omega)\int_0^t\int_0^r |X_r - Y_r^\delta||Y_r^\delta - Y_u^\delta|^\kappa(r-u)^{-\beta-1}du\,dr,$$
$$I_{12} = C(\omega)\int_0^t\int_0^r \Delta_{u,r}(X,Y^\delta)(r-u)^{-\beta-1}du\,dr.$$

Taking into account that $\beta > H > \alpha$, we obtain that
$$I_9 \leq C(\omega)\int_0^t Z_u^\delta du \qquad (3.4.24)$$

It follows from Theorem 3.1.4 that under assumptions (i)–(v) for any $0 < \rho < H$ there exists a constant $C(\omega)$ such that
$$\sup_{0\leq t\leq T}|X_t| \leq C(\omega), \quad \sup_{0\leq s\leq t\leq T}|X_t - X_s| \leq C(\omega)|t-s|^{H-\rho}. \qquad (3.4.25)$$

Moreover, we can choose $\rho > 0$ and $\beta > 1 - H$ such that $\kappa(H-\rho) > \beta$ and $H - \rho > 2\beta$, because $\kappa H > 1 - H$. In this case
$$I_{10} \leq C(\omega)\int_0^t Z_r^\delta \int_0^r (r-u)^{\kappa(H-\rho)-\beta-1}du\,dr \leq C(\omega)\int_0^T Z_r^\delta dr. \qquad (3.4.26)$$

Evidently, on the corresponding set $\Omega_{\varepsilon,\delta_0,\rho}$ the same estimate holds for I_{11}. Now estimate I_5.
$$I_5 \leq C(\omega)\int_0^t |\sigma(u,Y_u^\delta) - \sigma(t_u,Y_u^\delta)|u^{-\beta}du$$
$$+ C(\omega)\int_0^t\int_0^r |\sigma(r,Y_r^\delta) - \sigma(t_r,Y_r^\delta) - \sigma(u,Y_u^\delta) + \sigma(t_u,Y_u^\delta)|$$
$$\times (r-u)^{-\beta-1}du\,dr =: I_{13} + I_{14}.$$

Obviously,
$$I_{13} \leq C(\omega)\delta^\gamma, \tag{3.4.27}$$

$$I_{14} \leq C(\omega)\Big(\int_0^t\int_0^{t_r} + \int_0^t\int_{t_r}^r\Big) \leq C(\omega)\Big(\int_0^t\int_0^{t_r}\delta^\gamma(r-u)^{-\beta-1}du\,dr$$

$$+\int_0^t\int_{t_r}^r \big((r-u)^\gamma + (r-u)^{H-\rho}\big)(r-u)^{-\beta-1}du\,dr\Big) \leq C(\omega)(\delta^{\gamma-\beta} + \delta^{H-\rho-\beta}). \tag{3.4.28}$$

Similarly,
$$I_6 \leq C(\omega)\int_0^t |\sigma(t_u, Y_u^\delta) - \sigma(t_u, Y_{t_u}^\delta)|\,u^{-\beta}du$$
$$+ C(\omega)\int_0^t\int_0^r |\sigma(t_r, Y_r^\delta) - \sigma(t_r, Y_{t_r}^\delta) - \sigma(t_u, Y_u^\delta) + \sigma(t_u, Y_{t_u}^\delta)| \tag{3.4.29}$$
$$\times (r-u)^{-\beta-1}du\,dr =: I_{15} + I_{16}.$$

Here
$$I_{15} \leq C(\omega)\int_0^t \delta^{H-\rho}u^{-\beta}du \leq C(\omega)\delta^{H-\rho}, \tag{3.4.30}$$

$$I_{16} \leq C(\omega)\Big(\int_0^t\int_0^{t_r} + \int_0^t\int_{t_r}^r\Big) \leq C(\omega)\Big(\delta^{H-\rho}\int_0^t\int_0^{t_r}(r-u)^{-\beta-1}du\,dr$$
$$+ \int_0^t\int_{t_r}^r (r-u)^{H-\rho-\beta-1}du\,dr\Big) \leq C(\omega)\delta^{H-\rho-\beta}. \tag{3.4.31}$$

Substituting (3.4.18)–(3.4.31) into (3.4.17), we obtain that on $\Omega_{\varepsilon,\delta_0,\rho}$
$$Z_t^\delta \leq C(\omega)\Big(\int_0^t Z_r^\delta r^{-\beta}dr + \delta^{H-\rho-\beta} + \delta^{H-\rho} + \int_0^t \theta_r dr\Big), \tag{3.4.32}$$

where $\theta_r = \int_0^r \Delta_{r,u}(X, Y^\delta)(r-u)^{-\beta-1}du$. Recall that $H - \rho > 2\alpha$, therefore
$$Z_t^\delta \leq C(\omega)\Big(\int_0^t (Z_r^\delta r^{-\alpha} + \theta_r)dr + \delta^{2\alpha-\rho}\Big).$$

Now we estimate θ_t. Evidently, for $t > u$
$$\Delta_{t,u}(X, Y^\delta) \leq \int_u^t |b(s, X_s) - b(t_s, Y_{t_s}^\delta)|\,ds$$
$$+ \sum_{i,j=1}^m \Big|\int_u^t (\sigma_{ji}(s, X_s) - \sigma_{ji}(t_s, Y_{t_s}^\delta))dB_s^{j,H}\Big|.$$

Therefore we obtain that $\theta_t \leq \sum_{k=1}^9 J_k$, where

3.4 The Rate of Convergence of Euler Approximations

$$J_1 = \int_0^t \int_u^t \left|b(s, X_s) - b(s, Y_s^\delta)\right| (t-u)^{-\beta-1} ds\, du,$$

$$J_2 = \int_0^t \int_u^t \left|b(s, Y_s^\delta) - b(t_s, Y_s^\delta)\right| (t-u)^{-\beta-1} ds\, du,$$

$$J_3 = \int_0^t \int_u^t \left|b(t_s, Y_s^\delta) - b(t_s, Y_{t_s}^\delta)\right| (t-u)^{-\beta-1} ds\, du,$$

$$J_4 = C(\omega) \int_0^t \int_u^t \left|\sigma(s, X_s) - \sigma(s, Y_s^\delta)\right| (s-u)^{-\beta}(t-u)^{-\beta-1} ds\, du,$$

$$J_5 = C(\omega) \int_0^t \int_u^t \left|\sigma(s, Y_s^\delta) - \sigma(t_s, Y_s^\delta)\right| (s-u)^{-\beta}(t-u)^{-\beta-1} ds\, du,$$

$$J_6 = C(\omega) \int_0^t \int_u^t \left|\sigma(t_s, Y_s^\delta) - \sigma(t_s, Y_{t_s}^\delta)\right| (s-u)^{-\beta}(t-u)^{-\beta-1} ds\, du,$$

$$J_7 = C(\omega) \int_0^t \int_u^t \int_u^r \left|\sigma(r, X_r) - \sigma(r, Y_r^\delta) - \sigma(v, X_v) + \sigma(v, Y_v^\delta)\right|$$
$$\times (r-v)^{-\beta-1}(t-u)^{-\beta-1} dv\, dr\, du,$$

$$J_8 = C(\omega) \int_0^t \int_u^t \int_u^r \left|\sigma(r, Y_r^\delta) - \sigma(t_r, Y_r^\delta) - \sigma(v, Y_v^\delta) + \sigma(t_v, Y_v^\delta)\right|$$
$$\times (r-v)^{-\beta-1}(t-u)^{-\beta-1} dv\, dr\, du,$$

$$J_9 = C(\omega) \int_0^t \int_u^t \int_u^r \left|\sigma(t_r, Y_r^\delta) - \sigma(t_r, Y_{t_r}^\delta) - \sigma(t_v, Y_v^\delta) + \sigma(t_v, Y_{t_v}^\delta)\right|$$
$$\times (r-v)^{-\beta-1}(t-u)^{-\beta-1} dv\, dr\, du.$$

It is clear that $J_1 \leq C \int_0^t Z_s^\delta \int_0^s (t-u)^{-\beta-1} du\, ds$, $J_2 \leq C\delta^\theta$, $J_3 \leq C(\omega)\delta^{H-\rho}$. Further,

$$J_4 \leq C \int_0^t Z_s^\delta \int_0^s (s-u)^{-\beta}(t-u)^{-\beta-1} du\, ds.$$

The inner integral $\int_0^s (s-u)^{-\beta}(t-u)^{-\beta-1} du \leq (t-s)^{-2\beta} \int_0^\infty (1+y)^{-\beta-1} y^{-\beta} dy$. Therefore

$$J_4 \leq C \int_0^t (t-s)^{-2\beta} Z_s^\delta ds.$$

Similarly to J_2, $J_5 \leq C(\omega)\delta^\gamma$, and similarly to J_3, $J_6 \leq C(\omega)\delta^{H-\rho}$. Estimating J_7, J_8 and J_9 is, of course, a bit more complicated, but not dramatically. Obviously, $J_8 \leq C(\omega)\delta^\gamma \int_0^t \int_u^t \int_u^r (r-v)^{-\beta-1}(t-u)^{-\beta-1} du$
$= C(\omega)\delta^\gamma \int_0^t (t-u)^{-2\beta} dv\, dr\, du \leq C(\omega)\delta^\gamma$; similarly $J_9 \leq C(\omega)\delta^{H-\rho}$. Now we apply to J_7 inequality (3.4.23) and obtain the following estimate of the integrand:

$$|\sigma(r,X_r) - \sigma(r,Y_r^\delta) - \sigma(v,X_v) + \sigma(v,Y_v^\delta)| \leq M\Big[\Delta_{r,v}(X,Y^\delta)$$
$$+ |X_r - Y_r^\delta|(r-v)^\gamma + |X_r - Y_r^\delta||X_r - X_v|^\kappa + |X_r - Y_r^\delta||Y_r^\delta - Y_v^\delta|^\kappa\Big]. \quad (3.4.33)$$

According to this, we write $J_7 \leq \sum_{k=10}^{13} J_k$, where, in turn,

$$J_{10} = C(\omega) \int_0^t \int_u^t \int_u^r \Delta_{r,v}(X,Y^\delta)(r-v)^{-\beta-1}(t-u)^{-\beta-1} dv\, dr\, du$$
$$= C(\omega) \int_0^t \int_0^r \int_0^v (t-u)^{-\beta-1} \Delta_{r,v}(X,Y^\delta)(r-v)^{-\beta-1} dv\, dr\, du$$
$$\leq C(\omega) \int_0^t (t-r)^{-\beta} \theta_r dr;$$

$$J_{11} = C(\omega) \int_0^t \int_u^t \int_u^r |X_r - Y_r^\delta|(r-v)^{\gamma-\beta-1} dv\, dr (t-u)^{-\beta-1} du$$
$$\leq C(\omega) \int_0^t Z_r^\delta \int_0^r (t-u)^{-\beta-1} \Big(\int_u^r (r-v)^{\gamma-\beta-1} dv\Big) du\, dr$$
$$\leq C(\omega) \int_0^t (t-r)^{-\beta} Z_r^\delta dr,$$

$$J_{12} = C(\omega) \int_0^t \int_u^t \int_u^r |X_r - Y_r^\delta||X_r - X_v|^\kappa (r-v)^{-\beta-1} dv\, dr (t-u)^{-\beta-1} du$$
$$\leq C(\omega) \int_0^t \int_0^r \int_u^r Z_r^\delta (r-v)^{\kappa(H-\rho)-\beta-1}(t-u)^{-\beta-1} dv\, dr\, du$$
$$\leq C(\omega) \int_0^t Z_r^\delta (t-r)^{-\beta} dr,$$

and $J_{13} \leq C(\omega) \int_0^t Z_r^\delta (t-r)^{-\beta} dr$ is obtained the same way. Summing up these estimates, we obtain that $J_7 \leq C(\omega) \int_0^t (t-r)^{-\beta}(Z_r^\delta + \theta_r) dr$, whence

$$\theta_t \leq C(\omega)\Big(\int_0^t (t-r)^{-2\beta}(Z_r^\delta + \theta_r) dr + \delta^{H-\rho} + \delta^\theta\Big). \quad (3.4.34)$$

Coupling together (3.4.32) and (3.4.34), and taking into account that $H - \rho > 2\alpha$, $\theta > 2\alpha$, we obtain

$$\begin{aligned} Z_t^\delta + \theta_t &\leq C(\omega)\Big(\delta^{2\alpha} + \int_0^t \big((t-r)^{-2\beta} + r^{-\beta}\big)\big(Z_r^\delta + \theta_r\big) dr\Big) \\ &\leq C(\omega)\Big(\delta^{2\alpha} + t^{2\beta}\int_0^t (t-r)^{-2\beta} r^{-2\beta}\big(Z_r^\delta + \theta_r\big) dr\Big). \end{aligned} \qquad (3.4.35)$$

The proof now follows immediately from (3.4.35) and (3.1.22)–(3.1.23). Statement 2 is obvious. □

3.4.2 Approximation of Quasilinear Skorohod-type Equations

Now we proceed with the problem of the numerical solution of Skorohod-type equations driven by fractional white noise. From now on, we assume that our probability space is the white noise space, i.e. $(\Omega, \mathcal{F}, P) = (S'(R), \mathcal{B}(S'(R)), \mu)$, the symbol $\diamond$ stands for the Wick product, $W_t = \langle \mathbf{1}_{[0,t]}, \omega \rangle$ is the standard Brownian motion, $\dot{W}$ is the white noise. (See also Sections 1.4, 1.5, 2.3 and Subsection 3.3.2.)

Consider the quasilinear Skorohod-type equation driven by fractional white noise that is the one-dimensional analog of equation (3.3.2):

$$X_t = X_0 + \int_0^t b(s, X_s, \omega) \, ds + \int_0^t \sigma(s) X_s \diamond \dot{B}_s^H \, ds, \qquad (3.4.36)$$

with nonrandom initial condition X_0. Suppose that the coefficients b and σ satisfy conditions (iii)–(iv) of Theorem 3.3.2 (in this subsection we always refer to them as to conditions (iii)–(iv)), and

(vi) "Smoothness" of b w.r.t. ω: for any $t \in [0, T]$ and for $h \in L_1(\mathbb{R})$

$$|b(t, x, \omega + h) - b(t, x, \omega)| \leq C(1 + |x|) \int_{\mathbb{R}} |h(s)| \, ds.$$

(vii) Hölder continuity of b w.r.t. t or order H with constant that grows linearly in x:

$$|b(t, x, \omega) - b(s, x, \omega)| \leq C(1 + |x|) |t - s|^H;$$

(viii) Hölder continuity of σ w.r.t. t or order H:

$$|\sigma(t) - \sigma(s)| \leq C |t - s|^H.$$

Remark 3.4.4. Condition (vii) holds if, for example, the coefficient b has the stochastic derivative growing at most linearly in x. It is obviously true if b is nonrandom and Hölder of order H.

Consider the fractional Wick exponent

$$J_\sigma(t) = \exp^\diamond\left\{-\int_{\mathbb{R}} M_-^H \sigma_t(s) dW_s\right\}$$

$$= \exp\left\{-\int_{\mathbb{R}} M_-^H \sigma_t(s) dW_s - \frac{1}{2}\|\sigma_t\|^2_{|R_H|,1}\right\}$$

It easily follows from Theorem 3.3.2 that for nonrandom X_0 under conditions (iii)–(iv) the equation (3.4.36) has a unique solution that belongs to all $L_p(\Omega)$ and can be represented in the form

$$Z_t = J_\sigma(t) \diamond X_t, \quad \text{or} \quad X_t = J_{-\sigma}(t) \diamond Z_t,$$

where the process Z_t solves (ordinary) differential equation

$$Z_t = X_0 + \int_0^t J_\sigma(s) b(s, J_\sigma^{-1}(s) Z_s, \omega + M_-^H \sigma_s) \, ds. \tag{3.4.37}$$

This gives the following idea of construction of time-discrete approximations of the solution of (3.4.36). Take the uniform partition $\{\tau_n = n\delta,\ n = 1,\ldots,N\}$ of $[0,T]$ and define first the approximations of Z in a recursive way:

$$\begin{aligned}\widetilde{Z}_0 &= X_0, \\ \widetilde{Z}_{\tau_{n+1}} &= \widetilde{Z}_{\tau_n} + \widetilde{J}(\tau_n) b(\tau_n, \widetilde{J}^{-1}(\tau_n)\widetilde{Z}_{\tau_n}, \omega + M\widetilde{\sigma}_n)\delta,\end{aligned} \tag{3.4.38}$$

where

$$\widetilde{J}(t) := \exp\left\{-\int_0^t \widetilde{\sigma}(s) dB_s^H - \frac{1}{2}\|\widetilde{\sigma}\mathbf{1}_{[0,t]}\|^2_{|R_H|,1}\right\},$$

$$\widetilde{\sigma}(s) := \sigma(t_s),\ \widetilde{\sigma}_n := \widetilde{\sigma}\mathbf{1}_{[0,\tau_n]},\ M := M_-^H.$$

Note that both $\|\widetilde{\sigma}_n\|_{|R_H|,1}$ and $M\widetilde{\sigma}_n$ are easily computable as finite sums of elementary integrals. Further, we interpolate continuously by

$$\widetilde{Z}_t = X_0 + \int_0^t \widetilde{J}(t_s) b(t_s, \widetilde{J}^{-1}(t_s)\widetilde{Z}_{t_s}, \omega + M\widetilde{\sigma}_{n_s}) \, ds, \tag{3.4.39}$$

where $n_s = \max\{n : \tau_n \leq s\}$, and set

$$\widetilde{X}_t = T_{-M(\widetilde{\sigma}\mathbf{1}_{[0,t]})} \widetilde{J}^{-1}(t) \widetilde{Z}_t, \tag{3.4.40}$$

where for $\omega_0 \in S'(\mathbb{R})$ T_{ω_0} is the shift operator, $T_{\omega_0} F(\omega) = F(\omega + \omega_0)$.

Lemma 3.4.5. *Under the assumption (vi), the following estimate is true:*

$$\left|e^{\alpha_1} b(t, e^{-\alpha_1} x, \omega) - e^{\alpha_2} b(t, e^{-\alpha_2} x, \omega)\right| \leq C(1 + e^{\alpha_1} + e^{\alpha_2} + |x|) |\alpha_1 - \alpha_2|.$$

Proof. Write

$$\left|e^{\alpha_1}b(t,e^{-\alpha_1}x,\omega) - e^{\alpha_2}b(t,e^{-\alpha_2}x,\omega)\right|$$
$$\leq \left|e^{\alpha_1}b(t,e^{-\alpha_1}x,\omega) - e^{\alpha_1}b(t,e^{-\alpha_2}x,\omega)\right|$$
$$+ \left|e^{\alpha_1}b(t,e^{-\alpha_2}x,\omega) - e^{\alpha_2}b(t,e^{-\alpha_2}x,\omega)\right|$$

and apply (vi). □

Lemma 3.4.6. *Let ξ_1 and ξ_2 be jointly Gaussian variables. Then for $q \geq 1$*

$$\mathsf{E}\left[\left|e^{\xi_1} - e^{\xi_2}\right|^{2q}\right] \leq C(L,q)\left(\mathsf{E}\left[(\xi_1 - \xi_2)^2\right]\right)^q,$$

where $L = \max\left\{\mathsf{E}\left[\xi_1^2\right], \mathsf{E}\left[\xi_2^2\right]\right\}$.

Proof. By the Lagrange theorem, Cauchy–Schwartz inequality and Gaussian property,

$$\mathsf{E}\left[\left|e^{\xi_1} - e^{\xi_2}\right|^{2q}\right] \leq \left(\mathsf{E}\left[e^{4q\xi_1} + e^{4q\xi_2}\right]\mathsf{E}\left[|\xi_1 - \xi_2|^{4q}\right]\right)^{1/2}$$
$$\leq C(L)C(q)\left(\mathsf{E}\left[(\xi_1 - \xi_2)^2\right]\right)^q,$$

as required. □

Our first result is about convergence of $\widetilde{Z}$ to Z.

Theorem 3.4.7. *Under conditions (iii)–(iv) and (vi)–(vii) for any $p \geq 1$ the following estimate holds:*

$$\mathsf{E}\left[\left|Z_t - \widetilde{Z}_t\right|^{2p}\right] \leq C(p)\delta^{2pH}. \qquad (3.4.41)$$

Proof. Firstly, we recall that Z_t belongs to all $L_q(\Omega)$ and $\mathsf{E}[|Z_t|^q] \leq C(q)$. Therefore equation (3.4.37) together with condition (vii) gives $\mathsf{E}[|Z_t - Z_s|^q] \leq C(q)|t-s|^q$. Equation (3.4.38) and conditions (iii)–(iv) allow us to write

$$\left|\widetilde{Z}_{\tau_{n+1}}\right| \leq (1+C\delta)\left|\widetilde{Z}_{\tau_n}\right| + C\delta\widetilde{J}(\tau_n) \leq e^{C\delta}\left|\widetilde{Z}_{\tau_n}\right| + C\delta\widetilde{J}(\tau_n).$$

This gives an estimate $\left|\widetilde{Z}_{\tau_n}\right| \leq C\sum_{k=0}^{N-1}\widetilde{J}(\tau_k)\delta$. Then for any $q \geq 1$ by the Jensen inequality, $\left|\widetilde{Z}_{\tau_n}\right|^q \leq C(q)\sum_{k=0}^{N-1}\widetilde{J}^q(\tau_k)\delta$. Taking expectations, we get

$$\mathsf{E}\left[\left|\widetilde{Z}_{\tau_n}\right|^q\right] \leq C(q)\sum_{k=0}^{N-1}\mathsf{E}\left[\widetilde{J}^q(\tau_k)\right]\delta.$$

Using that each $\widetilde{J}$ is exponent of Gaussian variable and σ is bounded on $[0,T]$, we obtain

$$\mathsf{E}\left[\left|\widetilde{Z}_{\tau_n}\right|^q\right] \le C(q) \sum_{k=0}^{N-1} \delta = C(q).$$

This through (3.4.39) and (iii)–(iv) implies $\mathsf{E}\left[\left|\widetilde{Z}_t\right|^q\right] \le C(q)$.

Now we can write

$$\left|Z_t - \widetilde{Z}_t\right| \le I_1 + I_2 + I_3 + I_4 + I_5,$$

where

$$I_1 = \left|\int_0^t \widetilde{J}(t_s)\big(b(t_s, \widetilde{J}^{-1}(t_s)Z_{t_s}, \omega + M\widetilde{\sigma}_{n_s}) - b(t_s, \widetilde{J}^{-1}(t_s)\widetilde{Z}_{t_s}, \omega + M\widetilde{\sigma}_{n_s})\big)\, ds\right|,$$

$$I_2 = \left|\int_0^t \big(\widetilde{J}(t_s)b(t_s, \widetilde{J}^{-1}(t_s)Z_{t_s}, \omega + M\widetilde{\sigma}_{n_s}) - J_\sigma(s)b(t_s, J_\sigma^{-1}(s)Z_{t_s}, \omega + M\widetilde{\sigma}_{n_s})\big)\, ds\right|,$$

$$I_3 = \left|\int_0^t J_\sigma(s)\big(b(s, J_\sigma^{-1}(s)Z_{t_s}, \omega + M\widetilde{\sigma}_{n_s}) - b(t_s, J_\sigma^{-1}(s)Z_{t_s}, \omega + M\widetilde{\sigma}_{n_s})\big)\, ds\right|,$$

$$I_4 = \left|\int_0^t J_\sigma(s)\big(b(s, J_\sigma^{-1}(s)Z_{t_s}, \omega + M\widetilde{\sigma}_{n_s}) - b(s, J_\sigma^{-1}(s)Z_{t_s}, \omega + M\sigma_s)\big)\, ds\right|,$$

$$I_5 = \left|\int_0^t J_\sigma(s)\big(b(s, J_\sigma^{-1}(s)Z_s, \omega + M\sigma_s) - b(s, J_\sigma^{-1}(s)Z_{t_s}, \omega + M\sigma_s)\big)\, ds\right|.$$

First we estimate I_2 by using Lemma 3.4.5:

$$I_2 \le C \int_0^t \left(1 + J_\sigma(s) + \widetilde{J}(t_s) + |Z_{t_s}|\right) \left(\left|\int_0^s (\sigma(u) - \widetilde{\sigma}(u))\, dB_u^H\right|\right.$$
$$\left. + |\sigma(t_s)(B_s^H - B_{t_s}^H)| + \frac{1}{2}\left|\|\sigma_s\|_{|R_H|,1}^2 - \|\widetilde{\sigma}_{n_s}\|_{|R_H|,1}^2\right|\right)\, ds$$

$$\le C \int_0^t \left(1 + J_\sigma(s) + \widetilde{J}(t_s) + |Z_{t_s}|\right)$$
$$\times \left(\left|\int_0^s (\sigma(u) - \widetilde{\sigma}(u))\, dB_u^H\right| + |B_s^H - B_{t_s}^H| + \delta^H\right)\, ds,$$

where the inequality $\left|\|\sigma_s\|_{|R_H|,1}^2 - \|\widetilde{\sigma}_{n_s}\|_{|R_H|,1}^2\right| < C\delta^H$ is due to (viii) and boundedness of σ on $[0,T]$. Applying the Cauchy–Schwartz inequality, we arrive at

3.4 The Rate of Convergence of Euler Approximations

$$I_2 \leq C \left(\int_0^T (1 + J_\sigma^2(s) + \tilde{J}^2(t_s) + Z_{t_s}^2) ds \right)^{1/2}$$
$$\times \left(\int_0^T \left(\left(\int_0^s (\sigma(u) - \tilde{\sigma}(u)) dB_u^H \right)^2 + (B_t^H - B_{t_s}^H)^2 + \delta^{2H} \right) ds \right)^{1/2}.$$

Further, from (vii) it follows that

$$I_3 \leq C\delta^H \int_0^T (J_\sigma(s) + |Z_s|) ds,$$

from (vi)

$$I_3 \leq C\delta^H \int_0^T (J_\sigma(s) + |Z_s|) ds.$$

Conditions (iii)–(iv) allow us to estimate $I_1 \leq C \int_0^t \left| Z_{t_s} - \tilde{Z}_{t_s} \right| ds$, $I_5 \leq C \int_0^t |Z_s - Z_{t_s}| ds$. Summing up these estimates yields

$$\left| Z_t - \tilde{Z}_t \right| \leq C \left(\int_0^T (1 + J_\sigma^2(s) + \tilde{J}^2(t_s) + Z_{t_s}^2) ds \right)^{1/2}$$
$$\times \left(\delta^{2H} + \int_0^T \left(\left(\int_0^s (\sigma(u) - \tilde{\sigma}(u)) dB_u^H \right)^2 + (B_t^H - B_{t_s}^H)^2 \right) ds \right)^{1/2}$$
$$+ C \int_0^T \left| Z_{t_s} - \tilde{Z}_{t_s} \right| ds + C \int_0^t |Z_s - Z_{t_s}| ds.$$

Then, using the (discrete) Gronwall inequality, we get

$$\left| Z_t - \tilde{Z}_t \right| \leq C \left(\int_0^T (1 + J_\sigma^2(s) + \tilde{J}^2(t_s) + Z_{t_s}^2) ds \right)^{1/2}$$
$$\times \left(\delta^{2H} + \int_0^T \left(\left(\int_0^s (\sigma(u) - \tilde{\sigma}(u)) dB_u^H \right)^2 + (B_t^H - B_{t_s}^H)^2 \right) ds \right)^{1/2}$$
$$+ C \int_0^t |Z_s - Z_{t_s}| ds.$$

Then we raise this to the 2pth power and use the Jensen inequality. The last term will be bounded by $C(p)\delta^{2p}$; to the first one we apply the Cauchy–Schwartz inequality for expectations and the Jensen inequality, and use uniform boundedness of moments for Z, J_σ and $\tilde{J}$ (for J_σ and $\tilde{J}$ it follows from the fact that the both are exponents of some Gaussian variables with bonded variance) to get

$$\mathsf{E}\left[\left|Z_t - \widetilde{Z}_t\right|^{2p}\right] \leq C(p)\left(\delta^{2pH} + \left(\mathsf{E}\left[\left|\int_0^T (\sigma(u) - \widetilde{\sigma}(u))\, dB_u^H\right|^{4p}\right]\right)^{1/2}\right.$$
$$\left. + \left(\mathsf{E}\left[\left|B_t^H - B_{t_s}^H\right|^{4p}\right]\right)^{1/2}\right).$$

Using again that $\mathsf{E}\left[|\cdot|^{4p}\right] = C(p)(\mathsf{E}\left[(\cdot)^2\right])^{2p}$ for Gaussian variables, we get

$$\mathsf{E}\left[\left|Z_t - \widetilde{Z}_t\right|^{2p}\right] \leq C(p)\left(\delta^{2pH} + \left(\mathsf{E}\left[\left|\int_0^T (\sigma(u) - \widetilde{\sigma}(u))\, dB_u^H\right|^2\right]\right)^p\right.$$
$$\left. + \left(\mathsf{E}\left[\left|B_t^H - B_{t_s}^H\right|^2\right]\right)^p\right)$$
$$\leq C(p)(\delta^{2pH} + \|\sigma - \widetilde{\sigma}\|_{|R_H|,1}^{2p}) \leq C(p)\delta^{2pH};$$

the last is due to (viii). This is the desired result. □

Now we are ready to state the main result of this subsection.

Theorem 3.4.8. *Under conditions (iii)–(iv) and (vi)–(viii) the approximations $\widetilde{X}$ defined by (3.4.40) converge to the solution X of (3.4.36) in the mean-square sense, and, moreover,*

$$\mathsf{E}\left[(X_t - \widetilde{X}_t)^2\right] \leq C\delta^{2H}.$$

Proof. First, estimate for h such that $\int_R |h(s)|\, ds < C$ has the difference

$T_h Z(t) - Z(t) \leq A_1 + A_2 + A_3$, where

$$A_1 = \int_0^t T_h J_\sigma(s)\Big|b(s, (T_h J_\sigma^{-1}(s))T_h Z_s, \omega + h + M\sigma_s)$$
$$- b(s, (T_h J_\sigma^{-1}(s))Z_s, \omega + h + M\sigma_s)\Big|\, ds,$$

$$A_2 = \int_0^t T_h J_\sigma(s)\Big|b(s, (T_h J_\sigma^{-1}(s))Z(s), \omega + h + M\sigma_s)$$
$$- b(t, (T_h J_\sigma^{-1}(s))Z_s, \omega + M\sigma_s)\Big|\, ds,$$

$$A_3 = \int_0^t \Big|T_h J_\sigma(s)b(t, (T_h J_\sigma^{-1}(s))Z_s, \omega + M\sigma_s)$$
$$- J_\sigma(s)b(t, J_\sigma^{-1}(s)Z_s, \omega + M\sigma_s)\Big|\, ds.$$

Conditions (iii)–(iv) give $A_1 \leq C \int_0^t |T_h Z_s - Z_s|\, ds$, condition (vi) gives

$$A_2 \leq C \int_0^T (1 + |Z_s|)\, ds \int_R |h(s)|\, ds,$$

3.4 The Rate of Convergence of Euler Approximations   261

and Lemma 3.4.5 together with the boundedness of σ and the assumptions on h yields

$$A_3 \leq C \int_0^T (1 + J_\sigma(s) + T_h J_\sigma(s) + |Z_s|) J_\sigma(s) \int_{\mathbb{R}} |M\sigma_s(u) h(u)| \, du \, ds$$

$$\leq C \int_0^T (1 + J_\sigma(s) + T_h J_\sigma(s) + |Z_s|) J_\sigma(s) \, ds \int_{\mathbb{R}} |h(s)| \, ds.$$

Applying the Gronwall lemma, we get

$$|T_h Z_t - Z_t| \leq C \int_0^T (1 + J_\sigma(s) + T_h J_\sigma(s) + |Z_s|) J_\sigma(s) \, ds \int_{\mathbb{R}} |h(s)| \, ds.$$

Raising this inequality to the $2p$th power, taking expectations and using the Jensen inequality and boundedness of moments of Z, J_σ and $T_h J_\sigma$ (the last follows from the Girsanov theorem, Cauchy–Schwartz inequality and assumptions on h), we get

$$\mathsf{E}\left[(T_h Z(t) - Z(t))^{2p}\right] \leq C(p) \left(\int_{\mathbb{R}} |h(s)| \, ds\right)^{2p}. \qquad (3.4.42)$$

Further,

$$\mathsf{E}\left[(X_t - \widetilde{X}_t)^2\right] \leq 3(B_1 + B_2 + B_3),$$

$$B_1 = \mathsf{E}\left[\left(\overline{J}(t) T_{-M\widetilde{\sigma} \mathbf{1}_{[0,t]}} (Z_t - \widetilde{Z}_t)\right)^2\right],$$

$$B_2 = \mathsf{E}\left[\left((J_{-\sigma}(t) - \overline{J}(t)) T_{-M\widetilde{\sigma} \mathbf{1}_{[0,t]}} Z_t\right)^2\right],$$

$$B_3 = \mathsf{E}\left[\left(J_{-\sigma}(t)(T_{-M\sigma}(1 - T_{-M(\widetilde{\sigma} \mathbf{1}_{[0,t]} - \sigma_t)})) Z_t\right)^2\right],$$

where

$$J_{-\sigma}(t) = \exp\left\{\int_{\mathbb{R}} M\sigma_t(s) dW_s - \frac{1}{2} \|\sigma_t\|^2_{|R_H|,1}\right\},$$

$$\overline{J}(t) = \exp\left\{\int_{\mathbb{R}} M(\widetilde{\sigma} \mathbf{1}_{[0,t]})(s) dW_s - \frac{1}{2} \|\widetilde{\sigma} \mathbf{1}_{[0,t]}\|^2_{|R_H|,1}\right\}.$$

Now estimate using the Cauchy–Schwartz inequality, Girsanov theorem (which can be applied as σ and $\widetilde{\sigma}$ are bounded on $[0,T]$) and Theorem 3.4.7

$$B_1 \leq \left(\mathsf{E}\left[\overline{J}^4(t)\right] \mathsf{E}\left[T_{-M\widetilde{\sigma} \mathbf{1}_{[0,t]}} (Z_t - \widetilde{Z}_t)^4\right]\right)^{1/2},$$

$$\leq C \left(\mathsf{E}\left[\widetilde{J}(t)(Z_t - \widetilde{Z}_t)^4\right]\right)^{1/2}$$

$$\leq C \left(\mathsf{E}\left[\widetilde{J}^2(t)\right] \mathsf{E}\left[(Z_t - \widetilde{Z}_t)^8\right]\right)^{1/4} \leq C\delta^{2H}.$$

Similar reasoning and Lemma 3.4.6 imply that

$$B_2 \leq C\mathsf{E}\bigg[\Big(\int_{\mathbb{R}} M(\tilde{\sigma}\mathbf{1}_{[0,t]} - \sigma_t)(s)\,dW_s$$
$$+ \frac{1}{2}(\|\sigma_t\|^2_{|R_H|,1} - \|\tilde{\sigma}\mathbf{1}_{[0,t]}\|^2_{|R_H|,1})\Big)^2\bigg].$$

Using condition (vii), we obtain $B_2 \leq C\delta^{2H}$. And for B_3, using the estimate (3.4.42), we get

$$B_3 \leq \Big(\int_0^t |M(\tilde{\sigma}\mathbf{1}_{[0,t]} - \sigma_t)(s)|\,ds\Big)^2 \leq C\delta^{2H}.$$

This concludes the proof. □

Remark 3.4.9. It is natural to assume that the coefficient b is expressed in terms of fBm B^H rather then in terms of the underlying Brownian motion W (or underlying "Brownian" white noise $\dot{W}$). This justifies the fact that it is σ not $M\sigma$ that is discretized in (3.4.38).

Remark 3.4.10. Similarly to the proof of Theorem 3.4.8 one can prove that for any $s \geq 1$

$$\mathsf{E}\Big[\big|X_t - \tilde{X}_t\big|^s\Big] \leq \delta^{sH}.$$

The case $s = 2$ is considered here to keep the classical "scent" of the results.

Remark 3.4.11. The results of this subsection can be generalized to a random initial condition X_0 in the following form: under conditions (iii)–(iv), (vi)–(viii) and L_p-integrability of the initial condition one has convergence in any $L_{p'}$ for $p' < p$ with

$$\mathsf{E}\Big[\big|X_t - \tilde{X}_t\big|^s\Big] \leq \delta^{sH}.$$

Proofs need some simple changes: the Hölder inequality for appropriate powers instead of the Cauchy–Schwartz inequality.

3.5 Stochastic Differential Equation with Additive Wiener Integral w.r.t. Fractional Noise

Consider the following scalar stochastic differential equation

$$X_t = X_0 + \int_0^t b(s, X_s)\,ds + \int_0^t f(s)\,dB_s^H, \qquad (3.5.1)$$

where $b : [0,T] \times \mathbb{R} \to \mathbb{R}$ is the measurable function, $H \in (0,1)$, $X_0 \in \mathbb{R}$ and $f \in L_2^H(\mathbb{R})$. Equation (3.5.1) generalizes the equation

$$X_t = X_0 + \int_0^t b(s, X_s)\,ds + B_s^H, \qquad (3.5.2)$$

that was considered in the papers (MN03), (NO02), (NO03b).

3.5.1 Existence of a Weak Solution for Regular Coefficients

Definition 3.5.1. By a weak solution to equation (3.5.1) we mean a couple of adapted continuous processes $(\widetilde{B}^H, X)$ on a filtered probability space $(\Omega, \mathcal{F}, P, \{\mathcal{F}_t, t \in [0,T]\})$ such that

(a) $\widetilde{B}^H$ is an $\mathcal{F}_t$ - fractional Brownian motion;
(b) X and $\widetilde{B}^H$ satisfy (3.5.1).

The general approach to existence of the weak solution of (3.5.1) is the following. Let the function f be nonzero on $\mathbb{R}$, so that $g(s) := \frac{1}{f(s)}$ is determined on $\mathbb{R}$. Consider the process $\widetilde{B}^H_t := B^H_t - \int_0^t g(s)b(s, x + I_s(f))ds$, where $I_t(f) = \int_0^t f(s)dB^H_s$.

According to Theorem 2.8.1, under the following conditions

$$E \exp \left\{ L_t - \frac{1}{2}\langle L \rangle_t \right\} = 1, \quad t \in [0,T] \qquad (3.5.3)$$

where $L_t = \int_0^t s^\alpha \delta_s dB_s$, B is the Wiener process, $B_t = \widehat{\alpha} \int_0^t s^\alpha dM^H_s$, $M^H_t = \int_0^t l_H(t,s)dB^H_s$ and

$$\int_0^t l_H(t,s)g(s)b(s, x + I_s(f))ds = \widetilde{\alpha} \int_0^t \delta_s ds, \qquad (3.5.4)$$

we have that $\widetilde{B}^H_t$ will be an fBm w.r.t. the measure Q such that

$$\left.\frac{dQ}{dP}\right|_{\mathcal{F}_t} = \exp\left\{L_t - \frac{1}{2}\langle L\rangle_t\right\}.$$

In this case it is very easy to check that the couple $(\widetilde{B}^H_t, X_0 + I_t(f))$ creates a weak solution of equation (3.5.1).

Due to the Novikov condition, the equality (3.5.3) holds if

$$E \exp\left\{\frac{1}{2}\langle L\rangle_T\right\} < \infty, \qquad (3.5.5)$$

where

$$\langle L\rangle_t = \int_0^t s^{2\alpha}\delta_s^2 ds. \qquad (3.5.6)$$

Therefore, we must check inequality (3.5.5) together with (3.5.4) and (3.5.6). Denote the stochastic process $h(s) := g(s)b(s, X_0 + I_s(f))$. Note that in this section we begin the new numeration of the conditions.

Theorem 3.5.2. *Let one of the following assumptions hold:*

(i) $H \in (0, 1/2)$, the coefficients b and g satisfy the condition: there exists $\lambda > 0$ such that

$$\sup_{0 \le t \le T} E \exp \left\{ \lambda t^{2\alpha} \left(\int_0^t s^{-\alpha}(t-s)^{-\alpha-1} h(s) ds \right)^2 \right\} < \infty; \quad (3.5.7)$$

(ii) $H \in (1/2, 1)$, the coefficients b and g satisfy the condition:

$$E_\lambda := E \exp \left\{ \lambda \int_0^T \left(s^{-\alpha} |h(s)| \right. \right.$$

$$\left. \left. + \alpha s^\alpha \int_0^s \frac{|s^{-\alpha} h(s) - r^{-\alpha} h(r)|}{(s-r)^{\alpha+1}} dr \right)^2 ds \right\} < \infty, \quad (3.5.8)$$

for any $\lambda > 0$.

Then equation (3.5.1) has a weak solution.

Proof. Let $H \in (0, 1/2)$. Then we obtain δ_t directly from (3.5.4) (recall that $l_H(t, s) = C_H^{(5)} s^{-\alpha}(t-s)^{-\alpha}$)

$$\delta_t = C_H^{(5)} (-\alpha) \widehat{a} \int_0^t s^{-\alpha}(t-s)^{-\alpha-1} h(s) ds. \quad (3.5.9)$$

It follows from Example 13.32 (Ell82) that the condition: there exists $\lambda > 0$ such that $\sup_{0 \le s \le T} E \exp\{\lambda v_s^2\} < \infty$, is sufficient for the Novikov condition, if it has the form $E \exp \left\{ \frac{1}{2} \int_0^T v_s^2 ds \right\} < \infty$. Therefore, the proof follows immediately from (3.5.6), (3.5.7) and (3.5.9). Let $H \in (1/2, 1)$. In this case δ_t is a fractional derivative of the form:

$$\delta_t = \frac{d}{dt} \left(C_H^{(5)} \int_0^t (t-s)^{-\alpha} s^{-\alpha} h(s) ds \right)$$

$$= C_H^{(5)} \left(t^{-2\alpha} h(t) + \alpha \int_0^t (t^{-\alpha} h(t) - r^{-\alpha} h(r))(t-r)^{-\alpha-1} dr \right) \mathbf{1}_{(0,T)}(t) ,$$

whence the proof follows. □

Now we establish more convenient conditions for the existence of a weak solution in terms of g and b.

Denote the function $h(s, x) := g(s) b(s, x)$.

Theorem 3.5.3. *Let $0 < |f(t)| < f^*$ for any $t \in [0, T]$ and one of the following assumptions holds:*

(iii) $H \in (0, 1/2)$ and $h(t, x)$ is of linear growth:

$$|h(t, x)| \le C(1 + |x|), \quad (t, x) \in [0, T] \times \mathbb{R}$$

(iv) $H \in (1/2, 1)$, f is essentially bounded on $[0, T]$ and $h(s, x)$ is Hölder continuous:
$$|h(t, x) - h(s, y)| \leq C \left(|x - y|^\rho + |t - s|^\gamma\right),$$
where $1 > rho > 1 - \frac{1}{2H}$ and $1 \geq \gamma > \alpha$.

Then equation (3.5.1) has a weak solution.

Proof. In both cases we must check the conditions of Theorem 3.5.2.

Let $H \in (0, 1/2)$. Then $t^{2\alpha} \left(\int_0^t s^{-\alpha}(t-s)^{-\alpha-1} h(s) ds\right)^2$
$\leq Ct^{2\alpha} \sup_{0 \leq s \leq t} |h(s)|^2 \, t^{-4\alpha} \leq CT^{-2\alpha}(1 + |X_0| + I_T^*(f))$. Note that now $\alpha < 0$). Furthermore, inequality (3.5.7) is transformed into
$$E \exp\{\lambda(I_T^*(f))^2\} < \infty \quad \text{for some} \quad \lambda > 0.$$

The last inequality follows from the Fernique theorem (Fer74) about exponential integrability of the square of the supremum norm of a Gaussian process (recall that the process $I_t(f)$ is Gaussian). For $H \in (1/2, 1)$
$$|h(s)| \leq |h(0,0)| + C\left(s^\gamma + |X_0|^\rho + |I_s(f)|^\rho\right), \tag{3.5.10}$$

$$s^\alpha \int_0^s \frac{s^{-\alpha} h(s) - r^{-\alpha} h(r)}{(s-r)^{\alpha+1}} dr = \int_0^s \frac{h(s) - h(r)}{(s-r)^{\alpha+1}} dr$$
$$+ s^\alpha \int_0^s \frac{(s^{-\alpha} - r^{-\alpha})(h(r) - h(s))}{(s-r)^{\alpha+1}} dr + s^\alpha h(s) \int_0^s \frac{s^{-\alpha} - r^{-\alpha}}{(s-r)^{\alpha+1}} dr.$$

Further,
$$|h(s) - h(r)| \leq |h(s, X_0 + I_s(f)) - h(r, X_0 + I_r(f))|$$
$$\leq C\left(|s - r|^\gamma + |I_s(f) - I_r(f)|^\rho\right),$$

therefore
$$\left|\int_0^s \frac{h(s) - h(r)}{(s-r)^{\alpha+1}} dr\right| \leq C \int_0^s (s-r)^{\gamma-\alpha-1} dr + C \int_0^s \frac{|I_s(f) - I_r(f)|^\rho}{(s-r)^{\alpha+1}} dr$$
$$\leq C + C \int_0^s \frac{|I_s(f) - I_r(f)|^\rho}{(s-r)^{\alpha+1}} dr.$$

Similarly to Lemma 1.17.1, it follows from the Garsia–Rodemich–Rumsey inequality that for any $0 < \varepsilon < H$
$$|I_s(f) - I_r(f)| \leq C_{H,\varepsilon} |r - s|^{H-\varepsilon} \xi_\varepsilon,$$

where

$$\xi_\varepsilon = \left(\int_0^T \int_0^T \frac{|I_x(f) - I_y(f)|^{2/\varepsilon}}{|x-y|^{2H/\varepsilon}} dx\, dy\right)^{\varepsilon/2}.$$

Further, according to Corollary 1.9.4, it holds that

$$E\xi_\varepsilon^{2/\varepsilon} \le C\left(H, \frac{2}{\varepsilon}\right) \int_0^T \int_0^T \frac{\|f\|_{L^{\frac{1}{H}}[x,y]}^{2/\varepsilon}}{|x-y|^{2H/\varepsilon}} dx\, dy$$

$$\le C\left(H, \frac{2}{\varepsilon}\right)(f^*)^{2/\varepsilon} T^2. \quad (3.5.11)$$

Therefore,

$$\int_0^s \frac{|I_s(f) - I_r(f)|^\rho}{(s-r)^{\alpha+1}} dr \le C_{H,\varepsilon}^\rho \xi_\varepsilon^\rho \int_0^s (s-r)^{\rho(H-\varepsilon)-\alpha-1} dr \le C\xi_\varepsilon^\rho$$

for some constant C and such ε that $\rho(H-\varepsilon) - \alpha > 0$, and

$$\left|\int_0^s \frac{h(s) - h(r)}{(s-r)^{\alpha+1}} dr\right| \le C\left(1 + \xi_\varepsilon^\rho\right). \quad (3.5.12)$$

The next term admits an estimate

$$s^\alpha \left|\int_0^s \frac{(s^{-\alpha} - r^{-\alpha})(h(s) - h(r))}{(s-r)^{\alpha+1}} dr\right| \le \int_0^s \frac{|h(s) - h(r)|}{(s-r)} r^{-\alpha} dr$$

$$\le C + C \int_0^s \frac{|I_s(f) - I_r(f)|^\rho}{(s-r)} r^{-\alpha} dr \le C\left(1 + \xi_\varepsilon^\rho\right); \quad (3.5.13)$$

the proof follows now from (3.5.10)–(3.5.13), because for $\rho < 1$

$$E_\lambda \le CE \exp\left\{\lambda \int_0^T |I_s(f)|^{2\rho} s^{-2\alpha} ds + \lambda T C \xi_\varepsilon^{2\rho}\right\} < \infty.$$

$\square$

3.5.2 Existence of a Weak Solution for SDE with Discontinuous Drift

Consider equation (3.5.1) for the case when $f \equiv 1$, $b(s,x) = b(x)$ and $b(x)$ is Hölder continuous of order $\rho \in (1 - 1/2H, 1)$ except on a finite number of points, where there is a jump discontinuity (MN04).

Theorem 3.5.4. *Suppose that the function $b(x)$ is Hölder continuous of order $\rho \in (1 - 1/2H, 1)$ in a finite number of intervals $(-\infty, a_1), (a_1, a_2), \ldots, (a_{N-1}, a_N), (a_N, +\infty)$ and there is a jump discontinuity in the points $a_i, 1 \le i \le N$, that is, $b(a_i-) \ne b(a_i+) = b(a_i)$. Let B_t^H be an fBm with Hurst parameter $H \in \left(\frac{1}{2}, \frac{1+\sqrt{5}}{4}\right)$. Then equation (3.5.1) with $f \equiv 1$ has a weak solution.*

3.5 SDE with the Additive Wiener Integral w.r.t. Fractional Noise 267

Remark 3.5.5. The case $H \in (0, 1/2)$ is not specific now; for example, if b is discontinuous but bounded we have a weak solution.

Proof. A function $b(x)$ satisfying the conditions of Theorem 3.5.4 can be decomposed as follows:

$$b(x) = d(x) + \sum_{i=1}^{N} c_i \operatorname{sign}(x - a_i),$$

where the function d is Hölder continuous of order $\rho \in (1 - \frac{1}{2H}, 1)$, and $c_i \in \mathbb{R}$. Then, in order to prove Theorem 3.5.4 it suffices to check that the function $\operatorname{sign}(x - a_i)$ satisfies condition (3.5.8) for all $\lambda > 0$.

We have now that $h(s) = b(X_0 + B_s^H) = \operatorname{sign}(X_0 + B_s^H)$.

Since

$$\int_0^T \left| \operatorname{sign}(X_0 + B_s^H) s^{-\alpha} \right|^2 ds \leq \frac{T^{1-2\alpha}}{1 - 2\alpha},$$

it suffices to consider the term

$$A_s = s^\alpha \int_0^s \frac{\left| s^{-\alpha} \operatorname{sign}(X_0 + B_s^H) - r^{-\alpha} \operatorname{sign}(X_0 + B_r^H) \right|}{(s-r)^{\alpha+1}} dr.$$

We have

$$A_s = \int_0^s \frac{\left| \operatorname{sign}(X_0 + B_s^H) - \left(\frac{s}{r}\right)^\alpha \operatorname{sign}(X_0 + B_r^H) \right|}{(s-r)^{\alpha+1}} dr$$

$$\leq \int_0^s \frac{\left| \operatorname{sign}(X_0 + B_s^H) - \operatorname{sign}(X_0 + B_r^H) \right|}{(s-r)^{\alpha+1}} dr$$

$$+ \int_0^s \frac{\left| \left(1 - \left(\frac{s}{r}\right)^\alpha\right) \operatorname{sign}(X_0 + B_r^H) \right|}{(s-r)^{\alpha+1}} dr$$

$$= A_s^1 + A_s^2.$$

The term A_s^2 can be easily bounded:

$$A_s^2 \leq \int_0^s \frac{\left(\frac{s}{r}\right)^\alpha - 1}{(s-r)^{\alpha+1}} dr = c,$$

where

$$c = \int_0^1 \frac{z^{-\alpha} - 1}{(1-z)^{\alpha+1}} dz < \infty.$$

For the term A_s^1 we can write

$$A_s^1 \leq 2 \int_0^s \mathbf{1}_{\{X_0 + B_s^H > 0, X_0 + B_r^H < 0\}} (s-r)^{-\alpha-1} dr$$

$$+ 2 \int_0^s \mathbf{1}_{\{X_0 + B_s^H < 0, X_0 + B_r^H > 0\}} (s-r)^{-\alpha-1} dr$$

$$= 2A_s^{11} + 2A_s^{12}.$$

We will only consider the term A_s^{11}, because the term A_s^{12} can be treated in the same way. Since X_0 is any point from $\mathbb{R}$, we shall denote it simply x. We have
$$A_s^{11} = \int_0^s \mathbf{1}_{\{B_r^H < -x < B_s^H\}} (s-r)^{-\alpha-1} dr.$$

Denote $T_s := \sup\{t \in [0,s] : B_t^H = -x\}$ and notice that T_s is not a stopping time. But $T_s < s$ on the set $\{-x < B_s\}$ and

$$\int_0^s \mathbf{1}_{\{B_r^H < -x < B_s^H\}} (s-r)^{-\alpha-1} dr \leq \mathbf{1}_{\{-x < B_s^H\}} \int_0^{T_s} (s-r)^{-\alpha-1} dr$$
$$= \mathbf{1}_{\{-x < B_s^H\}} \frac{(s-T_s)^{-\alpha}}{\alpha}.$$

According to the Garsia–Rodemich–Rumsey inequality, for any $T > 0$, $p \geq 1$, $\gamma > \frac{1}{p}$ there exists a constant $C = C_{\gamma,p} > 0$ such that

$$|B_s^H - B_t^H|^p \leq C|t-s|^{\gamma p-1} \int_0^T \int_0^T \frac{|B_u - B_r|^p}{|u-r|^{\gamma p+1}} dr\, du \tag{3.5.14}$$

for any $s,t \in [0,T]$. Taking $t = T_s$ in (3.5.14) we obtain

$$|B_s^H + x|^p \leq C|s-T_s|^{\gamma p-1} \int_0^T \int_0^T \frac{|B_u - B_r|^p}{|u-r|^{\gamma p+1}} dr\, du. \tag{3.5.15}$$

Fix $0 < \varepsilon < H$ and take $p = \frac{2}{\varepsilon}$, $\gamma = H - \frac{\varepsilon}{2}$. Set

$$\xi_\varepsilon := \left(\int_0^T \int_0^T \frac{|B_u^H - B_r^H|^{\frac{2}{\varepsilon}}}{|u-r|^{\frac{2H}{\varepsilon}}} dr\, du \right)^{\frac{\varepsilon}{2}}.$$

The random variable ξ_ε verifies $E \exp(\lambda \xi_\varepsilon^\beta) < \infty$ for any $\lambda > 0$, $0 < \beta < 2$, due to (Fer74) and we obtain from (3.5.15) that on the set $\{B_s^H > -x\}$

$$|B_s^H + x| \leq C^{\frac{\varepsilon}{2}} |T_s - s|^{H-\varepsilon} \xi_\varepsilon.$$

Hence,
$$|T_s - s|^{-2\alpha} \leq C^{\frac{\varepsilon(2\alpha)}{2(H-\varepsilon)}} |B_s^H + x|^{\frac{-2\alpha}{H-\varepsilon}} \xi_\varepsilon^{\frac{2\alpha}{H-\varepsilon}}.$$

Therefore, in order to show (3.5.8) it suffices to prove the estimate

$$E \exp\left(\lambda \xi_\varepsilon^{\frac{2\alpha}{H-\varepsilon}} \int_0^T |B_s^H + x|^{\frac{-2\alpha}{H-\varepsilon}} ds \right) < \infty$$

for any $\lambda > 0$, $T > 0$, $x > 0$ and for some fixed $0 < \varepsilon < H$. Set $S_\varepsilon := \int_0^T |B_s^H + x|^{\frac{-2\alpha}{H-\varepsilon}} ds$. We can write, assuming $\varepsilon < \frac{1}{3}$

3.5 SDE with the Additive Wiener Integral w.r.t. Fractional Noise

$$E \exp\left(\lambda \xi_\varepsilon^{\frac{2\alpha}{H-\varepsilon}} S_\varepsilon\right) = E\left(\exp\left(\lambda \xi_\varepsilon^{\frac{2\alpha}{H-\varepsilon}} S_\varepsilon\right) \mathbf{1}_{\left\{S_\varepsilon < \xi_\varepsilon^{\frac{1-3\varepsilon}{H-\varepsilon}}\right\}}\right)$$

$$+ E\left(\exp\left(\lambda \xi_\varepsilon^{\frac{2\alpha}{H-\varepsilon}} S_\varepsilon\right) \mathbf{1}_{\left\{S_\varepsilon \geq \xi_\varepsilon^{\frac{1-3\varepsilon}{H-\varepsilon}}\right\}}\right)$$

$$\leq E \exp\left(\lambda \xi_\varepsilon^{2-\frac{\varepsilon}{H-\varepsilon}}\right) + E \exp\left(\lambda S_\varepsilon^{\frac{2H-3\varepsilon}{1-3\varepsilon}}\right).$$

We know that $E \exp\left(\lambda \xi_\varepsilon^{2-\frac{\varepsilon}{H-\varepsilon}}\right) < \infty$, so it suffices to show that $E \exp\left(\lambda S_\varepsilon^{\frac{2H-3\varepsilon}{1-3\varepsilon}}\right) < \infty$. By the Hölder inequality, assuming $\frac{-2\alpha}{H-\varepsilon} > -1 + \varepsilon$, we obtain that

$$S_\varepsilon \leq C_{T,\varepsilon} \left(\int_0^T |B_s^H + x|^{-1+\varepsilon} ds\right)^{\frac{2\alpha}{(H-\varepsilon)(1-\varepsilon)}}.$$

Hence,

$$S_\varepsilon^{\frac{2H-3\varepsilon}{1-3\varepsilon}} \leq C_{T,\varepsilon} \left(\int_0^T |B_s^H + x|^{-1+\varepsilon} ds\right)^\rho,$$

where $\rho = \frac{(2\alpha)(2H-3\varepsilon)}{(H-\varepsilon)(1-\varepsilon)(1-3\varepsilon)}$ can be expressed as $\rho = 4\alpha + \delta$, where $\delta > 0$ tends to zero as ε tends to zero. Therefore, it suffices to show that

$$E \exp\left(\lambda \psi_\varepsilon^{4\alpha+\delta}\right) < \infty, \tag{3.5.16}$$

where

$$\psi_\varepsilon = \int_0^T \mathbf{1}_{\{|B_s^H + x| < 1\}} |B_s^H + x|^{-1+\varepsilon} ds.$$

Lemma 3.5.6 below provides a proof for the estimate (3.5.16), provided $4\alpha H < 1$, and this leads to the condition $H < \frac{1+\sqrt{5}}{4}$. □

Lemma 3.5.6. *Fix $\nu < 1$ and define*

$$\widetilde{G} := \int_0^T \mathbf{1}_{\{|B_s^H + x| < 1\}} |B_s^H + x|^{-\nu} ds.$$

Then for any $p > 0$ such that $pH < 1$ we have

$$E\left(\exp \widetilde{G}^p\right) < \infty.$$

Proof. We need to estimate the moments of the random variable $\widetilde{G}$. Denote by Δ_n the simplex $\{0 < s_1 < \cdots < s_n < T\}$. We have

$$E(G^n) = n! E \int_{\Delta_n} \prod_{i=1}^{n} \mathbf{1}_{\{|B^H_{s_i}+x|<1\}} |B^H_{s_i}+x|^{-\alpha} ds_1 \cdots ds_n.$$

According to (Ber70) the joint density of the random vector $(B^H_{s_1}, B^H_{s_2} - B^H_{s_1}, \ldots, B^H_{s_n} - B^H_{s_{n-1}})$ can be estimated as follows:

$$p(y_1, \ldots, y_n) \leq \frac{2^{\frac{3}{2}n}}{(2\pi)^{n/2}} \prod_{i=1}^{n} (s_i - s_{i-1})^{-H},$$

where $s_0 = 0$, since

$$\det \left(E\left[\left(B^H_{s_i} - B^H_{s_{i-1}}\right) \left(B^H_{s_j} - B^H_{s_{j-1}}\right) \right] \right)_{1 \leq i,j \leq n} \geq 2^{-3n} \prod_{i=1}^{n} (s_i - s_{i-1})^{2H}.$$

Then

$$E \left(\prod_{i=1}^{n} \mathbf{1}_{\{|B^H_{s_i}+x|<1\}} |B^H_{s_i}+x|^{-\alpha} \right)$$

$$\leq c^n \prod_{i=1}^{n} (s_i - s_{i-1})^{-H} \int_{\mathbb{R}^n} \prod_{i=1}^{n} \mathbf{1}_{\{|\sum_{l=1}^{i} y_l + x|<1\}} \left| \sum_{l=1}^{i} y_l + x \right|^{-\alpha} dy_1 \cdots dy_n$$

$$= c^n \prod_{i=1}^{n} (s_i - s_{i-1})^{-H} \int_{\mathbb{R}^n} \prod_{i=1}^{n} \mathbf{1}_{\{|z_i+x|<1\}} |z_i + x|^{-\alpha} dz_1 \cdots dz_n$$

$$= d^n \prod_{i=1}^{n} (s_i - s_{i-1})^{-H},$$

where $c = \frac{2^{\frac{3}{2}}}{(2\pi)^{1/2}}$ and $d = \frac{2c}{1-\alpha}$. Finally,

$$E(\widetilde{G}^n) \leq n! d^n \int_{\Delta_n} \prod_{i=1}^{n} (s_i - s_{i-1})^{-H} ds_1 \cdots ds_n$$

$$= n! d^n \frac{1}{1-H} \frac{\Gamma(1-H)^{n-1} \Gamma(2-H)}{\Gamma(n(1-H)+1)} T^{n(1-H)}.$$

As a consequence we obtain

$$E\left(\exp \widetilde{G}^p \right) \leq e + 1 + \sum_{k=1}^{\infty} \frac{1}{k!} E(\widetilde{G}^{[pk]+1})$$

$$\leq C_1 + C_2 \sum_{k=1}^{\infty} \frac{1}{k!} \frac{([pk]+1)! C_3^{[pk]+1}}{\Gamma(([pk]+1)(1-H)+1)},$$

for some constants C_i, $i = 1, 2, 3$. Using the Stirling formula we finally obtain that this sum is finite provided $pH < 1$. $\square$

3.5.3 Uniqueness in Law and Pathwise Uniqueness for Regular Coefficients

We return to the case of subsection 3.5.1 when the conditions of Theorem 3.5.3 are fulfilled.

Lemma 3.5.7 ((NO02)). *Let the conditions of Theorem 3.5.3 hold for the coefficients of equation (3.5.1). Then any weak solution of this equation has the same distribution under the measure P.*

Proof. Let the pair (B^H, X) creates a weak solution of equation (3.5.1). Consider our function $h(s, X_s) := g(s)b(s, X_s)$. In the case $H \in (0, 1/2)$ we have by Gronwall inequality

$$X_T^* \leq (|X_0| + I_T^*(f) + CT)e^{CT}$$

and

$$|h(s, X_s)| \leq C(1 + X_T^*),$$

therefore the derivative

$$\left| \frac{d}{dt} \int_0^t l_H(t, s) h(s, X_s) ds \right| \leq C \int_0^t s^{-\alpha}(t-s)^{-\alpha-1} |h(s, X_s)| \, ds,$$

evidently, satisfies the condition, similar to (3.5.7):

$$\sup_{0 \leq t \leq T} E \exp \left\{ \lambda t^{2\alpha} \left(\int_s^t s^{-\alpha}(t-s)^{-\alpha-1} |h(s, X_s)| ds \right)^2 \right\} < \infty$$

for some $\lambda > 0$. For $H \in (1/2, 1)$ the condition similar to (3.5.8) can be easily checked similarly to (iv) in Theorem 3.5.3. So,

$$E \exp \left\{ -L_t - \frac{1}{2} \langle L \rangle_t \right\} = 1$$

for $t \in [0, T]$, $L_t = \int_0^t s^\alpha \delta_s dB_s$ with such Wiener process B w.r.t. the measure P that $\widetilde{\alpha} \int_0^t \delta_s ds = \int_0^t l_H(t, s) h(s, X_s) ds$, $\int_0^t l_H(t, s) dB_s^H = \widetilde{\alpha} \int_0^t s^{-\alpha} dB_s$. By Theorem 2.8.1, the process $\widehat{B}_t^H := B_t^H + \int_0^t h(s, X_s) ds$ is an fBm w.r.t. measure $\widehat{P}$ such that

$$\left. \frac{d\widehat{P}}{dP} \right|_t = \exp \left\{ -L_t - \frac{1}{2} \langle L \rangle_t \right\}. \qquad (3.5.17)$$

It means that $X_t - X_0 = \int_0^t f(s) d\widehat{B}_s^H$. Let $\widehat{B}_t$ be a Wiener process such that $\widehat{B}_t = B_t + \int_0^t s^\alpha \delta_s ds$. Also, let Ψ be a bounded measurable functional on $C[0, T]$. Then

$$E_P(\Psi(X - X_0)) = E_{\widehat{P}}\left(\Psi(X - X_0) \exp \left\{ L_t + \frac{1}{2} \langle L \rangle_t \right\} \right)$$

272   3 Stochastic Differential Equations Involving Fractional Brownian Motion

$$= E_{\widehat{P}}\Big(\Psi(X - X_0)\exp\{\int_0^t s^\alpha \delta_s d\widehat{B}_s - 1/2\int_0^t s^{2\alpha}\delta_s^2 ds\}\Big)$$

$$= E_P\Big(\Psi\Big(\int_0^t f(s)dB_s^H\Big)\exp\{\int_0^t s^\alpha \delta_s dB_s - 1/2\int_0^t s^{2\alpha}\delta_s^2 ds\}\Big).$$

The last relation demonstrates that the distribution of X is the same for any weak solution. □

Suppose now that X^1 and X^2 are two weak solutions defined on the same filtered probability space $(\Omega, \mathcal{F}, P, \{\mathcal{F}_t, t \in [0, T]\})$ with respect to the same fBm. Then $\max(X^1, X^2)$ and $\min(X^1, X^2)$ are also solutions and have the same distributions, whence $X^1 = X^2$. We proved the following result.

Theorem 3.5.8. *Under conditions of Theorem 3.5.3 any two weak solutions defined on the same filtered probability space coincide almost surely.*

3.5.4 Existence of a Strong Solution for the Regular Case

Let $H \in (1/2, 1)$, the function f be Hölder continuous of order $\beta > 1 - H$, and the function b be Lipschitz continuous. Then the conditions of Theorem 3.1.4 are fulfilled, therefore equation (3.5.1) has unique strong solution. In the case when $b(s, x) = b(x)$, according to Remark 3.1.11, equation (3.5.1) has a strong solution for $f \in C^\beta[0, T]$, $\beta > 1 - H$, and it is unique due to Theorem 3.5.8. So, the case $H \in (1/2, 1)$ is not hard or interesting.

Now, let $H \in (0, 1/2)$. Consider a Krylov-type inequality as an auxiliary result.

Lemma 3.5.9. *Let functions $g(s)$ and $b(s, x)$ are bounded, so $h(s, x)$ is bounded, X is a weak solution of (3.5.1), and for some $r > 1$ the integral $\int_0^T \psi_r(t)dt < \infty$, where $\psi_r(t) = \|f\|_{L_{\frac{1}{H}}[0,t]}^{-r}$. Then there exists the constant C depending on $h := \sup_{t \in [0,T], x \in \mathbb{R}} |h(t, x)|$ such that for any nonnegative measurable function $g(t, x) : [0, T] \times \mathbb{R} \to \mathbb{R}$*

$$E\int_0^T g(t, X_t)dt \leq C\left(\int_0^T \int_\mathbb{R} g^2(t, x)dx\, dt\right)^{1/2}. \quad (3.5.18)$$

Proof. Let X be a weak solution of (3.5.1) and consider the measure $\widehat{P}$ determined by (3.5.17). Then $X_t - X_0$ under measure $\widehat{P}$ has the Gaussian distribution with zero mean and covariance $\sigma_t^2 := \|I_t(f)\|_{L_2(P)}^2 = \|f\|_{L_2^H(0,t)}^2$. Denote $Z_t := \exp\{-L_t - \frac{1}{2}\langle L \rangle_t\}$. Then from the Hölder inequality with $\beta', \beta > 1$ and $1/\beta' + 1/\beta = 1$ we have that

$$E\int_o^T g(t, X_t)dt \leq \Big(\widehat{E}Z_T^{-\beta'}\Big)^{1/\beta'}\Big(\widehat{E}\int_o^T g(t, X_t)^\beta dt\Big)^{1/\beta}.$$

3.5 SDE with the Additive Wiener Integral w.r.t. Fractional Noise

The mathematical expectation

$$\widehat{E}Z_T^{-\beta'} = \widehat{E}\exp\left\{\beta' L_T + \frac{\beta'}{2}\langle L\rangle_T\right\} < \infty,$$

which follows from the boundedness of $\langle L\rangle_T$. Further, let $\gamma, \gamma' > 1$, $1/\gamma + 1/\gamma' = 1$ and $\gamma\beta = 2$. Then

$$\widehat{E}\int_0^T g(t, X_t)^\beta dt = \int_0^T \frac{1}{\sqrt{2\pi}\sigma_t}\int_{\mathbb{R}} g(t, y)^\beta e^{-\frac{(y-x)^2}{2\sigma_t^2}} dy\, dt$$

$$\leq \frac{1}{\sqrt{2\pi}}\left(\int_0^T\int_{\mathbb{R}} g(t,y)^{\gamma\beta} dy\, dt\right)^{1/\gamma}\left(\int_0^T\int_{\mathbb{R}} e^{-\frac{\gamma'(y-x)^2}{2\sigma_t^2}}\sigma_t^{-\gamma'} dy\, dt\right)^{1/\gamma'}$$

$$= \frac{1}{\sqrt{2\pi}}\left(\int_0^T\int_{\mathbb{R}} g(t,y)^2 dy\, dt\right)^{1/\gamma}\left(\int_0^T\int_{\mathbb{R}} e^{-\gamma' z^2}\sigma_t^{1-\gamma'} dz\, dt\right)^{1/\gamma'}$$

$$\leq C\left(\int_0^T\int_{\mathbb{R}} g(t,y)^2 dy\, dt\right)^{1/\gamma}\left(\int_0^T \sigma_t^{-\frac{\beta}{2-\beta}} dt\right)^{\frac{2-\beta}{2}}.$$

Finally, put $\frac{\beta}{2-\beta} = r > 1$, which means that $\beta = \frac{2r}{1+r}$, $\frac{1}{\alpha} = 1 - \frac{1}{\beta}$, $\gamma = 1 + \frac{1}{r}$. From inequality (1.9.1), $\|f\|_{L_2^H(0,t)} \geq C(H)\|f\|_{L_{\frac{1}{H}}(0,t)}$, so $\int_0^T \sigma_t^{-r} dt < \infty$, whence the proof follows. □

Lemma 3.5.10. *Let $b_n(t, x) = b_n(t, x)\mathbf{1}\{|x| \leq C_1\}$ be a sequence of measurable functions, $|b_n(t,x)| \leq C_2$, $\lim_{n\to\infty} b_n(t, x) = b(t, x)$, for all $(t, x) \in [0, T]\times\mathbb{R}$, and the conditions of Lemma 3.5.9 hold. Let also the corresponding solutions $X_t^{(n)}$ of the equations*

$$X_t^{(n)} = X_0 + \int_0^t b_n(s, X_s^{(n)})ds + I_t(f), t \in [0, T],$$

converge a.s. to some process X_t for all $t \in [0,T]$. Then the process X is a solution of equation (3.5.1).

Proof. It is sufficient to prove that $\lim_{n\to\infty} I_n := \lim_{n\to\infty} E\int_0^T |b_n(s, X_s^{(n)}) - b(s, X_s)|ds = 0$. But $I_n \leq I_n^{(1)} + I_n^{(2)}$, where

$$I_n^{(1)} = E\int_0^T |b_n(s, X_s^{(n)}) - b(s, X_s^{(n)})|ds,$$

$$I_n^{(2)} = E\int_0^T |b(s, X_s^{(n)}) - b(s, X_s)|ds.$$

Evidently, from (3.5.18) and finiteness of b_n and b, $I_n^{(1)} \leq C(\int_0^T\int_{\mathbb{R}} |b_n(t,x) - b(t,x)|^2 dt\, dx)^{1/2} \to 0, n\to\infty$, and also $I_n^{(2)} \to 0, n\to\infty$. □

Theorem 3.5.11. *Let both the functions $h(t, x)$ and $b(t, x)$ satisfy the linear growth condition. Then equation (3.5.1) has the unique strong solution.*

Remark 3.5.12. The next condition is sufficient for both the functions $h(t,x)$ and $b(t,x)$ to be of linear growth:

$$|b(s,x)| \leq C(|f(s)| \wedge 1)(1+|x|). \tag{3.5.19}$$

Proof. For any $R > 0$ denote $b^{(R)}(t,x) := b(t,x)\mathbf{1}_{\{|x|\leq R\}}$. Let φ be a smooth nonnegative function with compact support in $\mathbb{R}$ such that $\int_{\mathbb{R}} \varphi(x)dx = 1$. Define $b_{R,j}(t,x) := j\int_{\mathbb{R}} b^{(R)}(t,y)\varphi(j(x-y))dy$. Let for $n \leq k$ $\widetilde{b}_{R,n,k} = \wedge_{j=n}^{k} b_{R,j}$, $\widetilde{b}_{R,n} = \wedge_{j=n}^{\infty} b_{R,j}$. The functions $\widetilde{b}_{R,n,k}$ are Lipschitz in x uniformly in t and $\widetilde{b}_{R,n,k} \downarrow \widetilde{b}_{R,n}$, $k \to \infty$, $\widetilde{b}_{R,n} \uparrow b^{(R)}$, $n \to \infty$, for a.a. x and any t. Equation (3.5.1) with $\widetilde{b}_{R,n,k}$ has the unique solution $\widetilde{X}_{R,n,k}$ as an ordinary differential equation with Lipschitz coefficient. By the comparison theorem for ODEs the sequence $\widetilde{X}_{R,n,k}$ decreases in k, hence it has a limit $\widetilde{X}_{R,n}$. The sequence $\widetilde{X}_{R,n}$ increases in n, hence it has a limit $X^{(R)}$. Applying Lemma 3.5.10 we obtain that $\{X_t^{(R)}, t \in [0,T]\}$ is a solution of (3.5.1) with drift $b^{(R)}(t,x)$. Then we apply standard techniques: all $X_t^{(R)}$ are bounded by $(I_T^*(f) + |x|)e^{CT}$, and (3.5.1) has a unique solution on any $[0, \tau_R]$, where $\tau_R = \inf\{t : |X_t^{(R)}| \geq R\}$. It means that (3.5.1) has a unique solution on the whole interval $[0,T]$. □

3.5.5 Existence of a Strong Solution for Discontinuous Drift

Let $\Omega = C_0([0,T],\mathbb{R})$ be the Banach space of continuous functions, null at time 0, equipped with the supremum norm, and P be the unique probability measure on Ω such that the canonical process is an fBm with Hurst parameter $H \in (1/2, 1)$. Assume also that the canonical filtration is augmented with the P-negligible sets. We consider the following partial case of equation (3.5.1):

$$X_t = X_0 + \int_0^t b(X_s)ds + B_t^H \tag{3.5.20}$$

with $b(x) = \operatorname{sign} x$, $H \in (1/2, H_0)$, $H_0 = \frac{1+\sqrt{5}}{4}$. According to Theorem 3.5.4, equation (3.5.20) has a weak solution. Now we intend to prove the existence of its strong solution. For this purpose consider the following approximations of the function $b(x) = \operatorname{sign} x$:

$$b_n(x) = \begin{cases} -1, & x \leq 0; \\ n^3 x^2 - 1, & 0 < x \leq \frac{1}{n^2}; \\ 2nx - 1, & 1/n^2 < x \leq \frac{1}{n} - \frac{1}{n^2}; \\ 1 - n^3(x - \frac{1}{n})^2, & 1/n - 1/n^2 < x \leq \frac{1}{n}; \\ 1, & x \geq \frac{1}{n}. \end{cases}$$

Then

$$b_n'(x) = \begin{cases} 0, & x \leq 0; \\ 2n^3 x, & 0 < x \leq \frac{1}{n^2}; \\ 2n, & 1/n^2 < x \leq \frac{1}{n} - \frac{1}{n^2}; \\ 2n^3(x - \frac{1}{n}), & 1/n - 1/n^2 < x \leq \frac{1}{n}; \\ 0, & x \geq \frac{1}{n}. \end{cases}$$

3.5 SDE with the Additive Wiener Integral w.r.t. Fractional Noise

Evidently, any $b'_n \in C(\mathbb{R})$; moreover, it is Lipschitz: $|b'_n(x_1) - b'_n(x_2)| \leq 2n^3|x_1 - x_2|$.

Lemma 3.5.13. *For any $x \in \mathbb{R}$ $b_{n+1}(x) > b_n(x)$, $n \geq 1$.*

Proof. It is sufficient to consider the interval $(0, \frac{1}{n+1})$.
(a) For $x \in (0, \frac{1}{(n+1)^2}]$ $b_{n+1}(x) = (n+1)^3 x^2 - 1 > n^3 x^2 - 1 = b_n(x)$.
(b) For $x \in (\frac{1}{(n+1)^2}, \frac{1}{n^2}]$ $b_n(x) = n^3 x^2 - 1$, $b_{n+1}(x) = 2(n+1)x - 1$. But the inequality $2(n+1)x - 1 > n^3 x^2 - 1$ holds for $x < \frac{2(n+1)}{n^3}$, and it is our case.
(c) For $x \in (\frac{1}{n^2}, \frac{1}{n+1} - \frac{1}{(n+1)^2}]$ $b_{n+1}(x) = 2(n+1)x - 1 > 2nx - 1 = b_n(x)$.
(d) For $x \in (\frac{1}{n+1} - \frac{1}{(n+1)^2}, \frac{1}{n} - \frac{1}{n^2}]$ $b_{n+1}(x) = 1 - (n+1)^3 (x - \frac{1}{n+1})^2$, $b_n(x) = 2nx - 1$. The function $\varphi(x) := (n+1)^3 (x - \frac{1}{n+1})^2 + 2nx - 2$ has $\varphi'(x) = 2(n+1)^3 (x - \frac{1}{n+1}) + 2n = 0$ for $x_0 = \frac{1}{n+1} - \frac{n}{(n+1)^3}$, it is the point of local minimum and $x_0 \in (\frac{1}{n+1} - \frac{1}{(n+1)^2}, \frac{1}{n} - \frac{1}{n^2}]$ for $n > 2$. So, we must check the inequality $\varphi(x) < 0$ for $x = \frac{1}{n+1} - \frac{1}{(n+1)^2}$ and $x = \frac{1}{n} - \frac{1}{n^2}$, and it evidently holds.
(e) Finally, for $x \in (\frac{1}{n} - \frac{1}{n^2}, \frac{1}{n+1})$ the inequality $b_{n+1}(x) = 1 - (n+1)^3 (x - \frac{1}{n+1})^2 > 1 - n^3 (x - \frac{1}{n})^2 = b_n(x)$ is equivalent to $(2n + 1 + \sqrt{n(n+1)})x > 1$ and it is sufficient to check it in the point $x = \frac{1}{n} - \frac{1}{n^2}$:

$$(2n + 1 + \sqrt{n(n+1)})\left(\frac{1}{n} - \frac{1}{n^2}\right) > (3n+1)\frac{n-1}{n^2} = \frac{3n^2 - 2n - 1}{n^2} > 1$$

for $n \geq 2$. Therefore, $b_{n+1}(x) \geq b_n(x)$, $x \in \mathbb{R}$. □

Consider the approximating equation

$$X_t^n = x + \int_0^t b_n(X_s^n)ds + B_t^H. \tag{3.5.21}$$

The functions b_n are Lipschitz, therefore equation (3.5.21) has a unique strong solution X_t^n on $[0, T]$, and $X_t^n \leq X_t^{n+1}$ for any $t \in [0, T]$ a.s. Moreover, for any $0 < \varepsilon < H$

$$|X_{t_1}^n(\omega) - X_{t_2}^n(\omega)| \leq C(\omega)|t_2 - t_1|^{H-\varepsilon} + |t_2 - t_1|,$$

so, the set $\{X_n(\cdot, \omega), n \geq 1\}$ is tight for any $\omega \in \Omega'$, $P(\Omega') = 1$. We obtain that $X_t^n(\omega) \uparrow X_t(\omega)$, $\omega \in \Omega'$, where the limit process X is continuous in t. Further,

$$\left|\int_0^t b_n(X_s^n)ds - \int_0^t b(X_s)ds\right| \leq$$

$$\int_0^t |b_n(X_s^n) - b_n(X_s)|ds + \int_0^t |b_n(X_s) - b(X_s)|ds. \tag{3.5.22}$$

Note that $|b_n(X_s^n) - b_n(X_s)| = b_n(X_s) - b_n(X_s^n) \leq 2$. Consider all the cases of mutual values of X_s, X_s^n.
(a) For $X_s^n < 0$, $X_s \in (0, \frac{1}{n}]$ $b_n(X_s) - b_n(X_s^n) \leq 2 \mathbf{1}_{\{X_s \in (0, \frac{1}{n}]\}}$.
(b) For $X_s^n < 0$, $X_s > \frac{1}{n}$ $b_n(X_s) - b_n(X_s^n) \leq 2 \mathbf{1}_{\{X_s > 0, X_s^n < 0\}} \to 0$ a.s., $n \to \infty$.
(c) For $X_s, X_s^n \in [0, \frac{1}{n}]$ $b_n(X_s) - b_n(X_s^n) \leq 2 \mathbf{1}_{\{X_s \in [0, \frac{1}{n}]\}}$.
(d) For $X_s^n \in [0, \frac{1}{n}]$, $X_s > \frac{1}{n}$ $|b_n(X_s) - b_n(X_s^n)| \leq 2 \mathbf{1}_{\{X_s > 0, X_s^n \in [0, \frac{1}{n}]\}} \to 0$ a.s., $n \to \infty$.
Further,

$$\int_0^t |b_n(X_s) - b(X_s)| ds \leq 2 \int_0^t \mathbf{1}_{\{X_s \in [0, \frac{1}{n}]\}} ds. \qquad (3.5.23)$$

We obtain from (3.5.22) – (3.5.23) and (a)–(d) that

$$\lim_{n \to \infty} \left| \int_0^t b_n(X_s^n) ds - \int_0^t b(X_s) ds \right| \leq 6 \lim_{n \to \infty} \int_0^t \mathbf{1}_{\{X_s \in (0, \frac{1}{n})\}} ds$$

$$= 6 \int_0^t \mathbf{1}_{\{X_s = 0\}} ds.$$

Therefore, to prove the existence of a strong solution of (3.5.20) it is sufficient to prove that $E \int_0^T \mathbf{1}_{\{X_s = 0\}} ds = 0$, and in turn it is sufficient to establish the existence of bounded density $p_s(x)$, $x \in \mathbb{R}$, $s > 0$ of the process X_s. For this purpose, return to X_s^n: since the functions b_n are continuously differentiable, then X_s^n has a stochastic derivative, and on our probability space

$$D_s X_t^n = 1 + \int_s^t D_s X_u^n b_n'(X_u^n) du,$$

whence $D_s X_t^n = \exp\{\int_s^t b_n'(X_u^n) du\} \geq 1$, since $b_n' \geq 0$.

Now we use the result of (Nua95): let the random variable $F \in \mathbb{D}_{1,2}$, $h \in \mathcal{H}$, $\langle DF, h \rangle_\mathcal{H} \neq 0$ a.s. and $\frac{h}{\langle DF, h \rangle_\mathcal{H}} \in \text{Dom } \delta$. Then F has a continuous and bounded density

$$f(x) = E\left(\mathbf{1}_{\{F > x\}} \delta\left(\frac{h}{\langle DF, h \rangle_\mathcal{H}}\right)\right).$$

Now we put $F := X_t^n$, $h_t(s) := \mathbf{1}_{\{0 \leq s \leq t\}}$. Then

$$\langle DF, h \rangle_\mathcal{H} = 2\alpha H \int_0^t \int_0^t \exp\left\{\int_s^t b_n'(X_u^n) du\right\}$$

$$\times \exp\left\{\int_v^t b_n'(X_u^n) du\right\} |v - s|^{2\alpha - 1} dv \, ds \geq C_H t^{2H} > 0.$$

Consider the function $\theta(s) = \frac{h_t(s)}{\langle DF, h_t \rangle_\mathcal{H}} = h_t(s) \xi$, where ξ is a bounded random variable, $\xi = \langle DF, h \rangle_\mathcal{H}^{-1}$, $E\xi^2 < \infty$. To establish that $\theta \in \text{Dom } \delta$, it is sufficient to verify that

3.5 SDE with the Additive Wiener Integral w.r.t. Fractional Noise

$$E \int_0^T (D_s\xi)^2 ds < \infty. \tag{3.5.24}$$

Indeed,

$$D_s\xi = D_s\left(\left(\int_0^t\int_0^t \exp\left\{\int_z^t b'_n(X^n_u)du\right\}\exp\left\{\int_v^t b'_n(X^n_u)du\right\}\right.\right.$$
$$\left.\left.\times |v-z|^{2\alpha-1}dvdz\right)^{-1}\right) = \langle DF, h\rangle_{\mathcal{H}}^{-2} \cdot \int_0^t\int_0^t \exp\left\{\int_z^t b'_n(X^n_u)du\right\}$$
$$\times \exp\left\{\int_v^t b'_n(X^n_u)du\right\}|v-z|^{2\alpha-1}\int_z^t b''_n(X^n_u)du\, dv\, dz,$$

where $|b''_n(x)| \leq 2n^3$ (since $|b'_n(x_1) - b'_n(x_2)| \leq 2n^3|x_1-x_2|$). Therefore, $|D_s\xi| \leq C_H^{-2}t^{-2H} \cdot 4n^3 \cdot C(H,n,t)$ (note that $|b'_n(x)| \leq 2n$), and (3.5.24) holds. We obtain that $\theta \in Dom\,\delta$, and the density $p_t^n(x) := p_{X_t^n}(x)$ equals

$$p_t^n(x) = E\left\{\mathbf{1}_{\{X_t^n > x\}}\delta\left(\frac{h}{\langle DX_t^n, h_t\rangle_{\mathcal{H}}}\right)\right\}.$$

Let $\psi(y) := \mathbf{1}_{[a,b]}(y)$. Then from Proposition 2.1.1 (Nua95)

$$P\{a \leq X_n(t) \leq b\} = \int_a^b p_t^n(x)dx$$
$$= \int_a^b E\left\{\mathbf{1}_{\{X_t^n > x\}}\delta\left(\frac{h}{\langle DX_t^n, h_t\rangle_{\mathcal{H}}}\right)\right\}dx$$
$$= E\left(\left(\int_{-\infty}^{X_t^n}\psi(x)dx\right)\cdot\delta\left(\frac{h}{\langle DX_t^n, h_t\rangle_{\mathcal{H}}}\right)\right)$$
$$= E\left(\varphi(X_t^n)\delta\left(\frac{h}{\langle DX_t^n, h_t\rangle_{\mathcal{H}}}\right)\right) = \left(\varphi(y) = \int_{-\infty}^y \psi(z)dz\right)$$
$$= E\left(\left\langle D\varphi(X_t^n), \frac{h}{\langle DX_t^n, h_t\rangle_{\mathcal{H}}}\right\rangle_{\mathcal{H}}\right)$$
$$\leq \frac{1}{C_H t^{2H}} E\left(\langle D\varphi(X_t^n), h\rangle_{\mathcal{H}}\right)$$
$$= C_{1,H} t^{-2H}\int_a^b E\left(\mathbf{1}_{\{X_t^n > x\}}\delta(h)\right)dx$$
$$\leq C_{1,H} t^{-2H} E|\delta(h)|\int_a^b dx.$$

Therefore, $p_n^t(x) \leq C_{2,H}t^{-2H}$, and $P\{a \leq X_t \leq b\} = \lim_{n\to\infty} P\{a \leq X_t^n \leq b\} = C_{2,H}t^{-2H}(b-a)$ for any continuous points of distribution function of X_t. Choosing $a \uparrow 0$, $b \downarrow 0$, we obtain density $p_t(0) \leq C_{2,H}t^{-2H}$. So, we have proved the following result:

Theorem 3.5.14. *Equation* (3.5.20) *with* $b(x) = \text{sign}\,x$ *has a strong solution.*

3.5.6 Estimates of Moments of Solutions for Regular Case and $H \in (0, 1/2)$

Now we consider the case, when $H \in (0,1/2)$ and condition (3.5.19) holds. Then equation (3.5.1) has a unique strong solution. Suppose, in addition, that $f \in L_p[0,T] \cap D_p^H[0,T]$ for some $p > \frac{1}{H}$. Then the integral $I_t = I_t(f)$ is continuous on $[0,T]$ (see Section 1.11). Evidently, the solution X_t is also continuous on $[0,T]$. Let $\tau_N = \inf\{t > 0 : |X_t| \geq N\} \wedge T$. Then $|X_{t \wedge \tau_N}| \leq N$. The solution admits the evident estimate

$$|X_{t \wedge \tau_N}| \leq |X_0| + |I_{t \wedge \tau_N}| + C \int_0^t (1 + |X_{s \wedge \tau_N}|) ds,$$

and for any $r > 1$

$$\begin{aligned}E|X_{t \wedge \tau_N}|^r &\leq 3^r \left(|X_0|^r + C^r E(\int_0^t (1 + |X_{s \wedge \tau_N}|)ds)^r + E|I_{t \wedge \tau_N}|^r\right)\\ &\leq 3^r |X_0|^r + (6C)^r t^r + (6C)^r E \int_0^t |X_{s \wedge \tau_N}|^r ds \cdot t^{r-1} + 3^r E|I_{t \wedge \tau_N}|^r \quad (3.5.25)\\ &\leq g(t) + (6C)^r t^{r-1} \int_0^t |X_{s \wedge \tau_N}|^r ds.\end{aligned}$$

Here
$$g(t) = 3^r |X_0|^r + (6C)^r t^r + 3^r E|I_t^*|^r. \quad (3.5.26)$$

From the Gronwall inequality we obtain that

$$E|X_{t \wedge \tau_N}|^r \leq g(t)(1 + C_1 t^r e^{\frac{C_1 t^r}{r}}),$$

where $C_1 = (6C)^r$.
Let $N \to \infty$, then it holds that

$$E|X_t|^r \leq g(t)(1 + C_1 t^r e^{\frac{C_1 t^r}{r}}). \quad (3.5.27)$$

Now, it follows from Theorem 1.10.6 and the part 2 of Remark 1.10.7, that there exists a constant $C(H,p)$ such that

$$\|I_t^*\|_r \leq C(H,p) \left(\Gamma\left(\frac{r+1}{2}\right)\right)^{1/r} G_p^1(0,t,f). \quad (3.5.28)$$

It follows from (3.5.25)–(3.5.28) that

$$E|X_t|^r \leq g(t)\left(1 + C_1 t^r e^{\frac{C_1 t^r}{r}}\right) \quad (3.5.29)$$

where $g(t) = 3^r |X_0|^r + (6C)^r t^r + 3^r C(H,p)^r \Gamma(\frac{r+1}{2})^r (G_p^1(0,t,f))^r$. Estimate (3.5.29) means that $E|X_t|^r < \infty$, $t \in [0,T]$, and this permits us to reduce the value of the multiplier $g(t)$. Indeed, if we know that $E|X_t|^r < \infty$, we can write the following inequality instead of (3.5.25):

3.5 SDE with the Additive Wiener Integral w.r.t. Fractional Noise 279

$$E|X_t|^r \leq E\left(|X_0| + |I_t| + C\left(\int_0^t (1+|X_s|)ds\right)\right)^r \quad (3.5.30)$$
$$\leq g_1(t) + C_1 t^{r-1} \int_0^t E|X_s|^r ds,$$

where from (1.9.10) and (1.10.4) $g_1(t) = (3|X_0|)^r + C_1 t^r + 3^r \sup_{0 \leq s \leq t} E|I_s|^r$
$\leq (3|X_0|)^r + C_1 t^r + \widetilde{C}_r(C(H,p))^r(G_p^1(0,t,f))^r$, $\widetilde{C}_r = 3^r C_r$, $C_r = \frac{2^{r/2}}{\pi^{1/2}} \Gamma(\frac{r+1}{2})$.
Hence, from the Gronwall inequality it follows that

$$E|X_t|^r \leq g_1(t)(1 + C_1 t^r e^{\frac{C_1 t^r}{r}}). \quad (3.5.31)$$

Let estimate similarly $E|X_t - X_{t'}|^r$, $0 \leq t < t' \leq T$.

$$E|X_t - X_{t'}|^r \leq (2C)^r E(\int_t^{t'}(1+|X_s|)ds)^r + 2^r E|\int_t^{t'} f(s)dB_s^H|^r \quad (3.5.32)$$
$$\leq (4C)^r(1 + g_1(T)(1 + C_1 T^r e^{\frac{C_1 T^r}{r}}))(t'-t)^r + 2^r C_r (G^p(t,t',f))^r,$$

where $G^p(t,t',f) = C(H,p)(\|f\|_{L_p(t,t')}(t'-t)^{H-1/p} + \|f\|_{D_H(t,t')})$,
$\|f\|_{D_H(t,t')} = (\int_t^{t'}(\int_x^{t'}|f(x)-f(t)|(t-x)^{\alpha-1}dt)^2 dx)^{1/2}$.
Let $f \in C^\beta[0,T]$ with $\alpha + \beta > 0, 0 < \beta < 1$. Then

$$\|f\|_{L_p(t,t')} \cdot (t'-t)^{H-1/p} \leq \|f\|_{C^\beta[0,T]}(t'-t)^H,$$

$$\|f\|_{D_H(t,t')} \leq \|f\|_{C^\beta[0,T]} \cdot C_{H,\beta}^1 (t'-t)^{H+\beta}$$

with $C_{H,\beta}^1 = (H+\beta-1/2)^{-1}(2H+2\beta)^{-1/2}$. Therefore

$$E|X_t - X_{t'}|^r \leq (4C)^r(1 + g_1(T)(1 + C_1 T^r e^{\frac{C_1 T^r}{r}}))(t'-t)^r \quad (3.5.33)$$
$$+ 2^r C_r (C_{H,\beta,T})^r (t'-t)^{rH},$$

where $C_{H,\beta,T,p} = C(H,p)(1 + C_{H,\beta}^1 T^\beta)\|f\|_{C^\beta[0,T]}$. Estimates (3.5.31) and (3.5.33) can be strengthened by appropriate choice of partitions of $[0,T]$. More exactly, take $t_0 := (6C)^{-1}$. Then for $0 \leq t \leq t_0$ it follows from (3.5.30) that $E|X_t|^r \leq g_1(t) + 6C \int_0^t E|X_s|^r ds$, and from the Gronwall inequality

$$E|X_t|^r \leq g_1 \cdot e^{6Ct} \leq e \cdot g_1, \quad 0 \leq t \leq t_0,$$

where $g_1 = (3|X_0|)^r + 1 + \widetilde{C}_r(G^1(0,T,f))^r$.
Further, for $t_0 \leq t_1 \leq 2t_0$

$$E|X_t|^r \leq 3^r |X_{t_0}|^r + 3^r E|\int_{t_0}^t f(s)dB_s^H|^r + (6C)^r(t-t_0)^r$$
$$+ (6C)^r(t-t_0)^{r-1} E \int_{t_0}^t |X_s|^r ds \leq 3^r g_1 e + \widetilde{C}_r(G^1(0,T,f))^r + 1 + 6C \int_{t_0}^t |X_s|^r ds,$$

whence
$$E|X_t|^r \le g_2 e^{6C(t-t_0)} \le g_2 e,$$
where $g_2 = 3^r g_1 e + \widetilde{C_r}(G^1(0,T,f))^r + 1$.

Further, by induction, for $kt_0 \le t \le (k+1)t_0$ we have that $E|X_t|^r \le g_{k+1}e$, where $g_{k+1} \le 3^r g_k e + B_r \le \cdots \le (3^r e)^k (g_1 + B_r)$ for $B_r = \widetilde{C_r}(G^1(0,T,f))^r + 1$.

The number of such steps on the interval $[0,T]$ does not exceed $k = \left[\frac{T}{t_0}\right] + 1 \le 6CT + 1$. It means that for any $0 \le t \le T$

$$E|X_t|^r \le (3^r e)^{6CT+1}(g_1 + B_r) \le (3e)^{(6CT+1)r}(3^r|x|^r + 2 + 2\widetilde{C_r}(G^1(0,T,f))^r), \quad (3.5.34)$$

and similarly to (3.5.34) we obtain that

$$E|X_t - X_{t'}|^r \le (4C)^r D_r(t' - t)^r + 2^r C_r(G^2(t,t',f))^r,$$

where

$$D_r = 1 + (3e)^{(6CT+1)r}(g_1 + B_r)$$

$$= 1 + (3e)^{(6CT+1)r}(2 + (3|X_0|)^r + 2\widetilde{C_r}(G^1(0,T,f))^r). \quad (3.5.35)$$

For $f \in C^\beta[0,T]$ with $0 < \beta < 1$, $H + \beta > 1/2$ we have that

$$E|X_t - X_{t'}|^r \le (4C)^r (1 + (3e)^{(6CT+1)r}(g_1 + B_r))(t' - t)^r +$$
$$+ 2^r C_r (C_{H,\beta,T,p})^r (t' - t)^{Hr}, \quad (3.5.36)$$

whence

$$E|X_t - X_{t'}|^r \le (4C)^r D_r(t' - t)^r + 2^r C_r (C_{H,\beta,T,p})^r (t' - t)^{Hr}. \quad (3.5.37)$$

3.5.7 The Estimates of the Norms of the Solution in the Orlicz Spaces

The results of Subsections 3.5.7– 3.5.9 were motivated by the papers (KM06) and (KM07).

Let the function $U(x) = \exp\{x^2\} - 1$, $(\Omega, \mathcal{F}, P)$ be some probability space.

Definition 3.5.15. *The Orlicz space $L_U(\Omega)$ generated by the function $U(x)$ is the space of random variables ξ on $(\Omega, \mathcal{F})$, such that for some constant $C_\xi > 0$ $EU(\frac{\xi}{C_\xi}) < \infty$.*

The next result is proved in the monograph (BK00).

Theorem 3.5.16. *The Orlicz space $L_U(\Omega)$ is the Banach space with respect to the Luxemburg norm*

$$\|\xi\|_U = \inf\{r > 0 : E\exp\left\{\frac{\xi^2}{r^2}\right\} \le 2\}.$$

3.5 SDE with the Additive Wiener Integral w.r.t. Fractional Noise

Let $\mathbb{T}$ be some set of parameters.

Definition 3.5.17. *The random process $Y = \{Y_t, t \in \mathbb{T}\}$ belongs to the space $L_U(\Omega)$, if for any $t \in \mathbb{T}$ the random variable Y_t belongs to this space.*

Introduce the notations $a := (3e)^{6CT+1}$, $b := 3|X_0|a$, $c := 3aG^1(0,T,f)$, $c_1 = c\sqrt{2}$, $d := \max\{c_1, a\sqrt{e}, b\sqrt{e}\}$, $h := (3 + 2\sqrt{2})\exp\{\frac{d^2}{2c^2}\}$.

Theorem 3.5.18. *Let the conditions of the Theorem 3.5.11 hold and $\{X_t, t \in [0,T]\}$ be the solution of equation (3.5.1). Then for any $\varepsilon > 0$*

$$P\{|X_t| \geq \varepsilon\} \leq h \exp\left\{-\frac{\varepsilon^2}{2c^2}\right\}. \tag{3.5.38}$$

Proof. The next inequality follows from (3.5.34):

$$E|X_t|^r \leq 2a^r + b^r + \frac{2c_1^r}{\sqrt{\pi}} \Gamma\left(\frac{r+1}{2}\right). \tag{3.5.39}$$

Furthermore, from the Stirling formula

$$\Gamma(u) = \sqrt{2\pi} u^{u-1/2} e^{-u} e^{\theta(u)} \quad \text{with} \quad \theta(u) < \frac{1}{2u}, \quad u \geq 1,$$

we obtain that

$$\Gamma\left(\tfrac{r+1}{2}\right) \leq \sqrt{2\pi} \left(\tfrac{r+1}{2}\right)^{r/2} \cdot \exp\left\{-\tfrac{r+1}{2}\right\} \exp\left\{\tfrac{1}{6(r+1)}\right\}$$
$$= \sqrt{2\pi} r^{r/2} (2e)^{-r/2} (1+1/r)^{r/2} \exp\left\{-\tfrac{1}{2} + \tfrac{1}{6(r+1)}\right\}.$$

It is easy to see that for $r \geq 1$

$$h(r) := (1+1/r)^{r/2} \exp\left\{-\frac{1}{2} + \frac{1}{6(r+1)}\right\} \leq 1.$$

Indeed,

$$\ln h(r) = \tfrac{r}{2} \ln(1 + \tfrac{1}{r}) - \tfrac{1}{2} + \tfrac{1}{6(r+1)}$$
$$\leq \tfrac{r}{2}\left(\tfrac{1}{r} - \tfrac{1}{2r^2} + \tfrac{1}{3r^3}\right) - \tfrac{1}{2} + \tfrac{1}{6(r+1)} = \tfrac{2-r-r^2}{12(r+1)r^2} \leq 0$$

for $r \geq 1$, i.e.

$$\Gamma\left(\frac{r+1}{2}\right) \leq \sqrt{2\pi}(2e)^{-r/2} r^{r/2}. \tag{3.5.40}$$

It follows from (3.5.39) and (3.5.40) that

$$E|X_t|^r \leq 2a^r + b^r + 2\sqrt{2} l^r r^{r/2}, \tag{3.5.41}$$

where $l = \frac{c}{\sqrt{e}}$.

It follows from (3.5.41) and the Chebyshov inequality that

$$P\{|X_t| \geq \varepsilon\} \leq \frac{E|X_t|^r}{\varepsilon^r} \leq 2\left(\frac{a}{\varepsilon}\right)^r + \left(\frac{b}{\varepsilon}\right)^r + 2\sqrt{2}\left(\frac{l}{\varepsilon}\right)^r r^{r/2}. \quad (3.5.42)$$

We put $r = \left(\frac{\varepsilon}{l}\right)^2 \frac{1}{e}$, where $\varepsilon > l\sqrt{e}$, and obtain the inequality

$$P\{|X_t| \geq \varepsilon\} \leq 2\left(\frac{a}{\varepsilon}\right)^{\left(\frac{\varepsilon}{l}\right)^2 \frac{1}{e}} + \left(\frac{b}{\varepsilon}\right)^{\left(\frac{\varepsilon}{l}\right)^2 \frac{1}{e}} + 2\sqrt{2}\exp\left\{-\left(\frac{\varepsilon}{l}\right)^2 \frac{1}{2e}\right\}$$
$$= \exp\left\{\left(\ln\frac{b}{\varepsilon}\right)\left(\frac{\varepsilon}{l}\right)^2 \frac{1}{e}\right\} + 2\exp\left\{\left(\ln\frac{a}{\varepsilon}\right)\left(\frac{\varepsilon}{l}\right)^2 \frac{1}{e}\right\} + 2\sqrt{2}\exp\left\{-\left(\frac{\varepsilon}{l}\right)^2 \frac{1}{2e}\right\}. \quad (3.5.43)$$

Let $\ln\frac{a}{\varepsilon} \vee \ln\frac{b}{\varepsilon} \leq -\frac{1}{2}$, i.e. $\varepsilon \geq (a \vee b)\sqrt{e}$.

Then

$$P\{|X_t| \geq \varepsilon\} \leq (3 + 2\sqrt{2}) \cdot \exp\left\{-\frac{\varepsilon^2}{2el^2}\right\} = (3 + 2\sqrt{2}) \cdot \exp\left\{-\frac{\varepsilon^2}{2c^2}\right\}. \quad (3.5.44)$$

Evidently, (3.5.44) holds for $\varepsilon \geq d$. But $\exp\left\{\frac{d^2}{2c^2}\right\} \geq 1$, so it follows from (3.5.44) that inequality (3.5.38) holds for any $\varepsilon > 0$. □

Theorem 3.5.19. *Let the conditions of Theorem 3.5.11 hold and $\{X_t, t \in [0,T]\}$ be the solution of equation (3.5.1). Then the random variable X_t belongs to the Orlicz space $L_U(\Omega)$, and its norm in this space admits an estimate*

$$\|X_t\|_U \leq \sqrt{2}(1+h)c.$$

Proof. The statement of this theorem follows from Theorem 3.5.18 and the next lemma, which is the partial case of Theorem 2.3.4 (BK00). □

Lemma 3.5.20. *Let ξ be a random variable such that for any $\varepsilon > 0$ $P\{|\xi| \geq \varepsilon\} \leq C_1 \exp\left\{-\frac{\varepsilon^2}{2C_2^2}\right\}$ for some $C_i > 0, i = 1, 2$. Then $\xi \in L_U(\Omega)$ and $\|\xi\|_U \leq \sqrt{2}(1+C_1)C_2$.*

Now introduce the notations
$B_1 := 2(\sqrt{e})^{-1/2} C_{H,\beta,T,p}$, $B_2 := 4C\frac{c}{\sqrt{e}}T^{1-H}$, $B_3 := 4C(1+2a+b)T^{1-H}$,
$B_4 := B_1 + B_2$, $B_5 := (2\sqrt{2}+1)\exp\left\{\frac{B_3 \vee B_4}{2B_4^2}\right\}$, $B_6 := B_4\sqrt{e}$,
$B_7 := \sqrt{2}(1+B_5)B_6$.

Theorem 3.5.21. *Let $\{X_t, t \in [0,T]\}$ be the solution of equation (3.5.1), the conditions of Theorem 3.5.11 hold and the function $f \in C^\beta[0,T]$ with $H + \beta > 1/2$. Then for any $\varepsilon > 0$ and $0 \leq t < t' \leq T$*

$$P\{|X_{t'} - X_t| \geq \varepsilon\} \leq B_5 \exp\left\{-\frac{\varepsilon^2}{2B_6^2(t'-t)^{2H}}\right\} \quad (3.5.45)$$

and

$$\|X_{t'} - X_t\|_U \leq B_7(t'-t)^H. \quad (3.5.46)$$

3.5 SDE with the Additive Wiener Integral w.r.t. Fractional Noise 283

Proof. Inequality (3.5.46) follows from (3.5.45) and Theorem 3.5.19. So we prove only (3.5.45). It follows from inequalities (3.5.37) and (3.5.40) that

$$E|X_{t'} - X_t|^r \leq \left(\sqrt{2}B_1^r r^{r/2} + 2\sqrt{2}B_2^r r^{r/2} + B_3^r\right)(t'-t)^{rH}.$$

So, for any $\varepsilon > 0$

$$P\{|X_{t'} - X_t| \geq \varepsilon\} \leq \left(\left(\sqrt{2}\left(\tfrac{B_1}{\varepsilon}\right)^r + 2\sqrt{2}\left(\tfrac{B_2}{\varepsilon}\right)^r\right)r^{r/2} + \left(\tfrac{B_3}{\varepsilon}\right)^r\right)(t'-t)^{rH}$$
$$\leq \left(2\sqrt{2}\left(\tfrac{B_4}{\varepsilon}\right)^r r^{r/2} + \left(\tfrac{B_3}{\varepsilon}\right)^r\right)(t'-t)^{rH}.$$

Now we substitute $r = \frac{1}{e}\left(\frac{\varepsilon}{(t'-t)^H B_4}\right)^2$ and obtain for $r \geq 1$, i.e. for $\varepsilon > (t'-t)^H B_6$, that for $q(\varepsilon) := \frac{\varepsilon^2}{2(t'-t)^{2H} B_6^2}$

$$P\{|X_{t'} - X_t| \geq \varepsilon\} \leq 2\sqrt{2}\exp\{-q\} + \exp\left\{\ln\left(\tfrac{B_3}{e}(t'-t)^H\right) \cdot q\right\}.$$

Also, let $\ln\left(\frac{B_3}{e}(t'-t)^H\right) \leq -\frac{1}{2}$, i.e. $\varepsilon \geq \sqrt{e}(t'-t)^H B_3$.
Then for $\varepsilon \geq \varepsilon_0$, where $\varepsilon_0 := (B_3 \vee B_4)\sqrt{e}(t'-t)^H$ we have an inequality

$$P\{|X_{t'} - X_t| \geq \varepsilon\} \leq (2\sqrt{2}+1)\exp\{-q(\varepsilon)\} \leq B_5 \exp\{-q(\varepsilon)\}.$$

If $0 < \varepsilon < \varepsilon_0$, then

$$P\{|X_{t'} - X_t| \geq \varepsilon\} \leq (2\sqrt{2}+1)\exp\{q(\varepsilon_0)\}\exp\{-q(\varepsilon)\} = B_5\exp\{-q(\varepsilon)\}.$$

□

Corollary 3.5.22. *Let $\{X_t, t \in [0,T]\}$ be a solution of equation (3.5.1) for which the conditions of Theorem 3.5.11 hold and the function $f \in C^\beta[0,T]$ with $H + \beta > 1/2$. Then for any $\lambda \in \mathbb{R}$*

$$E\exp\{\lambda|X_{t'} - X_t|\} \leq 2\exp\left\{\frac{\lambda^2}{4}B_7^2(t'-t)^{2H}\right\}.$$

This statement follows directly from (3.5.46) and the following lemma, which is a partial case of Lemma 2.3.4 (BK00).

Lemma 3.5.23. *If the random variable ξ belongs to the space $L_U(\Omega)$, where $U(x) = = \exp\{x^2\} - 1$, then for any $\lambda \in \mathbb{R}$*

$$E\exp\{\lambda|\xi|\} \leq 2\exp\left\{\frac{\lambda^2\|\xi\|_U^2}{4}\right\}.$$

3.5.8 The Distribution of the Supremum of the Process X on $[0,T]$

First we present some facts from the theory of stochastic processes that belong to the Orlicz spaces.

Let $\mathbb{T}$ be some infinite set of parameters, $Y = \{Y_t, t \in \mathbb{T}\}$ be some real-valued process from the space $L_U(\Omega)$, where $U(x) = \exp\{x^2\} - 1$, $\sup_{t \in \mathbb{T}} \|Y_t\|_U < \infty$, $\rho_Y(t,s) = \|Y_t - Y_s\|_U$ be a semi-metric on $\mathbb{T}$.

Let the space $(\mathbb{T}, \rho_Y)$ be separable and the process Y_t be a separable process on $(\mathbb{T}, \rho_Y)$. Also, let $\mathcal{N}(\varepsilon) = \mathcal{N}(\mathbb{T}, \varepsilon)$ be the metric capacity of $(\mathbb{T}, \rho_Y)$, i.e. the minimal number of closed balls of radius ε that cover $(\mathbb{T}, \rho_Y)$. Note that $\mathcal{N}(\varepsilon) \to \infty$ as $\varepsilon \to 0$. (See also the beginning of Section 1.10, where similar questions are discussed for Gaussian processes.)

The next theorem is a partial case of Theorem 3.3.4 (BK00).

Theorem 3.5.24. *Let the following assumption holds:*

$$\int_0^{\varepsilon_0} (\ln(1 + \mathcal{N}(\varepsilon)))^{1/2} d\varepsilon < \infty,$$

where $\varepsilon_0 := \sup_{t,s \in \mathbb{T}} \rho_Y(t,s)$. *Then the random variable* $\sup_{t \in \mathbb{T}} |Y_t|$ *belongs to the space* $L_U(\Omega)$ *and*

$$\|\sup_{t \in \mathbb{T}} |Y_t|\|_U \leq K := \inf_{t \in \mathbb{T}} \|Y_t\|_U + \frac{e^2}{\theta(1-\theta)} \int_0^{\theta \varepsilon_0} (\ln(1 + \mathcal{N}(\varepsilon)))^{1/2} d\varepsilon < \infty, \quad (3.5.47)$$

where $0 < \theta < 1$ *and* $\mathcal{N}(\theta \varepsilon_0) > e^2 - 1$.

Remark 3.5.25. The statement of the theorem remains true if we replace $\mathcal{N}(\varepsilon)$ by any function $\mathcal{N}_1(\varepsilon) \geq \mathcal{N}(\varepsilon)$.

Remark 3.5.26. Under the assumption of Theorem 3.5.24 for any $\varepsilon > 0$ we have that

$$P\{\sup_{t \in \mathbb{T}} |Y_t| \geq \varepsilon\} \leq 2 \exp\left\{-\frac{\varepsilon^2}{K^2}\right\}, \quad (3.5.48)$$

where K was defined in (3.5.47).

Inequality (3.5.48) is implied by the following one: if $\xi \in L_U(\Omega)$, then for any $\varepsilon > 0$

$$P\{|\xi| \geq \varepsilon\} \leq 2 \exp\left\{-\frac{\varepsilon^2}{\|\xi\|_U}\right\}. \quad (3.5.49)$$

In turn, inequality (3.5.49) is a partial case of Theorem 3.3.4 (BK00).

Theorem 3.5.27. *Let* $\{Y_t, t \in \mathbb{T} = [a,b]\}$ *be the separable process from the space* $L_U(\Omega)$, *and let there exist* $\sigma = \sigma(h) : [0, b-a] \to \mathbb{R}_+$, *increasing and continuous in* h, *and such that* $\sigma(0) = 0$. *Also, let*

$$\sup_{|t-s| \leq h} \|Y_t - Y_s\|_U \leq \sigma(h), \quad (3.5.50)$$

3.5 SDE with the Additive Wiener Integral w.r.t. Fractional Noise

and
$$\int_0^{\widehat{\varepsilon}_0} \left(\ln\left(1 + \frac{3(b-a)}{2\sigma^{(-1)}(u)}\right)\right)^{1/2} du < \infty,$$

where $\sigma^{(-1)}(u)$ is the inverse function to $\sigma(u)$, and $\widehat{\varepsilon}_0 = \sigma(b-a)$.
Then $\sup\limits_{t\in[a,b]} |Y_t| \in L_U(\Omega)$ and the following estimate holds:

$$\left\|\sup_{t\in[a,b]} |Y_t|\right\|_U \leq K_1 := \inf_{t\in\mathbb{T}} \|Y_t\|_U$$

$$+ \frac{e^2}{\theta(1-\theta)} \int_0^{\theta\widehat{\varepsilon}_0} \left(\ln\left(1 + \frac{3}{2}\frac{b-a}{\sigma^{(-1)}(u)}\right)\right)^{1/2} du. \qquad (3.5.51)$$

Here θ is any number from the interval

$$\left(0, 1 \wedge \frac{\sigma\left(\frac{3(b-a)}{2(e^2-1)}\right)}{\sigma(b-a)}\right). \qquad (3.5.52)$$

Moreover, for any $\varepsilon > 0$, we have the estimate

$$P\{\sup_{t\in[a,b]} |Y_t| \geq \varepsilon\} \leq 2\exp\left\{-\frac{\varepsilon^2}{K_1^2}\right\}. \qquad (3.5.53)$$

Proof. The claim follows from Theorem 3.5.24 with $\mathbb{T} = [a,b]$. Indeed, according to (3.5.50), the process Y is separable in the space $([a,b], \rho_Y)$, where $\rho_Y(t,s) = \|Y_t - Y_s\|_U$. It is easy to see that $\mathcal{N}(u) \leq \frac{b-a}{2\sigma^{(-1)}(u)} + 1$, and for $0 < u \leq \widehat{\varepsilon}_0$, i.e. for $\frac{b-a}{\sigma^{(-1)}(u)} \geq 1$, we have that $\mathcal{N}(u) \leq \frac{3}{2}\frac{b-a}{\sigma^{(-1)}(u)}$. Therefore

$$\int_0^{\theta\widehat{\varepsilon}_0} (\ln(1 + \mathcal{N}(u)))^{1/2} du \leq \int_0^{\theta\widehat{\varepsilon}_0} \left(\ln\left(1 + \frac{3}{2}\frac{b-a}{\sigma^{(-1)}(u)}\right)\right)^{1/2} du.$$

The inequality $\mathcal{N}(\theta\widehat{\varepsilon}_0) > e^2 - 1$ can be reduced, according to Remark 3.5.25, to the inequality $\frac{3(b-a)}{2\sigma^{(-1)}(\theta\widehat{\varepsilon}_0)} > e^2 - 1$, i.e. to (3.5.52). Inequality (3.5.53) follows now from (3.5.48). □

Theorem 3.5.28. *Let the condition of Theorem 3.5.11 hold, $\{X_t, t \in \mathbb{T} = [0,T]\}$ be the solution of equation (3.5.1) and $0 \leq t_1 < t_2 \leq T$. Then the random variable* $\sup\limits_{t_1 \leq t \leq t_2} |X_t| \in L_U(\Omega)$, *and*

$$\left\|\sup_{t_1\leq t\leq t_2} |X_t|\right\|_U \leq (h+1)c_1 + e^2 C_{H,\gamma} \theta^{-\frac{\gamma}{2H}} \frac{(t_2-t_1)^H}{1-\theta} =: L, \qquad (3.5.54)$$

where $0 < \theta < \left(\frac{3}{2(e^2-1)}\right)^H$, $0 < \gamma < 2H$, $C_{H,\gamma} = \frac{\left(\frac{3}{2}\right)^{\frac{\gamma}{2}} H B_\gamma}{\gamma(H-\frac{\gamma}{2})}$.

Moreover, for any $\varepsilon > 0$

$$P\{\sup_{t_1 \leq t \leq t_2} |X_t| \geq \varepsilon\} \leq 2\exp\left\{-\frac{\varepsilon^2}{L^2}\right\}. \quad (3.5.55)$$

Proof. We use Theorem 3.5.27 with $[a,b] = [t_1, t_2]$. The process X_t is continuous with probability 1, hence is separable. It follows from (3.5.46) that $\sigma(h) = B_7 h^H$. It is easy to see that in this case $\widehat{\varepsilon}_0 = \sigma(t_2 - t_1)$ and

$$\begin{aligned} I(\theta\widehat{\varepsilon}_0) &:= \int_0^{\theta\widehat{\varepsilon}_0} \left(\ln\left(1 + \frac{3}{2}\frac{t_2-t_1}{\sigma^{(-1)}(u)}\right)\right)^{1/2} du \\ &= HB_7 \int_0^{\sigma^{(-1)}(\theta\widehat{\varepsilon}_0)} \left(\ln\left(1 + \frac{3}{2}\frac{t_2-t_1}{v}\right)\right)^{1/2} v^{H-1} dv. \end{aligned} \quad (3.5.56)$$

Since for $0 < \gamma \leq 1$ and $x > 0$

$$\ln(1+x) = \frac{1}{\gamma}\ln((1+x)^\gamma) \leq \frac{1}{\gamma}\ln(1+x^\gamma) \leq \frac{x^\gamma}{\gamma},$$

we obtain from (3.5.56) the following estimate for any $0 < \gamma < 2H$:

$$\begin{aligned} I(\theta\widehat{\varepsilon}_0) &\leq \left(\tfrac{3}{2}\right)^{\frac{\gamma}{2}} HB_7 \cdot \frac{1}{\gamma} \int_0^{\sigma^{(-1)}(\theta\widehat{\varepsilon}_0)} v^{H-1-\frac{\gamma}{2}} dv \cdot (t_2-t_1)^{\frac{\gamma}{2}} \\ &= C_{H,\gamma}(t_2-t_1)^{\frac{\gamma}{2}}(\sigma^{(-1)}(\theta\widehat{\varepsilon}_0))^{H-\frac{\gamma}{2}}. \end{aligned}$$

Evidently,

$$\sigma^{(-1)}(\theta\widehat{\varepsilon}_0) = \theta^{\frac{1}{H}}\sigma^{(-1)}(\widehat{\varepsilon}_0) = \theta^{\frac{1}{H}}(t_2-t_1).$$

Therefore

$$I(\theta\widehat{\varepsilon}_0) \leq C_{H,\gamma}\theta^{1-\frac{\gamma}{2H}}(t_2-t_1)^H. \quad (3.5.57)$$

Now the proof follows from (3.5.56)–(3.5.57) and Theorems 3.5.19 and 3.5.27. □

Remark 3.5.29. Estimate (3.5.54) demonstrates that up to constants the estimates for distribution of the supremum of the process X are of the same form as similar estimates for the Gaussian process (see (Fer74), for example).

Corollary 3.5.30. *Let $\{X_t, t \in [0,T]\}$ be a solution of equation (3.5.1) under the conditions of Theorem 3.5.11 and $0 \leq t_1 < t_2 \leq T$. Then for any $p \geq 1$ we have an estimate*

$$\left(\mathsf{E}\left(\sup_{t_1 \leq t \leq t_2}|X_t|\right)^p\right)^{\frac{1}{p}} \leq C_p \cdot L, \quad (3.5.58)$$

where L is defined in (3.5.54) and $C_p = 2^{\frac{1}{p}}\frac{\sqrt{p}}{2}$.

Proof. This statement follows from Theorem 3.5.28. Indeed, it was established in Lemma 2.33 (BK00), that for the random variable $\xi \in L_U(\Omega)$, $U(x) = \exp\{x^2\}-1$ and $p \geq 1$ $(\mathsf{E}|\xi|^p)^{\frac{1}{p}} \leq C_p\|\xi\|_U$. Now (3.5.58) follows from (3.5.54). □

Corollary 3.5.31. *Let $\{X_t, t \in [0,T]\}$ be the solution of equation (3.5.1), $0 \le t_1 < t_2 \le T$. Then for any $\lambda \in \mathbb{R}$*

$$\mathsf{E}\exp\left\{\lambda \sup_{t_1 \le t \le t_2} |X_t|\right\} \le 2\exp\left\{\frac{\lambda^2 L^2}{4}\right\}.$$

This estimate follows from Theorem 3.5.27 and Lemma 3.5.23.

3.5.9 Modulus of Continuity of Solution of Equation Involving Fractional Brownian Motion

Definition 3.5.32. We say that the C-function $U(x)$ (C-function is continuous, even, convex function that increases in $x > 0$ and is zero at the zero point) satisfies the Δ^2-condition if there exist such constants $x_0 > 0$ and $L_0 > 1$, that $U^2(x) \le U(L_0 x)$ for $x \ge x_0$.

Example 3.5.33. The function $U(x) = \exp\{x^2\} - 1$ satisfies Δ^2-condition with $x_0 := 0$ and $L_0 := \sqrt{2}$.

Theorem 3.5.34. *Let $\{Y_t, t \in \mathbb{T}\}$ be a stochastic process from the Orlicz space $L_U(\Omega)$, where the function $U(x)$ satisfies the Δ^2-condition with constants x_0, L_0, and let $Z_0 := x_0 \vee L_0$. Let $\rho_Y(t,s) = \|Y_t - Y_s\|_U$, $t,s \in T$ be a semi-metric generated by Y. Also, let $(\mathbb{T}, \rho_Y)$ be the separable space and the process Y be the separable process in the space $(\mathbb{T}, \rho_Y)$. Put $\varepsilon_0 := \sup_{t,s \in \mathbb{T}} \rho_Y(t,s)$, let $\mathcal{N}(u)$ be the minimal number of closed u-balls covering $(\mathbb{T}, \rho_Y)$, $\mathcal{N}_1(u) \ge \mathcal{N}(u), u > 0$ and let $\mathcal{N}_1(u)$ increase in u. If for any $\varepsilon > 0$*

$$q(\varepsilon) := \int_0^\varepsilon U^{(-1)}(\mathcal{N}_1(u))\, du < \infty, \qquad (3.5.59)$$

then for any $\varepsilon \in (0, \varepsilon_0)$ such that $\mathcal{N}_1(\varepsilon_0) \ge U(Z_0)$, and for any $x \ge Z_0$

$$P\left\{\sup_{0 < \rho_Y(t,s) \le \varepsilon} \frac{|Y_t - Y_s|}{C_0 q(\rho_Y(t,s))} \ge x\right\} \le \frac{3+\sqrt{2}}{U(x)}. \qquad (3.5.60)$$

Moreover, with probability 1

$$\limsup_{\varepsilon \downarrow 0} \frac{\Delta Y_\varepsilon}{C_0 Z_0 q(\varepsilon)} \le 1,$$

where $\Delta Y_\varepsilon = \sup_{0 < \rho_Y(t,s) \le \varepsilon} |Y_t - Y_s|$, $C_0 = 3L_0(5 + 4L_0)$.

Proof. For $\mathcal{N}_1(u) = \mathcal{N}(u)$ the theorem is proved in the book (BK00). If we replace $\mathcal{N}(u)$ for $\mathcal{N}_1(u)$, the proof will not change substantially. □

Corollary 3.5.35. Let $\{Y_t, t \in \mathbb{T} = [a,b]\}$ be the separable stochastic process from the space $L_U(\Omega)$, $U(x) = \exp\{x^2\} - 1$. Let for some $D_0 > 0$ and $0 < \beta \leq 1$

$$\sup_{s,t\in[a,b], |t-s|\leq h} \|Y_t - Y_s\|_U \leq D_0 h^\beta. \tag{3.5.61}$$

Then for any $x > \sqrt{2}$, $0 < \delta \leq \frac{b-a}{2(e^2-2)}$ the inequality holds

$$P\left\{\sup_{\substack{0 < |t-s| \leq \delta, \\ s,t \in [a,b]}} \frac{|Y_t - Y_s|}{D_1 g(D_0|t-s|^\beta)} \geq x\right\} \leq \frac{3+\sqrt{2}}{U(x)}, \tag{3.5.62}$$

where $g(\varepsilon) := \int_0^\varepsilon \left(\ln\left(2 + \frac{(b-a)D_0^{\frac{1}{\beta}}}{2u^{\frac{1}{\beta}}}\right)\right)^{\frac{1}{2}} du$, $D_1 = 3\sqrt{2}\left(5 + 4\sqrt{2}\right)$.

Moreover, with probability 1

$$\limsup_{\delta \downarrow 0} \frac{\widehat{\Delta} Y_\delta}{D_1 \sqrt{2} g(D_0 \cdot \delta^\beta)} \leq 1, \tag{3.5.63}$$

where $\widehat{\Delta} Y_\delta = \sup_{0 < |t-s| \leq \delta} |Y_t - Y_s|$.

Proof. As we have seen from Example 3.5.33, the C-function $U(x) = \exp\{x^2\} - 1$ satisfies the Δ^2-condition with $x_0 = 0$, $L_0 = \sqrt{2}$ (so, $Z_0 = \sqrt{2}$). Moreover, $U^{(-1)}(x) = (\ln(1+x))^{\frac{1}{2}}$, $x > 0$, and $q(\varepsilon) = \int_0^\varepsilon (\ln(1 + \mathcal{N}_1(u)))^{\frac{1}{2}} du$. Since in this case $\mathcal{N}(u) \leq \frac{D_0^{\frac{1}{\beta}}(b-a)}{2u^{\frac{1}{\beta}}} + 1$, we can put $\mathcal{N}_1(u) = \frac{D_0^{\frac{1}{\beta}}(b-a)}{2u^{\frac{1}{\beta}}} + 1$. It means that $q(\varepsilon) = \int_0^\varepsilon \left(\ln\left(2 + \frac{(b-a)D_0^{\frac{1}{\beta}}}{2u^{\frac{1}{\beta}}}\right)\right)^{\frac{1}{2}} du = g(\varepsilon)$. Therefore,

$$\sup_{0<|t-s|\leq \delta} \frac{|Y_t - Y_s|}{D_1 g(D_0|t-s|^\beta)} \leq \sup_{0<\rho_Y(t,s)\leq D_0 \delta^\gamma} \frac{|Y_t - Y_s|}{D_1 g(\rho_Y(t,s))}.$$

Now (3.5.62) follows from (3.5.60), since separability of Y and (3.5.61) imply its separability in the space $(\mathbb{T}, \rho_Y)$ with $\mathbb{T} = [a,b]$. Inequality (3.5.63) is proved similarly. The restriction on ε follows from the inequality $\mathcal{N}_1(\varepsilon) \geq U(Z_0) = e^2 - 1$. □

The next result follows from Corollary 3.5.35.

Theorem 3.5.36. Let $\{X_t, t \in [0,T]\}$ be the solution of equation (3.5.1) under the condition of Theorem 3.5.21, $f(y) := \int_0^y \left(\ln\left(2 + \frac{1}{2}v^{-\frac{1}{H}}\right)\right)^{\frac{1}{2}} dv$, $y > 0$. Then for any $x \geq \sqrt{2}$, $0 \leq t_1 < t_2 \leq T$, $0 < \delta \leq \frac{t_2 - t_1}{2(e^2 - 2)}$

3.5 SDE with the Additive Wiener Integral w.r.t. Fractional Noise

$$P\left\{\sup_{\substack{0 < |t-s| \le \delta \\ t, s \in [t_1, t_2]}} \frac{|X_t - X_s|}{B_7 D_1 (t_2 - t_1)^H f\left(\frac{|t-s|^H}{(t_2-t_1)^H}\right)} \ge x\right\} \le \frac{3 + \sqrt{2}}{U(x)}. \quad (3.5.64)$$

Moreover, with probability 1

$$\limsup_{\delta \downarrow 0} \frac{\sup_{\substack{|t-s| \le \delta \\ t, s \in [t_1, t_2]}} |X_t - X_s|}{B_7 D_1 (t_2 - t_1)^H f\left(\frac{\delta^H}{(t_2-t_1)^H}\right)} \le 1. \quad (3.5.65)$$

Proof. It follows from Theorems 3.5.19, 3.5.21 and Corollary 3.5.35. Indeed, in this case $\mathbb{T} = [t_1, t_2]$, $\beta = H$, $D_0 = B_7$,
$g(D_0|t-s|^H) = f\left(\frac{|t-s|^H}{(t_2-t_1)^H}\right) D_0 (t_2 - t_1)^H$. □

Definition 3.5.37. Let $(\mathbb{T}, \rho)$ be a metric space, $\Theta = \{\Theta(u), u \ge 0\}$ be a modulus of continuity (see Definition 1.16.1). The family of functions $\{y_t, t \in \mathbb{T}\}$ such that

$$\sup_{\substack{t, s \in \mathbb{T} \\ t \ne s}} \frac{|y_t - y_s|}{\Theta(\rho(t,s))} < \infty$$

is called the Lipschitz space $\Lambda_\Theta(\mathbb{T}, \rho)$.

(Compare with the Definition 1.16.3; note that now our process is not Gaussian.)

Remark 3.5.38. Theorem 3.5.36 states that the solution of equation (3.5.1) under the conditions of this theorem with probability 1 belongs to the space $\Lambda_\Theta(\mathbb{T}, \rho)$, where $\mathbb{T} = [t_1, t_2]$, $\rho(t, s) = |t - s|$, $\Theta(x) = f\left(\frac{x^H}{(t_2-t_1)^H}\right)$, and inequality (3.5.64) gives the estimates of the distribution of the norm of X_t in this space.

Corollary 3.5.39. *Let $\{X_t, t \in [0, T]\}$ be the solution of equation (3.5.1) under the conditions of Theorem 3.5.36. Then for any $0 < \gamma < 2H$ with probability 1 the trajectories of X_t belong to the space $\Lambda_\Theta(\mathbb{T}, \rho)$, where $\mathbb{T} = [t_1, t_2] \subset [0, T]$,*

$$\rho(s, u) = |s - u|, \Theta(x) = C_{H,\gamma,1} x^{H - \frac{\gamma}{2}},$$

$C_{H,\gamma,1} = B_7 D_1 C_{H,\gamma} (t_2 - t_1)^{\frac{\gamma}{2}}, C_{H,\gamma} = \gamma^{-1/2}(2H - \gamma + 2^{1+\gamma/2} H)(2H - \gamma)^{-1}.$
Moreover, for $x > \sqrt{2}$ and $\delta < (t_2 - t_1) \wedge \delta_0$

$$P\left\{\sup_{\substack{0 < |t-u| < \delta, \\ t, u \in [t_1, t_2]}} \frac{|X_t - X_u|}{C_\gamma |t - u|^{H - \frac{\gamma}{2}}} > x\right\} \le \frac{3 + \sqrt{2}}{U(x)}. \quad (3.5.66)$$

Proof. From the inequality $\ln(1+x) \leq \frac{1}{\gamma}x^\gamma$, $x > 0$, $0 < \gamma \leq 1$ it is easy to obtain for $\delta < (t_2 - t_1)$

$$f\left(\delta^H (t_2 - t_1)^{-H}\right) \leq \int_0^{\left(\frac{\delta}{t_2-t_1}\right)^H} \left(\ln\left(\frac{1}{2}v^{-\frac{1}{H}} + 2\right)\right)^{\frac{1}{2}} dv$$

$$\leq \int_0^{\left(\frac{\delta}{t_2-t_1}\right)^H} \frac{1 + 2^{\gamma/2} v^{-\frac{\gamma}{2H}}}{\gamma^{\frac{1}{2}}} dv \leq \gamma^{-\frac{1}{2}} \left(\frac{\delta}{t_2 - t_1}\right)^H$$

$$+ 2^{\gamma/2} \frac{\gamma^{-\frac{1}{2}}}{1 - \frac{\gamma}{2H}} \left(\frac{\delta}{t_2 - t_1}\right)^{H-\frac{\gamma}{2}} \leq C_{H,\gamma} \left(\frac{\delta}{t_2 - t_1}\right)^{H-\frac{\gamma}{2}}, \quad (3.5.67)$$

and the proof immediately follows from (3.5.64) and (3.5.65). □

4
Filtering in Systems with Fractional Brownian Noise

4.1 Optimal Filtering of a Mixed Brownian–Fractional-Brownian Model with Fractional Brownian Observation Noise

Consider the real-valued signal process X_t and the observation process Y_t defined by the following system of equations:

$$\begin{cases} X_t = \eta + \int_0^t a(s, X_s)ds + \sum_{i=1}^N \int_0^t b_i(s, X_s)\, dW_s^i \\ \quad + \sum_{j=1}^M \int_0^t c_j(s) dB_s^{H_j}, \qquad\qquad\qquad t \in [0,T], \\ Y_t = \xi + \int_0^t A(s, X_s)\, ds + \int_0^t C(s) dB_s^H, \end{cases} \qquad (4.1.1)$$

where $\{W^i, 1 \leq i \leq N\}$ are independent Wiener processes, $\{B^{H_j}, 1 \leq j \leq M\}$ are independent fractional Brownian motions with Hurst indices $H_j \in (\frac{1}{2}, 1)$, B^H is an fBm with Hurst index $H \in (\frac{1}{2}, 1)$, all the processes are mutually independent, random initial conditions (η, ξ) are independent of each other and independent of all the processes (W^i, B^{H_j}, B^H), the functions $a, b, A : [0,T] \times \mathbb{R} \to \mathbb{R}$, $c_j, C : [0,T] \to \mathbb{R}$ are measurable in their variables and satisfy the conditions that are sufficient for the existence of pathwise integrals w.r.t. corresponding fBms.

The problem is to construct the optimal filter of the signal X according to the observation Y, which will be expressed in terms of the conditional expectation $\pi_t(X) := E(X_t/\mathcal{F}_t^Y)$, where $\mathcal{F}_t^Y := \sigma\{Y_s, 0 \leq s \leq t\}$.

Note that the partial cases of this problem were considered in (KLeBR99), (KLeBR00), where $N = 1, c_j = 0$ (see also (KKA98b), (LeB98)), and in (Pos05), where $b_i = 0$. Suppose that the following condition holds:

  (i) the function $C \in L_2^H(\mathbb{R})$, does not vanish and $1/C(s)$ is bounded on $[0,T]$, $c_j \in L_2^H(\mathbb{R})$.

4 Filtering in Systems with Fractional Brownian Noise

Here we use the approach to the solution of optimal filtering problem developed in (KLeBR00) but simplify it and modify it in accordance with our model (4.1.1).

Introduce the following processes, connected with fBm B^H:

$$Z_t^* := \int_0^t l_H(t,s)C^{-1}(s)dY_s = \int_0^t l_H(t,s)D(s,X_s)ds$$

$$+ \int_0^t l_H(t,s)dB_s^H = J_t(D) + M_t^H, \qquad (4.1.2)$$

where $J_t(D) = \int_0^t l_H(t,s)D(s,X_s)ds$, M_t^H is the Molchan martingale, introduced in (1.8.5), $D(s,X_s) = A(s,X_s)/C(s)$. Recall that

$$l_H(t,s) = C_H^{(5)} s^{-\alpha}(t-s)^{-\alpha} \mathbf{1}_{\{0<s<t\}}, \; \alpha = H - \frac{1}{2}.$$

Suppose that the functional D satisfies the condition

(ii) $\int_0^t s^{-\alpha}(t-s)^{-\alpha}|D(s,X_s)|ds < \infty$ P-a.s., so the integral $J_t(D)$ exists.

Moreover, suppose that

(iii) $D(s,x_s)s^{-\alpha} \in I_{0+}^\alpha(L_1[0,T])$, i.e. there exists the fractional derivative

$$\frac{d}{dt}\int_0^t (t-s)^{-\alpha}s^{-\alpha}D(s,x_s)ds = \Gamma(1-\alpha)D_{0+}^\alpha(D(u,x_u)u^{-\alpha})(t)$$

$$= \Gamma(1-\alpha)t^{-2\alpha}\left(D(t,x_t) + \alpha\int_0^t \frac{D(t,x_t)t^\alpha - D(u,x_u)t^{2\alpha}u^{-\alpha}}{(t-u)^{1+\alpha}}du\right)$$

$$=: \Gamma(1-\alpha)t^{-2\alpha}E(t,x) \in L_1[0,T],$$

where x_t is any Hölder function from $C^{1/2-}[0,T]$; for example, sufficient condition is

(iii') $D(t,x_t)t^{-\alpha} \in C^{\alpha+\varepsilon}[0,T]$ for some $\varepsilon > 0$.

Then the integral

$$J_t(D) = \Gamma(1-\alpha)C_H^{(5)} \int_0^t E(s,X)s^{-2\alpha}ds$$

has a.s. bounded variation, so, it follows from (4.1.2) that Z_t^* is the semimartingale w.r.t. the σ-field $\mathcal{F}_t := \sigma\{\eta, \xi, X_s, W_s^i, 1 \leq i \leq N, B_s^{H_j}, 1 \leq j \leq M, Y_s, 0 \leq s \leq t\}$ and admits the representation

$$Z_t^* = M_t^H + C_H \int_0^t E(s,X)s^{-2\alpha}ds, \; C_H = \Gamma(1-\alpha)C_H^{(5)}, \qquad (4.1.3)$$

and, in addition, Z_t^* is $\mathcal{F}_t^Y$-adapted.

Further, let

4.1 Optimal Filtering of a Mixed Model in Fractional Brownian Observation

$$\nu_t := Z_t^* - C_H \int_0^t \pi_s(E(s,X)) s^{-2\alpha} ds, \tag{4.1.4}$$

where $\pi_s(Q) := E(Q/\mathcal{F}_s^Y)$, It follows from (4.1.4) and (4.1.3) that

$$\nu_t = C_H \int_0^t \Big(E(s,X) - \pi_s(E(s,X))\Big) s^{-2\alpha} ds + M_t^H.$$

Moreover, for $0 \le s < t \le T$

$$E(\nu_t - \nu_s/\mathcal{F}_s^Y) = E(M_t^H - M_s^H/\mathcal{F}_s/\mathcal{F}_s^Y) = 0,$$

and the integral $C_H \int_0^t \Big(E(s,X) - \pi_s(E(s,X))\Big) s^{-2\alpha} ds$ is continuous and has a bounded variation. Hence $\langle \nu \rangle_t = t^{1-2\alpha}$, where $\langle \nu \rangle_t$ is calculated w.r.t. the filtration $\{\mathcal{F}_t^Y, 0 \le t \le T\}$. So, ν is a continuous Gaussian martingale w.r.t. this filtration. (Evidently, ν is adapted to this filtration since Z_t^* is adapted.)

Further we need the following evident result.

Lemma 4.1.1. *Any square-integrable martingale* $\{M_t, \mathcal{F}_t^Y, t \in [0,T]\}$ *with* $M_0 = 0$ *admits the representation*

$$M_t = \int_0^t \varphi_s d\nu_s, t \in [0, T],$$

where the process φ_t *is* $\mathcal{F}_t^Y$*-adapted and* $E \int_0^t \varphi_s^2 s^{-2\alpha} ds < \infty$.

The next statement is proved similarly to Theorem 18.11 (Ell82); see also Theorem 3 (KLeBR00). For any integrable process X, let $\widehat{X}_t := E(X_t/\mathcal{F}_t^Y)$.

Theorem 4.1.2. *Let* $\{S_t, \mathcal{F}_t, t \in [0,T]\}$ *be the semimartingale of the form*

$$S_t = S_0 + \int_0^t \alpha_s ds + m_t, \quad t \in [0, T],$$

where $ES_0^2 < \infty$, $E \int_0^T \alpha_s^2 ds < \infty$ *and* $\{m_t, \mathcal{F}_t, t \in [0,T]\}$ *be a square integrable martingale with mutual bracket* $\langle m, M^H \rangle_t = \int_0^t \lambda_s s^{-2\alpha} ds$.

Then the process $\{\widehat{S}_t, t \in [0,T]\}$ *satisfies the following stochastic differential equation:*

$$\widehat{S}_t = \widehat{S}_0 + \int_0^t \widehat{\alpha}_s ds + \int_0^t \Big(\widehat{\lambda}_s + C_H \Big(\widehat{S_s E(s,X)} - \widehat{S}_s \pi_s(E(s,X))\Big)\Big) d\nu_s, \quad t \in [0,T].$$

Proof. If we define the $\mathcal{F}_t^Y$-adapted process

$$M_t := \widehat{S}_t - \widehat{S}_0 - \int_0^t \widehat{\alpha}_s ds, \tag{4.1.5}$$

then for $s \leq t$ $E(M_t - M_s/\mathcal{F}_s^Y) = E(\widehat{S}_t - \widehat{S}_s/\mathcal{F}_s^Y) - \int_s^t \widehat{\alpha}_u du$
$= E(\int_s^t \alpha_u du/\mathcal{F}_s^Y) - \int_s^t \widehat{\alpha}_u du = 0$.

Therefore, M_t is a $\mathcal{F}_t^Y$-square-integrable martingale. By Lemma 4.1.1, M_t admits the representation

$$M_t = \int_0^t \varphi_s d\nu_s, t \in [0, T], \tag{4.1.6}$$

whence

$$\widehat{S}_t = \widehat{S}_0 + \int_0^t \widehat{\alpha}_s ds + \int_0^t \varphi_s d\nu_s.$$

Now we use the same reasonings as in Theorem 18.11 (Ell82). On the one hand, with the help of (4.1.3) the product $S_t Z_t^*$ can be decomposed by the Itô formula as

$$S_t Z_t^* = \int_0^t S_s(dM_s^H + C_H E(s, X)s^{-2\alpha} ds) + \int_0^t Z_s^*(\alpha_s ds + dm_s)$$

$$+ \int_0^t \lambda_s s^{-2\alpha} ds,$$

whence

$$\widehat{S_t Z_t^*} = \widehat{S_t Z_t^*} = \int_0^t (C_H S_s \widehat{E(s, X)} s^{-2\alpha}$$

$$+ \widehat{\alpha}_s Z_s^* + \widehat{\lambda}_s s^{-2\alpha}) ds + N_t^1, \tag{4.1.7}$$

where N_t^1 is continuous $\mathcal{F}_t^Y$-martingale.

On the other hand, using (4.1.4) and (4.1.5)–(4.1.6) we obtain the following decomposition for $\widehat{S}_t Z_t^*$:

$$\widehat{S}_t Z_t^* = \int_0^t (Z_s^* \widehat{\alpha}_s + C_H \widehat{S}_s \pi_s(E(s, X))s^{-2\alpha} + \varphi_s s^{-2\alpha}) ds + N_t^2, \tag{4.1.8}$$

where N_t^2 is a continuous $\mathcal{F}_t^Y$- martingale. It follows from (4.1.7)–(4.1.8) that $N^1 = N^2$ and $\widehat{\lambda}_s + C_H S_s \widehat{E(s, X)} = C_H \widehat{S}_s \pi_s(E(s, X)) + \varphi_s$, whence the proof follows. □

Now we can establish the form of the optimal filter in our model. In this order we rewrite all the integrals $\int_0^t c_j(s) dB_s^{H_j}, 1 \leq j \leq M$, by using Theorem 1.8.3, in the form

$$\int_0^t c_j(s) dB_s^{H_j} = \int_0^t K_{H_j}^{c_j}(t, s) dM_s^{H_j},$$

where

$$K_H^C(t, s) = C_H^{(7)} \int_s^t C(u) u^\alpha (u - s)^{\alpha - 1} du.$$

Further, consider for any $t \in [0, T]$ the process

$$X_u^t := \eta + \int_0^u a(s, X_s)ds + \sum_{i=1}^N \int_0^u b_i(s, X_s)dW_s^i$$

$$+ \sum_{j=1}^M \int_0^u K_{H_j}^{c_j}(t, s)dM_s^{H_j}, 0 \leq u \leq t, \quad (4.1.9)$$

so that $X_t^t = X_t$ from (4.1.1).

Evidently, $\{X_u^t, 0 \leq u \leq t\}$ is the semimartingale with respect to the filtration $\{\mathcal{F}_t, 0 \leq t \leq T\}$. Therefore we can use Theorem 4.1.2 to establish the following result.

Theorem 4.1.3. Let $\phi \in C_b^2(\mathbb{R}), \pi_t(\phi) = E(\phi(X_t)/\mathcal{F}_t^Y)$,

$$\mathcal{L}_s^t \phi_y(x) = a(s,y)\phi'(x) + \frac{1}{2}\sum_{i=1}^N b_i^2(s,y)\phi''(x) + \sum_{j=1}^M \beta_j (K_{H_j}^{c_j}(t,s))^2 s^{-2\alpha_j}\phi''(x),$$

$0 \leq s \leq t \leq T$, $\beta_j = 1 - 2\alpha_j$, and the conditions (i)–(iii) hold.

Then the equation for the optimal filter π_t has the form:

$$\pi_t(\phi) = \pi_0(\phi) + \int_0^t \pi_s(\mathcal{L}_s^t \phi_{X_s}(X_s^t))ds + C_H \int_0^t (\pi_s^t(\phi E) - \pi_s^t(\phi)\pi_s(E))d\nu_s,$$

where $\pi_s^t(\phi E) = E(\phi(X_s^t)E(s,X)/\mathcal{F}_s^Y)$, $\pi_s^t(\phi) = E(\phi(X_s^t)/\mathcal{F}_s^Y)$, $\pi_s(E) = \pi_s(E(s,X))$.

Proof. It follows from (4.1.9) that X_t is a "boundary" value of the semimartingale $X_u^t, 0 \leq u \leq t$.

Since $\phi(X_u^t) = \phi(\eta) + \int_0^u (\phi'(X_s^t)a(s,X_s) + \frac{1}{2}\sum_{i=1}^N \phi''(X_s^t)b_i^2(s,X_s)$

$+\sum_{j=1}^M (1-2\alpha_j)(K_{H_j}^{c_j}(t,s))^2 s^{-2\alpha_j}\phi''(X_s^t))ds + \sum_{i=1}^N \int_0^u \phi'(X_s^t)b_i(s,X_s)dW_s^i$

$+\sum_{j=1}^M \int_0^u \phi'(X_s^t)K_{H_j}^{c_j}(t,s)dM_s^{H_j}$ and $\widehat{\lambda}_s = 0$ in our case, the proof immediately follows from Theorem 4.1.2. $\square$

4.2 Optimal Filtering in Conditionally Gaussian Linear Systems with Mixed Signal and Fractional Brownian Observation Noise

Now we suppose that the real-valued signal process $\{X_t, t \in [0,T]\}$ and the observation process $\{Y_t, t \in [0,T]\}$ satisfy the following system of equations:

4 Filtering in Systems with Fractional Brownian Noise

$$\begin{cases} X_t = \eta + \int_0^t a(s)X_s ds + \sum_{i=1}^N \int_0^t b_i(s)dW_s^i \\ \quad + \sum_{j=1}^M \int_0^t c_j(s)dB_s^{H_j}, \\ Y_t = \xi + \int_0^t A(s)X_s ds + \int_0^t C(s)dB_s^H, \quad t \in [0,T] \end{cases} \quad (4.2.1)$$

where $\{W^i, 1 \leq i \leq N\}$ are independent Wiener processes, $\{B^{H_j}, 1 \leq j \leq M\}$ are independent fBms with Hurst indices $H_j \in (\frac{1}{2}, 1)$, B^H is an fBm with Hurst index $H \in (\frac{1}{2}, 1)$, W^i, B^{H_j}, B^H are mutually independent, random initial conditions (η, ξ) are independent of all the processes (W^i, B^{H_j}, B^H), a, b_i, c_j, A, $C : [0,T] \to \mathbb{R}$, are bounded measurable functions which satisfy the conditions sufficient for the existence of Lebesgue integrals, corresponding pathwise integrals w.r.t. fBms and Itô integrals w.r.t. Wiener processes.

As before we suppose that $C(s)$ does not vanish and $1/C(s)$ is a bounded function on $[0,T]$. Suppose also that the conditional distribution $\pi_0 := E(\eta/\xi)$ is Gaussian. Under these assumptions the mutual distribution of the pair (X,Y) is well-defined, and this pair is conditionally Gaussian pair, i.e., for any $0 \leq t_1 \leq t_2 \leq \cdots \leq t_n \leq t \leq T$ the joint conditional distribution of $(X_{t_1}, \ldots, X_{t_n})$ given $\mathcal{F}_t^Y$ is Gaussian. The same is obviously true for the system $((X, E(\cdot, X)), Y)$. Then for any $t \in [0,T]$ the optimal filter π_t has a Gaussian distribution, which can be completely characterized by its conditional mean value $\widehat{X}_t := E(X_t/\mathcal{F}_t^Y)$ and conditional variance $\widehat{\sigma}_t^2 := E((X_t - \widehat{X}_t)^2/\mathcal{F}_t^Y), t \in [0,T]$. Denote $\mathcal{D}(s) := A(s)/C(s)$ and note that now $E(s, x_s) = \mathcal{D}(s)x_s$ for any $x \in C^{1/2-}[0,T]$. Suppose that the following version of the condition (iii) is now fulfilled:

(iii") $\mathcal{D}(s)x_s s^{-\alpha} \in I_{0+}^\alpha(L_1[0,T])$ for any $x \in C^{1/2-}[0,T]$. Evidently, the set of such $\mathcal{D}(s)$ is nonempty.

Consider for any $t \in [0,T]$ the semimartingale that is similar to (4.1.9):

$$X_u^t := \eta + \int_0^u a(s)X_s ds + \sum_{i=1}^N \int_0^u b_i(s)dW_s^i + \sum_{j=1}^M \int_0^u K_{H_j}^{c_j}(t,s)dM_s^{H_j},$$

$0 \leq u \leq t$, so that $X_t^t = X_t$ from (4.2.1). Denote $\widehat{\sigma}_0^2 := E(\eta^2/\xi) - (\pi_0)^2$.

Lemma 4.2.1. *For all $t \in [0,T]$*

$$\widehat{X}_t = \pi_0 + \int_0^u a(s)\widehat{X}_s ds + C_H \int_0^u \mathcal{D}(s)\widehat{\sigma}_s^2 d\nu_s, \quad (4.2.2)$$

$$\widehat{\sigma}_t^2 = \widehat{\sigma}_0^2 + 2\int_0^t a(s)(\widehat{\sigma}_s^t)^2 ds + \sum_{i=1}^N \int_0^t b_i^2(s)ds$$

4.2 Optimal Filtering in Conditionally Gaussian Linear Systems 297

$$+ \sum_{j=1}^{M}(1-2\alpha_j)\int_0^t (K_{H_j}^{c_j}(t,s))^2 s^{-2\alpha}ds - (1-2\alpha)C_H^2 \int_0^t \mathcal{D}(s)\widehat{\sigma}_s^2 s^{-2\alpha}ds$$

$$+ C_H \int_0^t \mathcal{D}(s)\Big(\widehat{(X_s^t)^2 X_s} - \widehat{(X_s^t)^2}\widehat{X}_s - 2\widehat{\sigma}_s^2(\widehat{X}_s^t)\Big)d\nu_s, \qquad (4.2.3)$$

where $(\widehat{\sigma}_s^t)^2 := E\Big((X_s^t - \widehat{X}_s^t)(X_s - \widehat{X}_s)/\mathcal{F}_s^Y\Big)$.

Proof. By using Theorem 4.1.2 and independence of $\{W^i, M^{H_j}\}$ of M^H we obtain that

$$\widehat{X}_u^t := E(X_u^t/\mathcal{F}_u^Y) = \pi_0 + \int_0^u a(s)\widehat{X}_s ds + C_H \int_0^u (\widehat{X_s E(s,X)} - \widehat{X}_s\pi_s(E))d\nu_s$$

$$= \pi_0 + \int_0^u a(s)\widehat{X}_s ds + C_H \int_0^u \mathcal{D}(s)\widehat{\sigma}_s^2 d\nu_s,$$

whence (4.2.2) follows. Now we apply the Itô formula to the semimartingale $\{\widehat{X}_u^t, 0 \le u \le t\}$:

$$(\widehat{X}_u^t)^2 = (\pi_0)^2 + \int_0^u 2a(s)(\widehat{X}_s^t)\widehat{X}_s ds + 2C_H \int_0^u \mathcal{D}(s)\widehat{\sigma}_s^2 \widehat{X}_s^t d\nu_s$$

$$+ C_H^2(1-2\alpha)\int_0^u \mathcal{D}(s)\widehat{\sigma}_s^2 s^{-2\alpha}ds, t \in [0,T]. \qquad (4.2.4)$$

On the other hand,

$$(X_u^t)^2 = \eta^2 + \int_0^u 2a(s)(X_s^t)X_s ds$$

$$+ \sum_{i=0}^{N}\int_0^u b_i^2(s)ds + \sum_{j=1}^{M}(1-2\alpha_j)\int_0^u (K_{H_j}^{c_j}(t,s))^2 s^{-2\alpha_j}ds$$

$$+ \sum_{i=1}^{N}\int_0^u 2b_i(s)X_s^t dW_s^i + \sum_{j=1}^{M}\int_0^u 2K_{H_j}^{c_j}(t,s)X_s^t dM_s^{H_j},$$

whence

$$\widehat{(X_u^t)^2} = E(\eta^2/\xi) + \int_0^u 2a(s)\widehat{(X_s^t)X_s}ds + \sum_{i=1}^{N}\int_0^u b_i^2(s)ds$$

$$+ \sum_{j=1}^{M}(1-2\alpha_j)\int_0^u (K_{H_j}^{c_j}(t,s))^2 s^{-2\alpha_j}ds$$

$$+ C_H \int_0^u \mathcal{D}(s)(\widehat{(X_s^t)^2 X_s}) - \widehat{(X_s^t)^2}\widehat{X}_s)d\nu_s, t \in [0,T]. \qquad (4.2.5)$$

Subtracting (4.2.4) from (4.2.5) for $u = t$, we obtain (4.2.3). □

4.3 Optimal Filtering in Systems with Polynomial Fractional Brownian Noise

In all previous filtering models the noises were presented as the integrals w.r.t. a Wiener process or w.r.t. an fBm, but everywhere with nonrandom integrands. In this section we consider the simple case of a random integrand.

Let the signal process $\{X_t, t \in [0,T]\}$ and the observation process $\{Y_t, t \in [0,T]\}$ are defined by the following system of equations:

$$X_t = \eta + \int_0^t a(s, X_s)ds + \sum_{n=1}^{N} b_n (B_t^{H_1})^n,$$

$$Y_t = \xi + \int_0^t A(s, X_s)ds + \int_0^t C(s)dB_s^{H_2}, t \in [0,T],$$

where $(B_t^{H_1}, B_t^{H_2}, t \in [0,T])$ are fBms with Hurst indexes $H_i \in (1/2, 1)$, $a, A : [0,T] \times \mathbb{R} \to \mathbb{R}$ are measurable functions and $b_n, 1 \le n \le N$ are real numbers. Suppose that the pair (η, ξ) does not depend on (B^{H_1}, B^{H_2}), condition (i) holds for the function C, and condition (ii) holds for $E(s, x_s)$ with any $x \in C^{H_1-}[0,T]$. First we try to present the power term $(B_t^{H_1})^n$ in the form

$$(B_t^{H_1})^n = \int_0^t M_n(t,s)dB_s + \int_0^t K_n(t,s)ds,$$

where B is the underlying Wiener process (it means that $B_t^{H_1} = \int_0^t m_{H_1}(t,s)dB_s$ with the kernel $m_{H_1}(t,s)$ from Section 1.8), $M_n(t,s)$ and $K_n(t,s)$ are some $\mathcal{F}_s$-adapted random functions. Evidently, for $n=1$

$$B_t^{H_1} = \int_0^t m_{H_1}(t,s)dB_s.$$

Therefore $M_1(t,s) = m_{H_1}(t,s)$, $K_1(t,s) = 0$. For arbitrary $n \ge 2$ $(B_t^{H_1})^n = \int_0^t n(B_s^{H_1})^{n-1} m_{H_1}(t,s)dB_s + \int_0^t \frac{n(n-1)}{2}(B_s^{H_1})^{n-2}(m_{H_1}(t,s))^2 ds.$

So, the signal process can be presented as

$$X_t = \eta + \int_0^t a(s, X_s)ds + \sum_{n=1}^{N} b_n \left(\int_0^t M_n(t,s)dB_s + \int_0^t K_n(t,s)ds \right)$$

$$= \eta + \int_0^t a(s, X_s)ds + \int_0^t M(t,s)dB_s + \int_0^t K(t,s)ds,$$

where

$$M_n(t,s) = n(B_s^{H_1})^{n-1} m_{H_1}(t,s),$$

$$K_n(t,s) = \frac{n(n-1)}{2}(B_s^{H_1})^{n-2}(m_{H_1}(t,s))^2,$$

4.3 Optimal Filtering with Polynomial Fractional Noise

$$M(t,s) = \sum_{n=1}^{N} b_n M_n(t,s), \quad K(t,s) = \sum_{n=1}^{N} b_n K_n(t,s).$$

Suppose that $\langle B, M^{H_2}\rangle_t = \int_0^t \lambda_s s^{-2\alpha} ds$, where $M_t^{H_2} = \int_0^t l_{H_2}(t,s) dB_s^{H_2}$.

Consider the family of semimartingales

$$X_s^t := \eta + \int_0^s a(u, X_u) du + \int_0^s M_n(t,u) dB_u + \int_0^s K_n(t,u) du,$$

$s \in [0, t], t \in [0, T]$. Let the function $\phi \in C^2(\mathbb{R})$. Then the process $\phi(X_s^t), s \in [0, t]$ is a semimartingale with the representation

$$\phi(X_s^t) = \phi(\eta) + \int_0^s \phi'(X_u^t) M(t,u) dB_u + \int_0^s L_u^t(\phi(\cdot)) du,$$

where $L_u^t(\phi(\cdot)) = (a(u, X_u) + K(t,u)) \phi'(\cdot) + \frac{1}{2}\phi''(\cdot)(M(t,u))^2$.

So, Theorem 4.1.3 gives the following representation for the optimal filter $\pi_s(\phi(X_s^t))$:

$$\pi_s(\phi(X_s^t)) = \pi(\phi(\eta)) + \int_0^s \pi_u(L_u^t(\phi(X_u^t))) du + \int_0^s \Big(\pi_u(\phi'(X_u^t) M(t,u) \lambda_u)$$

$$+ C_H (\pi_u(\phi(X_u^t) E(u, X)) - \pi_u(\phi(X_u^t)) \pi_u(E))\Big) d\nu_u.$$

If we put $s = t$ then the equation for the optimal filter $\pi_t(\phi(X_t))$ receives the form:

$$\pi_t(\phi(X_t)) = \pi(\phi(\eta)) + \int_0^t \pi_u(L_u^t(\phi(X_u^t))) du + \int_0^t \Big(\pi_u(\phi'(X_u^t) M(t,u) \lambda_u)$$

$$+ C_H (\pi_u(\phi(X_u^t) E(u, X)) - \pi_u(\phi(X_u^t)) \pi_u(E))\Big) d\nu_u.$$

5

Financial Applications of Fractional Brownian Motion

5.1 Discussion of the Arbitrage Problem

5.1.1 Long-range Dependence in Economics and Finance

As mentioned in the paper (WTT99), long-range dependence in economics and finance has a long history and is an area of active research (e.g., see (Lo91), (CKW95)). The importance of long-range dependent processes as stochastic models lies in the fact that they provide an explanation and interpretation of an empirical law that is commonly referred to as the Hurst law or Hurst effect. In short, for a given set of observations $\{X_i, i \geq 1\}$ with partial sum $Y(n) = \sum_{i=1}^{n} X_i$, $n \geq 1$, and sample variance $S^2(n) = n^{-1} \sum_{i=1}^{n} (X_i - n^{-1} Y(n))^2$, $n \geq 1$, the rescaled adjusted range statistic or R/S-statistic is defined by

$$\frac{R}{S}(n) = \frac{1}{S(n)} \left(\max_{0 \leq t \leq n} \left(Y(t) - \frac{t}{n} Y(n) \right) - \min_{0 \leq t \leq n} \left(Y(t) - \frac{t}{n} Y(n) \right) \right), \quad n \geq 1.$$

Hurst in (Hur51) found that many naturally occurring empirical records appear to be well represented by the relation $E\left((R/S)(n)\right) \sim c_1 n^H$ as $n \to \infty$, with typical values of the Hurst parameter $H \in (1/2, 1)$, and c_1 a finite positive constant not depending on n. But in the case when the observations come from a short-range dependent model, then $E\left(R/S(n)\right) \sim c_2 n^{1/2}$ as $n \to \infty$, where c_2 does not depend on n. The discrepancy between these two relations is called the Hurst effect or Hurst phenomenon. The analysis of the R/S-statistic, provided in (WTT99), (TTW95) and (TT97), leads to the recommendation to use a diverse portfolio of time-domain-based and frequency-domain-based graphics and statistical methods, including the graphical R/S-method, the modified R/S-statistic (Lo91) and Whittle's approach. Also, another (possibly, surprising) recommendation is: in the case when statistical analysis cannot be expected to provide a definitive answer concerning the presence or absence of long-range dependence in asset price returns, a more revealing and also much more challenging approach to tackle this problem consists of providing a mathematically rigorous physical "explanation" for the presence or absence of the long-range dependence phenomenon in stock returns.

5.1.2 Arbitrage in "Pure" Fractional Brownian Model. The Original Rogers Approach

Suppose that we establish that the existence of long-range dependence on the financial market in which we operate, and we must model a share price process using long-range dependence of returns. So, we can try to replace the clasical log-Brownian model (BlSc73)

$$dS_t = S_t(\mu dt + \sigma dW_t), \; t \geq 0$$

involving some Brownian motion W by the model involving fBm B^H:

$$dS_t = S_t(\mu dt + dB_t^H), \; t \geq 0, \tag{5.1.1}$$

where $H \in (1/2, 1)$.

Three main problems arise immediately: what will be the class of financial strategies, what will be the kind of stochastic integral w.r.t. the fBm used in the model and is such a model arbitrage-free or not? There has been wide discussion on these topics and we will present here the main (in our opinion) results and conclusions. It seems that the first attempt to construct arbitrage on the financial market that is modeled with fBm, was made by Rogers (Rog97). He did not use *geometric fBm* like (5.1.1) but fBm itself, and exploits its stationary properties, obtains an arbitrage possibility and immediately concludes that fBm is an absurd model for finance markets (as we shall see later, the situation is not so dramatic).

The notion of arbitrage that will be used (only in this subsection) is the following: we say that an arbitrage exists on the interval $[a, b]$ if there is some trading strategy whose gains process $\{\eta_t, a \leq t \leq b\}$ satisfies the following conditions: (a) $\eta_a = 0$; (b) $\eta_t \geq -\beta$ for all $a \leq t \leq b$ and some $\beta > 0$; (c) $P\{\eta_b > 0\} > 0$.

The brief description of the Rogers construction is the following. Suppose that $(\Omega, \mathcal{F}, \{\mathcal{F}_t, t \in \mathbb{R}_+\}, P)$ is a filtered probability space and $\{X_t, t \in \mathbb{R}_+\}$ is a continuous integrable adapted process. For any $a > 0$ and $0 \leq t < b$ define

$$\tau(t, b, a) := \inf\{u > t : X_u - X_t \notin [-a, a]\} \wedge b.$$

Lemma 5.1.1. *Let, for any rational a, b, t with $t < b$,*

$$E\left(X_{\tau(t,b,a)} - X_t / \mathcal{F}_t\right) = 0 \;\; a.s. \tag{5.1.2}$$

Then X is a local martingale.

5.1 Discussion of the Arbitrage Problem

Proof. For any stopping time $T \leq c$ equality (5.1.2) can be extended to

$$E\left(X_{\tau(T,b,a)} - X_T/\mathcal{F}_T\right) = 0 \quad a.s. \tag{5.1.3}$$

Indeed, we can approximate T by a sequence of stopping times $T^{(n)} = 2^{-n}\left([2^n T] + 1\right)$, taking discretely many rational values and decreasing to T. Now fix $N \in \mathbb{N}$, define $\tau := \tau(0, N, N)$, fix $\varepsilon > 0$ and define the stopping times $\sigma_0^\varepsilon = 0$,

$$\sigma_{n+1}^\varepsilon := \inf\{u > \sigma_n^\varepsilon : X_u - X_{\sigma_n^\varepsilon} \notin (-\varepsilon, \varepsilon)\} \wedge \tau,$$

$n \geq 0$. Evidently, $\sigma_n^\varepsilon \uparrow \tau$ as $n \to \infty$. From (5.1.3) it follows that

$$E\left(X_{\sigma_{n+1}^\varepsilon}/\mathcal{F}_{\sigma_n^\varepsilon}\right) = X_{\sigma_n^\varepsilon}.$$

Since $|X_{\sigma_n^\varepsilon}| \leq N + |X_0|$, we have that for any $n \geq 0$ $X_{\sigma_n^\varepsilon} = E\left(X_\tau/\mathcal{F}_{\sigma_n^\varepsilon}\right)$, and as $\varepsilon \to \infty$ we obtain that for any $t < N$

$$X_{t \wedge \tau} = E(X_\tau/\mathcal{F}_t),$$

which means that $X_{t \wedge \tau}$ is a martingale, and this is sufficient. □

Now, as we have seen in Section 1.15, fBm B^H is not a semimartingale (in particular, it is not a local martingale) unless $H = 1/2$. As a conclusion, we obtain from Lemma 5.1.1 that for fBm $\{B_t^H, t \in \mathbb{R}\}$ the following is true: if we define for any $n \in \mathbb{N}$ the process

$$Y_n(t) := (B_{t \cdot 2^{-n} - 2^{1-n}}^H - B_{-2^{1-n}}^H) 2^{nH},$$

$0 \leq t \leq 1, H \in (1/2, 1)$, and $\mathcal{Y}_n := \mathcal{F}_{-2^{-n}}^{B^H}$, then there exist $a > 0$ and $\varepsilon > 0$, such that

$$P\{E(Y_n(\tau_n)/\mathcal{Y}_{n-1}) \geq \varepsilon\} \geq \varepsilon$$

where $\tau_n = \inf\{t > 0 : Y_n(t) \in [-a, a]\} \wedge 1$. Note that by the scaling properties of B^H the sequence $\{Y_n, n \in \mathbb{Z}\}$ of $C[0,1]$-valued random variables is stationary and even ergodic since $\bigcap_n \sigma\{Y_k, k \leq -n\}$ is trivial. The ergodic theorem guarantees that

$$P\{E(Y_n(\tau_n)/\mathcal{Y}_{n-1}) \geq \varepsilon \text{ for infinitely many } n \geq 0\} = 1.$$

Consider the period $(-2^{1-n}, -2^{-n}]$ and call this period "promising" if $E(Y(\tau_n)/\mathcal{Y}_{n-1}) \geq \varepsilon$. There will be infinitely many "promising" periods. The investment strategy is the following one: we invest a unit amount in each "promising" period but immediately sell our holding and wait until the end of the period if Y_n goes out of $[-a, a]$ during the promising period. So, the gain ζ_n made during a "promising" period satisfies the relations $-a \leq \zeta_n \leq a$,

$E(\zeta_n/\mathcal{Y}_{n-1}) \geq \varepsilon$, and for the "nonpromising" period $\zeta_n = 0$. Denote accumulated gain by $\eta_n = \sum_{k \leq n} \zeta_k$. Then we can take $\lambda > 0$ sufficiently small such that

$$E(e^{-\lambda \eta_n}/\mathcal{Y}_{n-1}) \leq e^{-\lambda \eta_{n-1}}.$$

Therefore, $e^{-\lambda \eta_n}$ is a nonnegative supermartingale convergent a.s. to 0. If we stop η_n at the first time ν when $\eta_n < -a$, then

$$P\{\nu < \infty\} \leq \exp\{-\lambda a\} < 1$$

and on the event $\{\nu = +\infty\}$ $\eta_n \to +\infty$. Finally, the arbitrage strategy can be described as follows: invest a unit amount in Y (which is the same as investing an amount 2^{nH} in B^H during period n) in each "promising" period until either η_n has risen to 1 or falls to below $-a$. The former happens at least with probability $1 - \exp\{-\lambda a\}$, and the resulting gain is 1, and if the latter happens we lose at most $2a$. If the latter happens we invest $1/2$ in each "promising" period until either η_n has risen to 1 or has fallen below $\frac{5a}{2}$. If the latter happens we lose at most $3a$, and invest $1/4$ in each "promising" period until either η has risen to 1 or has fallen below $\frac{13a}{4}$ and so on. To continue in this way, successively halving the stake when things go badly, we shall eventually be successful and make a net gain of at least 1, and the worst that can happen is that our wealth meantime could fall to 4α, so we have arbitrage in our definition.

5.1.3 Arbitrage in the "Pure" Fractional Model. Results of Shiryaev and Dasgupta

Consider a $(B(r), S(r))$-market with

$$\begin{aligned} B_t(r) &= e^{rt}, \\ S_t(r) &= e^{\mu t + \sigma B_t^H}, \quad t \geq 0, \end{aligned} \quad (5.1.4)$$

$H \in (1/2, 1)$. Let for simplicity $\mu = r$, $\sigma = 1$. We construct a portfolio $\pi = (\beta, \gamma)$ with $\beta_t = 1 - e^{2B_t^H}$, $\gamma_t = 2(B_t^H - 1)$. For such a portfolio we have that the corresponding capital X_t^π equals

$$X_t^\pi = \beta_t B_t(r) + \gamma_t S_t(r) = e^{rt}\left(e^{B_t^H} - 1\right)^2.$$

From the Itô formula (2.7.5) for a pathwise integral w.r.t. fBm,

$$\begin{aligned} X_t^\pi &= \int_0^t r e^{rs} \left(e^{B_s^H} - 1\right)^2 ds + 2 \int_0^t e^{rs + B_s^H}\left(e^{B_s^H} - 1\right) dB_s^H \\ &= \int_0^t \beta_s dB_s(r) + \int_0^t \gamma_s dS_s(r), \end{aligned} \quad (5.1.5)$$

and (5.1.5) exactly means that the strategy π is self-financing strategy in usual sense. So, for this portfolio $X_0^\pi = 0$ and $X_t^\pi > 0$ a.s. for any $t > 0$, and everyone understands that it is an arbitrage possibility (in any appropriate definition). This is Shiryaev's example (Shi01).

A very close result was obtained by Dasgupta (Das98). He considered a one-dimensional portfolio $\pi_t, 0 \le t \le 1$, the same model as in (5.1.4), defined discounted gain as

$$G_t = \int_0^t \pi(s) B_s^{-1}(r) \left(\sigma dB_s^H + (\mu - r) ds \right),$$

and determined arbitrage as the following possibility:
(a) there exists $\alpha \in \mathbb{R}$ such that $P\{G_t \ge \alpha, 0 \le t \le 1\} = 1$;
(b) $P\{G_t \ge 0\} = 1$, (c) $P\{G_1 > 0\} > 0$.

Now, consider the particular case $\mu = r$ and the particular portfolio

$$\pi_t = 2e^{rt + \sigma B_t^H} \left(e^{\sigma B_t^H} - 1 \right). \tag{5.1.6}$$

With portfolio (5.1.6) the gain process equals

$$G_t = \int_0^t 2e^{\sigma B_s^H} \left(e^{\sigma B_s^H} - 1 \right) \sigma dB_s^H = \int_0^t e^{2\sigma B_s^H} \left(2\sigma dB_s^H \right)$$

$$-2 \int_0^t e^{\sigma B_s^H} \left(2\sigma dB_s^H \right) = e^{2\sigma B_t^H} - 1 - 2e^{\sigma B_t^H} + 2 = \left(e^{\sigma B_t^H} - 1 \right)^2.$$

Of course, we obtain arbitrage possibility. As a conclusion, we see that the "pure" continuous-time model based on fBm is not arbitrage-free, if the arbitrage possibility is defined in any appropriate terms. The same fact is emphasized in PhD thesis of Cheridito (Che01b), the paper of Salopek (Sal98); see also an early discussion on arbitrage with fBm in finance in (MS93).

Now we can discuss discrete-time models and "mixed" models (the latter ones are much more promising).

5.1.4 Mixed Brownian–Fractional-Brownian Model: Absence of Arbitrage and Related Topics

Let $\{W_t, t \ge 0\}$ be a standard Wiener process and $\{B_t^H, t \ge 0\}$ be an fBm with the Hurst index $H \in (1/2, 1)$, both defined on a filtered probability space $(\Omega, \mathcal{F}, \{\mathcal{F}_t, t \ge 0\}, P)$.

Consider a mixed version of the Black–Merton–Scholes model, i.e. a (B, S)-market with a bond B and a stock S, where

$$B_t = e^{rt}, \quad S_t = e^{aW_t + bB_t^H + ct}, \quad r, a, b, c \in \mathbb{R}, \ t \in \mathbb{R}_+. \tag{5.1.7}$$

For a given strategy (or a portfolio) $\pi = \{\beta_t, \gamma_t, t \ge 0\}$ the capital $\{X_t, t \ge 0\}$ corresponding to this portfolio equals

$$X_t = B_t \cdot \beta_t + S_t \cdot \gamma_t. \qquad (5.1.8)$$

We make the following assumptions about the strategy π:
1) π is a self-financing strategy, i.e.

$$X_t = X_0 + \int_0^t \beta_s \, dB_s + \int_0^t \gamma_s \, dS_s; \qquad (5.1.9)$$

2) π is a Markov-type strategy, i.e.

$$\beta_t = \beta(S_t, t), \quad \gamma_t = \gamma(S_t, t). \qquad (5.1.10)$$

One needs to be accurate with condition (5.1.9), for it to reflect the real economic concept of "self-financing". This entails that the meaning of the second integral in (5.1.9) should be specified clearly. We understand it now in the pathwise sense, i.e. as the following limit with probability 1:

$$\int_0^t \gamma_s \, dS_s = \lim_{\max|s_{k+1}-s_k| \to 0} \sum_{k=0}^{n-1} \gamma_{s_k} \left(S_{s_{k+1}} - S_{s_k} \right).$$

Here, the sum $\sum_{k=0}^{n-1} \gamma_{s_k} \left(S_{s_{k+1}} - S_{s_k} \right)$ is an obvious formula for the capital, earned on the price variation of S with a piecewise buy-and-hold strategy $\{\tilde{\gamma}_t, t \in \mathbb{R}_+\} = \{\gamma_{s_k}, \; s_k \le t < s_{k+1}, \; t \ge 0\}$. Hence, the integral $\int_0^t \gamma_s \, dS_s$, as the capital earned on S with the continuous strategy $\{\gamma_t, t \in \mathbb{R}_+\}$, agrees with the "fundamental moral" in the definition of self-financing conditions (for discussion on this topic see Section 5.2.2).

We say that the strategy π has an arbitrage opportunity if there exists $T > 0$ such that

$$X_0 = 0, \quad X_T \ge 0 \; (P - a.s.), \quad P(X_T > 0) > 0.$$

In the mixed model (5.1.7) with $a \ne 0$ and $b \ne 0$, some results in this direction have been obtained in the papers of (Ku99), (Che01b), (MV02), (Zah02a). More exactly, Kuznetsov (Ku99) established the absence of arbitrage under the condition of independence of processes W and B^H. As we mentioned in Subsection 3.4.2, Cheridito (Che01b) proved that, for $H \in (3/4, 1)$, the mixed model with independent W and B^H is equivalent to the one with Brownian motion and hence it is arbitrage-free. Zähle (Zah02a) proved the absence of arbitrage in the general mixed model with independent Wiener process and the process of zero quadratic variation (Dirichlet processes, see, for example, (Fol81b)). In the mixed model, studied in the paper (MV02), there is no requirement of independence. Conversely, the absence of arbitrage is demonstrated under the condition that the process B^H is connected with the process W as in formula (1.8.17).

The main result of this subsection is that the mixed market is arbitrage-free without any conditions on the dependence of W and B^H, if we restrict ourselves to the self-financing Markov-type strategies with smooth β and γ.

Conditions of Self-Financing and Their Consequences

Note that in the case of the Markov-type strategy (5.1.10), the process of capital X_t can be written as a function of price of the stock S at the moment t:
$$X_t = \Phi(S_t, t), \tag{5.1.11}$$
where
$$\Phi(x, t) = e^{rt} \cdot \beta(x, t) + x \cdot \gamma(x, t). \tag{5.1.12}$$

We prove in this section that the self-financing assumption strongly restricts the class of possible functions Φ in (5.1.11).

In the case of $\gamma_t = \gamma(S_t, t)$ with smooth $\gamma(\cdot, \cdot)$, the integral $\int_0^t \gamma_s \, dS_s$ exists and it can be presented in the form

$$\int_0^t \gamma_s \, dS_s = \int_0^t a\gamma_s S_s \, dW_s + \int_0^t b\gamma_s S_s \, dB_s^H + \int_0^t \left(c + \frac{a^2}{2}\right) \gamma_s S_s \, ds, \tag{5.1.13}$$

where the first integral on the right-hand side is the Itô integral, the second integral is the pathwise Riemann–Stieltjes integral and the third one is the Riemann integral. Formula (5.1.13) gives the Itô formula for an exponent of the mixed process. In addition, we shall refer in this subsection to the Itô formula for processes with generalized quadratic variation (see Subsection 2.7.2).

The Itô integral in (5.1.13) appears due to the choice of the left endpoint s_k in the expression under the summation sign in (5.1.12). Such a choice is crucial for condition (5.1.9) to have the economic sense of self-financing. The second integral $\int_0^t b\gamma_s S_s \, dB_s^H$ does not depend on the choice of inner points of the intervals.

Theorem 5.1.2. *Let the (B, S)-market be given by (5.1.7) with $a \neq 0$. Suppose also that for all $t > 0$ the support of the distribution of S_t coincides with*

$$\mathrm{supp}(S_t) = [0, +\infty). \tag{5.1.14}$$

Then in the class of Markov-type strategies (5.1.10) with

$$\{\beta(x, t), \gamma(x, t)\} \subset C^2((0, +\infty)) \times C^1([0, +\infty))$$

the condition of self-financing (5.1.9) is equivalent to the following one:

(i) There exists a function $\phi(x, t) \in C^2((0, +\infty)) \times C^1([0, +\infty))$, which satisfies the equation

$$\phi'_t(x, t) + \frac{a^2}{2} x^2 \phi''_{xx}(x, t) + r x \phi'_x(x, t) - r \phi(x, t) = 0, \tag{5.1.15}$$

and the strategy (β, γ) can be expressed in terms of ϕ:

308   5 Financial Applications of Fractional Brownian Motion

$$\begin{cases} \beta(x,t) = e^{-rt}(\phi(x,t) - x \cdot \phi'_x(x,t)); \\ \gamma(x,t) = \phi'_x(x,t). \end{cases} \quad (5.1.16)$$

Remark 5.1.3. Condition (5.1.14) holds, for example, in the case when the processes W and B^H are jointly Gaussian, and, hence, $\log(S_t) = aW_t + bB_t^H + ct$, $t \geq 0$ is a Gaussian process.

Remark 5.1.4. Under condition (i) we have the identity $\Phi(x,t) = \phi(x,t)$.

Proof of Theorem 5.1.2. Below we use the Itô formula for processes with generalized quadratic variation; see (3.1.25), (3.1.26). Firstly, the Itô formula holds for continuous processes with generalized bracket. Secondly, if the process Z has the usual bracket, then it has the same generalized bracket.

Let consider the process S_t and prove that it has usual bracket. Indeed,

$$\sum_{k=0}^{n} (\Delta S_{t_k})^2 = \sum_{k=0}^{n} \left(e^{aW_{t_{k+1}} + bB^H_{t_{k+1}} + ct_{k+1}} - e^{aW_{t_k} + bB^H_{t_k} + ct_k} \right)^2$$

$$= \sum_{k=0}^{n} e^{2aW_{t_{k+1}}} \left(e^{bB^H_{t_{k+1}} + ct_{k+1}} - e^{bB^H_{t_k} + ct_k} \right)^2$$

$$+ \sum_{k=0}^{n} \left(e^{aW_{t_{k+1}}} - e^{aW_{t_k}} \right)^2 e^{2bB^H_{t_k} + 2ct_k}$$

$$+ 2 \sum_{k=0}^{n} e^{aW_{t_{k+1}}} e^{bB^H_{t_k} + ct_k} \left(e^{aW_{t_{k+1}}} - e^{aW_{t_k}} \right) \left(e^{bB^H_{t_{k+1}} + ct_{k+1}} - e^{bB^H_{t_k} + ct_k} \right)$$

$$=: I_1^n + I_2^n + I_3^n.$$

Evidently, $I_2^n \to \int_0^t S_u^2 a^2 \, du$, a.s. and in $L_2(P)$. Further,

$$\left| e^{bB^H_{t_{k+1}} + ct_{k+1}} - e^{bB^H_{t_k} + ct_k} \right| \leq e^{bB^H_{t_k} + ct_k} \left| b \Delta B^H_{t_k} + c \Delta t_k \right|$$

and the trajectories of B^H belong to the class $C^{H-}[0,T]$ with $H > 1/2$. Therefore $I_1^n \to 0$ a.s., and the same is true for I_3^n. It means that the bracket of S has the form

$$[S]_t = \int_0^t a^2 S_u^2 \, du. \quad (5.1.17)$$

Let us apply the Itô formula (2.7.8) to the processes $B_t \beta(S_t, t)$ and $S_t \gamma(S_t, t)$ from (5.1.8). We obtain the equalities

$$B_t \beta(S_t, t) - \beta(1,0) = \int_0^t d(B_u \beta(S_u, u))$$

$$= \int_0^t \beta(S_u, u) \, dB_u + \int_0^t B_u \beta'_t(S_u, u) \, du + \int_0^t B_u \beta'_x(S_u, u) \, dS_u \quad (5.1.18)$$

$$+ \frac{1}{2} \int_0^t B_u \beta''_{xx}(S_u, u) \, d[S]_u,$$

5.1 Discussion of the Arbitrage Problem   309

and

$$S_t \gamma(S_t, t) - \gamma(1, 0) = \int_0^t d(S_u \gamma(S_u, u))$$

$$= \int_0^t \gamma(S_u, u) \, dS_u + \int_0^t S_u \gamma'_t(S_u, u) \, du + \int_0^t S_u \gamma'_x(S_u, u) \, dS_u \quad (5.1.19)$$

$$+ \frac{1}{2} \int_0^t \left(2\gamma'_x(S_u, u) + S_u \gamma''_{xx}(S_u, u) \right) d[S]_u.$$

Combining equations (5.1.18) and (5.1.19), we obtain:

$$X_t - X_0 - \int_0^t \beta(S_u, u) \, dB_u - \int_0^t \gamma(S_u, u) \, dS_u$$

$$= \int_0^t \left(B_u \beta'_t(S_u, u) + S_u \gamma'_t(S_u, u) \right) du + \int_0^t \left(B_u \beta'_x(S_u, u) + S_u \gamma'_x(S_u, u) \right) dS_u$$

$$+ \frac{1}{2} \int_0^t \left(B_u \beta''_{xx}(S_u, u) + 2\gamma'_x(S_u, u) + S_u \gamma''_{xx}(S_u, u) \right) d[S]_u. \quad (5.1.20)$$

Comparing equations (5.1.20) and (5.1.9), we conclude that the condition of self-financing of the strategy $\pi = \{\beta_t, \gamma_t, t \in \mathbb{R}_+\}$ is equivalent to the equation

$$\int_0^t \left(B_u \beta'_t(S_u, u) + S_u \gamma'_t(S_u, u) \right) du + \int_0^t \left(B_u \beta'_x(S_u, u) + S_u \gamma'_x(S_u, u) \right) dS_u$$

$$+ \frac{1}{2} \int_0^t \left(B_u \beta''_{xx}(S_u, u) + 2\gamma'_x(S_u, u) + S_u \gamma''_{xx}(S_u, u) \right) d[S]_u = 0, \quad t > 0. \quad (5.1.21)$$

From the same Itô formula and definition of the process S, we obtain that

$$S_t = S_0 + \int_0^t S_u \, d(aW_u + bB_u^H + cu) + \int_0^t \frac{1}{2} a^2 S_u \, du,$$

where the integral $\int_0^t S_u \, dW_u$ exists as the usual Itô integral, and the integral $\int_0^t S_u \, dB_u^H$ exists as the limit of the Riemann–Stieltjes sums, because $S \in C^{1/2-}[0, T]$, $B^H \in C^{H-}[0, T]$, and $1/2 + H > 1$.

Substituting equation (5.1.17) into equation (5.1.21), we obtain that equation (5.1.21) can be rewritten as

$$\int_0^t \left(B_u \beta'_t(S_u, u) + S_u \gamma'_t(S_u, u) \right) du$$

$$+ \int_0^t \left(B_u \beta'_x(S_u, u) + S_u \gamma'_x(S_u, u) \right) S_u \, d\left(aW_u + bB_u^H + (c + a^2/2) u\right)$$

$$+ \frac{a^2}{2} \int_0^t \Big(B_u \beta''_{xx}(S_u, u) + 2\gamma'_x(S_u, u) + S_u \gamma''_{xx}(S_u, u) \Big) S_u^2 \, du = 0. \quad (5.1.22)$$

Let us take the quadratic variation of the both sides of (5.1.22). Evidently, the usual bracket of all Lebesgue integrals in (5.1.22) vanishes, and the bracket of the Itô integral equals

$$\left[\int_0^{\cdot} \Big(B_u \beta'_x(S_u, u) + S_u \gamma'_x(S_u, u) \Big) S_u \, d(aW_u) \right]_t =$$
$$= a^2 \int_0^t \Big(B_u \beta'_x(S_u, u) + S_u \gamma'_x(S_u, u) \Big)^2 S_u^2 \, du.$$

Establish now that the usual bracket of the process $\int_0^t \Big(B_u \beta'_x(S_u, u) + S_u \gamma'_x(S_u, u) \Big) S_u \, d(bB_u^H)$ a.s. equals 0. In this order denote $f_u := b \Big(B_u \beta'_x(S_u, u) + S_u \gamma'_x(S_u, u) \Big)$. Evidently, the trajectories of this process belong to the class $C^{1/2-}[0, T]$. Further, from the estimate in Proposition 22 (FdP99), it follows that

$$\left| \int_{t_k}^{t_{k+1}} f_u \, dB_u^H - f_{t_k} \Delta B_{t_k}^H \right| \leq C \, \|f\|_{C^{1/2-\delta}} \, \|B^H\|_{C^{H-\delta}} \, (\Delta t_k)^{1/2+H-2\delta},$$

with constant C not depending on f and B^H and such δ that $1/2+H-2\delta > 1$, i.e. $\delta < \alpha/2$. Therefore,

$$\sum_{k=0}^n \Big(\int_{t_k}^{t_{k+1}} f_u \, dB_u^H \Big)^2 \leq 2 \sum_{k=0}^n \Big(\int_{t_k}^{t_{k+1}} f_u \, dB_u^H - f_{t_k} \Delta B_{t_k}^H \Big)^2$$
$$+ 2 \sum_{k=0}^n (f_{t_k})^2 (\Delta B_{t_k}^H)^2 \leq 2 C^2 \, \|f\|_{C^{1/2-\delta}}^2 \, \|B^H\|_{C^{H-\delta}}^2 \sum_{k=0}^n (\Delta t_k)^{1+2H-4\delta}$$
$$+ 2 \sum_{k=0}^n (f_{t_k})^2 (\Delta B_{t_k}^H)^2 \to 0 \quad \text{a.s.}$$

From all these estimations and (5.1.22) we obtain

$$a^2 \int_0^t \Big(B_u \beta'_x(S_u, u) + S_u \gamma'_x(S_u, u) \Big)^2 S_u^2 \, du = 0. \quad (5.1.23)$$

Since (5.1.23) holds for all $t > 0$, we easily deduce that

$$B_u \beta'_x(S_u, u) + S_u \gamma'_x(S_u, u) = 0 \quad (5.1.24)$$

for all $u > 0$ and almost all (a.a.) $\omega \in \Omega$.

5.1 Discussion of the Arbitrage Problem   311

Substituting (5.1.24) into (5.1.22) we obtain another equation for all $t > 0$:

$$\int_0^t \left(B_u \, \beta'_t(S_u, u) + S_u \, \gamma'_t(S_u, u) \right) du + \frac{a^2}{2} \int_0^t \left(B_u \, \beta''_{xx}(S_u, u) + 2\gamma'_x(S_u, u) \right.$$
$$\left. + S_u \, \gamma''_{xx}(S_u, u) \right) S_u^2 \, du = 0.$$

This means that the equality

$$B_u \, \beta'_t(S_u, u) + S_u \, \gamma'_t(S_u, u) \quad (5.1.25)$$
$$+ \frac{a^2}{2} \left(B_u \, \beta''_{xx}(S_u, u) + 2\gamma'_x(S_u, u) + S_u \, \gamma''_{xx}(S_u, u) \right) S_u^2 = 0$$

holds for all $u > 0$ and a.a. $\omega \in \Omega$.

Condition (5.1.14) of the theorem ensures that equations (5.1.24) and (5.1.25) may hold if and only if

$$B_t \, \beta'_x(x, t) + x \, \gamma'_x(x, t) = 0; \quad (5.1.26)$$

$$B_t \, \beta'_t(x, t) + x \, \gamma'_t(x, t) + \frac{a^2}{2} \left(B_t \, \beta''_{xx}(x, t) + 2\gamma'_x(x, t) + x \, \gamma''_{xx}(x, t) \right) x^2 = 0, \quad (5.1.27)$$

for all $t > 0$, $x > 0$.

The last relations mean that the strategy $(\beta(S_t, t), \gamma(S_t, t))$ is self-financing if and only if the pair $(\beta(x, t), \gamma(x, t))$ satisfies equations (5.1.26), (5.1.27).

Now assume that condition (i) of the theorem holds. Substituting β and γ from (5.1.16) into (5.1.26) and (5.1.27) we obtain an identity $0 = 0$ in the first equation and identity (5.1.15) in the second one.

Conversely, if (5.1.26) and (5.1.27) hold, we set

$$\phi(x, t) := B_t \cdot \beta(x, t) + x \cdot \gamma(x, t).$$

For such function ϕ we obtain from (5.1.26) that

$$\phi'_x(x, t) = B_t \cdot \beta'_x(x, t) + \gamma(x, t) + x \cdot \gamma'_x(x, t) = \gamma(x, t),$$
$$\beta(x, t) = B_t^{-1} \left(\phi(x, t) - x \cdot \gamma(x, t) \right) = e^{-rt} \left(\phi(x, t) - x \cdot \phi'_x(x, t) \right),$$

i.e. we come to (5.1.16). Substituting β and γ from (5.1.16) into identity (5.1.27), we obtain that $\phi(x, t)$ satisfies equation (5.1.15). □

Remark 5.1.5. Let the process $\{Z_t, t \geq 0\}$ be defined on $(\Omega, \mathcal{F}, \{\mathcal{F}_t, t \geq 0\}, P)$ with $Z_0 = 0$ and $[Z] \equiv 0$, where $[Z]$ stands for usual bracket, i.e. quadratic variation. Then it is not hard to see that Theorem 5.1.2 is valid for the $(B, \tilde{S})$-market with

$$B_t = e^{rt}, \quad \tilde{S}_t = e^{aW_t + Z_t + ct},$$

if only condition (5.1.14) holds for the process $\tilde{S}$.

Absence of Arbitrage

Theorem 5.1.6. *Let the (B, S)-market be given by (5.1.7) with $a \neq 0$. Let the support of the distribution of S_t coincides with*

$$\operatorname{supp}(S_t) = [0, +\infty) \qquad (5.1.28)$$

for all $t > 0$.

Then there is no arbitrage strategy in the class of self-financing Markov-type strategies (5.1.10) with

$$\{\beta(x, t), \gamma(x, t)\} \subset C^2((0, +\infty)) \times C^1([0, +\infty)).$$

Proof. Theorem 5.1.2 states that for any strategy in the class, described in the theorem, the process of capital X_t is given by

$$X_t = \phi(S_t, t),$$

where ϕ satisfies the equation

$$\phi'_t(x,t) + \frac{a^2}{2} x^2 \phi''_{xx}(x,t) + r x \phi'_x(x,t) - r \phi(x,t) = 0. \qquad (5.1.29)$$

Suppose that an arbitrage strategy exists. So, there exists $T > 0$ such that

$$X_0 = 0, \quad X_T \geq 0 \ (P-a.s.). \qquad (5.1.30)$$

Together with (5.1.28) conditions (5.1.30) are equivalent to the following ones:

$$\phi(1, 0) = 0, \quad \phi(x, T) \geq 0 \quad \forall x > 0. \qquad (5.1.31)$$

We are going to prove that $\phi \equiv 0$ is the only function that satisfies (5.1.29) and (5.1.31) simultaneously. Hence, it would mean that there is no arbitrage strategies in the given class.

Let us use the standard approach in solving equation (5.1.29). Suppose the function ϕ satisfies equation (5.1.29) with boundary conditions (5.1.31). Then a new function $\eta(z, t)$, given by

$$\eta(z, t) = \theta(az, T - t), \quad z \in \mathbb{R}, \ t \in [0, T],$$

where

$$\theta(z, t) = e^{-(\alpha z + \beta t)} \phi(e^z, t), \quad \alpha = \frac{1}{2} - \frac{r}{a^2}, \quad \beta = -\frac{a^2}{8} + \frac{r^2}{2a^2},$$

satisfies a heat equation

$$\eta'_t(z, t) = \frac{1}{2} \eta''_{zz}(z, t) \qquad (5.1.32)$$

with additional conditions
$$\forall z \in \mathbb{R} \quad \eta(z,0) \geq 0, \quad \eta(0,T) = 0. \tag{5.1.33}$$
Here, an inverse change is given by
$$\phi(x,t) = x^{\left(\frac{1}{2} - \frac{r}{a^2}\right)} \cdot e^{\left(-\frac{a^2}{8} + \frac{r^2}{2a^2}\right)} \cdot \eta\left(\frac{\ln(x)}{a}, T - t\right).$$
The continuous solution of equation (5.1.32) is well known and has the form
$$\eta(z,t) = \int_{\mathbb{R}} \eta(\xi,0) \cdot (2\pi t)^{-\frac{1}{2}} \cdot \exp\left(-\frac{(z-\xi)^2}{2t}\right) d\xi,$$
which together with boundary conditions (5.1.33) gives $\eta \equiv 0$ and, therefore, $\phi \equiv 0$. □

Convergence of Lebesgue–Stieltjes Integrals to the Integral w.r.t. fBm

In this subsection, we use Theorem 1.15.3 and prove a theorem which establishes the convergence in probability of integrals with respect to $B^{H,\beta}$ from (1.15.15) to the integral with respect to fBm.

Theorem 5.1.7. *Let the process f be such that for some $\varepsilon > 0$ and for a.a. $\omega \in \Omega$*
$$f(\cdot, \omega) \in C^{2(1-H)+\varepsilon}[0,T]. \tag{5.1.34}$$
Then
$$\int_0^T f(u) \, dB_u^{H,\beta} \xrightarrow{P} \int_0^T f(u) \, dB_u^H \quad \text{as} \quad \beta \to 0+,$$
where $\xrightarrow{P}$ denotes the convergence in probability.

Proof. For any $N > 0$ we introduce the step process of the form
$$f_N(u) = \sum_{k=1}^N f(u_{k-1}) \mathbf{1}_{[u_{k-1},\, u_k)}(u), \quad u \in [0,T), \quad f_N(T) = f(u_N),$$
where
$$u_k = \frac{kT}{N}, \quad 0 \leq k \leq N.$$
Then the following obvious inequality holds:
$$\left| \int_0^T f(u) \, dB_u^{H,\beta} - \int_0^T f(u) \, dB_u^H \right|$$
$$\leq \left| \int_0^T (f(u) - f_N(u)) \, dB_u^{H,\beta} \right| + \left| \int_0^T f_N(u) \, d(B_u^{H,\beta} - B_u^H) \right|$$
$$+ \left| \int_0^T (f_N(u) - f(u)) \, dB_u^H \right| =: I_1(N,\beta) + I_2(N,\beta) + I_3(N).$$

314   5 Financial Applications of Fractional Brownian Motion

We shall establish that for the subsequence N_β such that $N_\beta = \left[\frac{T}{\beta^{1/2}}\right]$ the following convergence holds:

$$I_1(N_\beta,\beta) \xrightarrow{P} 0, \quad I_2(N_\beta,\beta) \xrightarrow{P} 0, \quad I_3(N_\beta) \xrightarrow{P} 0 \quad \text{as} \quad \beta \to 0+.$$

Condition (5.1.34) is equivalent to the relation: there exists a finite random variable $K = K(\omega)$ such that P-a.s. $\forall\, 0 \leq x < y \leq T$ we have

$$|f(x) - f(y)| \leq K|x-y|^\lambda \qquad (5.1.35)$$

with $\lambda = 2(1-H) + \varepsilon$.

Consider $I_1(N_\beta,\beta)$. We use (5.1.33), (5.1.34), (5.1.35) to obtain:

$$I_1(N_\beta,\beta) = \left|\int_0^T (f(u) - f_{N_\beta}(u))\,dB_u^{H,\beta}\right|$$

$$= C\left|\sum_{k=1}^N \int_{u_{k-1}}^{u_k} (f(u) - f(u_{k-1}))\left(u^{H-\frac{1}{2}}\int_0^{(u-\beta)_+} (u-y)^{\alpha-1} y^{\frac{1}{2}-H}\,d\tilde W_y\right) du\right|$$

$$\leq CK \sum_{k=1}^N (u_k - u_{k-1})^\lambda \int_{u_{k-1}}^{u_k} u^{H-\frac{1}{2}}\left|\int_0^{(u-\alpha)_+} (u-y)^{\alpha-1} y^{\frac{1}{2}-H}\,d\tilde W_y\right| du$$

$$=: CK\,\zeta_1(N,\beta).$$

where $\tilde W$ is now the underlying Wiener process (before it was denoted B, but now B is bond process). From now on C means a constant, the value of which is not interesting for us. Without loss of generality we may assume that $\beta < T/2$. Let estimate the mathematical expectation of $\zeta_1(N_\beta,\beta)$:

$$E\zeta_1(N_\beta,\beta) \leq \beta^{\frac{\lambda}{2}} \sum_{k=1}^N \int_{u_{k-1}}^{u_k} u^{H-\frac{1}{2}} E\left|\int_0^{(u-\beta)_+} (u-y)^{\alpha-1} y^{\frac{1}{2}-H}\,d\tilde W_y\right| du$$

$$\leq \beta^{\frac{\lambda}{2}} \sum_{k=1}^N \int_{u_{k-1}}^{u_k} u^\alpha \left(\int_0^{(u-\beta)_+} (u-y)^{2H-3} y^{1-2H}\,dy\right)^{1/2} du$$

$$\leq \beta^{\frac{\lambda}{2}} \left(\int_0^{1-\beta/T} (1-y)^{2H-3} y^{1-2H}\,dy\right)^{1/2} \sum_{k=1}^N \int_{u_{k-1}}^{u_k} u^{H-1}\,du$$

$$\leq C\beta^{\frac{\lambda}{2}} \left(\int_0^{1/2} (1-y)^{2H-3} y^{1-2H}\,dy + 2^{2\alpha}\int_{1/2}^{1-\beta/T} (1-y)^{2H-3}\,dy\right)^{1/2}$$

$$\leq C\beta^{\frac{\lambda}{2}}\left(1 + \beta^{2\alpha-1}\right)^{\frac{1}{2}}, \qquad (5.1.36)$$

Substituting $\lambda = 2(1-H) + \varepsilon$ in (5.1.36) we obtain

$$E\zeta_1(N_\beta,\beta) \le C\alpha^{1-H+\varepsilon/2}\left(1+\beta^{2\alpha-1}\right)^{\frac{1}{2}} = O(\beta^{\varepsilon/2}) \to 0, \quad \beta \to 0+.$$

Hence, $I_1(N_\beta,\beta) \xrightarrow{P} 0$ as $\beta \to 0+$.

Let consider $I_2(N_\beta,\beta)$.

$$I_2(N_\beta,\beta) = \left|\sum_{k=1}^{N} f(u_{k-1})\left(\left(B_{u_k}^{H,\beta} - B_{u_k}^{H}\right) - \left(B_{u_{k-1}}^{H,\beta} - B_{u_{k-1}}^{H}\right)\right)\right|$$

$$\le \sum_{k=1}^{N} |f(u_k) - f(u_{k-1})| \cdot \left|B_{u_k}^{H,\beta} - B_{u_k}^{H}\right| + \left|f(T)\left(B_T^{H,\beta} - B_T^{H}\right)\right|$$

$$\le K \sum_{k=1}^{N} (u_k - u_{k-1})^\lambda \cdot \left|B_{u_k}^{H,\beta} - B_{u_k}^{H}\right| + \left|f(T)\left(B_T^{H,\beta} - B_T^{H}\right)\right|.$$

The term $\left|f(T)\left(B_T^{H,\beta} - B_T^H\right)\right| \xrightarrow{P} 0$ because $B_T^{H,\beta} \xrightarrow{P} B_T^H$ as $\beta \to 0+$. Denote $\zeta_2(N_\beta,\beta) := \sum_{k=1}^{N}(u_k - u_{k-1})^\lambda \left|B_{u_k}^{H,\beta} - B_{u_k}^{H}\right|$. With the help of Theorem 1.15.3, the mathematical expectation of $\left|B_t^{H,\beta} - B_t^H\right|$ can be estimated in the following way:

$$E\left|B_t^{H,\beta} - B_t^H\right| \le C \begin{cases} t^H, & t < \beta \\ \beta^\alpha \sqrt{t\left(1 + \ln\frac{t}{\beta}\right)}, & \beta \le t \end{cases}$$

$$\le C \max\left(\beta^H, \beta^\alpha \sqrt{T\left(1 + \ln\frac{T}{\beta}\right)}\right) = o\left(\beta^{\alpha-\rho}\right), \quad \beta \to 0+, \quad (5.1.37)$$

for any fixed $\rho > 0$. For $N = \left[\frac{T}{\beta^{1/2}}\right]$, $\rho = \varepsilon/2$ and $\lambda = 2(1-H) + \varepsilon$ we obtain from (5.1.37) that

$$E\zeta_2(N_\beta,\beta) \le \beta^{\frac{\lambda}{2}} \sum_{k=1}^{N} \mathbb{E}\left|B_{u_k}^{H,\beta} - B_{u_k}^{H}\right| \le \beta^{\frac{\lambda}{2}}\left([N_\beta]+1\right) o\left(\beta^{\alpha-\rho}\right) =$$

$$= o\left(\beta^{\frac{2(1-H)+\varepsilon-1}{2} + \left(\alpha - \frac{\varepsilon}{2}\right)}\right) = o(1) \to 0, \quad \beta \to 0+.$$

Hence, $I_2(N_\beta,\beta) \xrightarrow{P} 0$ as $\beta \to 0+$.

Finally, it follows from Theorem 2.1.7 that

$$I_3(N_\beta) = \left|\int_0^T f_{N_\beta}(u)\,dB_u^H - \int_0^T f(u)\,dB_u^H\right| \to 0$$

a.s., and hence in probability, as $\beta \to 0+$. $\square$

The Capital Process as a Limit of Semimartingales

Let the (B, S)-market be given by (5.1.7) and a Markov-type strategy $(\tilde{\beta}, \tilde{\gamma})$ be self-financing for this market. Then the capital, based on this strategy, is given by

$$X_t = X_0 + \int_0^t \tilde{\beta}(S_s, s) \, dB_s + \int_0^t \tilde{\gamma}(S_s, s) \, dS_s.$$

For $\beta > 0$ and the given $(\beta(\cdot, \cdot), \gamma(\cdot, \cdot))$ consider the processes

$$S_t^\beta = e^{aW_t + bB_t^{H, \beta} + ct}$$

and

$$X_t^\beta = X_0 + \int_0^t \tilde{\beta}(S_s^\beta, s) \, dB_s + \int_0^t \tilde{\gamma}(S_s^\beta, s) \, dS_s^\beta. \tag{5.1.38}$$

The Itô formula and definition of $B^{H, \beta}$ imply that the process X^β can be rewritten as

$$X_t^\beta = X_0 + \int_0^t \left(r \, B_s \, \tilde{\beta}(S_s^\beta, s) + \left(b \, (B_s^{H, \beta})'_s + c\right) S_s^\beta \, \gamma(S_s^\beta, s) \right) ds$$

$$+ a \int_0^t S_s^\beta \, \gamma(S_s^\beta, s) \, dW_s \tag{5.1.39}$$

with

$$(B_s^{H, \beta})'_s = C_H^{(6)} \alpha s^\alpha \int_0^{(s-\beta)_+} (s-u)^{\alpha-1} u^{-\alpha} d\tilde{W}_u,$$

which means that X^β is a semimartingale at least if the following condition holds:

$$\int_0^T E \left(S_s^\beta \, \tilde{\gamma}(S_s^\beta, s) \right)^2 ds < \infty. \tag{5.1.40}$$

Theorem 5.1.8. *Let* $H \in (3/4, 1)$ *and the pair* $(\tilde{\beta}(\cdot, \cdot), \tilde{\gamma}(\cdot, \cdot))$ *satisfy the assumptions:*

(ii) $\forall t \geq 0 \quad \tilde{\beta}(\cdot, t), \, \tilde{\gamma}(\cdot, t) \in C^1(\mathbb{R})$

(iii) $\forall T, L > 0$ *there exists* $K = K(T, L) > 0$ *such that*

$$\left|\tilde{\beta}(x, t) - \tilde{\beta}(x, s)\right| + |\tilde{\gamma}(x, t) - \tilde{\gamma}(x, s)| \leq K \, |t - s|^{\frac{1}{2}}, \quad \forall \, |x| \leq L, \, t, s \in [0, T].$$

(iv) $\forall T > 0$ *there exist* $M = M(T) > 0$ *and* $N = N(T) > 0$ *such that*

$$\left|\tilde{\beta}'_x(x, t)\right| + |\tilde{\gamma}'_x(x, t)| \leq M \, (1 + |x|^N), \quad \forall \, t \in [0, T].$$

Then $X_t^\beta \xrightarrow{P} X_t$ *as* $\beta \to 0+$ *for any* $t \in [0, T]$.

5.1 Discussion of the Arbitrage Problem   317

Remark 5.1.9. Evidently, conditions (ii)–(iv) imply (5.1.40) and the pair (B, S^β) can be regarded as a new stock market with a price of the stock being a semimartingale. It follows from Theorem 1.15.2 that $S_t^\beta \xrightarrow{P} S_t$ as $\beta \to 0+$ at any moment $t \geq 0$. If, additionally, condition (5.1.14) holds for S^β and $\tilde{\beta}, \tilde{\gamma} \in (C^2 \times C^1)(\mathbb{R}_+)$, then the strategy $(\tilde{\beta}(S_s^\beta, s), \tilde{\gamma}(S_s^\beta, s))$ is self-financing and the market (B, S^β) is arbitrage-free. In this case the process X^β is a process of capital in this market.

Proof of Theorem 5.1.8. Using (5.1.38), (5.1.39) and (5.1.9), we may write

$$X_t^\beta - X_t$$
$$= \int_0^t \left(\tilde{\beta}(S_s^\beta, s) - \tilde{\beta}(S_s, s) \right) dB_s + \int_0^t \gamma(S_s^\beta, s) \, dS_s^\beta - \int_0^t \gamma(S_s, s) \, dS_s$$
$$= r \int_0^t \left(f^\beta(s) - f(s) \right) ds + a \int_0^t \left(g^\beta(s) - g(s) \right) dW_s$$
$$+ b \left(\int_0^t g^\beta(s) \, dB_s^{H,\beta} - \int_0^t g(s) \, dB_s^H \right) + \left(c + \frac{a^2}{2} \right) \int_0^t \left(g^\beta(s) - g(s) \right) ds,$$

where

$$f^\beta(s) = e^{rs} \tilde{\beta}(S_s^\beta, s), \quad f(s) = e^{rs} \tilde{\beta}(S_s, s),$$
$$g^\beta(s) = S_s^\beta \tilde{\gamma}(S_s^\beta, s), \quad g(s) = S_s \tilde{\gamma}(S_s, s).$$

To prove that $X_t^\beta \to X_t$, it is enough to establish that

$$\int_0^t \left(f^\beta(s) - f(s) \right) ds \xrightarrow{P} 0; \tag{5.1.41}$$

$$\int_0^t \left(g^\beta(s) - g(s) \right) dW_s \xrightarrow{P} 0; \tag{5.1.42}$$

$$\int_0^t g^\beta(s) \, dB_s^{H,\beta} \xrightarrow{P} \int_0^t g(s) \, dB_s^H; \tag{5.1.43}$$

$$\int_0^t \left(g^\beta(s) - g(s) \right) ds \xrightarrow{P} 0, \quad \text{as} \quad \beta \to 0+. \tag{5.1.44}$$

The convergence in (5.1.41), (5.1.42) and (5.1.44) holds if $\int_0^t (f^\beta(s) - f(s))^2 \, ds \xrightarrow{P} 0$ and $\int_0^t (g^\beta(s) - g(s))^2 \, ds \xrightarrow{P} 0$ as $\beta \to 0+$, which, in turn, follows immediately from the relations

$$E\left(f^\beta(s) - f(s) \right)^2 \leq C \beta^{2\alpha}, \tag{5.1.45}$$
$$E\left(g^\beta(s) - g(s) \right)^2 \leq C \beta^{2\alpha}, \tag{5.1.46}$$

which will be proved in Lemma 5.1.10.

Let us prove (5.1.43). Obviously, the following inequality holds:

$$\left| \int_0^t g^\beta(s) \, dB_s^{H,\beta} - \int_0^t g(s) \, dB_s^H \right|$$
$$\leq \left| \int_0^t g(s) \, dB_s^{H,\beta} - \int_0^t g(s) \, dB_s^H \right| + \left| \int_0^t (g^\beta(s) - g(s)) \, dB_s^{H,\beta} \right|. \quad (5.1.47)$$

The trajectories of the process $\eta(t) = aW_t + bB_t^H + ct$ a.s. belong to the space $C^{\frac{1}{2}-}[0,T]$. It means that for any $\rho > 0$ there exists $K_1(\delta, \omega) > 0$ such that

$$|\eta(t) - \eta(s)| \leq K_1(\delta, \omega) |t-s|^{\frac{1}{2}-\rho}, \quad \forall t, s \in [0,T]. \quad (5.1.48)$$

Let us prove that the process $g(s) =: \psi(\eta(s), s)$ also belongs to $C^{\frac{1}{2}-}[0,T]$ P-a.s. Indeed, it follows from (iii) that $\forall L > 0$ there exists $K_2(L) > 0$ such that

$$|\psi(x,t) - \psi(x,s)| \leq K_2(L) |t-s|^{\frac{1}{2}}, \quad \forall |x| \leq L, \; t, s \in [0,T]. \quad (5.1.49)$$

It follows from the definition of $\psi(x,s)$ and (iv) that $\exists \tilde{M}, \tilde{N} > 0$

$$|\psi'_x(x,s)| \leq \tilde{M} \exp\{\tilde{N} |x|\}, \quad \forall s \in [0,T]. \quad (5.1.50)$$

Now we use (5.1.48)–(5.1.50) to obtain

$$\left| \psi(\eta(t), t) - \psi(\eta(s), s) \right| \leq \left| \psi(\eta(t), t) - \psi(\eta(s), t) \right| + \left| \psi(\eta(s), t) - \psi(\eta(s), s) \right|$$
$$\leq \sup_{|x| \leq |\eta(s)| \vee |\eta(t)|} |\psi'_x(x,t)| \cdot |\eta(t) - \eta(s)| + \left| \psi(\eta(s), t) - \psi(\eta(s), s) \right|$$
$$\leq \tilde{M} \exp\left\{ \tilde{N} \sup_{t \in [0,T]} |\eta(t)| \right\} K_1(\delta, \omega) |t-s|^{\frac{1}{2}-\delta} + K_2 \left(\sup_{t \in [0,T]} |\eta(t)| \right) |t-s|^{\frac{1}{2}}$$
$$\leq K_3(\delta, \omega) |t-s|^{\frac{1}{2}-\delta},$$

where

$$K_3(\delta, \omega) = \tilde{M} \exp\left\{ \tilde{N} \sup_{t \in [0,T]} |\eta(t)| \right\} K_1(\delta, \omega) + T^\delta K_2 \left(\sup_{t \in [0,T]} |\eta(t)| \right).$$

For any $H \in (3/4, 1)$ it is possible to find $\varepsilon = \varepsilon(H) > 0$ such that $C^{\frac{1}{2}-}[0,T] \subset C^{2(1-H)+\varepsilon}[0,T]$. So, we can apply Theorem 5.1.7 to the first term on the right-hand side of (5.1.47) and obtain its convergence to 0 in probability.

Consider the second term on the right-hand side of (5.1.47). Using (5.1.46) we obtain, as in (5.1.36), that

5.1 Discussion of the Arbitrage Problem

$$E\left|\int_0^t (g^\beta(s) - g(s)) \, dB_s^{H,\beta}\right|$$

$$\leq E \int_0^t |g^\beta(s) - g(s)| \cdot C s^\alpha \left|\int_0^{(s-\beta)_+} (s-y)^{\alpha-1} y^{-\alpha} \, d\tilde{W}_y\right| ds$$

$$\leq C \int_0^t s^\alpha \left(E(g^\beta(s) - g(s))^2 \cdot \int_0^{(s-\beta)_+} (s-y)^{2\alpha-2} y^{-2\alpha} \, dy \right)^{\frac{1}{2}} ds$$

$$\leq C \beta^\alpha \left(\int_0^{(1-\beta/T)_+} (1-y)^{2\alpha-2} y^{-2\alpha} \, dy \right)^{\frac{1}{2}} \int_0^t s^{H-1} ds$$

$$\leq C \beta^\alpha \left(1 + \beta^{2\alpha-1}\right)^{\frac{1}{2}} = O\left(\beta^{2\alpha-\frac{1}{2}}\right), \quad \beta \to 0+,$$

which means that $E\left|\int_0^t (f^\beta(s) - f(s)) \, dB_s^{H,\beta}\right| \to 0$ if $H \in \left(\frac{3}{4}, 1\right)$. □

Lemma 5.1.10. *Inequalities (5.1.45) and (5.1.46) are true for every $s \in [0, T]$ and $\beta \in (0, 1)$ with a constant C that does not depend on β and s.*

Proof. We prove only inequality (5.1.46) since (5.1.45) can be established similarly.

Denote a function $\psi(x, s) := \exp\{x\} \cdot \gamma(\exp\{x\}, s)$. Then the processes $g^\beta(s)$ and $g(s)$ are given by

$$g^\beta(s) = \psi(aW_s + bB_s^{H,\beta} + cs, \, s), \quad g(s) = \psi(aW_s + bB_s^H + cs, \, s).$$

We obtain from the Hölder inequality that

$$E(g^\beta(s) - g(s))^2 \leq E \left(\sup_{x \in I_1(s, \beta, \omega)} \left|\frac{\partial \psi(x, s)}{\partial x}\right| \cdot b \left(B_s^{H,\beta} - B_s^H\right) \right)^2$$

$$\leq b^2 \left(E \sup_{x \in I_1(s, \beta, \omega)} \left|\frac{\partial \psi(x, s)}{\partial x}\right|^{2p} \right)^{\frac{1}{p}} \left(E(B_s^{H,\beta} - B_s^H)^{2q} \right)^{\frac{1}{q}}, \quad (5.1.51)$$

where $p, q > 1$, $1/p + 1/q = 1$ and

$$I_1(s, \beta, \omega)$$
$$= \left\{ x : \, aW_s + \min(bB_s^{H,\beta}, bB_s^H) < x - cs < aW_s + \max(bB_s^{H,\beta}, bB_s^H) \right\}.$$

In the case when $2q < \frac{1}{1-H}$ (which is equivalent to the inequality $p > \frac{1}{2\alpha}$), we can use Theorem 1.15.2 and derive the following estimation:

$$\left(E(B_s^{H,\beta} - B_s^H)^{2q} \right)^{\frac{1}{q}} \leq C \begin{cases} (s^{2qH})^{\frac{1}{q}}, & s < \beta \\ (\beta^{2q\alpha} s^q + \beta^{2q(H-1)+1} s^{2q-1})^{\frac{1}{q}}, & \beta \leq s \end{cases}$$

$$\leq C \max\left(\beta^{2H}, \, \beta^{2\alpha}T + \beta^{2\alpha-1/p} T^{1+1/p}\right)$$

$$\leq \tilde{C} \beta^{2\alpha-1/p}, \quad \beta \in (0, 1), \quad (5.1.52)$$

where $\tilde{C} = C \max(1, 2T, 2T^{2H})$.

To estimate the first expectation in (5.1.51) note that

$$I_1(s, \beta, \omega) \subset \left\{ x : |x| \leq |aW_s| + |bB_s^H| + |bB_s^{H,\beta}| + |c|\,s \right\}$$
$$\subset \left\{ x : |x| \leq |aW_s| + 2|bB_s^H| + |b(B_s^{H,\beta} - B_s^H)| + |c|\,s \right\}$$
$$=: I_2(s, \beta, \omega). \tag{5.1.53}$$

We use (5.1.50), (5.1.53) and the Hölder inequality to obtain

$$\left(E \sup_{x \in I_1(s,\beta,\omega)} \left| \frac{\partial \psi(x,s)}{\partial x} \right|^{2p} \right)^{\frac{1}{p}} \leq \tilde{M}^2 \left(E \sup_{x \in I_2(s,\beta,\omega)} \exp\{2p\tilde{N}|x|\} \right)^{\frac{1}{p}}$$
$$\leq \tilde{M}^2 \left(E \exp\left\{ 2p\tilde{N}\left(|aW_s| + 2|bB_s^H| + |b(B_s^{H,\alpha} - B_s^H)| + |c|\,s \right) \right\} \right)^{\frac{1}{p}} \tag{5.1.54}$$
$$\leq \tilde{M}^2 \exp\{Ls\}$$
$$\times \left(E \exp\{3L|W_s|\} \cdot E \exp\{3L|B_s^H|\} \cdot E \exp\{3L|B_s^{H,\beta} - B_s^H|\} \right)^{\frac{1}{3}},$$

where $L = 2\tilde{N} \max(|c|, |a|, 2|b|)$. For a Gaussian random variable with zero mean $\xi \sim \mathcal{N}(0, \sigma^2)$, the following bound is well-known:

$$E \exp\{a|\xi|\} \leq 2 \exp\left\{ \frac{a^2 \sigma^2}{2} \right\}. \tag{5.1.55}$$

We use (5.1.55) and Theorem 1.15.2 to deduce from (5.1.54) that

$$\left(E \sup_{x \in I_1(s,\beta,\omega)} \left| \frac{\partial \psi(x,s)}{\partial x} \right|^{2p} \right)^{\frac{1}{p}}$$
$$\leq 2\tilde{M}^2 \exp\left\{ Ls + 3L^2/2 \left(E(W_s)^2 + E(B_s^H)^2 + E(B_s^{H,\beta} - B_s^H)^2 \right) \right\}$$
$$\leq 2\tilde{M}^2 \exp\left\{ LT + 3L^2/2 \left(T + T^{2H} + \right. \right.$$
$$\left. \left. + C \max(\beta^{2H}, \beta^{2\alpha} T(1 + \ln T - \ln \beta)) \right) \right\} \leq C < \infty, \tag{5.1.56}$$

for some $C > 0$ and all $\beta \in (0,1)$. Summarizing (5.1.51), (5.1.52) and (5.1.56) we obtain that for any $p > \frac{1}{2\alpha}$

$$\mathbb{E}\left(g^\beta(s) - g(s) \right)^2 \leq C \beta^{2\alpha - 1/p}, \quad s \in [0,T],\ \beta \in (0,1), \tag{5.1.57}$$

where constant C does not depend on p or β. Since p is arbitrary, inequality (5.1.46) follows from (5.1.57). $\square$

5.1.5 Equilibrium of Financial Market. The Fractional Burgers Equation

Definition 5.1.11. The financial market described by equation (5.1.7) is in equilibrium on $[0, T]$ if both the kernel φ_t and likelihood ratio $\frac{dQ}{dp}\big|_{\mathcal{F}_t}$ are the functions of t and W_t, twice differential in both the variables, and do not depend on the path of $\{W_s, 0 \leq s < t\}$ (for the corresponding notations see Subsection 3.2.3).

This definition generalizes the usual definition of equilibrium of the financial market involving only the Wiener process (see (HC93)), where the path's independence of $\frac{dQ}{dp}\big|_{\mathcal{F}_t}$ is declared, and the kernel φ_t equals simply $e(t, W_t)$, up to a constant multiplier.

Theorem 5.1.12. *If the financial market is in equilibrium, then φ_t satisfies the Burgers equation*

$$-\varphi(s,x)\varphi'_x(s,x) = \varphi'_t(s,x) + \frac{1}{2}\varphi''_{xx}(s,x).$$

Proof. Let $\varphi_t = g(t, W_t)$, and $\int_0^t \varphi_s dW_s - \frac{1}{2}\int_0^t \varphi_s^2 ds = G(t, W_t)$, where $g, G \in C^2(\mathbb{R}_+ \times \mathbb{R})$. Then

$$\int_0^t g(s, W_s) dW_s - \frac{1}{2}\int_0^t g^2(s, W_s) ds = G(t, W_t), \quad t \in [0.T].$$

From the Itô formula,

$$G(t, W_t) = \int_0^t (G'_t(s, W_s) + \frac{1}{2}G''_{xx}(s, W_s)) ds + \int_0^t G'_x(s, W_s) dW_s.$$

From here $g(s, W_s) = G'_x(s, W_s)$, $-\frac{1}{2}g^2(s, W_s) = G'_t(s, W_s) + \frac{1}{2}G''_{xx}(s, W_s)$, or, simply, $g(s,x) = G'_x(s,x)$, $-\frac{1}{2}g^2(s,x) = G'_t(s,x) + \frac{1}{2}G''_{xx}(s,x)$. Further, $g'_2(s,x) = G''_{22}(s,x)$, $-\frac{1}{2}g^2(s,x) = G'_t(s,x) + \frac{1}{2}g'_x(s,x)$. Therefore,

$$g'_t(s,x) = G'''_{tx}(s,x), -g(s,x)g'_x(s,x) = G'''_{tx}(s,x) + \frac{1}{2}g''_{xx}(s,x),$$

whence the proof follows. □

Remark 5.1.13. It is easy to see that the "principal" kernel $\theta_t = \varphi_t t^{-\alpha}$ satisfies the equation

$$s^{\alpha+1}\theta(s,x)\theta'_x(s,x) = \alpha\theta(s,x) + s\theta'_t(s,x) + s\frac{1}{2}\theta''_{xx}(s,x),$$

$s > 0$, $x \in \mathbb{R}$, and $\alpha = H - 1/2$, which can be called, in this connection, the fractional analog of the Burgers equation. (Recall that the usual Burgers equation has the form $u'_t = u''_{xx} + uu'_x$.)

5.2 The Different Forms of the Black–Scholes Equation on the Fractional Market

5.2.1 The Black–Scholes Equation for the Mixed Brownian–Fractional-Brownian Model

Consider a mixed version of the Black–Merton–Scholes model (5.1.7) with the value process X_t, described by (5.1.8), and self-financing strategies, defined by (5.1.9)–(5.1.10). Consider $C(t, S_t)$, the price of a European call option with striking price K at time $t \in [0, T]$. Suppose that $C \in C^1[0, T] \times C^2(\mathbb{R})$, then we can present the function $\widetilde{C}(t, S(t)) := C(T - t, S(t))$ according to the Itô formula from Theorem 2.7.2 as

$$\widetilde{C}(t, S(t)) = \widetilde{C}(0, x) + \int_0^t \left(\widetilde{C}'_t(u, S_u) + c\widetilde{C}'_S(u, S_u)S_u + \widetilde{C}'_S \frac{a^2}{2} S_u \right.$$

$$\left. + C''_{ss} \frac{a^2}{2} S_u^2 \right) du + a \int_0^t \widetilde{C}'_S(u, S_u) S_u dW_u + b \int_0^t C'_S(u, S_u) S_u dB_u^H. \quad (5.2.1)$$

Now, let the portfolio on value process consist of one option and an amount of $-\delta$ of underlying assets. The number $-\delta$ will be specified later. The value of this portfolio equals $X = \widetilde{C} - \delta S$.

The jump in the value of this portfolio in one-step time equals

$$dX = d\widetilde{C} - \delta dS = \left(\widetilde{C}'_t + c\widetilde{C}'_S + \frac{a^2}{2} \widetilde{C}''_{SS} S^2 \right) du + a\widetilde{C}'_S S dW_u + bC'_S S dB_u^H$$

$$- \delta \left(aSdW_u + bSdB_u^H + \frac{a^2 S}{2} du + cSdu \right). \quad (5.2.2)$$

If we choose $\delta = \frac{\partial \widetilde{C}}{\partial S}$ to eliminate the stochastic noise, then

$$dX = \left(\widetilde{C}'_t + \frac{a^2}{2} \widetilde{C}''_{SS} \cdot S^2 \right) du.$$

The return of an amount X invested in bank account equals $rXdt$ at time dt. For absence of arbitrage, these values must be the same. Hence we obtain the traditional Black–Scholes equation

$$\widetilde{C}'_t + \frac{1}{2} a^2 S^2 \frac{\partial^2 \widetilde{C}}{\partial S^2} - r\widetilde{C} + rS\widetilde{C}'_S = 0,$$

or, in terms of $C(t, S_t)$,

$$-C'_t + \frac{1}{2} a^2 S^2 \frac{\partial^2 C}{\partial S^2} - rC + rSC'_S = 0.$$

Remark 5.2.1. The same equation was obtained by Zähle (Zah02a) for the process $\widetilde{\mathcal{Z}}_t$ instead of $aW_t + bB_t^H$, where $\widetilde{\mathcal{Z}}_t = aW_t + b\mathcal{Z}_t$, and $\mathcal{Z}$ is continuous process with vanishing generalized quadratic variation.

5.2.2 Discussion of the Place of Wick Products and Wick–Itô–Skorohod Integral in the Problems of Arbitrage and Replication in the Fractional Black–Scholes Pricing Model

This section appears as a result of the interesting discussion of the related problems contained in the papers (SV03) and (BH05).

The fact of the existence of arbitrage in the "pure" fractional Brownian model is, to some degree, the consequence of the fact that the mathematical expectation of the stochastic integral w.r.t. fBm defined in the pathwise sense is nonzero (and you immediately obtain such an integral as a limit of the portfolio value created by step buy-and-hold strategies; we discussed this topic in Subsection 5.1.4). Note, however, that the arbitrage opportunity constructed by Rogers (Rog97) does not depend on any particular notion of integration. The same is true for the pre-limit arbitrage of the fractional Black–Scholes model considered in (Sot01). Nevertheless, many efforts were made to create the "pure" fractional model which will be "free of arbitrage", with the help of the stochastic integral constructed by Wick products. We mention in this connection the papers (HO03), (EvH03), (Ben03), (BO03), (BHOS02), (Mis04). Now we present the corresponding list of propositions for alternative definitions of portfolio values and self-financial conditions:

(i) the price of risky asset S is modeled by a geometric fBm and is the solution of the equation

$$dS_t = S_t \Diamond dB_t^H, \quad S_0 = s_0, \tag{5.2.3}$$

where $H \in (1/2, 1)$ everywhere. In this case

$$S_t = s_0 \exp^{\Diamond}(B_t^H) = s_0 \exp\left\{B_t^H - \frac{1}{2}t^{2H}\right\} \tag{5.2.4}$$

(see Section 2.3.1 for the definition of the Wick integral and recall that $\exp^{\Diamond}(X) = \sum_{n=0}^{\infty} X^{\Diamond n}$). Such an approach was developed in (EvH03) and (HO03). The portfolio value is defined in (EvH03). The standard way is $V_t = f_t B_t + g_t S_t$, where f and g are the respective numbers of units of the riskless and the risky asset held in the portfolio. However, in (HO03) the portfolio value is defined as

$$V_t = f_t B_t + g_t \Diamond S_t.$$

The standard Itô-type self-financing condition $dV_t = g_t dS_t$ is replaced by $dV_t = g_t S_t \Diamond dB_t^H$ in (EvH03) and by $dV_t = g_t \Diamond dS_t$ in (HO03).

The paper (BH05) claims that the definition of V_t as $V_t = f_t B_t + g_t S_t$ together with $dV_t = g_t S_t \Diamond dB_t^H$ (where we put $B_t \equiv 1$) has no economic interpretation as a self-financing condition. Here are the brief arguments. Consider a buy-and-hold portfolio. It must satisfy

$$V_t - V_u = g_u(S_t - S_u), \tag{5.2.5}$$

from intuitive point of view. However, in our case $V_t - V_u = \int_u^t g_u S_z \diamond dB_z^H$, where the last integral, in general, does not coincide with $g_u \int_u^t S_z \diamond dB_z^H$ and does not coincide with the right-hand side of (5.2.5). To be precise with this statement, consider the following example from (BH05): let the initial capital $x > 0$; at time $t = 0$ we put our money into the bank account and wait until $t = 1$. Since $B_t \equiv 1$ we receive x at time $t = 1$. At this moment we put our money into the risky asset, i.e., buy x/S_1 shares at the price S_1 and hold this position until $t = 2$. The value of this portfolio at time $t = 2$ is $V_2 = \frac{x}{S_1} S_2$. Evidently, such a strategy must be considered as self-similar since nothing was added or subtracted. Nevertheless, $\frac{x}{S_1} S_2 \neq x + \int_0^2 g_u S_u \diamond dB_u^H$ with $g_u = \frac{x}{S_1} \mathbf{1}_{(1,2]}(u)$. Indeed, $E(x + \int_0^2 g_u S_u \diamond dB_u^H)$ exists and equals x, but

$$xE \frac{S_2}{S_1} = xE \exp\left\{B_2^H - B_1^H - \frac{1}{2} 2^{2H} + \frac{1}{2} 1^{2H}\right\} = x \exp\left\{\frac{1}{2}(1 - 2^{2H})\right\}$$

$$\times E \exp\{B_2^H - B_1^H\} = x \exp\left\{\frac{1}{2}(1-2^{2H})\right\} \exp\left\{\frac{1}{2} \cdot (2-1)^{2H}\right\} = x \exp\{1 - 2^{2\alpha}\},$$

which is not x unless $H \neq 1/2$. There are some other objections concerning this model, see (BH05).

As to the model with $dV_t = g_t \diamond dS_t$, simple buy-and-hold strategies will be self-financing in this case. However, the objection in this case is that such a definition of portfolio $V_t = f_t dB_t + g_t \diamond dS_t$ is hard to motivate from the economic point of view. The reasoning in (BH05) is more moral and practical than mathematical: indeed, to calculate the value of portfolio in this case one needs to know Wick calculus and it is hard to instruct the broker how to do it. But there are also some mathematical reasonings against this model, because it can be proved that there exists a portfolio $f = 0$, $g_1 > 0$ such that $g_1 \diamond S_1 < 0$ with positive probability (index 1 stands for the moment of time here). It is sufficient to put $\Omega' = \{\omega \in \Omega | B_1^H(\omega) \in (1/2, 3/2)\}$, $g_1 = S_1 - 1$, where $S_1 = \exp\{B_1^H - 1/2\}$. Then $g_1 > 0$ on Ω', $P(\Omega') > 0$, $g_1 \diamond S_1 = S_1 \diamond S_1 - S_1 = \exp\{2B_1^H - 2\} - \exp\{B_1^H - \frac{1}{2}\} < 0$ on Ω'.

In spite of all this criticism, we can say some positive words about Wick (and Skorohod) models with fBm in finances. For other interesting facts and approaches to these topics see, for example, (AOPU00),(Oks07).

First, we mention that geometric fBm can be written in two forms:

$$S_t^{(1)} = S_0 e^{\mu t + \sigma B_t^H} \quad \text{or} \quad S_t^{(2)} = S_0 e^{\mu t + \sigma B_t^H - \frac{\sigma^2}{2} t^{2H}}. \tag{5.2.6}$$

The first form is very simple to understand but the second one is similar to usual geometrical Brownian model $S_t = S_0 e^{\mu t + \sigma B_t - \frac{1}{2}\sigma^2 t}$, because $ES_t^{(2)} = S_0$ for $\mu = 0$. (In Section 6.1 we shall consider the null hypothesis $H : S = S_t^{(2)}$ against $A : S = S_t^{(1)}$, but in a more complex form, see below.)

As mentioned in (SV03), if we consider it in the Riemann–Stieltjes sense, the geometric fBm $S_t^{(2)}$ with $\mu = 0$ is the solution of the equation

5.2 The Different Forms of the Black–Scholes Equation

$$dS_t^{(2)} = S_t^{(2)}(dB_t^H - Ht^{2\alpha}dt), \qquad (5.2.7)$$

and in the Wick–Skorohod sense $\delta S_t^{(2)} = S_t^{(2)} \delta B_t^H$ or $dS_t^{(2)} = S_t^{(2)} \diamond dB_t^H$, i.e. we obtain the model (5.2.3). Nevertheless, due to the Riemann–Stieltjes interpretation, we can consider self-financing condition as

$$V_t = V_0 + \int_0^t g_s S_s^{(2)} d(B_s^H - Hs^{2\alpha}ds),$$

and it has a clear economic meaning. Indeed, one can consider the Riemann–Stieltjes integral as an almost sure limit of simple predictable trading strategies.

Now we use the Itô formula (Theorem 2.7.6) for $m = 1$, $S_t := S_t^{(2)}$, $Y_t = \sigma B_t^H + \mu t - \frac{\sigma^2}{2} t^{2H}$, $H \in (1/2, 1)$ and $\widetilde{F}(t, x) = F(t, S_0 e^x)$, take (5.2.7) into account and obtain

$$\widetilde{F}(t, Y_t) := F(t, S_t) = F(0, S_0) + \int_0^t \frac{\partial F}{\partial t}(u, S_u) du$$

$$+ \int_0^t \frac{\partial F}{\partial x}(u, S_u) S_u(\mu - H\sigma^2 u^{2\alpha}) du + \sigma \int_0^t \frac{\partial F}{\partial x}(u, S_u) d(B_u^H - Hu^{2\alpha} du)$$

$$+ H\sigma^2 \int_0^t u^{2\alpha} \left(\frac{\partial^2 F}{\partial x^2}(u, S_u) S_u^2 + \frac{\partial F}{\partial x}(u, S_u) S_u \right) du$$

$$= F(0, S_0) + \int_0^t \frac{\partial F}{\partial t}(u, S_u) du + \mu \int_0^t \frac{\partial F}{\partial x}(u, S_u) S_u du$$

$$+ \sigma \int_0^t \frac{\partial F}{\partial x}(u, S_u) d(B_u^H - Hu^{2\alpha} du) + H\sigma^2 \int_0^t u^{2\alpha} \frac{\partial^2 F}{\partial x^2}(u, S_u) S_u^2 du.$$

Consider the assumption

$$E \sup_{0 \le s \le t} \left(\frac{\partial F}{\partial x}(s, S_s) S_s \right)^2 + E \sup_{0 \le s \le t} \left(\frac{\partial^2 F}{\partial x^2}(s, S_s) S_s^2 \right)^2 < \infty. \qquad (5.2.8)$$

Let $F(t, S_t) := \widetilde{C}(t, S_t) := C(T - t, S_t)$, where $C(t, x)$ is the price of some European option with $C(T, x) = c(x)$, and S satisfying assumption (5.2.8). Then, similarly to (5.2.2), we can present $d\widetilde{C}$ in differential form as

$$d\widetilde{C}_t = \sigma \frac{\partial \widetilde{C}}{\partial S} \cdot S(dB_t^H - Ht^{2\alpha}dt)$$

$$+ \left(\mu S \frac{\partial \widetilde{C}}{\partial S} + \frac{\partial \widetilde{C}}{\partial t} + \sigma^2 Ht^{2\alpha} \frac{\partial^2 \widetilde{C}}{\partial S^2} S^2 \right) dt.$$

Now, if the portfolio of value process V consists of one option and an amount of $-\delta$ of underlying assets, then the value $V = \widetilde{C} - \delta \cdot S$, the jump in the value of this portfolio in one time step equals

$$dV_t = d\widetilde{C}_t - \delta \cdot dS_t$$

$$= \sigma \frac{\partial \widetilde{C}}{\partial S} \cdot S_t(dB_t^H - Ht^{2\alpha}dt) - \delta(\sigma S_t(dB_t^H - Ht^{2\alpha}dt))$$

$$+ \left(\mu S_t \frac{\partial \widetilde{C}}{\partial S} + \frac{\partial \widetilde{C}}{\partial t} + \sigma^2 Ht^{2\alpha}\frac{\partial^2 \widetilde{C}}{\partial S^2}S_t^2 - \mu S_t \delta\right)dt.$$

If we choose $\delta := \frac{\partial \widetilde{C}}{\partial S}$ to eliminate the stochastic noise, then

$$dV = \left(\frac{\partial \widetilde{C}}{\partial t} + \sigma^2 Ht^{2\alpha}\frac{\partial^2 \widetilde{C}}{\partial S^2}S^2\right)dt.$$

The return on an amount V_t invested in the bank account equals $rV dt$ at time dt. For absence of arbitrage they must be equal, whence we obtain the fractional Black–Scholes equation ("Wick" version):

$$\frac{\partial \widetilde{C}}{\partial t} + \sigma^2 Ht^{2\alpha}\frac{\partial^2 \widetilde{C}}{\partial S^2}S^2 + rS\frac{\partial \widetilde{C}}{\partial S} - r\widetilde{C} = 0.$$

We can solve this equation on the segment $[0,T]$ with boundary condition $c(x) = (x-K)^+$, where $K > 0$ is strike price, and obtain

$$C(t,S) = \widetilde{C}(T-t,S) = S\Phi\left(\frac{\ln \frac{S}{K} + r(T-t) + (T^{2H} - t^{2H})\frac{\sigma^2}{2}}{\sigma\sqrt{T^{2H} - t^{2H}}}\right)$$

$$- Ke^{-r(T-t)}\Phi\left(\frac{\ln \frac{S}{K} + r(T-t) - (T^{2H} - t^{2H})\frac{\sigma^2}{2}}{\sigma\sqrt{T^{2H} - t^{2H}}}\right),$$

where $\Phi(\cdot)$ is a function of standard normal distribution. Note that it coincides with the solution of usual Black–Scholes equation for $H = 1/2$.

6

Statistical Inference with Fractional Brownian Motion

6.1 Testing Problems for the Density Process for fBm with Different Drifts

As we have seen in Subsection 5.2.2, the form of geometric fBm (5.2.6) depends on the kind of integral that is used in its calculations: if we use the Riemann–Stieltjes integral,

$$S_t^{(1)} = S_0^{(1)} + \mu \int_0^1 S_s^{(1)} ds + \sigma \int_0^t S_s^{(1)} dB_s^H, \text{ then}$$

$S_t^{(1)} = S_0^{(1)} \exp\{\mu t + \sigma B_t^H\}$, and if the behavior of geometric process is guided by the Wick integral,

$$S_t^{(2)} = S_0^{(2)} + \mu \int_0^1 S_s^{(2)} ds + \sigma \int_0^t S_s^{(2)} \diamond dB_s^H, \text{ then}$$

$S_t^{(2)} = S_0^{(2)} \exp\{\mu t + \sigma B_t^H - \frac{1}{2}\sigma^2 t^{2H}\}$. So, the natural question arises: what trend actually has geometric fBm? This question was considered in the paper (KMV05), and here we present a solution of this problem. In what follows the notation $X_n = o_P(1)$ means that $X_n \xrightarrow{P} 0$, $X_n = O_P(1)$ means that $\lim_{C \to \infty} \limsup_n P\{|X_n| \geq C\} = 0$. Assume that $H \in (1/2, 1)$. For a fixed $\mu \in \mathbb{R}$ let $P_{\mu,\sigma,\sigma}$ be the distribution of the process

$$X_t := \sigma B_t^H + \mu t - \frac{\sigma^2}{2} t^{2H}, \ 0 \leq t \leq T \tag{6.1.1}$$

in the space $C_{[0,T]}$ of continuous functions. Similarly, $P_{\mu,\sigma}$ is the distribution of the process

$$X_t := \sigma B_t^H + \mu t, \ 0 \leq t \leq T \tag{6.1.2}$$

in the space $C_{[0,T]}$.

Suppose now that we observe a trajectory of the process $\{X_t,\ 0 \leq t \leq T\}$ in the space $C_{[0,T]}$. Denote by P_X the law of X. We want to test the following complex hypothesis:

$$H : P_X \in \{P_{\mu,\sigma,\sigma} : \mu \in \mathbb{R}, \sigma \in \mathbb{R}_+\}$$

against the complex alternative

$$A : P_X \in \{P_{\mu,\sigma} : \mu \in \mathbb{R}, \sigma \in \mathbb{R}_+\}.$$

From the point of view of the general theory, models of observation (6.1.1) and (6.1.2) are equivalent to the classical model

$$\widetilde{X}_t = \int_0^t l_H(t,s)dX_s = \sigma M_t^H + \mu B_1 t^{1-2\alpha} - \sigma^2 H B_2 t \qquad (6.1.3)$$

and

$$\widetilde{X}_t = \sigma M_t^H + \mu B_1 t^{1-2\alpha}, \qquad (6.1.4)$$

where

$$\widetilde{X}_t = \int_0^t l_H(t,s)dX_s, \quad M_t^H = \int_0^t l_H(t,s)dB_s^H,$$

the kernel l_H is defined in Section 1.8, $B_1 := C_H^{(5)} B(1-\alpha, 1-\alpha)$, $B_2 := C_H^{(5)} B(1+\alpha, 1+\alpha)$.

Introduce the following density processes (Radon–Nikodym derivatives) based on the observed trajectory of X:

$$f_1(X : \mu, \sigma, \sigma) := \frac{dP_{\mu,\sigma,\sigma}}{dP_{0,\sigma}}(X) \qquad (6.1.5)$$

and

$$f_2(X : \mu, \sigma) := \frac{dP_{\mu,\sigma}}{dP_{0,\sigma}}(X). \qquad (6.1.6)$$

Theorem 6.1.1. *Assume we observe X on the interval $[0,T]$. We have*

$$f_1(X : \mu, \sigma, \sigma) = \exp\left\{a\frac{\mu}{\sigma^2}\widetilde{X}_T - b\widetilde{X}_T^1 - c\frac{\mu^2}{\sigma^2}T^{1-2\alpha} + d\mu T - k\sigma^2 T^{2H}\right\} \qquad (6.1.7)$$

and

$$f_2(X : \mu, \sigma) = \exp\left\{a\frac{\mu}{\sigma^2}\widetilde{X}_t - c\frac{\mu^2}{\sigma^2}T^{1-2\alpha}\right\}, \qquad (6.1.8)$$

where

$$\widetilde{X}_t^1 = \int_0^t s^{2\alpha}d\widetilde{X}_s, \quad a = B_1, \quad b = \frac{HB_2}{2(1-H)},$$
$$c = \tfrac{1}{2}B_1^2, \quad d = B_1 B_2 H, \quad k = \frac{HB_2^2}{8(1-H)}. \qquad (6.1.9)$$

Proof. Follows immediately from (6.1.1) to (6.1.6) and the classical Girsanov theorem. □

6.1.1 Observations Based on the Whole Trajectory with σ and H Known

In this section we demonstrate how to test the hypothesis **H** against the alternative **A**, when σ is known and the whole trajectory $\{X_t : t \in [0,T]\}$ is observed. We can use the likelihood ratio to test this (for the likelihood ratio see (Bor84), p. 319). In our problem the likelihood ratio $l(X.) = l(X.|\sigma)$ has the form

$$l(X.|\sigma) := \frac{\sup_{\mu \in R} f_1(X; \mu, \sigma, \sigma)}{\sup_{\mu \in R} f_2(X; \mu, \sigma)}. \qquad (6.1.10)$$

Note that in (6.1.10) both upper bounds are attained, since the densities f_1 and f_2 are the quadratic functions of μ. More precisely, we have

$$\sup_{\mu \in R} f_1(X; \mu, \sigma, \sigma) = \exp\Big\{ \tfrac{a^2}{4\sigma^2 c}(\widetilde{X}_t)^2 T^{2\alpha - 1} - 2\alpha b \widetilde{X}_t \cdot T^{2\alpha} \\ + 2\alpha b \int_0^T s^{2\alpha - 1} \widetilde{X}_s ds - 4\alpha^2 k \sigma^2 T^{2H} \Big\}, \qquad (6.1.11)$$

and the value of μ giving the maximal value in (6.1.11) is

$$\widehat{\mu}_H := \frac{a\widetilde{X}_t + dT\sigma^2}{2cT^{1-2\alpha}}. \qquad (6.1.12)$$

Similarly, for the denominator in (6.1.10) we have

$$\sup_{\mu \in R} f_2(X; \mu, \sigma) = \exp\left\{ \frac{a\widetilde{X}_T^2 T^{2\alpha-1}}{4\sigma^2 C} \right\} \qquad (6.1.13)$$

an the maximum in (6.1.13) is achieved by

$$\widehat{\mu}_A := \frac{a\widetilde{X}_T T^{2\alpha - 1}}{2c}. \qquad (6.1.14)$$

We obtain the following theorem as a direct consequence of (6.1.10) – (6.1.14):

Theorem 6.1.2. *The likelihood $l(X.|\sigma)$ from (6.1.10) admits the representation*

$$l(X.|\sigma) = \exp\Big\{ -2\alpha b \widetilde{X}_T T^{2\alpha} + 2\alpha b \int_0^T s^{2\alpha - 1} \widetilde{X}_s ds - 4\alpha^2 k \sigma^2 T^{2H} \Big\}.$$

Remark 6.1.3. Note that in the case when $H = \tfrac{1}{2}$ we have $l(X.|\sigma) = 1$. It means that our method does not work in this case, because the drift $(-\sigma^2 \tfrac{t}{2})$ has the same order in t as μt, and we cannot distinguish them. Therefore our method works worse if H is close to $\tfrac{1}{2}$.

Next we describe the testing procedure. Given a confidence level $1 - \rho$, $\rho \in (0, 1/2)$, consider the critical areas defined by $K_1 := \{X_. : l(X_.|\sigma) \geq K_\rho\}$ and $K_2 := \{X_. : l(X_.|\sigma) < k_\rho\}$. The critical values $0 < k_\rho \leq K_\rho$ are chosen in such a way that we have

$$\sup_{\mu \in R} P_{\mu,\sigma}(K_1) \leq \rho, \quad \sup_{\mu \in R} P_{\mu,\sigma,\sigma}(K_2) \leq \rho. \tag{6.1.15}$$

The test is now clear: if $X_. \in K_1$ we accept **H**, if $X_. \in K_2$ we accept **A**. If $l(X_.|\sigma) \in [k_\rho, K_\rho)$ then no hypothesis is accepted. Inequalities (6.1.15) show that the probabilities of so-called errors of the first and of the second kind will not exceed the level ρ.

Next we compute the critical values K_ρ, k_ρ. To compute k_ρ recall that under **A** the process X has the same distribution as the process $\sigma Z_t + \mu t$. Similarly, to compute K_ρ we use the fact that under **H** the process X has the same distribution as the process $\sigma Z_t + \mu t - \frac{\sigma^2}{2} t^{2H}$.

We have that

$$l(\sigma Z_. + \mu_. | \sigma) = \exp\left\{ -2\alpha b\sigma M_T^H \cdot T^{2\alpha} + 2\alpha b\sigma \int_0^T s^{2\alpha-1} M_s^H ds - 4\alpha^2 k\sigma^2 T^{2H} \right\}$$

and

$$l\left(\sigma Z_. + \mu_. - \frac{\sigma^2}{2} \cdot |\sigma\right) = \exp l(\sigma Z_. + \mu_. | \sigma) \exp\left\{ 8\alpha^2 k\sigma^2 T^{2H} \right\}.$$

Hence, we have that

$$P_{\mu,\sigma}(K_1) = P\Big\{ -2\alpha b\sigma M_T^H \cdot T^{2\alpha} \\ + 2\alpha b\sigma \int_0^T s^{2\alpha-1} M_s^H ds \geq \log K_\rho + 4\alpha^2 k\sigma^2 T^{2H} \Big\} \tag{6.1.16}$$

The random variable in the above expression is Gaussian with zero mean and variance

$$v^2 = \frac{\alpha^2 H B_2^2 \sigma^2 T^{2H}}{1 - H}.$$

Therefore, by (6.1.15)

$$P_{\mu,\sigma}(K_1) = 1 - \Phi\left(\frac{\log K_\rho}{v} + \frac{v}{2} \right), \tag{6.1.17}$$

where Φ is the distribution function of standard normal distribution. If ξ_ρ is such that $1 - \Phi(\xi_\rho) = \rho$, then $K_\rho \geq \exp\{v\xi_\rho - \frac{v^2}{2}\}$.

Similarly,

$$P_{\mu,\sigma,\sigma}(K_2) = 1 - \Phi\left(\frac{1}{v} \log\left(\frac{1}{k_\rho}\right) + \frac{v}{2} \right), \tag{6.1.18}$$

that is $k_\rho \leq \exp\{-v\xi_\rho + \frac{v^2}{2}\}$. Finally, we can choose $K_\rho = \max(1, \exp\{v\xi_\rho - \frac{v^2}{2}\})$, $k_\rho = K_\rho^{-1}$.

6.1.2 Discretely Observed Trajectory and σ Unknown

Assume now that we observe the process X discretely and the intensity σ of the fractional noise is unknown. We replace the parameter σ in $l(X.|\sigma)$ with a consistent estimate $\hat{\sigma}_n$, where n is the number of time points, and instead of the stochastic integrals w.r.t. X we will use sums in terms of the increments of X. We obtain a quasi-likelihood ratio, which is constructed from the observations. The critical values will be computed uniformly w.r.t. all possible values of μ and σ. We will give an asymptotic description of the critical levels.

First, choose the critical values independently of the parameter σ. For $K_\rho \geq 1$ we have that

$$\frac{1}{v}\log K_\rho + \frac{v}{2} \geq 2\sqrt{\frac{1}{2}\log K_\rho} = \sqrt{2\log K_\rho},$$

and from (6.1.17)

$$P_{\mu,\sigma}(K_1) \leq 1 - \Phi(\sqrt{2\log K_\rho}).$$

Take $K_\rho^* := e^{\frac{\xi_\rho^2}{2}}$ and put $K_1^* := \{X. : l(X.) \geq K_\rho^*\}$. Then we have

$$\sup_{\mu,\sigma > 0} P_{\mu,\sigma}(K_1^*) \leq \rho. \tag{6.1.19}$$

Similarly, using (6.1.18) and taking $k_\rho^* = e^{-\frac{\xi_\rho^2}{2}}$ and if $K_2^* := \{X. : l(X.) \leq k_\rho^*\}$ we will have

$$\sup_{\mu,\sigma > 0} P_{\mu,\sigma,\sigma}(K_2^*) \leq \rho. \tag{6.1.20}$$

Put

$$K_0^* := \{X. : k_\rho^* < l(X.) < K_\rho^*\};$$

note that K_0^* is (a conservative variant of) the region, where neither the hypothesis **H** nor the hypothesis **A** is accepted. Let $C_1 := \frac{\alpha\sqrt{H}B_2\sigma}{\sqrt{1-H}}$.

Theorem 6.1.4. *Assume that* $T > \left(\frac{\sqrt{2}}{C_1}\xi_\rho\right)^{1/H}$. *Then we have that*

$$\sup_{\mu,\sigma} P_{\mu,\sigma}(K_0^*) \leq \frac{4}{C_1}T^{-H}\exp\left\{-\frac{C_1^2 T^{2H}}{32}\right\} \tag{6.1.21}$$

and

$$\sup_{\mu,\sigma,\sigma} P_{\mu,\sigma,\sigma}(K_0^*) \leq \frac{4}{C_1}T^{-H}\exp\left\{-\frac{C_1^2 T^{2H}}{32}\right\}. \tag{6.1.22}$$

Proof. We have that

$$P_{\mu,\sigma}(K_0^*) \leq P_{\mu,\sigma}(\{X. : l(X.) > k_\rho^*\}) = 1 - \Phi\left(\frac{v}{2} + \frac{1}{v}\log k_\rho^*\right). \tag{6.1.23}$$

We have the following inequality for $x > 0$:

$$1 - \Phi(x) = \frac{1}{\sqrt{2\pi}} \int_x^\infty e^{-\frac{u^2}{2}} du \le \frac{1}{\sqrt{2\pi} x} e^{-\frac{x^2}{2}}.$$

Apply this to (6.1.22) with $x = \frac{C_1 T^H}{2} - \frac{1}{C_1 T^H} \frac{\xi_\rho^2}{2}$, and if $T > \left(\frac{\sqrt{2}\xi_\rho}{C_1}\right)^{1/4}$ we obtain (6.1.21). The estimate (6.1.22) is obtained similarly. □

Corollary 6.1.5.
$$\lim_{T \to \infty} \sup_\mu P_{\mu,\sigma}\{K_0^*\} = 0$$

and

$$\lim_{T \to \infty} \sup_\mu P_{\mu,\sigma,\sigma}\{K_0^*\} = 0.$$

Assume that we observe the process X at points $0 \le t_{n,1} < \cdots t_{n,n} \le T$, where $t_{n,k} \in \pi^n$. Put $\Delta^n = \max\{t_{n,1}, |\pi^n|, T - t_{n,n}\}$ and assume that

$$\lim_{n \to \infty} \Delta^n = 0. \tag{6.1.24}$$

We will introduce a discrete version of the functional $l(X.)$. Put $s_k = t_{n,k}$, $\Delta s_k = s_{k+1} - s_k$, $x_k = X_{t_{n,k}}$ and $\Delta x_k = x_{k+1} - x_k$. Assume that $\hat{\sigma}_n^2$ is some consistent estimator of σ^2. Put

$$l_n(x_1, \ldots, x_n) = \exp\Bigg(-2\alpha b T^{2\alpha} C_H^{(5)} \sum_{k=0}^{n-1} s_{k+1}^{-\alpha}(T - s_k)^{-\alpha} \Delta x_k$$

$$+ 2\alpha b C_H^{(5)} \sum_{k=1}^n s_{k+1}^{2\alpha-1} \left(\sum_{i=0}^{k-1} s_{i+1}^{-\alpha}(s_k - s_i)^{-\alpha} \Delta x_i \right) \Delta s_k - 4\alpha^2 k \hat{\sigma}_n^2 T^{2H} \Bigg).$$

With the help of constants K_ρ^* and k_ρ^* from (6.1.19) and (6.1.20) define the critical domains

$$K_{1n}^* := \{(x_{n,1}, \ldots, x_{n,n}) \in \mathbb{R}^n | l_n(x_{n,1}, \ldots, x_{n,n}) \ge K_\rho^*\}$$

and

$$K_{2n}^* := \{(x_{n,1}, \ldots, x_{n,n}) \in \mathbb{R}^n | l_n(x_{n,1}, \ldots, x_{n,n}) < k_\rho^*\}.$$

If the observations belong to K_{1n}^* then **H** is accepted and if the observations belong to K_{2n}^* then **A** is accepted.

Theorem 6.1.6. *Assume that we have (6.1.24) as $n \to \infty$. Then for any $\mu \in \mathbb{R}$, $\sigma > 0$ we have that*

$$l_n(x_{n,1}, \ldots, x_{n,n}) \xrightarrow{P_{\mu,\sigma,\sigma}} l(X.|\sigma) \tag{6.1.25}$$

and

$$l_n(x_{n,1}, \ldots, x_{n,n}) \xrightarrow{P_{\mu,\sigma}} l(X.|\sigma). \tag{6.1.26}$$

6.1 Testing Problems for the Density Process for fBm with Different Drifts

Proof. We prove the claim (6.1.25) (the claim (6.1.26) is proved similarly). Denote by $l(\widehat{X}.|\sigma)$ the random variable $l(X.|\sigma)$, when the process X_t is replaced by the process $\sigma Z_t + \mu t - \frac{\sigma^2 t^{2H}}{2}$, $0 \leq t \leq T$, and by $l_n(\widehat{x}_{n,1}, \ldots, \widehat{x}_{n,n})$ the variable where we replace $\Delta X_{n,k}$ by $\widehat{\sigma}_n \Delta Z_{n,k} + \mu \Delta t_k - \frac{\sigma^2 (\Delta t_k)^{2H}}{2}$. Then for any $\varepsilon > 0, C > 0$ we have that

$$P_{\mu,\sigma,\sigma}\{|l_n(x_{n,1}, \ldots, x_{n,n}) - l(\widehat{X}.|\sigma)| > \varepsilon\}$$
$$\leq P_{\mu,\sigma,\sigma}\{|l_n(x_{n,1}, \ldots, x_{n,n})| \geq C\} + P_{\mu,\sigma,\sigma}\{|l(\widehat{X}.|\sigma)| \geq C\}$$
$$+ P_{\mu,\sigma,\sigma}\{|\log l_n(\widehat{x}_{n,1}, \ldots, \widehat{x}_{n,n}) - \log l(\widehat{X}.|\sigma)| > \varepsilon e^{-C}\}. \quad (6.1.27)$$

The first two probabilities can be chosen sufficiently small for large $C > 0$. The structure of the functionals $l(\widehat{X}.|\sigma)$ and $l_n(\widehat{x}_{n,1}, \ldots, \widehat{x}_{n,n})$, the facts that $\widehat{\sigma}_n \xrightarrow{P_{\mu,\sigma,\sigma}} \sigma$, $C_H^{(5)} \sum_{k=1}^{n-1} s_{k+1}^{-\alpha}(T - s_k)^{-\alpha} \Delta s_k \to \int_0^t l_H(t,s) ds$ and

$$C_H^{(5)} \sum_{k=1}^{n} s_k^{2\alpha-1} \sum_{i=1}^{k-1} s_{i+1}^{-\alpha}(s_k - s_i)^{-\alpha} \Delta s_i \Delta s_k \to \int_0^T s^{2\alpha-1} \int_0^s l_H(s,u) du \, ds,$$

supply that it is sufficient to prove that

$$C_H^{(5)} \sum_{k=1}^{n} s_{k+1}^{-\alpha}(T - s_k)^{-\alpha} \Delta B_k^H \xrightarrow{P_{\mu,\sigma,\sigma}} M_T^H, \quad (6.1.28)$$

where $\Delta Y_k := Y_{(k+1)T/n} - Y_{kT/n}$ for any process Y, and that

$$C_H^{(5)} \sum_{k=1}^{n} s_{k+1}^{2\alpha-1} \left(\sum_{i=1}^{k-1} s_{i+1}^{-\alpha}(s_k - s_i)^{-\alpha} \Delta B_i^H \right) \Delta s_k$$
$$\xrightarrow{P_{\mu,\sigma,\sigma}} \int_0^T s^{2\alpha-1} M_s^H ds. \quad (6.1.29)$$

To prove (6.1.28) consider for $f_{n,T}(s) = C_H^{(5)} s_{k+1}^{-\alpha}(T - s_k)^{-\alpha} \mathbf{1}_{\{s \in [s_k, s_{k+1})\}}$

$$E\left(M_T^H - C_H^{(5)} \sum_{k=1}^{n-1} s_{k+1}^{-\alpha}(T - s_k)^{-\alpha} \Delta B_k^H \right)^2$$
$$= 2H\alpha \int_0^T \int_0^T (l_H(T,s) - f_{n,T}(s))(l_H(T,u) - f_{n,T}(u)) |u - s|^{2\alpha-2} du \, ds.$$

We have that $f_{n,T}(s) \uparrow l_H(T,s)$ for $s \in (0,T)$, and $\int_0^T \int_0^T l_H(T,s) l_H(T,u) \times |u-s|^{2\alpha-1} du \, ds < \infty$. Therefore, by monotone convergence,

$$\int_0^T \int_0^T (l_H(T,s) - f_{n,T}(s))(l_H(T,u) - f_{n,T}(u)) |u-s|^{2\alpha-1} du \, ds \to 0$$

as $n \to \infty$ and (6.1.28) follows.

To finish, we prove (6.1.29). Denote $g(s) := s^{2\alpha-1} M_s^H$ and

$$g_n(s) := C_H^{(5)} s_{k+1}^{2\alpha-1} \sum_{i=1}^{k-1} s_{i+1}^{-\alpha} (s_k - s_i)^{-\alpha} \Delta B_i^H \mathbf{1}_{\{s \in [s_k, s_{k+1})\}}.$$

Then, for any $s \in (0, T]$,

$$E|g(s) - g_n(s)| \leq |s^{2\alpha-1} - s_{k+1}^{2\alpha-1}| E|M_s^H| + 2H\alpha s^{2\alpha-1}$$
$$\times \left(E \int_0^s \int_0^s (l_H(s,u) - f_{n,s}(u))(l_H(s,r) - f_{n,s}(r))|u - r|^{2\alpha-1} du\, dr \right)^{1/2},$$
(6.1.30)

and as in previous inequalities, the second term on the right-hand side of (6.1.30) goes to zero, moreover the left-hand side can be dominated, according to Remark 1.9.5, by

$$\widetilde{C}_H^{(1)} s^{2\alpha-1} T^{1-H} + \widetilde{C}_H^2 \|l_H(s, \cdot)\|_{L_{1/H}[0,s]} \leq \widetilde{C}_H^{(3)} T^{1-H},$$

where $\widetilde{C}_H^{(i)}$ are some constants, $i = 1, 2, 3$. From here $E|\int_0^T (g(s) - g_n(s)) ds| \to 0$, as $n \to \infty$ and we obtain (6.1.29).

Corollary 6.1.7. *Assume that (6.1.24) holds. Then*

$$\limsup_n P_{\mu,\sigma}(K_{1n}^*) \leq \rho, \quad \limsup_n P_{\mu,\sigma,\sigma}(K_{2n}^*) = 0$$

and

$$\lim_T \limsup_n (P_{\mu,\sigma} + P_{\mu,\sigma,\sigma})(K_{0n}^*) = 0,$$

where $K_{0n}^* := \{(x_{n,1}, \ldots, x_{n,n}) : k_\rho^* < \log l_n(x_{n,1}, \ldots, x_{n,n}) < K_\rho^*\}$.

Proof. By Theorem 6.1.6 we have, as $n \to \infty$:

$$P_{\mu,\sigma}(K_{1n}^*) \to P_{\mu,\sigma}(K_1^*), \quad P_{\mu,\sigma,\sigma}(K_{2n}^*) \to P_{\mu,\sigma}(K_2^*)$$

and

$$(P_{\mu,\sigma} + P_{\mu,\sigma,\sigma})(K_{0n}^*) \to (P_{\mu,\sigma} + P_{\mu,\sigma,\sigma})(K_0^*).$$

Hence the statements of the corollary follow from Theorem 6.1.6. $\square$

Note that according to Corollary 6.1.7 the proposed test procedure has asymptotically the level of errors less than or equal to ρ for both kinds of errors. Note also that the probability not to make a decision goes to zero as $T \to \infty$. It is also easy to see from the proof of Theorem 6.1.6 and Corollary 6.1.7 that this convergence is uniform for all μ and all $\sigma \geq \sigma_0 > 0$, where σ_0 is fixed.

6.2 Goodness-of-fit Test

6.2.1 Introduction

Suppose that **H** was tested against **A**, and we conclude that, e.g. **A** is true. Consider a certain functional depending on the trajectory of the observed process $\{X_t, 0 \le t \le T\}$. If the distribution of this functional under **A** is known we can construct the corresponding goodness-of-fit test. For a given confidence level we either reject **A** or do not reject **A**. If we reject **A** it means that the observed trajectory does not fit the model described by **A**, and we conclude finally in this case that both **A** and **H** are wrong.

If the parameters in the models are unknown we propose an asymptotical test which provides a given confidence level as $T \to +\infty$.

6.2.2 The Whole Trajectory Is Observed and the Parameters μ and σ Are Known

Introduce a functional which depends on the whole observed trajectory $\{x(t), t \in [0,T]\}$, in a linear way:

$$Q_T := \int_0^T Z(T,s) dX_s,$$

where

$$Z(T,s) = s^{1/4-H}(T-s)^{3/4-H}.$$

We choose here the exponents $\frac{1}{4} - H$ and $\frac{3}{4} - H$ different from $\frac{1}{2} - H$ in order to obtain the functional which is essentially different from M_T^H. The reason for that will be clear from Theorem 6.2.3. The integral exists in both cases when $X_t = \sigma B_t^H + \mu t$ and $X_t = \sigma B_t^H + \mu t - \frac{\sigma^2}{2} t^{2H}$.

Denote

$$B_3 = B\left(\frac{5}{4} - H, \frac{7}{4} - H\right), \quad B_4 = B\left(H + \frac{1}{4}, \frac{7}{4} - H\right).$$

Theorem 6.2.1. *Let the parameters μ and σ be known.*

(i) *Assume that we have **H**: $X_t = \sigma B_t^H + \mu t - \frac{\sigma^2}{2} t^{2H}$. Then*

$$R_T^{\mathbf{H}} := T^{H-1} Q_T - \mu B_3 \cdot T^{1-H} + \sigma^2 H \cdot B_4 \cdot T^H \sim N(0, C_2 \sigma^2);$$

(ii) *Assume that we have **A**: $X_t = \sigma B_t^H + \mu t$. Then*

$$R_T^{\mathbf{A}} := T^{H-1} Q_T - \mu B_3 \cdot T^{1-H} \sim N(0, C_2 \sigma^2),$$

where

$$C_2 = 2H\alpha \int_0^1 \int_0^1 (us)^{\frac{1}{4}-H}((1-u)(1-s))^{\frac{3}{4}-H} \cdot |u-s|^{2\alpha-1} du\, ds.$$

Proof. Assume **H**. Then we have

$$Q_T = \sigma \int_0^T Z(T,s)dB_s^H + \mu T^{1-2\alpha} B_3 - \sigma^2 H B_4 T \qquad (6.2.1)$$

and so

$$R_T^{\mathbf{H}} = T^{H-1} \sigma \int_0^T Z(T,s)dB_s^H.$$

Obviously, $R_T^{\mathbf{H}}$ is normally distributed with mean zero and with variance

$$E(R_T^{\mathbf{H}})^2 = \sigma^2 T^{2\alpha-1} 2\alpha H \int_0^T \int_0^T (us)^{\frac{1}{4}-H}((T-u)(T-s))^{\frac{3}{4}-H}|s-u|^{2\alpha-1} du\, ds,$$

i.e. $E(R_T^{\mathbf{H}})^2 = \sigma^2 C_2$ and the first claim now follows.

Assume **A**. Then we can write Q_T as

$$Q_T = \sigma \int_0^T Z(T,s)dB_s^H + \mu T^{1-2\alpha} B_3, \qquad (6.2.2)$$

and the second claim follows from (6.2.2) as above. $\square$

The goodness-of-fit tests are based on the statistics

$$\overline{R}_T^{\mathbf{H}} := \frac{R_T^{\mathbf{H}}}{\sigma(C_2)^{\frac{1}{2}}}, \quad \overline{R}_T^{\mathbf{A}} := \frac{R_T^{\mathbf{A}}}{\sigma(C_2)^{\frac{1}{2}}}.$$

Fix a confidence level $1-\rho, \rho \in (0,\frac{1}{2})$, and let $\xi_{\frac{\rho}{2}}$ be a $\frac{\rho}{2}$-fractile of a standard normal law, i.e $P\{N(0,1) \geq \xi_{\frac{\rho}{2}}\} = \frac{\rho}{2}$. We reject **H** if $|\overline{R}_T^{\mathbf{H}}| > \xi_{\frac{\rho}{2}}$, and reject **A** if $|\overline{R}_T^{\mathbf{A}}| > \xi_{\frac{\rho}{2}}$.

Note that under **H**, $\overline{R}_T^{\mathbf{A}} \xrightarrow{P_{\mu,\sigma,\sigma}} -\infty$, $T \to +\infty$, therefore the inequality $\overline{R}_T^{\mathbf{A}} < -\xi_{\frac{\rho}{2}}$ is an additional argument in favor of **H**.

Also, if **A** is true, then $\overline{R}_T^{\mathbf{H}} \xrightarrow{P_{\mu,\sigma}} +\infty, T \to +\infty$, therefore the inequality $\overline{R}_T^{\mathbf{A}} > \xi_{\frac{\rho}{2}}$ is an additional argument in favor of **A**.

Remark 6.2.2. Suppose that in reality we have the model $X_t = \sigma B_t^{H_1} + \mu t$, $H_1 > H$, not $\sigma B_t^H + \mu t$. Denote the law of X in this case by P. Then

$$\overline{R}_T^{\mathbf{H}} = \frac{T^{H-1}}{(C_2)^{\frac{1}{2}}} \int_0^T s^{\frac{1}{4}-H}(T-s)^{\frac{3}{4}-H} dB_s^{H_1},$$

and $E(\overline{R}_T^{\mathbf{H}})^2$ has the order $T^{2(H_1-H)}$ for large T, thus $\overline{R}_T^{\mathbf{H}} \xrightarrow{P} \infty, T \to \infty$, and

$$\overline{R}_T^{\mathbf{H}} = \overline{R}_T^{\mathbf{A}} + \frac{\sigma H B_4 T^H}{(C_2)^{\frac{1}{2}}} = T^{H_1-H} O_P(1) + \frac{\sigma H B_4 T^H}{(C_2)^{\frac{1}{2}}} \xrightarrow{P} +\infty, \quad T \to \infty.$$

Therefore our statistics can distinguish this case, too.

6.2.3 Goodness-of-fit Tests with Discrete Observations

Asymptotic Behavior of Discrete Statistics for μ Unknown and σ Known

Suppose for simplicity that we observe the values $X_{\frac{kT}{n}}, k = 0, 1, \ldots, n$. We substitute in $R_T^{\mathbf{A}}, R_T^{\mathbf{H}}$ a discretization of Q_T,

$$\widehat{Q}_T := \sum_{k=0}^{n-1} \left(\frac{(k+1)T}{n}\right)^{\frac{1}{4}-H} \left(T - \frac{kT}{n}\right)^{\frac{3}{4}-H} \triangle X_k.$$

Instead of μ we substitute the estimates (6.1.12) and (6.1.14), respectively. Thus we define

$$\widehat{R}_T^{\mathbf{H}} := T^{H-1}\widehat{Q}_T - \widehat{\mu}_{\mathbf{A}} B_3 T^{1-H} + \sigma^2 H B_4 T^H$$

and

$$\widehat{R}_T^{\mathbf{A}} := T^{H-1}\widehat{Q}_T - \widehat{\mu}_{\mathbf{H}} B_3 T^{1-H}.$$

Under hypothesis **H** we have

$$\widehat{R}_T^{\mathbf{H}} = \sigma T^{-H} \sum_{k=0}^{n-1} \left(\frac{k+1}{n}\right)^{\frac{1}{4}-H} \left(1 - \frac{k}{n}\right)^{\frac{3}{4}-H} \triangle B_{\frac{kT}{n}}^H \quad (6.2.3)$$

$$+ \mu T^{1-H} \sum_{k=0}^{n-1} \left(\frac{k+1}{n}\right)^{\frac{1}{4}-H} \left(1 - \frac{k}{n}\right)^{\frac{3}{4}-H} \cdot \frac{1}{n} - \frac{\sigma^2}{2} T^H \sum_{k=0}^{n-1} \left(\frac{k+1}{n}\right)^{\frac{1}{4}-H}$$

$$\times \left(1 - \frac{k}{n}\right)^{\frac{3}{4}-H} \left(\left(\frac{k+1}{n}\right)^{2H} - \left(\frac{k}{n}\right)^{2H}\right) - \widehat{\mu}_{\mathbf{A}} B_3 T^{1-H} + \sigma^2 H B_4 T^H,$$

and under hypothesis **A**

$$\widehat{R}_T^{\mathbf{A}} = \sigma T^{-H} \sum_{k=0}^{n-1} \left(\frac{k+1}{n}\right)^{\frac{1}{4}-H} \left(1 - \frac{k}{n}\right)^{\frac{3}{4}-H} \triangle B_{\frac{kT}{n}}^H \quad (6.2.4)$$

$$+ \mu T^{1-H} \sum_{k=0}^{n-1} \left(\frac{k+1}{n}\right)^{\frac{1}{4}-H} \left(1 - \frac{k}{n}\right)^{\frac{3}{4}-H} \cdot \frac{1}{n} - \widehat{\mu}_{\mathbf{H}} B_3 T^{1-H}.$$

To begin we find the rate of convergence of the integral sums in (6.2.3) and (6.2.4) to the corresponding integrals.

Define $\widetilde{R}_T^{\mathbf{A}}$ by

$$\widetilde{R}_T^{\mathbf{A}} := \frac{\sigma}{T^{1-H}} \int_0^T s^{1/4-H} (T-s)^{3/4-H} dB_s^H + B_3 T^{1-H} (\mu - \widehat{\mu}_{\mathbf{A}})$$

and $\widetilde{R}_T^{\mathbf{H}}$ similarly, with $\widehat{\mu}_{\mathbf{H}}$ replacing $\widehat{\mu}_{\mathbf{A}}$.

We study the differences $\widehat{R}_T^{\mathbf{A}} - \widetilde{R}_T^{\mathbf{A}}$ and $\widehat{R}_T^{\mathbf{H}} - \widetilde{R}_T^{\mathbf{H}}$. Put

$$q_n(T,s) := \sum_{k=0}^{n-1} \left(\frac{(k+1)T}{n}\right)^{\frac{1}{4}-H} \left(T - \frac{kT}{n}\right)^{\frac{3}{4}-H} \cdot \mathbf{1}_{[\frac{kT}{n}, \frac{(k+1)T}{n})}(s),$$

$$I(\delta, \beta) := B(\delta+1, \beta+1) = \int_0^1 s^\delta (1-s)^\beta ds,$$

and

$$I_n(\delta, \beta) = \sum_{k=0}^{n-1} \left(\frac{k+1}{n}\right)^\delta \left(1 - \frac{k}{n}\right)^\beta \frac{1}{n}.$$

We have that

$$\widehat{R}_T^{\mathbf{A}} - \widetilde{R}_T^{\mathbf{A}} = T^{H-1} \int_0^T (q_n(T,s) - q(T,s)) \, dB_s^H$$
$$- T^{1-H} \mu \left(I_n(1/4 - H, 3/4 - H) - I(1/4 - H, 3/4 - H) \right) \quad (6.2.5)$$

and

$$\widehat{R}_T^{\mathbf{H}} - \widetilde{R}_T^{\mathbf{H}} = \widehat{R}_T^{\mathbf{A}} - \widetilde{R}_T^{\mathbf{A}} - \frac{\sigma^2}{2} T^H 2H \Big(I_n(H - 3/4, 3/4 - H)$$
$$- I(H - 3/4, 3/4 - H) \Big) - \frac{\sigma^2}{2} T^H \left(\sum_{k=0}^{n-1} \left(\left(\frac{k+1}{n}\right)^{1/4-H} \left(1 - \frac{k}{n}\right)^{3/4-H} \right. \right. \quad (6.2.6)$$
$$\left. \left. \times \left(\left(\frac{k+1}{n}\right)^{2H} - \left(\frac{k}{n}\right)^{2H} \right) - 2H \left(\frac{k+1}{n}\right)^{H-3/4} \left(1 - \frac{k}{n}\right)^{3/4-H} \frac{1}{n} \right) \right).$$

Using self-similarity, we obtain that

$$E\left(T^{H-1} \int_0^T (q_n(T,s) - q(T,s)) \, dB_s^H \right)^2$$
$$= E\left(\int_0^1 (q_n(1,s) - q(1,s)) \, dB_s^H \right)^2. \quad (6.2.7)$$

According to Remark 1.9.5 we have

$$E\left(\int_0^1 (q_n(1,s) - q(1,s)) \, dB_s^H \right)^2 \leq c_H \|q_n(1,s) - q(1,s)\|_{L_{1/H}[0,1]}^2. \quad (6.2.8)$$

Now we use these preliminary calculations to prove the next result. Let $n = n(T)$ be the number of approximation points.

Theorem 6.2.3. *Assume*

(iii) *For $\frac{1}{2} < H \leq \frac{3}{4}$,*

$$\frac{T^\beta}{n(T)} \to 0, \quad T \to \infty, \quad \text{with } \beta = \frac{H}{H + \frac{1}{4}}.$$

(iv) For $\frac{3}{4} < H < 1$,
$$\frac{T^\beta}{n(T)} \to 0, T \to \infty, \text{ with } \beta = H.$$

Then under **H**
$$\widehat{R}_T^{\mathbf{H}} - \widetilde{R}_T^{\mathbf{H}} = o_P(1), \quad T \to \infty \tag{6.2.9}$$

and under **A**
$$\widehat{R}_T^{\mathbf{A}} - \widetilde{R}_T^{\mathbf{A}} = o_P(1), \quad T \to \infty. \tag{6.2.10}$$

Moreover, under **H** $\widetilde{R}_T^{\mathbf{H}} \sim N(0, r^2)$, *and under* **A** $\widetilde{R}_T^{\mathbf{A}} \sim N(0, r^2)$, *where*

$$r^2 := 2\sigma^2 \alpha H \int_0^1 \int_0^1 \varphi(s)\varphi(u) \cdot |u-s|^{2\alpha-1} \, du \, ds,$$

with
$$\varphi(s) := s^{\frac{1}{4}-H}(1-s)^{\frac{3}{4}-H} - \frac{B_3}{B_1} s^{-\alpha}(1-s)^{-\alpha}.$$

Proof. To prove the claims note first that using Lemmas B.0.1 and B.0.2 from Appendix B we have that $\widehat{R}_T^{\mathbf{H}} - \widetilde{R}_T^{\mathbf{H}} = o_P(1)$ under **H** and $\widehat{R}_T^{\mathbf{A}} - \widetilde{R}_T^{\mathbf{A}} = o_P(1)$ under **A**. Next, we substitute (6.1.12) into $\widetilde{R}_T^{\mathbf{H}}$ and obtain

$$\widetilde{R}_T^{\mathbf{H}} = \frac{\sigma}{T^{1-H}} \int_0^T \left(s^{\frac{1}{4}-H}(T-s)^{\frac{3}{4}-H} - \frac{B_3}{B_1} s^{-\alpha}(T-s)^{-\alpha} \right) dB_s^H.$$

This implies that under **H** $\widetilde{R}_T^{\mathbf{H}} \sim N(0, r^2)$. Similarly, one shows that under **A** $\widetilde{R}_T^{\mathbf{A}} \sim N(0, r^2)$. □

Remark 6.2.4. For the kernel $l_H(t,s)$ instead of $Z(t,s)$ we obtain the degenerate distribution of $\widetilde{R}_T^{\mathbf{H}}$ and $\widetilde{R}_T^{\mathbf{A}}$. This is the reason why we take the kernel $Z(t,s)$.

Goodness-of-fit Test

Based on Theorem 6.2.3, we construct the goodness-of-fit test similarly to the one from Subsection 6.2.2. Choose $\xi_{\frac{\rho}{2}}$ as there. We reject **H** if $\left|\widehat{R}_T^{\mathbf{H}}\right| > r\xi_{\frac{\rho}{2}}$, and we reject **A** if $\left|\widehat{R}_T^{\mathbf{A}}\right| > r\xi_{\frac{\rho}{2}}$. The test is applicable for large T only, contrary to the test from Subsection 6.2.2, because for the probability $p_{\mathbf{H}}(T)$ that **H** is rejected when **H** is true, we have now

$$\lim_{T \to \infty} p_{\mathbf{H}}(T) = \rho$$

and similarly for **A** and $p_{\mathbf{A}}(T)$.

6.2.4 On Volatility Estimation

In this subsection we construct an estimator for the parameter σ. We end this subsection by giving the goodness-of-fit test for the case where both μ and σ are unknown.

Introductory Computations for Volatility Estimation

Assume **H**. Then the background process is $X_t = \sigma B_t^H + \mu t - \frac{\sigma^2}{2} t^{2H}, t \geq 0$. We make observations at time points $t_k = \frac{kT}{n}, k = 0, 1, \ldots, n$. Put, as before, $\Delta X_k = X_{\frac{k+1}{n}T} - X_{\frac{k}{n}T}, k = 0, \ldots, n-1$. Then we have, with obvious notation, that

$$\Delta X_k = \sigma \Delta B_k^H + \mu \Delta t_k - \frac{\sigma^2}{2} \Delta(t^{2H})_k,$$

$k = 0, \ldots, n-1$. Consider now $\frac{\Delta X_k}{T^H}$ and write this as

$$\frac{\Delta X_k}{T^H} = \sigma \frac{1}{n^H} \varepsilon_k + \frac{\mu \Delta t_k}{T^H} - \frac{\sigma^2}{2T^H} \Delta(t^{2H})_k. \quad (6.2.11)$$

In (6.2.11) we used the notation $\varepsilon_k = \frac{\Delta B_k^H n^H}{T^H}$. By self-similarity the distribution of the vector $(\varepsilon_0, \ldots, \varepsilon_{n-1})$ is the same as of the vector

$$\frac{B_{\frac{1}{n}}^H - B_0^H}{\frac{1}{n^H}}, \ldots, \frac{B_1^H - B_{\frac{(n-1)}{n}}^H}{\frac{1}{n^H}} \stackrel{d}{=} B_1^H - B_0^H, B_2^H - B_1^H, \ldots, B_n^H - B_{n-1}^H,$$

where we again used self-similarity. Simple computation gives $E\varepsilon_k = 0$, $E\varepsilon_k^2 = 1$ and

$$E\varepsilon_k \varepsilon_l = \frac{1}{2} \left(|k - l + 1|^{2H} - 2|k - l|^{2H} + |k - l - 1|^{2H} \right).$$

If $k > l \geq 1$ and $\frac{1}{2} \leq H < 1$, then, applying the mean value theorem twice gives

$$0 \leq E\varepsilon_k \varepsilon_l \leq 2H\alpha(k - l)^{2\alpha - 1}. \quad (6.2.12)$$

Denote $\mu_1 := \frac{n^H \mu \Delta t}{T^H}$, $y_t := \frac{n^H \Delta X_t}{T^H}$ and rewrite (6.2.11):

$$y_t = \sigma \varepsilon_t + \mu_1 - \frac{\sigma^2}{2} T^{-H} n^H \Delta t^{2H}.$$

To simplify the notation put

$$y_k := \sigma \varepsilon_k + \mu_1 - \frac{\sigma^2}{2} T^H n^H \Delta \tau_k^{2H}, k = 0, 1, \ldots, n-1, \quad (6.2.13)$$

where $\Delta \tau_k^{2H} = \left(\frac{k+1}{n}\right)^{2H} - \left(\frac{k}{n}\right)^{2H}$. We use a sample variance to estimate σ:

$$\hat{\sigma}_n^2 := \frac{n}{n-1} (\overline{y_n^2} - \overline{y}_n^2) \text{ with } \overline{y}_n := \frac{y_1 + \cdots + y_n}{n}. \quad (6.2.14)$$

Let
$$z_k = \sigma\varepsilon_k - \frac{\sigma^2}{2}T^H n^H \Delta\tau_k^{2H}, \quad k = 0, 1, \ldots, n-1. \text{ Then}$$

$$\bar{z}_n = \sigma\bar{\varepsilon}_n - \frac{\sigma^2}{2}T^H n^{H-1} \tag{6.2.15}$$

and

$$\widehat{\sigma}^2 = \frac{n}{n-1}(\overline{z_n^2} - \bar{z}_n^2) = \frac{n\sigma^2}{n-1}\left(\overline{\varepsilon_n^2} - \sigma T^H n^H \overline{\varepsilon_n \Delta\tau_n^{2H}}\right.$$
$$+ \frac{\sigma^2}{4}T^{2H}n^{2H}\overline{(\Delta\tau_n^{2H})^2} - \bar{\varepsilon}_n^2 \tag{6.2.16}$$
$$\left. + \sigma T^H n^H \bar{\varepsilon}_n \overline{\Delta\tau_n^{2H}} - \frac{\sigma^2}{4}T^{2H}n^{2H}(\overline{\Delta\tau_n^{2H}})^2\right).$$

Again we have a problem with the rate of the discretization with respect to the observation interval. We start with one lemma:

Lemma 6.2.5. *Assume that X, Y are two standard normal random variables:*
$$EX = EY = 0 \text{ and } Var(X) = Var(Y) = 1.$$
Assume that $EXY = q$. Then
$$E((X^2 - 1)(Y^2 - 1)) = 2q^2. \tag{6.2.17}$$

Lemma 6.2.6. *With the notation above:*
(v)   *If $H < \frac{3}{4}$, then $E|\overline{\varepsilon_n^2} - 1| \leq C\frac{1}{\sqrt{n}}$.*
(vi) *If $H = \frac{3}{4}$, then $E|\overline{\varepsilon_n^2} - 1| \leq C\sqrt{\frac{\log n}{n}}$.*
(vii) *If $\frac{3}{4} < H < 1$, then $E|\overline{\varepsilon_n^2} - 1| \leq Cn^{2\alpha - 1}$.*

Proof. We have that
$$\overline{\varepsilon_n^2} - 1 = \frac{1}{n}\sum_{i=0}^{n-1}(\varepsilon_i^2 - 1).$$

From Lemma 6.2.5 and (6.2.6):
$$E(\overline{\varepsilon_n^2} - 1)^2 = \frac{1}{n^2}\sum_{i=0}^{n-1} E(\varepsilon_i^2 - 1)^2 + \frac{2}{n^2}\sum_{0 \leq j < i \leq n-1} E(\varepsilon_i^2 - 1)(\varepsilon_j^2 - 1)$$
$$\leq \frac{C}{n} + \frac{C}{n^2}\sum_{0 \leq j < i \leq n-1}(i - j)^{4H - 4}. \tag{6.2.18}$$

Note that
$$\sum_{0 \leq j < i \leq n-1}(i - j)^{4H-4} = \sum_{j=1}^{n-1}(n - j)j^{4H-4}.$$
This and inequality (6.2.18) give the result. □

We have
$$\bar{z}_n \stackrel{d}{=} \sigma n^{H-1}\left(B_1^H - \frac{\sigma}{2}T^H\right) \tag{6.2.19}$$
so that
$$0 \leq E\bar{z}_n^2 \leq \sigma^2 n^{2\alpha-1}\left(2E(B_1^H)^2 + \frac{\sigma^2}{2}T^{2H}\right). \tag{6.2.20}$$

Estimation of σ

Theorem 6.2.7. *Assume* **H**. *If $n(T)$ is such that $\frac{T^{3H}}{n(T)^{1-2\alpha}} \to 0$, then $T^H(\hat{\sigma}_n^2 - \sigma^2) = o_P(1)$.*

Assume **A**. *Then*

(viii) If $\frac{1}{2} < H < \frac{3}{4}$ and $n(T)$ is such that $\frac{T^{2H}}{n(T)} \to 0$, then $T^H(\hat{\sigma}_n^2 - \sigma^2) = o_P(1)$.

(ix) If $H = \frac{3}{4}$ and $n(T)$ is such that $\frac{T^{\frac{3}{2}}\log(n(T))}{n(T)} \to 0$, then $T^H(\hat{\sigma}_n^2 - \sigma^2) = o_P(1)$.

(x) If $\frac{3}{4} < H < 1$ and $n(T)$ is such that $\frac{T^H}{n(T)^{1-2\alpha}} \to 0$, then $T^H(\hat{\sigma}_n^2 - \sigma^2) = o_P(1)$.

Proof. Using Lemma 6.2.5 and (6.2.16) we obtain that
$$T^H E|\hat{\sigma}_n^2 - \sigma^2| \leq C\left(T^H E|\overline{\varepsilon_n^2} - 1| + \frac{T^H + T^{2H} + T^{3H}}{n^{1-2\alpha}}\right).$$
where C depends on σ^2. Under **H** the statement follows from Lemma 6.2.6. Under **A** we have
$$\hat{\sigma}^2 = \frac{n\sigma^2}{n-1}\left(\overline{\varepsilon_n^2} - (\bar{\varepsilon}_n)^2\right),$$
and
$$T^H \cdot E|\hat{\sigma}_n^2 - \sigma^2| \leq C\left(T^H E|\overline{\varepsilon_n^2} - 1| + \frac{T^H}{n^{1-2\alpha}}\right).$$
The claims (viii)–(x) follow from Lemma 6.2.6. □

6.2.5 Goodness-of-fit Test with Unknown μ and σ

If the parameter σ is unknown, then using the observation $X_{\frac{kT}{n}}$, $k = 0, 1, \ldots, n$, with $n = n(T)$, an estimator $\hat{\sigma}^2 = \hat{\sigma}_n^2$ is constructed. The construction of this estimator is explained in Subsection 6.2.4.
If
$$\frac{T^{\frac{3H}{1-2\alpha}}}{n(T)} \to 0, \quad T \to \infty \tag{6.2.21}$$
we have $(\hat{\sigma}^2 - \sigma^2) \cdot T^H \xrightarrow{P_{\mu,\sigma,\sigma}} 0$, when **H** is true.

If conditions (viii)–(x) of Theorem 6.2.7 hold, we have the same convergence for $(\hat{\sigma}^2 - \sigma^2) \cdot T^H$, then **A** is true. Define the statistics

$$\widehat{S}_T^{\mathbf{A}} := r^{-1}\widehat{R}_T^{\mathbf{A}}\bigg|_{\sigma=\widehat{\sigma}}, \quad \widehat{S}_T^{\mathbf{H}} := r^{-1}\widehat{R}_T^{\mathbf{H}}\bigg|_{\sigma=\widehat{\sigma}}. \tag{6.2.22}$$

Consider the model with unknown μ and σ.

Theorem 6.2.8. *(a) Assume that* **H** *is true, and that* $\frac{T^{3H}}{n(T)^{1-2\alpha}} \to 0$, $T \to \infty$.
Then
$\widehat{S}_T^{\mathbf{H}} \to N(0,1)$ *in distribution.*
(b) Assume that **A** *is true and conditions (vii)–(x) of Theorem 6.2.7 hold.
Then* $\widehat{S}_T^{\mathbf{A}} \to N(0,1)$ *in distribution.*

Proof. (a) Suppose that **H** is true. By Theorem 6.2.7 we have $\widehat{\sigma}^2 - \sigma T^H \xrightarrow{P} 0$. Rewrite (6.2.22) as

$$\widehat{S}_T^{\mathbf{H}} = r^{-1}\widehat{R}_T^{\mathbf{H}}|_{\sigma=\widehat{\sigma}} \cdot \frac{\sigma}{\widehat{\sigma}}$$
$$= r^{-1}\frac{\sigma}{\widehat{\sigma}}\left(\widehat{R}_T^{\mathbf{H}} + B_3 T^{1-H}(\widehat{\mu_{\mathbf{H}}} - \mu_{\mathbf{H}}|_{\sigma=\widehat{\sigma}}) + HB_4 T^H(\widehat{\sigma}^2 - \sigma^2)\right).$$

Now, $\widehat{\sigma} \xrightarrow{P} \sigma$ and $HB_4 T^H(\widehat{\sigma} - \sigma) \xrightarrow{P} 0$, as $T \to \infty$. From (6.1.12)

$$B_3 T^{1-H}(\widehat{\mu_{\mathbf{H}}} - \mu_{\mathbf{H}}|_{\sigma=\widehat{\sigma}}) \text{ equals } B_5 T^H(\widehat{\sigma}^2 - \sigma^2) \xrightarrow{P} 0, \tag{6.2.23}$$

where B_5 is some constant.

And now from (6.2.23) and Theorem 6.2.7 the convergence $\widehat{S}_T^{\mathbf{A}} \to N(0,1)$ follows.

(b) If **A** is true then $T^H(\widehat{\sigma}^2 - \sigma^2) \xrightarrow{P} 0$ holds under the conditions of the Theorem 6.2.7. The proof now follows in the same way. □

The goodness-of-fit test is now organized in such a way. We reject **H** if $|\widehat{S}_T^{\mathbf{H}}| > \xi_{\frac{\rho}{2}}$, and we reject **A** if $|\widehat{S}_T^{\mathbf{A}}| > \xi_{\frac{\rho}{2}}$. Asymptotic relations for the errors $p_{\mathbf{A}}(T)$ and $p_{\mathbf{H}}(T)$ are the same in Section 6.2.4.

6.3 Parameter Estimates in the Models Involving fBm

In this section we consider very simple diffusion models involving fBm and in some cases the Wiener process. Our goal is to demonstrate the properties of drift parameter estimates depending on the form of the model. We follow but slightly modify an approach of (MR01).

6.3.1 Consistency of the Drift Parameter Estimates in the Pure Fractional Brownian Diffusion Model

First we consider the "pure" fractional diffusion (nonlinear) model and establish strong consistency and asymptotic normality of the maximum likelihood drift parameter estimate.

The Girsanov Theorem for the Pure Fractional Diffusion Model and Likelihood Ratio for Drift Parameter

We assume that the fBm B_t^H with $H \in (1/2, 1)$ is defined on a probability space $(\Omega, \mathcal{F}, P)$ and denote by $(\mathcal{F}_t)_{t\geq 0}$ the filtration generated by B_t^H. Consider a diffusion equation containing a stochastic differential driven by B^H:

$$dX_t = \theta a(t, X_t)dt + b(t, X_t)dB_t^H, \quad X_{t=0} = X_0 \in \mathbb{R}, \qquad (6.3.1)$$

$$\theta \in \mathbb{R}, \ 0 \leq t \leq T, \ T > 0.$$

Differential equation (6.3.1) can be rewritten in the integral form

$$X_t = X_0 + \theta \int_0^t a(s, X_s)ds + \int_0^t b(s, X_s)dB_s^H, \ t \in [0, T]. \qquad (6.3.2)$$

Here we use pathwise construction of the integral w.r.t. fBm. Suppose that equation (6.3.2) has unique pathwise solution. (Sufficient conditions of existence and uniqueness of the solution on the interval $[0, T]$ are presented in Theorem 3.1.4.)

Now, let $T > 0$ be fixed. We are in a position to find the likelihood ratio $\frac{dP_\theta(t)}{dP_0(t)}$ for the probability measure $P_\theta(t)$ corresponding to our model and the probability measure $P_0(t)$ corresponding to the model with zero drift. Suppose that the following assumption holds:

(i) $b(t, X_t) \neq 0, t \in [0, T]$ and $\frac{a(t,X_t)}{b(t,X_t)}$ is a.s. Lebesgue integrable on $[0, T]$.

Denote $\varphi_t := \frac{a(t,X_t)}{b(t,X_t)}$ and introduce the new process

$$\widehat{B}_t^H := B_t^H + \theta \int_0^t \varphi_s ds. \qquad (6.3.3)$$

Let also the following conditions hold (recall that $\widetilde{\alpha} = (1 - 2\alpha)^{1/2}$, $\widehat{\alpha} = (1 - 2\alpha)^{-1/2}$):

(ii) $\int_0^t l_H(t, s)|\varphi(s)|ds < \infty, \ t \in [0, T]$

(iii) $\theta \int_0^t l_H(t, s)\varphi(s)ds = \widetilde{\alpha} \int_0^t \delta_s ds, \ t \in [0, T]$

and

(iv) $E \int_0^t s^{2\alpha} \delta_s^2 ds < \infty, t \in [0, T]$.

Then $L_t = \int_0^t s^\alpha \delta_s d\widehat{B}_s$ is a square-integrable martingale for the Wiener process $\widehat{B}$ w.r.t. the measure $P_0(t)$ such that $\int_0^t l_H(t,s) d\widehat{B}_s^H = \widetilde{\alpha} \int_0^t s^{-\alpha} d\widehat{B}_s$. According to the Girsanov theorem for fBm (Theorem 2.8.1), under the assumptions (i)–(iv) and

(v) $E \exp\left\{L_t - \frac{1}{2}\langle L \rangle_t\right\} = 1,$

the process $\widehat{B}_t^H$ is an fBm on $[0,T]$ w.r.t. the measure Q defined via the relation

$$\frac{dP_\theta(t)}{dP_0(t)} = \exp\left\{L_t - \frac{1}{2}\langle L \rangle_t\right\}, \; t \in [0,T]. \tag{6.3.4}$$

Remark 6.3.1. We can try to present the likelihood ratio (6.3.4) as a function of the observed process X_t, according to statistical tradition. Toward this end recall that

$$\int_0^t l_H(t,s) d\widehat{B}_s^H = \widetilde{\alpha} \int_0^t s^{-\alpha} d\widehat{B}_s = \int_0^t l_H(t,s) b^{-1}(s, X_s) dX_s. \tag{6.3.5}$$

Suppose that the process $J_t := \int_0^t l_H(t,s) b^{-1}(s, X_s) dX_s$ admits a differential of the form $dJ_t = F(t, X_t) dX_t$; then, evidently,

$$L_t = \int_0^t s^\alpha \delta_s d\widehat{B}_s = \widehat{\alpha} \int_0^t s^{2\alpha} \delta_s F(s, X_s) dX_s,$$

and δ_s is a functional of the process X under the conditions of Lemma 6.3.2 (see below). In turn, the existence of the differential dJ_t can be established separately for $\int_0^t l_H(t,s) \varphi_s ds$ (it is realized in Lemma 6.3.2) and for $\int_0^t l_H(t,s) dB_s^H = M_t^H$, but the last problem is of the same complexity as the original one. Another possibility is to establish, similarly to Lemma 6.3.2, the conditions of the existence of the derivative $(s^\alpha \delta_s)'$, in general, this problem is solvable; then we can rewrite

$$L_t = t^\alpha \delta_t \widehat{B}_t - \int_0^t \widehat{B}_s (s^\alpha \delta_s)' ds,$$

and, of course, $\widehat{B}$ is an adapted functional of X. Indeed, we can present $\widehat{B}$ via X with the help of $\widehat{B}^H$ (see (6.3.5)), relation (6.3.3) and the equality $B_t^H = \int_0^t b^{-1}(s, X_s) dX_s - \int_0^t \varphi_s ds$.

Consistency of the Drift Parameter Estimates

In order to find the maximum likelihood estimate of the parameter θ, we use likelihood ratio (6.3.4), which can be rewritten as

$$\frac{dP_\theta(t)}{dP_0(t)} = \exp\left\{\int_0^t s^\alpha \delta_s d\widehat{B}_s - \frac{1}{2}\int_0^t s^{2\alpha} \delta_s^2 ds\right\},$$

where δ_s is defined according to the integral representation (iii). First we establish sufficient conditions ensuring the existence of representation (iii).

Denote $\psi(t,x) = \frac{a(t,x)}{b(t,x)}$, so that $\psi(t, X_t) = \varphi(t)$, $I(t) := \int_0^t l_H(t,s)\varphi(s)ds$.

Lemma 6.3.2. Let $\psi(t,x) \in C^1[0,T] \cap C^2(\mathbb{R})$. Then for $t > 0$

$$I'(t) = C(H)\psi(0,0)t^{-2\alpha} + \int_0^t l_H(t,s)\left(\psi_t'(s,X_s) + \theta\psi_x'(s,X_s)a(s,X_s)\right)ds$$

$$- \alpha C_H^{(5)} \int_0^t s^{-1-\alpha}(t-s)^{-\alpha} \int_0^s \left(\psi_t'(u,X_u) + \theta\psi_x'(u,X_u)a(u,X_u)\right) du\, ds$$

$$+ (1-2\alpha)C_H^{(5)} t^{-2\alpha} \int_0^t s^{2\alpha-2} \int_0^s u^{1-\alpha}(s-u)^{-\alpha}\psi_x'(u,X_u)b(u,X_u)dB_u^H\, ds$$

$$+ C_H^{(5)} t^{-1} \int_0^t u^{1-\alpha}(t-u)^{-\alpha}\psi_x'(u,X_u)b(u,X_u)dB_u^H,$$

where $C(H) = (1-2\alpha)B(1-\alpha, 1-\alpha)C_H^{(5)}$.

Proof. According to the Itô formula (2.7.3),

$$\varphi_s = \phi(0,0) + \int_0^s (\psi_t'(u,X_u) + \psi_x'(u,X_u)\theta a(u,X_u))du$$

$$+ \int_0^s \psi_x'(u,X_u)b(u,X_u)dB_u^H. \qquad (6.3.6)$$

Substituting (6.3.6) into the integral $I(t) = \int_0^t l_H(t,s)\varphi_s ds$, we obtain

$$I(t) = C(H,1)\psi(0,0)t^{1-2\alpha} + \int_0^t l_H(t,s) \int_0^s \psi_t'(u,X_u)du\, ds$$

$$+ \theta \int_0^t l_H(t,s) \int_0^s \psi_x'(u,X_u)a(u,X_u)du\, ds$$

$$+ \int_0^t l_H(t,s) \int_0^s \psi_x'(u,X_u)b(u,X_u)dB_u^H\, ds, \qquad (6.3.7)$$

$C(H,1) = C_H^{(5)} B(1-\alpha, 1-\alpha)$ and now our aim is to differentiate $I(t)$. The first term on the right-hand side of (6.3.7) is obviously differentiable, i.e. can be presented as $C(H)\psi(0,0) \int_0^t s^{-2\alpha}ds$. The second and the third terms can be transformed using integration by parts:

$$s^{-\alpha} \int_0^s \psi_t'(u,X_u)du = \int_0^s u^{-\alpha}\psi_t'(u,X_u)du - \alpha \int_0^s u^{-1-\alpha} \int_0^u \psi_t'(v,X_v)dv\, du,$$

and

6.3 Consistency of Drift Parameter Estimates

$$s^{-\alpha}\int_0^s \psi_x'(u, X_u)a(u, X_u)du = \int_0^s u^{-\alpha}\psi_x'(u, X_u)a(u, X_u)du$$

$$-\alpha \int_0^s u^{-1-\alpha}\int_0^u \psi_x'(u, X_u)a(v, X_v)dv\, du. \quad (6.3.8)$$

According to representation (6.3.8), there exist a.s. the fractional derivatives of order α, i.e. the derivatives of fractional integrals:

$$\frac{d}{dt}\int_0^t l_H(t, s)\int_0^s \psi_t'(u, X_u)du\, ds = \int_0^t l_H(t, s)\psi_t'(s, X_s)ds$$

$$-\alpha C_H^{(5)}\int_0^t s^{-1-\alpha}(t-s)^{-\alpha}\int_0^s \psi_t'(u, X_u)du\, ds. \quad (6.3.9)$$

$$\frac{d}{dt}\int_0^t l_H(t, s)\int_0^s \psi_x'(u, X_u)a(u, X_u)du\, ds = \int_0^t l_H(t, s)\psi_x'(s, X_s)a(s, X_s)ds$$

$$-\alpha C_H^{(5)}\int_0^t s^{-1-\alpha}(t-s)^{-\alpha}\int_0^s \psi_x'(u, X_u)a(u, X_u)du\, ds. \quad (6.3.10)$$

Further, it follows from Lemma 2.8.2 that

$$\int_0^t l_H(t, s)\int_0^s \psi_x'(u, X_u)b(u, X_u)dB_u^H ds$$

$$= C_H^{(5)}t^{1-2\alpha}\int_0^t s^{2\alpha-2}\int_0^s u^{1-\alpha}(s-u)^{-\alpha}\psi_x'(u, X_u)b(u, X_u)dB_u^H ds. \quad (6.3.11)$$

The proof follows immediately from relations (6.3.9)–(6.3.11). □

Now, we can rewrite (6.3.6) as

$$\frac{dP_\theta(t)}{dP_0(t)} = \exp\left\{\frac{\theta}{\widetilde{\alpha}}\int_0^T s^\alpha I'(s)d\widehat{B}_s - \frac{\theta^2}{2(1-2\alpha)}\int_0^T s^{2\alpha}(I'(s))^2 ds\right\}. \quad (6.3.12)$$

It follows from (6.3.12) that the maximum likelihood estimate is achieved under the condition

$$\int_0^T s^\alpha I'(s)dB^s - \frac{\theta}{\widetilde{\alpha}}\int_0^T s^{2\alpha}(I'(s))^2 ds = 0,$$

whence

$$\widehat{\theta}_t = \frac{\widetilde{\alpha}\int_0^t s^\alpha I'(s)d\widehat{B}_s}{\int_0^t s^{2\alpha}(I'(s))^2 ds}. \quad (6.3.13)$$

Using Lemma 6.3.2, we obtain

$$B_t + \theta\widehat{\alpha}\int_0^t s^\alpha I'(s)ds = \widehat{B}_t, \qquad (6.3.14)$$

where $\widehat{B}_t$ is a Wiener process under measure Q. Substituting (6.3.14) into (6.3.13) we obtain

$$\widehat{\theta}_t = \theta + \frac{\widetilde{\alpha}\int_0^t s^\alpha I'(s)dB_s}{\int_0^t s^{2\alpha}(I'(s))^2 ds}. \qquad (6.3.15)$$

Recall that under condition (iv) $\int_0^t s^\alpha I'(s)dB_s$ is the square-integrable P_θ-martingale with angle bracket $\int_0^t s^{2\alpha}(I'(s))^2 ds$.

Theorem 6.3.3. *Let the conditions of Theorem 3.1.4 and (i)–(v) hold for any $T > 0$ and, moreover,*

$$\text{(vi)} \quad \int_0^\infty s^{2\alpha}(I'(s))^2 ds = \infty \quad a.s.$$

Then the maximum likelihood estimate $\widehat{\theta}_T$ is strongly consistent as $T \to \infty$.

Proof follows immediately from representation (6.3.15) and from Theorem 6.10 (LS86). This theorem establishes that $\frac{X_t}{\langle X\rangle_t} \to 0$ a.s. if X_t is a square-integrable martingale and $\langle X \rangle_\infty \to \infty$ a.s. In other words,

$$\frac{\int_0^t s^\alpha I'(s)dB_s}{\int_0^t s^{2\alpha}(I'(s))^2 ds} \to 0, \ t \to \infty,$$

with P_θ-probability 1.

Example 6.3.4. Consider the linear model of the form

$$dX_t = \theta X_t dt + X_t dB_t^H.$$

In this case $\varphi_t = 1$, so $\int_0^t \delta_s ds = \int_0^t l_H(t,s)ds = C(H,1)t^{1-2\alpha}$, $\delta_t = C(H)t^{-2\alpha}$. Hence

$$\widehat{\theta}_t = \theta + \frac{\widetilde{\alpha}\int_0^t s^{-\alpha}dB_s}{C(H)t^{1-2\alpha}}.$$

Since $\widetilde{\alpha}\int_0^t s^{-\alpha}dB_s$ is the square-integral martingale with the angle bracket $t^{1-2\alpha} \to \infty$ when $t \to \infty$, then, according to Theorem 6.3.3, $\frac{\widetilde{\alpha}\int_0^t s^{-\alpha}dB_s}{C(H)t^{1-2\alpha}} \to 0$, a.s. as $t \to \infty$.

So, the estimate $\widehat{\theta}_t$ is consistent with probability 1.

6.3.2 Consistency of the Drift Parameter Estimates in the Mixed Brownian–fractional-Brownian Diffusion Model with "Linearly" Dependent W_t and B_t^H

Now we consider the linear mixed Brownian–fractional-Brownian diffusion model represented by the stochastic differential equation of the form

$$dX_t = \theta X_t dt + \sigma_1 X_t dB_t + \sigma_2 X_t dB_t^H, \qquad (6.3.16)$$

$X_{t=0} = X_0 \in \mathbb{R}$, $0 \le t \le T$, $T > 0$, $\{\theta, \sigma_1, \sigma_2\} \subset \mathbb{R}$, $\sigma_1 \sigma_2 < 0$, θ is a parameter that we need to estimate.

We suppose that the Wiener process B and the fBm B^H in (6.3.16) are connected via the relations (1.8.3), (1.8.5). The integral form of equation (6.3.16) is

$$X_t = X_0 + \theta \int_0^t X_s ds + \sigma_1 \int_0^t X_s dB_s + \sigma_2 \int_0^t X_s dB_s^H, \; 0 \le t \le T. \quad (6.3.17)$$

The existence and the uniqueness of the solution of the equation (6.3.17) was established in Theorem 3.2.1.

The Girsanov Theorem for the Mixed Fractional Diffusion Model

First we try to change the probability measure P_θ for the another measure P_0, $P_\theta(T) \sim P_0(T)$ in order to exclude the drift $\theta X_t dt$ from equations (6.3.16) and (6.3.17).

We introduce probability measures $P_{0,i}, i = 1, 2$ and $P_{\theta,i}, i = 1, 2$ as follows. The probability measures $P_{0,1}(t)$ and $P_{\theta,1}(t)$ are determined by the following condition:

$$\frac{dP_{\theta,1}(t)}{dP_{0,1}(t)} = \exp\left\{\int_0^t \psi_s \, dB_s^{(1)} - \frac{1}{2}\int_0^t \psi_s^2 \, ds\right\}$$

for a nonrandom function ψ_s such that $\int_0^t \psi_s^2 \, ds < \infty$ and

$$E \exp\left\{\int_0^t \psi_s \, dB_s^{(1)} - \frac{1}{2}\int_0^t \psi_s^2 \, ds\right\} = 1.$$

Here the process $B_t^{(1)}$ is created according to the Girsanov theorem,

$$B_t^{(1)} := B_t + \int_0^t \psi_s \, ds, \qquad (6.3.18)$$

and $B_t^{(1)}$ is a standard Wiener process with respect to the probability measure $P_{0,1}(t)$. The probability measures $P_{0,2}$ and $P_{\theta,2}(t)$ satisfy the relation

$$\frac{dP_{\theta,2}(t)}{dP_{0,2}(t)} = \exp\left\{\int_0^t s^\alpha \delta_s \, dB_s^{(2)} - \frac{1}{2}\int_0^t s^{2\alpha} \delta_s^2 \, ds\right\},$$

where δ_s satisfies the relation $\int_0^t l_H(t,s)|\delta_s|\,ds < \infty$, $t \in [0,T]$ and admits the following integral representation:

$$\int_0^t l_H(t,s)\varphi_s\,ds = \tilde{\alpha}\int_0^t \delta_s\,ds, \tag{6.3.19}$$

the Wiener process $B_t^{(2)}$ is defined from the equation

$$\int_0^t l_H(t,s)\,dB_s^{H,2} = \tilde{\alpha}\int_0^t s^{-\alpha}\,dB_s^{(2)}.$$

Moreover, the process

$$B_t^{H,2} := B_t^H + \int_0^t \varphi_s\,ds \tag{6.3.20}$$

is a fractional Brownian motion on $[0,T]$ with respect to the measure $P_{0,2}(t)$. So, the total drift coefficient equals

$$\sigma_1\int_0^t \psi_s\,ds + \sigma_2\int_0^t \varphi_s\,ds = \theta t,$$

and if we suppose that the functions ψ and φ are continuous, we obtain that

$$\sigma_1\psi_t + \sigma_2\varphi_t = \theta. \tag{6.3.21}$$

Obviously, from (6.3.18)–(6.3.20) and since the likelihood ratios $\frac{dP_{\theta,i}(t)}{dP_{0,i}(t)}$ must coincide, we obtain that

$$\widehat{M}_t^H = M_t^H + \tilde{\alpha}\int_0^t s^{-\alpha}\psi_s\,ds$$

and

$$\widehat{M}_t^H = M_t^H + \tilde{\alpha}\int_0^t \delta_s\,ds,$$

whence $t^\alpha \delta_t = \psi_t$, $t \in [0,T]$. Moreover,

$$\int_0^t l_H(t,s)\varphi_s\,ds = \tilde{\alpha}\int_0^t s^{-\alpha}\psi_s\,ds.$$

Multiplying by $(t-s)^{\alpha-1}$ and integrating, we obtain

$$C_H^{(5)}\int_0^t (t-s)^{\alpha-1}\int_0^s u^{-\alpha}(s-u)^{-\alpha}\varphi_u\,du\,ds$$

$$= \tilde{\alpha}\int_0^t (t-s)^{\alpha-1}\int_0^s \delta_u\,du\,ds, \tag{6.3.22}$$

6.3 Consistency of Drift Parameter Estimates   351

and the Fubini theorem applied to both sides of (6.3.22) gives

$$C(H,2)\int_0^t u^{-\alpha}\varphi_u du = \frac{\widetilde{\alpha}}{\alpha}\int_0^t (t-u)^\alpha \delta_u du,$$

whence

$$\varphi_t = \frac{1}{C(H,3)} t^\alpha \int_0^t (t-u)^{\alpha-1} u^{-\alpha}\psi_u du. \qquad (6.3.23)$$

Here $C(H,2) = C_H^{(5)} B(\alpha, 1-\alpha), C(H,3) = \frac{\widetilde{\alpha}}{C(H,2)}$. Substituting (6.3.23) into (6.3.21), we obtain a Volterra equation of the second kind, with weak singularity, of the form

$$\sigma_1\psi_t + \frac{\sigma_2}{C(H,3)} t^\alpha \int_0^t (t-u)^{\alpha-1} u^{-\alpha}\psi_u du = \theta,$$

or

$$\rho_t + \frac{\sigma_2}{\sigma_1}\frac{1}{C(H,3)}\int_0^t (t-u)^{\alpha-1}\rho_u du = \frac{e_t}{\sigma_1}, \qquad (6.3.24)$$

where $\rho_t = t^{-\alpha}\psi_t$, $e_t = \theta t^{-\alpha}$. We solve (6.3.24) using successive approximations

$$\rho_t^{(n+1)} + \frac{\sigma_2}{\sigma_1}\frac{1}{C(H,3)}\int_0^t (t-u)^{\alpha-1}\rho_u^{(n)} du = \frac{e_t}{\sigma_1}. \qquad (6.3.25)$$

Denote for simplicity $C := \frac{\sigma_2}{\sigma_1 C(H,3)}$ and start with $\rho_t^{(0)} = 0$, $\rho_t^{(1)} = \frac{e_t}{\sigma_1}$. Then we obtain from (6.3.25) that

$$\rho_t^{(2)} = (-1)\frac{C}{\sigma_1}\int_0^t (t-u)^{\alpha-1} e_u du + \frac{e_t}{\sigma_1}.$$

It is very simple now to prove by induction that for $n > 1$

$$\rho_t^{(n)} = \frac{1}{\sigma_1}\sum_{k=1}^{n-1}(-C)^k \int_0^t e_s(t-s)^{k\alpha-1}\frac{\Gamma^k(\alpha)}{\Gamma(k\alpha)} ds + \frac{e_t}{\sigma_1},$$

and the solution $\rho_t = \lim_{n\to\infty} \rho_t^{(n)}$ evidently can be represented as a series

$$\rho_t = \frac{1}{\sigma_1}\sum_{n=1}^\infty (-C)^n \int_0^t e_s(t-s)^{n\alpha-1}\frac{\Gamma^n(\alpha)}{\Gamma(n\alpha)} ds + \frac{e_t}{\sigma_1}.$$

Hence

$$\psi_t = t^\alpha \rho_t = \frac{t^\alpha \theta}{\sigma_1}\sum_{n=1}^\infty (-C)^n \frac{\Gamma^n(\alpha)}{\Gamma(n\alpha)}\int_0^t s^{-\alpha}(t-s)^{n\alpha-1} ds + \frac{\theta}{\sigma_1}$$

$$= \frac{\theta}{\sigma_1}\Gamma(1-\alpha)\sum_{n=1}^\infty (-C)^n \frac{\Gamma^n(\alpha)}{\Gamma((n-1)\alpha+1)} t^{n\alpha} + \frac{\theta}{\sigma_1}. \qquad (6.3.26)$$

The series on the right-hand side of (6.3.26) can be expressed in terms of the Mittag–Leffler function $E_\rho(z) := \sum_{n=0}^{\infty} \frac{z^n}{\Gamma(n/\rho+1)}$ (see, for example, (Po99)),

$$A_t = -Ct^\alpha \pi (\sin \pi\alpha)^{-1} E_{1/\alpha}(-C\Gamma(\alpha)t^\alpha),$$

and in these terms

$$\psi_t = \frac{\theta}{\sigma_1}(A_t + 1).$$

Therefore, the likelihood ratio for the mixed fractional Brownian model equals

$$\frac{dP_{\theta,1}(t)}{dP_{0,1}(t)} = \exp\left\{\int_0^t \psi_s dB_s^{(1)} - \frac{1}{2}\int_0^t \psi_s^2 ds\right\}$$

$$= \exp\left\{\frac{\theta}{\sigma_1}\int_0^t (A_s + 1)dB_s^{(1)} - \frac{1}{2}\frac{\theta^2}{\sigma_1^2}\int_0^t (A_s + 1)^2 ds\right\},$$

whence the maximum likelihood estimate for θ equals

$$\widehat{\theta}_T^1 = \sigma_1 \frac{\int_0^T (A(s)+1) dB_s^{(1)}}{\int_0^t (A(s)+1)^2 ds}$$

$$= \sigma_1 \frac{\int_0^T (A(s)+1) dB_s + \frac{\theta}{\sigma_1}\int_0^T (A(s)+1)^2 ds}{\int_0^t (A(s)+1)^2 ds} = \theta + \sigma_1 \frac{\int_0^t (A(s)+1) dB_s}{\int_0^T (A(s)+1)^2 ds}.$$

For the demonstration of the consistency of the estimate $\widehat{\theta}_T^1$ with probability 1, it is sufficient to prove the divergence of the integral $\int_0^t (A(s)+1)^2 ds$ when $t \to \infty$. Note that $C < 0$ since $\frac{\sigma_1}{\sigma_2} < 0$, and

$$\sum_{n=1}^{\infty}(-C\Gamma(\alpha))^n \frac{t^{n\alpha}}{\Gamma((n-1)\alpha+1)} > \sum_{n=1}^{\infty}(-C\Gamma(\alpha))^n \frac{t^{n\alpha}}{\Gamma(n+1)} =$$

$$= \sum_{n=1}^{\infty}(-C\Gamma(\alpha))^n \frac{t^{n\alpha}}{n!} = \exp\{-C\Gamma(\alpha)t^\alpha\} \to \infty,$$

when $t \to \infty$ because $\alpha > 0$ and $-C\Gamma(\alpha) > 0$. Note that $\delta_t = t^{-\alpha}\psi_t$ satisfies conditions (ii)–(vi). So we have proved the following result.

Theorem 6.3.5. *The drift parameter maximum likelihood estimate of the linear Brownian–fractional-Brownian model (6.3.16) is consistent with probability 1.*

The Asymptotic Normality of the Maximum Likelihood Estimates

First, consider one of the limit theorems for the stochastic integrals w.r.t. the Wiener process $\{W_t, \mathcal{F}_t, t > 0\}$. Let $\{h(s), s \geq 0\}$ be an $\mathcal{F}_s$-adapted predictable function such that $\mathbb{E}\int_0^t h^2(s)ds$ is finite for any $t > 0$ and

6.3 Consistency of Drift Parameter Estimates   353

$\mathcal{F}_n(t) = \sigma\{h(s), W(s), s \leq nt\}$. Consider the sequence $Y_n(t) := \int_0^{nt} h(s)dW_s$. Evidently, $Y_n(t)$ are $\mathcal{F}_n(t)$-square-integrable martingales, $t \in [0,T]$, and their angle brackets equal $\langle Y_n \rangle (t) = \int_0^{nt} h^2(s)ds$. Suppose that the following conditions hold:

(vii) there exists an increasing real-valued sequence $\{A_n, n \geq 1\}$ such that $A_n \uparrow \infty$, $n \to \infty$ and for some constant $c_0 > 0$ we have that

$$\int_0^n h^2(s)ds \cdot A_n^{-2} \xrightarrow{P} c_0,$$

Consider the sequence of normalized square-integrable martingales $X_n(t) := A_n^{-1} \cdot \int_0^{nt} h(s)dW_s$. Then $\langle X_n \rangle (1) = \int_0^n h_2(s)ds \cdot A_n^{-2} \xrightarrow{P} c_0$, therefore X_n satisfy conditions of Theorem 4.1 (LS86), if we consider the set of convergence points consisting of one point $t = 1$. By using this theorem we obtain the following result:

Lemma 6.3.6. *Let condition (vii) holds. Then the random variable*

$$Z_n := \int_0^n h(s)dW_s \cdot \left(\int_0^n h^2(s)ds\right)^{-1/2}$$

weakly converges to the random variable $c_0^{-1/2} N(0,1)$.

Proof. From the Theorem 4.1 (LS86) and the condition (vii) we obtain that $X_n(1)$ weakly converges to the value $Z(1)$ of the Gaussian martingale Z with independent increments such that $\langle Z \rangle (t) = c_0 t$. Evidently, $Z(1) \sim c_0^{1/2} N(0,1)$. Moreover, from the same condition, the weak convergence holds:

$$Z_n = \frac{A_n}{\left(\int_0^n h^2(s)ds\right)^{1/2}} \cdot X_n(1) \to c_0^{-1/2} N(0,1).$$

$\square$

Consider the estimate $\widehat{\theta}_n^1$ satisfying relation (6.3.15). We see that for the pure fractional diffusion model $h(s) = A(s) + 1$ and is nonrandom. Therefore we obtain from Lemma 6.3.6 that

$$\left(\int_0^n (A(s)+1)^2 ds\right)^{1/2} (\widehat{\theta}_n^1 - \theta) \to N(0,1).$$

Moreover, under the assumption

(viii) there exists an increasing real-valued sequence $\{A_n, n \geq 1\}$ such that $A_n \uparrow \infty$, $n \to \infty$ and

$$\int_0^n s^{2\alpha}(I_s')^2 ds \cdot A_n^{-2} \xrightarrow{P} \varphi_0, n \to \infty,$$

we have a weak convergence

$$\varphi_0^{1/2} A_n(\widehat{\theta}_n - \theta) \to N(0,1).$$

In this sense we say that the estimates $\widehat{\theta}_n$ and $\widehat{\theta}_n^1$ are asymptotically normal.

6.3.3 The Properties of Maximum Likelihood Estimates in Diffusion Brownian–Fractional-Brownian Models with Independent Components

Now we consider an "opposite" situation when the components of the diffusion model are independent, more exactly, the processes B^H and B are independent, where B^H is a fBm and B is a Wiener process.

The Estimates of the Drift Parameter in the Mixed Brownian–Fractional-Brownian Diffusion Model Where B_t and B_t^H are Independent

Let the diffusion equation contain stochastic differentials with respect to fBm and the Wiener process,

$$dX_t = \theta X_t\, dt + \sigma_1 X_t\, dB_t + \sigma_2 X_t\, dB_t^H,$$

$X_{t=0} = X_0 \in \mathbb{R}$, $0 \le t \le T$, $T > 0$, $\{\theta, \sigma_1, \sigma_2\} \subset \mathbb{R} \setminus \{0\}$, where the processes B_t and B_t^H are independent. Evidently, we can rewrite the solution of our simple linear equation as

$$X_t = X_0 \exp\{\theta t + \sigma_1 B_t + \sigma_2 B_t^H - 1/2\sigma_1^2 t\}.$$

It was mentioned by B.L.S. Prakasa Rao in the private conversation that we cannot prove the equivalence of the observation of the whole process X_t and the observation of its two independent components, B_t and B_t^H, i.e., we cannot separate these components (note that the measures corresponding to these processes are singular). So, we suppose that we observe both the components. Let, as before, θ be the parameter to be estimated. We shall try to represent the estimate of θ via the components B_t and B_t^H because it seems to be impossible to represent it via the whole process X_t. Let P_θ be the basic probability measure corresponding to the process X. We introduce probability measures $P_{0,i}$, $i = 1, 2$ and $P_{\theta,i}$, $i = 1, 2$ as follows. The probability measures $P_{0,1}(t)$ and $P_{\theta,1}(t)$ are determined by the following condition:

$$\frac{dP_{\theta,1}(t)}{dP_{0,1}(t)} = \exp\left\{\int_0^t \psi_s\, dB_s^{(1)} - \frac{1}{2}\int_0^t \psi_s^2\, ds\right\}$$

for a nonrandom function ψ_s such that $\int_0^t \psi_s^2\, ds < \infty$ and

$$E \exp\left\{\int_0^t \psi_s\, dB_s^{(1)} - \frac{1}{2}\int_0^t \psi_s^2\, ds\right\} = 1.$$

6.3 Properties of Maximum Likelihood Estimates

Here the process $B_t^{(1)}$ is created according to the Girsanov theorem,

$$B_t^{(1)} := B_t + \int_0^t \psi_s \, ds \qquad (6.3.27)$$

and $B_t^{(1)}$ is a standard Wiener process with respect to the probability measure $P_{0,1}(t)$. The probability measures $P_{0,2}$ and $P_{\theta,2}(t)$ satisfy the relation

$$\frac{dP_{\theta,2}(t)}{dP_{0,2}(t)} = \exp\left\{\int_0^t s^\alpha \delta_s \, dB_s^{(2)} - \frac{1}{2}\int_0^t s^{2\alpha} \delta_s^2 \, ds\right\},$$

where δ_s satisfies the relation $\int_0^t l_H(t,s)|\delta_s|\, ds < \infty$, $t \in [0,T]$, admits the following integral representation:

$$\int_0^t l_H(t,s)\varphi_s \, ds = \tilde{\alpha}\int_0^t \delta_s \, ds, \qquad (6.3.28)$$

the Wiener process $B_t^{(2)}$ is defined from the equation

$$\int_0^t l_H(t,s) \, dB_s^{H,2} = \tilde{\alpha}\int_0^t s^{-\alpha} \, dB_s^{(2)},$$

and the process

$$B_t^{H,2} := B_t^H + \int_0^t \varphi_s \, ds$$

is a fractional Brownian motion on $[0,T]$ with respect to the measure $P_{0,2}(t)$. So, the total drift coefficient equals

$$\sigma_1 \int_0^t \psi_s \, ds + \sigma_2 \int_0^t \varphi_s \, ds = \theta t,$$

or, if we suppose that the functions ψ and φ are continuous,

$$\sigma_1 \psi_t + \sigma_2 \varphi_t = \theta. \qquad (6.3.29)$$

Since B_t and B_t^H are independent, the final probability measure $P_0(t)$ is the product of the measures $P_{0,1}(t)$ and $P_{0,2}(t)$. Thus the final likelihood ratio is

$$\frac{dP_\theta(t)}{dP_0(t)} = \exp\left[\left\{\int_0^t \psi_s \, dB_s^{(1)} - \frac{1}{2}\int_0^t \psi_s^2 \, ds\right\}\right.$$
$$\left.\times \left\{\int_0^t s^\alpha \delta_s \, dB_s^{(2)} - \frac{1}{2}\int_0^t s^{2\alpha} \delta_s^2 \, ds\right\}\right]$$

$$= \exp\left\{\int_0^t \psi_s \, dB_s^{(1)} + \int_0^t s^\alpha \delta_s \, dB_s^{(2)} - \frac{1}{2}\int_0^t (\psi_s^2 + s^{2\alpha}\delta_s^2) \, ds\right\}. \qquad (6.3.30)$$

Solving equations (6.3.28) and (6.3.29) with respect to the functions ψ_t and δ_t, respectively, we obtain

$$\psi_t = \frac{1}{\sigma_1}(\theta - \sigma_2 \varphi_t), \tag{6.3.31}$$

$$\delta_t = \widehat{\alpha} \left(\int_0^t l_H(t,s) \varphi_s ds \right)'_t. \tag{6.3.32}$$

Substituting equalities (6.3.31) and (6.3.32) into likelihood ratio (6.3.30), we get at the point $t = T$ that

$$\frac{dP_\theta(T)}{dP_0(T)} = \exp\left\{ \frac{1}{\sigma_1} \int_0^T (\theta - \sigma_2 \varphi_s)\, dB_s^{(1)} + \widehat{\alpha} \int_0^T s^\alpha \left(\int_0^s l_H(s,u)\varphi_u\, du \right)'_s dB_s^{(2)} \right.$$

$$\left. - \frac{1}{2} \int_0^T \left[\frac{1}{\sigma_1^2}(\theta - \sigma_2 \varphi_s)^2 + s^{2\alpha} \widehat{\alpha} \left(\left(\int_0^s l_H(s,u)\varphi_u\, du \right)'_s \right)^2 \right] ds \right\}. \tag{6.3.33}$$

If follows from (6.3.33) that the maximum likelihood estimate $\widehat{\theta}_T^1$ of the parameter θ satisfies the equality

$$\frac{1}{\sigma_1} \int_0^T dB_s^{(1)} - \frac{1}{\sigma_1^2} \int_0^T (\theta - \sigma_2 \varphi_s)\, ds = 0,$$

which can be rewritten as follows:

$$\sigma_1 B_T^{(1)} + \sigma_2 \int_0^T \varphi_s\, ds - \theta T = 0.$$

This gives us the following estimate of the parameter θ:

$$\widehat{\theta}_T^1 = \frac{\sigma_1 B_T^{(1)}}{T} + \frac{\sigma_2 \int_0^T \varphi_s\, ds}{T}. \tag{6.3.34}$$

Now we solve equation (6.3.29) with respect to the function φ_t and substitute it into equation (6.3.34):

$$\widehat{\theta}_T^1 = \theta + \frac{\sigma_1}{T}\left(B_T^{(1)} - \int_0^T \psi_s\, ds \right). \tag{6.3.35}$$

Substituting (6.3.27) into (6.3.35) yields

$$\widehat{\theta}_T^1 = \theta + \sigma_1 \frac{B_T}{T}. \tag{6.3.36}$$

It is evident that the estimate (6.3.36) of parameter $\widehat{\theta}_T^1$ is strongly consistent.

We can construct another estimate of the parameter θ. The function δ_t is expressed via φ_t by equality (6.3.28). Denote also $\zeta_t := \left(\int_0^t l_H(t,s)\psi_s\, ds\right)'_t$. Then

$$\delta_t = \widehat{\alpha}\left(\int_0^t l_H(t,s)\varphi_s\, ds\right)'_t = \frac{1}{\sigma_2}\widehat{\alpha}\left(\int_0^t l_H(t,s)(\theta - \sigma_1\psi_s)\, ds\right)'_t$$

$$= \widehat{\alpha}\left(\frac{\theta}{\sigma_2}\left(\int_0^t l_H(t,s)\, ds\right)'_t - \frac{\sigma_1}{\sigma_2}\zeta_t\right)$$

$$= \widehat{\alpha}\left(\frac{\theta}{\sigma_2}C(H)t^{-2\alpha} - \frac{\sigma_1}{\sigma_2}\zeta_t\right), \tag{6.3.37}$$

where $C(H) = C_H^{(5)}(1 - 2\alpha)B_1 C_H^{(5)}$, $B_1 = B(1-\alpha, 1-\alpha)$. Using equality (6.3.37) for likelihood ratio (6.3.30), taking the logarithms, differentiating with respect to θ, and equating the derivative to zero, we obtain at the point $t = T$

$$\int_0^T s^{-\alpha}\, dB_s^{(2)} - \widehat{\alpha}\int_0^T \left(\frac{\theta C(H)}{\sigma_2}s^{-2\alpha} - \frac{\sigma_1}{\sigma_2}\zeta_s\right) ds = 0,$$

or

$$\int_0^T s^{-\alpha}\, dB_s^{(2)} - \widehat{\alpha}^3\theta\frac{C(H)}{\sigma_2}T^{1-2\alpha} + \widehat{\alpha}\frac{\sigma_1}{\sigma_2}\int_0^T l_H(T,s)\psi_s\, ds = 0.$$

This implies another estimate for the parameter θ:

$$\widehat{\theta}_T^2 = \frac{\sigma_2\widetilde{\alpha}\int_0^T s^{-\alpha}\, dB_s^{(2)} + \sigma_1\int_0^T l_H(T,s)\psi_s\, ds}{C_H^{(5)} B_1 T^{1-2\alpha}}. \tag{6.3.38}$$

Now we substitute the expression (6.3.31) for the function ψ_t into relation (6.3.38) and obtain with $C(H,1) = C_H^{(5)} B_1$ that

$$\widehat{\theta}_T^2 = \theta - \frac{\sigma_2}{C(H,1)T^{1-2\alpha}}\left[\int_0^T l_H(T,s)\varphi_s\, ds - \widetilde{\alpha}\int_0^T s^{-\alpha}\, dB_s^{(2)}\right].$$

Recall that $\widetilde{\alpha}\int_0^T s^{-\alpha}\, dB_s^{(2)} = \int_0^T l_H(T,s)\, dB_s^{H,2}$.

Further,

$$\int_0^T l_H(T,s)\varphi_s\, ds - \int_0^T l_H(T,s)\, dB_s^{H,2} = -\int_0^T l_H(T,s)\, dB_s^H.$$

So, the second estimate of the parameter θ is given by

$$\widehat{\theta}_T^2 = \theta + \frac{\sigma_2}{C(H,1)T^{1-2\alpha}}\int_0^T l_H(T,s)\, dB_s^H,$$

or

$$\widehat{\theta}_T^2 = \theta + \frac{\sigma_2 \widetilde{\alpha}}{C(H,1)} \frac{\int_0^T s^{-\alpha} d\widetilde{B}_s}{T^{1-2\alpha}}, \qquad (6.3.39)$$

where $\widetilde{B}_s$ is some Wiener process. The strong consistency of the estimate $\widehat{\theta}_T^2$ is also clear.

Now we compare the estimates $\widehat{\theta}_T^1$ and $\widehat{\theta}_T^2$. First we compute the variances of the remainder terms in formulae (6.3.36) and (6.3.39) and compare $\sigma_1^2 T^{-1}$ and $\sigma_2^2 C(H,1)^{-2} T^{2\alpha-1}$. Since $H \in (\frac{1}{2}, 1)$, it is obvious that there exists a number N such that $\sigma_1^2 T^{-1} < \sigma_2^2 C(H,1)^{-2} T^{2\alpha-1}$ for all $T > N$. It means that the variance of the deviation between the estimate $\widehat{\theta}_T^1$ and true value is smaller than that of the corresponding deviation between the estimate $\widehat{\theta}_T^2$ and the true value. It this sense, the estimate $\widehat{\theta}_T^1$ is better than $\widehat{\theta}_T^2$.

Local Asymptotic Normality and Asymptotic Efficiency of the Estimate of the Drift Parameter in a Linear Brownian Diffusion Model

Consider (only for comparison with the fractional case, see below) a pure linear Brownian model

$$dX_t^\theta = \frac{1}{T^\beta}\theta X_t \, dt + c X_t \, dB_t, \; X_{t=0} = X_0, \; c \in \mathbb{R} \setminus \{0\}, \; t \in [0,T], \; \beta \in \left(\frac{1}{2},1\right].$$

Put $\Theta = (0,\infty)$ and let $\theta \in \Theta$. According to Definition 2.1 (IK81), a family of measures $P_\theta(t)$ has the property of local asymptotic normality (LAN) at the point $\theta \in \Theta$ as $t \to \infty$, if

$$Z_{t,\theta}(u) := \frac{dP_{\theta+A(t,\theta)u}(t)}{dP_\theta(t)} = \exp\left\{u\xi_{t,\theta} - \frac{1}{2}u^2 + \zeta_t(u,\theta)\right\} \qquad (6.3.40)$$

for some function $A(t,\theta)$ and any number $u \in \mathbb{R}$, where $\xi_{t,\theta} \Rightarrow N(0,1)$ as $t \to \infty$ with respect to the measure $P_\theta(t)$, and $\zeta_t(u,\theta) \stackrel{P_\theta(t)}{\to} 0$, $t \to \infty$, for all numbers $u \in \mathbb{R}$. We say in this case that the LAN property holds for the family of measures $P_\theta(t)$ as $t \to \infty$ at the point θ.

Theorem 6.3.7. *The LAN property holds for the family of measures $P_\theta(t)$ as $t \to \infty$ at any point $\theta \in \Theta$.*

Proof. We change the probability measure $P_\theta(t)$, which corresponds to the process X_t^θ for the measure $P_0(t)$. Then the drift $\theta X_t \, dt$ disappears and we obtain

$$X_t^0 = X_0 + c \int_0^t X_s^0 \, d\widehat{B}_t,$$

where $\widehat{B}_t = B_t + t\theta/(cT^\beta)$ is a Wiener process w.r.t. the measure $P^0(t)$.

Consider the likelihood ratio corresponding to this change of measure with $\varphi_s = \theta/(cT^\beta)$:

6.3 Properties of Maximum Likelihood Estimates

$$\frac{dP_\theta(t)}{dP_0(t)} = \exp\left\{\int_0^t \frac{\theta}{cT^\beta}\, d\widehat{B}_s - \frac{1}{2}\int_0^t \frac{\theta^2}{(cT^\beta)^2}\, ds\right\}$$

$$= \exp\left\{\frac{\theta}{cT^\beta}\widehat{B}_t - \frac{1}{2}\frac{\theta^2}{(cT^\beta)^2}t\right\}.$$

Now we consider the linear model with a parameter θ shifted by $A(t)u$. The likelihood ratio for such a change of measure is of the form

$$\frac{P_{\theta+A(t)u}(t)}{dP_0(t)} = \exp\left\{\frac{1}{cT^\beta}(\theta+A(t)u)\widehat{B}_t - \frac{1}{2(cT^\beta)^2}(\theta+A(t)u)^2 t\right\}$$

and

$$\frac{dP_{\theta+A(t,\theta)u}(t)}{dP_\theta(t)} = \frac{dP_{\theta+A(t,\theta)u}(t)}{dP_0(t)}\cdot\left(\frac{dP_\theta(t)}{dP_0(t)}\right)^{-1}$$

$$= \exp\left\{\frac{1}{cT^\beta}(\theta+A(t)u)\widehat{B}_t - \frac{1}{2(cT^\beta)^2}(\theta+A(t)u)^2 t - \frac{\theta}{cT^\beta}\widehat{B}_t - \frac{1}{2}\frac{\theta^2}{(cT^\beta)^2}t\right\}$$

$$= \exp\left\{\frac{uA(t)}{cT^\beta}\widehat{B}_t - \frac{1}{2}u^2\frac{A^2(t)}{(cT^\beta)^2}t - \frac{A(t)u\theta}{(cT^\beta)^2}t\right\}.$$

Set $A(t) := cT^\beta/\sqrt{t}$. Then

$$\frac{dP_{\theta+A(t,\theta)u}(t)}{dP_\theta(t)} = \exp\left\{u\frac{\widehat{B}_t}{\sqrt{t}} - \frac{1}{2}u^2 - \frac{u\theta\sqrt{t}}{cT^\beta}\right\}.$$

Since $\widehat{B}_t/\sqrt{t} \Rightarrow N(0,1)$ under both the measures $P_0(t)$ and $P_\theta(t)$ and, in addition, $u\theta\sqrt{t}/(cT^\beta) \to 0$ as $t \to \infty$ for $T \geq t$ and $\alpha > \frac{1}{2}$, the above definition implies the LAN property for the family $P_\theta(t)$ as $t \to \infty$ and at any point $\theta \in \Theta$. □

Consider now the asymptotic efficiency of the estimate of parameter θ. According to definition (11.3), introduced in the monograph (IK81), an estimate $\{\theta_t, t > 0\}$ of a parameter θ is asymptotically efficient under the LAN property for the cost function $w(A^{-1}(t,\theta)x)$ at the point θ if

$$\lim_{\delta\to 0}\lim_{t\to\infty}\sup_{|\theta'-\theta|<\delta} E_{P_{\theta'}(t)} w\left(A^{-1}(t,\theta)(\theta_t - \theta')\right) = Ew(N(0,1)).$$

Let $w \in W$, where W is the class of functions defined on Θ and satisfying the conditions:

1) $w(u) \geq 0$, $w(0) = 0$, w is a Borel function, continuous at zero and not identically zero;
2) $w(u) = w(-u)$,
3) the set $\{u: w(u) < c\}$ is convex for any $c > 0$.

Further we consider the cost function $w\left(A^{-1}(t,\theta)x\right) \in W_p$, where $W_p \subset W$ is the class of functions of W that have a dominant polynomial.

Consider the maximum likelihood estimate of the parameter θ in a linear Brownian model

$$\widehat{\theta}_t = \frac{cT^\beta}{t}\widehat{B}_t = \frac{cT^\beta}{t}\left(B_t + \frac{1}{cT^\beta}\theta t\right) = \theta + \frac{cT^\beta}{t}B_t.$$

To prove the asymptotic efficiency of the estimate $\widehat{\theta}_t$ we use Theorem 1.3 of Chapter III from (IK81). According to this theorem, the estimate $\widehat{\theta}_t$ is asymptotically efficient in the sense mentioned above if the following conditions hold:

(a) $\lim_{t\to\infty} A^{-1}(t,\theta_2)A(t,\theta_1) = B(\theta_1,\theta_2)$ exists, the convergence is uniform in $\theta_i \in \Theta$ and $B(\theta_1,\theta_2)$ is continuous in θ_1;

(b) $\zeta_t(\theta) := A^{-1}(t,\theta)(\widehat{\theta}_t - \theta) \Rightarrow N(0,1)$ uniformly in $\theta_i \in \Theta$ as $t \to \infty$ with respect to the measure $P_\theta(t)$;

(c) for any $N > 0$ random variables $|A^{-1}(t,\theta)(\widehat{\theta}_t-\theta)|^N$, are $P_\theta(t)$-integrable for any $\theta \in \Theta$ uniformly in $t > t_0(N)$.

Condition (a) holds in our case because $A(t) = \frac{cT^\beta}{\sqrt{t}}$ does not depend on θ. Now we check condition (b):

$$\zeta_t(\theta) = A^{-1}(t,\theta)(\widehat{\theta}_t - \theta) = \frac{\sqrt{t}}{cT^\beta}\frac{cT^\beta}{t}B_t = B_t\frac{1}{\sqrt{t}} \Rightarrow N(0,1)$$

under both the measures $P_0(t)$ and $P_\theta(t)$. Condition (c) now is evident. Thus the estimate $\widehat{\theta}_t$ is asymptotically efficient as $t \to \infty$.

Local Asymptotic Normality and Asymptotic Efficiency of the Estimate of the Drift Parameter in a Linear Fractional Brownian Diffusion Model

Now consider a pure linear fractional Brownian model

$$dX_t = \frac{1}{T^\beta}\theta X_t \, dt + X_t \, dB_t^H, X_{t=0} = X_0, \theta \in \Theta, t \in [0,T], \beta \in (1-H,1].$$

It will be clear later that in this model it is sufficient to consider $\beta \in \left(1-H,\frac{1}{2}\right)$. Now $\varphi_t = \theta/T^\beta$. Then

$$\widetilde{\alpha}\int_0^t \delta_s \, ds = \int_0^t l_H(t,s)\frac{\theta}{T^\beta} \, ds = \frac{\theta}{T^\beta}C(H,1)t^{1-2\alpha}, \delta_t = (\theta/T^\beta)C(H,1)t^{-2\alpha}\widetilde{\alpha}.$$

Therefore $\widehat{\theta}_t = T^\beta \widetilde{\alpha}\int_0^t s^{-\alpha}\,d\widehat{B}_s C(H,1)^{-1}t^{2\alpha-1}$, where

$$\widetilde{\alpha}\int_0^t s^{-\alpha}\,d\widehat{B}_s = \widetilde{\alpha}\int_0^t s^{-\alpha}\,dB_s + \frac{\theta}{T^\beta}C(H,1)t^{1-2\alpha}.$$

In other words,

$$\widehat{\theta}_t = \theta + \frac{T^\beta \widetilde{\alpha}\int_0^t s^{-\alpha}\,dB_s}{C(H,1)t^{1-2\alpha}}.$$

6.3 Properties of Maximum Likelihood Estimates 361

Theorem 6.3.8. *The LAN property holds for the family $P_\theta(t)$ as $t \to \infty$ at any point $\theta \in \Theta$.*

Proof. We change the probability measure $P_\theta(t)$ for the measure $P_0(t)$. As a result, the drift $\theta X_t \, dt$ disappears. The corresponding likelihood ratio is given by

$$\frac{dP_\theta(t)}{dP_0(t)} = \exp\left\{ \int_0^t s^\alpha \delta_s \, d\widehat{B}_s - \frac{1}{2}\int_0^t s^{2\alpha} \delta_s^2 \, ds \right\}$$

$$= \exp\left\{ \frac{\theta C(H,1)\widetilde{\alpha}}{T^\beta} \int_0^t s^{-\alpha} \, d\widehat{B}_s - \frac{1}{2T^{2\beta}}(\theta C(H,1))^2 t^{1-2\alpha} \right\}.$$

Now we consider the linear model with parameter θ shifted by $A(t)u$ and denote for simplicity $K = C(H)$.

$$\frac{P_{\theta+A(t)u}(t)}{dP_0(t)} = \exp\left\{ \frac{(\theta + A(t)u)K}{T^\beta} \int_0^t s^{-\alpha} \, d\widehat{B}_s \right.$$

$$\left. - \frac{1}{2T^{2\beta}}((\theta + A(t)u)K)^2 \frac{t^{1-2\alpha}}{1-2\alpha} \right\}.$$

The likelihood ratio for this model is of the form

$$\frac{dP_{\theta+A(t,\theta)u}(t)}{dP_\theta(t)} = \frac{dP_{\theta+A(t,\theta)u}(t)}{dP_0(t)} \cdot \left(\frac{dP_\theta(t)}{dP_0(t)}\right)^{-1}$$

$$= \exp\left\{ \frac{K}{T^\beta} A(t)u \left(\int_0^t s^{-\alpha} \, d\widehat{B}_s - \frac{1}{2}A(t)u \frac{K}{T^\beta} \frac{t^{1-2\alpha}}{1-2\alpha} - \theta \frac{K}{T^\beta} \frac{t^{1-2\alpha}}{1-2\alpha} \right) \right\}.$$

Set $A(t) := T^\beta \widetilde{\alpha} / K t^{1-H}$. Then the likelihood ratio obtains the form

$$\frac{dP_{\theta+A(t,\theta)u}(t)}{dP_\theta(t)} = \exp\left\{ \widetilde{\alpha} u \frac{\int_0^t s^{-\alpha} d\widehat{B}_s}{t^{1-H}} - \frac{1}{2}u^2 - \frac{u\theta K t^{1-H}}{T^\beta \widetilde{\alpha}} \right\}.$$

Since

$$\widetilde{\alpha} \frac{\int_0^t s^{-\alpha} \, d\widehat{B}_s}{t^{1-H}} \Rightarrow N(0,1)$$

and

$$\frac{u\theta K t^{1-H}}{T^\beta \widetilde{\alpha}} \to 0 \quad \text{as } t \to \infty,$$

the LAN property holds for the family $P_\theta(t)$ as $t \to \infty$ at any point $\theta \in \Theta$. □

Now we check the asymptotic efficiency of the estimate $\widehat{\theta}_t$. Consider conditions (a)-(c). Two of them, (a) and (c), are evident. To check (b) we use the following relations:

$$\zeta_t(\theta) = A^{-1}(t,\theta)(\widehat{\theta}_t - \theta) = \frac{C(H)t^{1-H}}{T^\beta \widetilde{\alpha}} \frac{T^\beta \int_0^t s^{-\alpha} \, dB_s}{C(H,1)t^{1-2\alpha}}$$

$$= \frac{\left(\int_0^t s^{-\alpha}dB_s\right)\widetilde{\alpha}}{t^{1-H}} \Rightarrow N(0,1).$$

Therefore, the estimate $\widehat{\theta}_t$ of the parameter θ is asymptotically efficient as $t \to \infty$.

Remark 6.3.9. The maximum likelihood estimators for the drift coefficient in the stochastic differential equations involving fBm were considered also in the paper (TV03), the estimate of the diffusion coefficient for diffusion driven by fBm is contained in the paper (LL00).

A

Mandelbrot–van Ness Representation: Some Related Calculations

Now we calculate the constant that appeared in the Mandelbrot–van Ness representation of fBm (see Section 1.3, Theorem 1.3.1).

Lemma A.0.1. *The following equalities hold:*

$$C_H^{(2)} := \left(\int_{\mathbb{R}_+} ((1+s)^\alpha - s^\alpha)^2 ds + \frac{1}{2H} \right)^{-1} = \frac{(2H \sin \pi H \Gamma(2H))^{1/2}}{\Gamma(1+\alpha)}.$$

Proof. Recall that the constant $C_H^{(2)}$ is chosen to normalize the fBm

$$\overline{B}_t^H = C_H^{(2)} \int_{\mathbb{R}} k_H(t,u) dW_u = C_H^{(2)} \Gamma(1+\alpha) \int_{\mathbb{R}} (I_-^\alpha \mathbf{1}_{(0,t)})(x) dW_x$$

(see Lemma 1.1.3). Therefore, the first equality is evident, since

$$\int_{\mathbb{R}} (k_H(t,u))^2 du = \int_{-\infty}^0 ((t-x)^\alpha - (-x)^\alpha)^2 dx + \int_0^t (t-x)^{2\alpha} dx$$

$$= t^{2H} \left(\int_0^\infty ((1+s)^\alpha - s^\alpha)^2 ds + \frac{1}{2H} \right).$$

We obtain the second equality if we note that

$$\int_{\mathbb{R}} (I_-^\alpha \mathbf{1}_{(0,t)})(x)^2 dx = \frac{1}{2\pi} \int_{\mathbb{R}} \left(\widehat{\mathcal{F}}(I_-^\alpha \mathbf{1}_{(0,t)})(x) \right)^2 dx$$

and according to Theorem 1.1.5

$$\widehat{\mathcal{F}}(I_-^\alpha \mathbf{1}_{(0,t)})(x)(\lambda) = \widehat{\mathbf{1}_{(0,t)}}(\lambda)|\lambda|^{-\alpha} \exp\left\{ \frac{\alpha \pi i}{2} \operatorname{sign} \lambda \right\}$$

$$= \frac{e^{it\lambda} - 1}{i\lambda} |\lambda|^{-\alpha} \exp\left\{ \frac{\alpha \pi i}{2} \operatorname{sign} \lambda \right\}.$$

Therefore,

$$\int_{\mathbb{R}} (I_-^\alpha \mathbf{1}_{(0,t)})(x)^2 dx = \frac{1}{2\pi} \int_{\mathbb{R}} |e^{it\lambda} - 1|^2 |\lambda|^{-2\alpha-2} d\lambda$$

$$= \frac{1}{2\pi} \int_{\mathbb{R}} (1 - \cos t\lambda)^2 |\lambda|^{-2\alpha-2} d\lambda + \frac{1}{2\pi} \int_{\mathbb{R}} \sin^2 t\lambda |\lambda|^{-2\alpha-2} d\lambda$$

$$= \frac{1}{\pi} \int_0^\infty \frac{(1 - \cos t\lambda)^2}{\lambda^{2\alpha+2}} d\lambda + \frac{1}{\pi} \int_0^\infty \frac{\sin^2 t\lambda}{\lambda^{2\alpha+2}} d\lambda$$

$$= t^{2H} \left(\frac{1}{\pi} \int_0^\infty \frac{(1 - \cos \lambda)^2}{\lambda^{2\alpha+2}} d\lambda + \frac{1}{\pi} \int_0^\infty \frac{\sin^2 \lambda}{\lambda^{2\alpha+2}} d\lambda \right) = \frac{t^{2H}}{2H \sin \pi H \Gamma(2H)},$$

whence the proof follows. □

B

Approximation of Beta Integrals and Estimation of Kernels

These results were obtained by E.Valkeila (KMV05).

Lemma B.0.1. *Assume that $-1 < \delta < 0$, $\beta > -1$ and $n \geq 2$. Then for $\beta \geq 0$*

$$|I(\delta, \beta) - I_n(\delta, \beta)| \leq C_1(\delta, \beta) n^{-\alpha - 1}, \qquad (B.0.1)$$

and with $-1 < \beta < 0$ we have

$$|I(\alpha, \beta) - I_n(\delta, \beta)| \leq C_2(\delta, \beta) n^{-\alpha - \beta - 1} \qquad (B.0.2)$$

(for the value of the constants, see the proof).

Proof. We start the proof with

$$I(\delta, \beta) - I_n(\delta, \beta) = \int_0^{\frac{1}{n}} s^\delta (1-s)^\beta ds - n^{-\delta - 1}$$
$$+ \sum_{k=1}^{n-2} \int_{\frac{k}{n}}^{\frac{k+1}{n}} \left(s^\delta (1-s)^\beta - \left(\frac{k+1}{n}\right)^\delta \left(1 - \frac{k}{n}\right)^\beta \right) ds$$
$$+ \int_{1-\frac{1}{n}}^1 s^\delta (1-s)^\beta ds - n^{-\beta - 1}.$$

We work first with the integral on $(0, 1/n)$. We have

$$\int_0^{\frac{1}{n}} s^\delta (1-s)^\beta ds - n^{-\delta - 1} = \int_0^{\frac{1}{n}} \left(s^\delta - n^{-\delta} \right) ds$$
$$+ \int_0^{\frac{1}{n}} s^\delta \left((1-s)^\beta - 1 \right) ds; \qquad (B.0.3)$$

here

$$0 \leq \int_0^{\frac{1}{n}} \left(s^\delta - n^{-\delta} \right) ds = -\delta/(\delta + 1) n^{-\delta - 1},$$

if $\beta \geq 0$, then

$$\left| \int_0^{\frac{1}{n}} s^\delta \left((1-s)^\beta - 1\right) ds \right| \leq \int_0^{\frac{1}{n}} s^\delta ds$$

and if $\beta < 0$ and $s \leq 1/n$, then $0 \leq (1-s)^\beta - 1 \leq 2^{-\beta} - 1$. Use these estimates in (B.0.3) to obtain

$$\left| \int_0^{\frac{1}{n}} s^\delta (1-s)^\beta ds - n^{-\delta-1} \right| \leq C_1(\delta, \beta) n^{-\delta-1}. \tag{B.0.4}$$

Next, we work with the integral on $(1 - 1/n, 1)$. We have

$$\int_{1-\frac{1}{n}}^1 s^\delta (1-s)^\beta ds - n^{-\beta-1} = \int_{1-\frac{1}{n}}^1 \left((1-s)^\beta - n^{-\beta}\right) ds$$
$$+ \int_{1-\frac{1}{n}}^1 (1-s)^\beta \left(s^\delta - 1\right) ds,$$

and this gives

$$\left| \int_{1-\frac{1}{n}}^1 s^\delta (1-s)^\beta ds - n^{-\beta-1} \right| \leq \frac{|\beta|}{1+\beta} n^{-\beta-1} + 2^{-\delta} n^{-\beta-1}. \tag{B.0.5}$$

We continue with the middle term. We have

$$\sum_{k=1}^{n-2} \left(\int_{\frac{k}{n}}^{\frac{k+1}{n}} s^\delta (1-s)^\beta ds - \left(\frac{k+1}{n}\right)^\delta \left(1 - \frac{k}{n}\right)^\beta \frac{1}{n} \right)$$

$$= \sum_{k=1}^{n-2} \left(\int_{\frac{k}{n}}^{\frac{k+1}{n}} \left(s^\delta - \left(\frac{k+1}{n}\right)^\delta \right) (1-s)^\beta ds \right)$$

$$+ \sum_{k=1}^{n-2} \left(\int_{\frac{k}{n}}^{\frac{k+1}{n}} \left(\frac{k+1}{n}\right)^\delta \left((1-s)^\beta - \left(1 - \frac{k}{n}\right)^\beta\right) ds \right). \tag{B.0.6}$$

The first term on the right-hand side of (B.0.6) is always positive, when $\delta < 0$. We use the estimate

$$s^\delta - ((k+1)/n)^\delta \leq (k/n)^\delta - ((k+1)/n)^\delta.$$

If $\beta \geq 0$, then $(1-s)^\beta \leq 1$ and so for the first term on the right-hand side of (B.0.6) we obtain

$$0 \leq \sum_{k=1}^{n-2} \left(\int_{k/n}^{(k+1)/n} \left(s^\delta - ((k+1)/n)^\delta\right) (1-s)^\beta ds \right)$$

B Approximation of Beta Integrals and Estimation of Kernels   367

$$\leq n^{-\delta-1} \sum_{k=1}^{n-2} \left(k^{\delta} - (k+1)^{\delta}\right) \leq n^{-\delta-1}. \tag{B.0.7}$$

If $\beta \leq 0$ then

$$\int_{k/n}^{(k+1)/n} \left(s^{\delta} - ((k+1)/n)^{\delta}\right)(1-s)^{\beta} ds \leq \frac{1}{1+\beta} n^{-\delta-\beta-1} \left(k^{\delta}\right.$$
$$\left. - (k+1)^{\delta}\right) \left((n-k)^{\beta+1} - (n-(k+1))^{\beta+1}\right) \leq n^{-\delta-\beta-1} \left(k^{\delta} - (k+1)^{\delta}\right),$$

and this gives the estimate

$$0 \leq \sum_{k=1}^{n-2} \left(\int_{k/n}^{(k+1)/n} (s^{\alpha} - ((k+1)/n)^{\alpha})(1-s)^{\beta} ds\right) \leq n^{-\alpha-\beta-1}. \tag{B.0.8}$$

Finally, the second part of the middle term is

$$J_n := \sum_{k=1}^{n-2} \left(\int_{k/n}^{(k+1)/n} ((k+1)/n)^{\delta} \left((1-s)^{\beta} - (1-k/n)^{\beta}\right) ds\right).$$

If $\beta \geq 0$, then with calculations similar to above

$$|J_n| \leq n^{-\delta-1}, \tag{B.0.9}$$

and if $\beta < 0$, then

$$|J_n| \leq -\frac{1}{\beta} 2^{\beta} n^{-\alpha-\beta-1}. \tag{B.0.10}$$

Combining the bounds (B.0.3)–(B.0.7) and (B.0.9) we get $C_1(\delta, \beta)$, and combining the bounds (B.0.3)–(B.0.6), (B.0.8) and (B.0.10) we get $C_2(\delta, \beta)$. □

Lemma B.0.2. *Put*

$$H_n := \sum_{k=0}^{n-1} \left(\left(\frac{k+1}{n}\right)^{1/4-H} \left(1-\frac{k}{n}\right)^{3/4-H} \left(\left(\frac{k+1}{n}\right)^{2H} - \left(\frac{k}{n}\right)^{2H}\right)\right.$$
$$\left. -2H \left(\frac{k+1}{n}\right)^{H-3/4} \left(1-\frac{k}{n}\right)^{3/4-H} \frac{1}{n}\right).$$

Then

$$|H_n| \leq C n^{-\min(1, \frac{1}{4}+H)}. \tag{B.0.11}$$

Proof. The proof of Lemma B.0.2 is similar to Lemma B.0.1. □

The proof of the following lemma is obvious.

B Approximation of Beta Integrals and Estimation of Kernels

Lemma B.0.3. *Consider the expression*

$$\bar{u}_n(H) := \frac{1}{n} \sum_{k=0}^{n-1} \left(\left(\frac{k+1}{n}\right)^{2H} - \left(\frac{k}{n}\right)^{2H} \right)^2.$$

Then

$$|\bar{u}_n(H)| \leq \frac{C}{n^2}. \tag{B.0.12}$$

References

[AOPU00] Aase, K., Øksendal, B., Privault, N., Ubøe, J.: White noise generalization of the Clark-Haussmann-Ocone theorem with applications to mathematical finance. Finance Stoch., **4**, 465–496 (2000)

[AS96] Abry, P., Sellan, F.: The wavelet-based synthesis for fractional Brownian motion proposed by F. Sellan and Y. Meyer: Remarks and fast implementation. Appl. Comp. Harmon. Analysis, **3**, 377–383 (1996)

[AS95] Adler, R.J.; Samorodnitsky, G.: Super fractional Brownian motion, fractional super Brownian motion and related self-similar (super) processes. Ann. Prob., **23**, 743–766 (1995)

[ALN01] Alòs, E., León, I.A., Nualart, D.: Stratonovich stochastic calculus with respect to fractional Brownian motion with Hurst parameter less than 1/2. Taiwanesse J. Math., **5**, 609–632 (2001)

[AMN00] Alòs, E., Mazet, O., Nualart, D.: Stochastic calculus with respect to fractional Brownian motion with Hurst parameter less than 1/2. Stoch. Proc. Appl., **86**, 121–139 (2000)

[AMN01] Alòs, E., Mazet, O., Nualart, D.: Stochastic calculus with respect to Gaussian processes. Ann. Prob., **29**, 766–801 (2001)

[AN02] Alòs, E., Nualart, D.: Stochastic integration with respect to the fractional Brownian motion. Stoch. Stoch. Rep., **75**, 129–152 (2002)

[And05] Androshchuk, T.: The approximation of stochastic integral w.r.t. fBm by the integrals w.r.t. absolutely continuous processes. Prob. Theory Math. Stat., **73**, 11–20 (2005)

[AM06] Androshchuk, T., Mishura Y.: Mixed Brownian–fractional Brownian model: absence of arbitrage and related topics. Stochastics: Intern. J. Prob. Stoch. Proc., **78**, 281–300 (2006)

[AG03] Anh, V., Grecksch, W.: A fractional stochastic evolution equation driven by fractional Brownian motion. Monte Carlo Methods Appl. **9**, 189–199 (2003)

[AHL01] Anh, V., Heyde, C., Leonenko, N.: Dynamic models of long-memory processes driven by Lévy noise with applications to finance and macroeconomics. J. Appl. Probab. **39**, 730–747 (2002)

[AI04] Anh, V., Inoue, A.: Prediction of fractional Brownian motion with Hurst index less than 1/2. Bull. Aust. Math. Soc., **70**, 321–328 (2004)

[AI05a] Anh, V., Inoue, A.: Financial markets with memory I: Dynamic models. Stoch. Anal. Appl., **23**, 275–300 (2005)

[AI05b] Anh, V., Inoue, A.: Financial markets with memory II: Innovation processes and expected utility maximization. Stoch. Anal. Appl., **23**, 301–328 (2005)

[ALN01] Anh, V., Leonenko, N., Nguyen, C.: Stochastic differential equations with fractional Riesz–Bessel input. Inform. Techn. Econom. Management, **1**, 1–16 (2001)

[ALP02a] Ayache, A., Leger, S., Pontier M.: Drap brownien fractionnaire. (Fractional Brownian sheet.) Potential Analysis, **17**, 31–43 (2002)

[ALP02b] Ayache, A., Leger, S., Pontier, M.: Les ondelettes a la conquite du drap brownien fractionnaire. (Wavelets conquering the fractional Brownian field.) C. R. Acad. Sci. Paris Sér. I Math., **335**, 1063–1068 (2002)

[BLOPST03] Bardet, J.M., Lang, G., Oppenheim, G., Philippe, A., Stoev, S., Taqqu, M.S.: Semi-parametric estimation of the long-range dependence parameter: a survey. In: Doukhan, P., Oppenheim, G., Taqqu, M. (eds) Theory and Applications of Long-Range Dependence. Birkhäuser, Boston, 557–579 (2003)

[BJ06] Bardina, X., Jolis, M.: Multiple fractional integral with Hurst parameter less than 1/2. Stoch. Proc. Appl., **116**, 463–479 (2006)

[BP88] Barton, R.J., Vincent Poor, H.: Signal detection in fractional Gaussian noise. IEEE Trans. Inform. Theory, **34**, 943–959 (1988)

[BaNu06] Baudoin, F., Nualart, D.: Notes on the two-dimensional fractional Brownian motion. Ann. Prob., **34**, 159–180 (2006)

[Ben03a] Bender, C.: Integration with respect to a fractional Brownian motion and related market models. PhD Thesis, Hartung-Gorre Verlag, Konstanz (2003)

[Ben03b] Bender, C.: An S-transform approach to integration with respect to a fractional Brownian motion. Bernoulli, **9**, 955–983 (2003)

[Ben03c] Bender, C.: An Itô formula for generalized functionals of a fractional Brownian motion with arbitrary Hurst parameter. Stoch. Proc. Appl., **104**, 81–106 (2003)

[BE03] Bender, C., Elliott, R.J.: On the Clark–Ocone theorem for fractional Brownian motions with Hurst parameter bigger than a half. Stoch. Stoch. Rep., **75**, 391–405 (2003)

[BE04] Bender, C., Elliott, R.J.: Arbitrage in a discrete version of the Wick-fractional Black-Scholes market. Math. Oper. Res., **29**, 935–945 (2004)

[BSV06] Bender, C., Sottinen, T., Valkeila, E.: No-arbitrage pricing beyond semimartingales. Preprint, Weierstrass-Institut Ang. Anal. Stoch., No. 1110, Berlin (2006)

[Ben03] Benth, F.E.: On arbitrage-free pricing of weather derivatives based on fractional Brownian motion. Appl. Math. Finance, **10**, 303–324 (2003)

[Ber94] Beran, J.: Statistics for Long-memory Processes. Chapman & Hall, New York (1994)

[Ber69] Berman, S.M.: Harmonic analysis of local times and sample functions of Gaussian processes. Trans. Amer. Math. Soc., **143**, 269–281 (1969)

[Ber70] Berman, S.M.: Gaussian processes with stationary increments: local times and sample function properties. Ann Math. Stat., **41**, 1260–1272 (1970)

[BGK06] Bhansali, R.J., Giraitis, L., Kokoszka, P.S.: Estimation of the memory parameter by fitting fractionally differenced autoregressive models. J. Multivariate Anal. **97**, 2101–2130 (2006).

[BHOS02] Biagini, F., Hu, Y., Øksendal, B., Sulem, A.: A stochastic maximum principle for processes driven by a fractional Brownian motion. Stoch. Proc. Appl., **100**, 233–254 (2002)

[BHOZ07] Biagini, F., Hu, Y., Øksendal, B., Zhang T.: Stochastic Calculus for Fractional Brownian Motion and Applications. Forthcoming. Probability and Its Applications, Springer (2007)

[BO02] Biagini, F., Øksendal, B.: Minimal variance hedging for fractional Brownian motion. Methods Appl. Analysis, **10**, 347–362 (2003)

[BO04] Biagini, F., Øksendal, B.: Forward integrals and an Itô formula for fractional Brownian motion. Dept. of Math., University of Oslo, No. 22 (2004)

[Bi81] Bichteler, K.: Stochastic integration and L^p-theory of semimartingales. Ann. Prob., **9**, 49–89 (1981)

[BG96] Bisaglia, L., Guegan, D.: A review of techniques of estimation in long memory processes: application to intraday data. Comp. Stat. Data Analysis, **26**, 61–81 (1997)

[BG98] Bisaglia, L., Guegan, D.: A comparison of techniques of estimation in long-memory processes. Comp. Stat. Data Analysis, **27**, 61–81 (1998)

[BH05] Björk, T., Hult, H.: A note on Wick products and the fractional Black–Scholes model. Finance Stoch., **9**, 197–209 (2005)

[BlSc73] Black, F., Scholes, M.: The pricing of options and corporate liabilities. J. Political Econ., **81**, 637–654 (1973)

[Bor84] Borovkov, A.A.: Mathematical Statistics. Estimates of Parameters. Testing Hypotheses. (Russian) Nauka, Moscow (1984)

[BL01] Boufoussi, B., Lakhel, El H.: Weak convergence in Besov spaces to fractional Brownian motion. C. R. Acad. Sci. Paris Sér. I Math., **333**, 39–44 (2001)

[BO03] Boufoussi, B., Ouknine, Y.: On a SDE driven by a fractional Brownian motion and with monotone drift. Electronic Comm. Prob., **8**, 122–134 (2003)

[BK00] Buldygin, V.V., Kozachenko, Yu.V.: Metric Characterization of Random Variables and Random Processes. AMS, Providence, Rhode Island (2000)

[CC98] Carmona, P., Coutin, L.: Fractional Brownian motion and the Markov property. Electronic Commun. Prob., **3**, 95–107 (1998)

[CC00] Carmona, P., Coutin, L.: Integrale stochastique pour le mouvement brownien fractionnaire. C. R. Acad. Sci. Paris Sér. I Math., **330**, 231–236 (2000)

[CCM98] Carmona, P., Coutin, L., Montseny, G.: Applications of a representation of long memory Gaussian processes. Preprint, Université de Toulouse. Laboratoire de Statistique et Probabilités (1998)

[CCM03] Carmona, P., Coutin, L., Montseny, G.: Stochastic integration with respect to fractional Brownian motion. Ann. Inst. Henri Poincaré, Prob. Stat., **39**, 27–68 (2003)

[Che01a] Cheridito, P.: Mixed fractional Brownian motion. Bernoulli, **7**, 913–934 (2001)

[Che01b] Cheridito, P.: Regularizing fractional Brownian motion with a view towards stock price modelling. PhD thesis, Zurich (2001)

[CKM03] Cheridito, P., Kawaguchi, H., Maejima, M.: Fractional Ornstein–Uhlenbeck processes. Electronic J. Prob., Paper no. 3, 1–14 (2003)

[CN02] Cheridito, P., Nualart, D.: Symmetric integration with respect to fractional Brownian motion. Preprint, University of Barcelona (2002)

[CN05] Cheridito, P., Nualart, D.: Stochastic integral of divergence type with respect to the fractional Brownian motion with Hurst parameter $H < 1/2$. Ann. Inst. Henri Poincaré, Prob. Stat. **41**, 1049–1081 (2005)

[CMS80] Chou, C.S., Meyer, P.A., Stricker, C.: Sur les intégrales stochastiques de processus prévisibles non bornées. Séminaire Prob. XIV, In: Lect. Notes Math. **784**, 128–139 (1980)

[ChW83] Chung, K.L., Williams, R.J.: Introduction to Stochastic Integration. Birkhäuser, Boston (1983)

[CKR93] Ciesielski, Z., Kerkyacharian, G., Roynette, B.: Quelques espaces fouctionnels associes á des processus gaussiens. Studia Mat., **107**, 171–204 (1993)

[CM95] Cioczek-Georges, R., Mandelbrot, B.: A class of micropulses and antipersistent fractional Brownian motion. Stoch. Proc. Appl., **60**, 1–18 (1995)

[CM96] Cioczek-Georges, R., Mandelbrot, B.: Alternative micropulses and fractional Brownian motion. Stochastic Processes Appl., **64**, 143–152 (1996)

[CNW06] Corcuera, J.M., Nualart, D., Woerner, J..C.: Power variation of some integral long-memory processes. Bernoulli, **12**, 713–735 (2006).

[Cou07] Coutin, L.: An introduction to (stochastic) calculus with respect to fractional Brownian motion. In: Donati-Martin, C. et al. (eds) Séminaire de Probabilités XL, Lecture Notes in Math. 1899, Springer (2007)

[CD99] Coutin, L., Decreusefond, L.: Abstract nonlinear filtering theory in the presence of fractional Brownian motion. Ann. Appl. Prob., **9**, 1058–1090 (1999)

[CD01] Coutin, L., Decreusefond, L.: Stochastic Volterra equations with singular kernels. In: Cruzeiro, Ana Bela et al. (eds), Stochastic Analysis and Mathematical Physics. Birkhäuser, Boston. Prog. Probab. 50, 39–50 (2001)

[CNT01] Coutin, L., Nualart, D., Tudor, C.A.: Tanaka formula for the fractional Brownian motion. Stoch. Proc. Appl., **94**, 301–315 (2001)

[CQ00] Coutin, L., Qian, Z.: Stochastic differential equations for fractional Brownian motions. C. R. Acad. Sci. Paris Sér. I Math., **331**, 75–80 (2000)

[CQ02] Coutin, L., Qian, Z.: Stochastic analysis, rough path analysis and fractional Brownian motion. Prob. Theory Rel. Fields, **122**, 108–140 (2002)

[CM96] Csörgö, S., Mielniczuk, J.: Extreme values of derivatives of smoothed fractional Brownian motions. Prob. Math. Stat., **16**, 211–219 (1996)

[CKW95] Cutland, N.G., Kopp, P.E., Willinger, W.: Stock price returns and the Josef effect: a fractional version of the Black–Scholes model. In: Bolthausen, E., Dozzi, M., Russo, F. (eds) Sem. Stoch. Anal., Random Fields Appl. Birkhäuser, Boston, 327–351 (1995)

[DH96] Dai, W., Heyde, C.C.: Itô's formula with respect to fractional Brownian motion and its application. J. Appl. Math. Stoch. Analysis, **9**, 439–448 (1996)

[Das98] Dasgupta, A.: Fractional Brownian motion. Its properties and applications to stochastic integration. PhD Thesis, University of North Carolina (1998)

[Daye03] Daye, Z.J.: Introduction to Fractional Brownian Motion. Advanced FE, www.stat.purdue.edu/~zdaye/Projects/fBM.doc (2003)

[Dec01] Decreusefond, L.: A Skorokhod-Stratonovitch integral for the fractional Brownian motion. In: Stochastic Analysis and Related Topics VII. Proc. 7th Silivri Workshop. Birkhäuser, Boston, 177–198 (2001)

[Dec03] Decreusefond, L.: Stochastic integration with respect to fractional Brownian motion. In: Doukhan, P., Oppenheim, G., Taqqu, M. (eds) Theory and Applications of Long-Range Dependence. Birkhäuser, Boston, 203–225 (2003)

[DU95] Decreusefond, L., Üstünel, A.S.: Application du calcul des variations stochastiques an mouvement brownien fractionnaire. C. R. Acad. Sci. Paris Sér. I Math., **321**, 1605–1608 (1995)

[DU98] Decreusefond, L., Üstünel, A.S.: Fractional Brownian motion: Theory and applications. ESAIM Proc., **5**, 75–86 (1998)

[DU99] Decreusefond, L., Üstünel, A.S.: Stochastic analysis of the fractional Brownian motion. Potential Analysis, **10**, 177–214 (1999)

[DS06] Delbaen, F., Schachermayer, W.: The Mathematics of Arbitrage. Springer, Berlin (2006)

[Del72] Dellacherie, C.: Capacités et Processus Stochastiques. Springer, Berlin (1972)

[DM82] Dellacherie, C., Meyer, P.-A.: Probabilities and Potential. B: Theory of Martingales. North-Holland, Amsterdam (1982)

[DH03] Deo, R.S., Hurvich, C.M.: Estimation of long-memory in volatility. In: Doukhan, P., Oppenheim, G., Taqqu, M. (eds) Theory and Applications of Long-Range Dependence. Birkhäuser, Boston, 313–325 (2003)

[DDM70] Doléans-Dade, C., Meyer, P.A.: Intégrales stochastiques par rapport aux martingales locales. Séminaire Prob. IV. In: Lect. Notes Math. **124**, 77–107 (1970)

[Do03] Doukhan, P.: Models, inequalities and limit theorems for stationary sequences. In: Doukhan, P., Oppenheim, G., Taqqu, M. (eds) Theory and Applications of Long-Range Dependence. Birkhäuser, Boston, 43–100 (2003)

[DKL03] Doukhan, P., Khezour, A., Lang,G.: Non-parametric estimation under long-range dependence. In: Doukhan, P., Oppenheim, G., Taqqu, M. (eds) Theory and Applications of Long-Range Dependence. Birkhäuser, Boston, 303–313 (2003)

[DN99] Dudley, R.M., Norvaiša, R.: An introduction to p-variation and Young integrals. Concentrated advanced course. Maphysto, CMPhS University of Aarhus, Denmark (1999)

[Dun00] Duncan, T.E.: Some applications of fractional Brownian motion to linear systems. In: System theory: modeling, analysis, and control. Kluwer Int. Ser. Eng. Comput. Sci, **518**, Kluwer Academic Publishers, Boston, 97–105 (2000)

[Dun01] Duncan, T.: Some aspects of fractional Brownian motion. Nonlinear Analysis Theory Methods Appl., **47**, 4775–4782 (2001)

[Dun04] Duncan, T.: Some processes associated with a fractional Brownian motion. In: Mathematics of Finance. Proc. AMS-IMS-SIAM Conf. Contemporary Mathematics, **351**, AMS, Providence, RI, 93–101 (2004)

[Dun06] Duncan, T.E.: Prediction for some processes related to a fractional Brownian motion. Stat. Prob. Lett., **76**, 128–134 (2006)

[DHP00] Duncan, T.E., Hu, Y., Pasik-Duncan, B.: Stochastic calculus for fractional Brownian motion, I. Theory. SIAM J. Control Optimization, **38**, 582–612 (2000)

[DPM02] Duncan, T., Pasik-Duncan, B., Maslowski, B.: Fractional Brownian motion and stochastic equations in Hilbert spaces. Stoch. Dyn., **2**, 225–250 (2002)

[DF02] Dzhaparidze, K., Ferreira, J.A.: A frequency domain approach to some results on fractional Brownian motion. Stat. Prob. Lett., **60**, 155–168 (2002)

[DvZ04] Dzhaparidze, K., van Zanten, H.: A series expansion of fractional Brownian motion. Prob. Theory Relat. Fields, **130**, 39–55 (2004)

[DvZ05] Dzhaparidze, K., van Zanten, H.: Krein's spectral theory and the Paley-Wiener expansion for fractional Brownian motion. Ann. Prob., **33**, 620–644 (2005)

[ElN93] El-Nouty, Ch.: A Hanson-Russo-type law of the iterated logarithm for fractional Brownian motion. Stat. Prob. Lett. **17**, 27–34 (1993)

[ElN99] El-Nouty, Ch.: On the large increments of fractional Brownian motion. Stat. Prob. Lett., **41**, 169–178 (1999)

[Ell82] Elliott R.J.: Stochastic Calculus and Applications. Applications of Mathematics, **18**. Springer-Verlag, New York (1982)

[EvH01] Elliott, R.J., van der Hoek, J.: Fractional Brownian motion and financial modelling. In: Kohlmann, M., Tang, S. (eds) Proc. Workshop Math. Finance, Konstanz, Birkhäuser, Basel, 140–150 (2001)

[EvH03] Elliott, R.J., van der Hoek, J.: A general fractional white noise theory and applications to finance. Math. Finance, **13**, 301–330 (2003)

[ENO02] Erraoui, M., Nualart, D., Ouknine, Y.: Hyperbolic stochastic partial differential equations with additive fractional Brownian sheet. Preprint No. 307, Univ.de Barcelona (2002)

[Fel36] Feller, W.: Zur Theorie der Stochastischen Prozesse (Existenz- und Eindeutigkeitssatze). Math. Ann., **113**, 113–160 (1936)

[Fer74] Fernique, X.M.: Regularitè des trajectories de fonctions alèatories gaussiennes. In: Lect. Notes Math., **480**, 2–95 (1974)

[FdP99] Feyel, D., de la Pradelle, A.: Fractional integrals and Brownian processes. Potential Analysis, **10**, 273–288 (1999)
[FdP01] Feyel, D., de la Pradelle, A.: The FBM Itô's formula through analytic continuation. Electronic J. Prob., paper **26**, 1–22 (2001)
[Fol81a] Föllmer H.: Calcul d'Itô sans probabilités. Séminaire Prob. XV, In: Lect. Notes Math. **850**, 144–150 (1981)
[Fol81b] Föllmer, H.: Dirichlet processes. In: Stochastic integrals, Proc. LMS Durham Symp. 1980, Lect. Notes Math. **851**, 476–478 (1981).
[Gap04] Gapeev, P.: On arbitrage and Markovian short rates in fractional bond markets. Stat. Prob. Lett., **70**, 211–222 (2004)
[GRR71] Garsia, A., Rodemich, E., Rumsey, H.: A real variable lemma and the continuity of paths of some Gaussian processes. Indiana Univ. Math. Journal, **20**, 565–578 (1970/71)
[GR03a] Giraitis, L., Robinson, P.M.: Edgeworth expansions for semi-parametric Whittle estimation of long memory. Ann. Stat. **31**, 1325–1375 (2003)
[GR03b] Giraitis, L., Robinson, P.M.: Parametric estimation under long-range dependence. In: Doukhan, P., Oppenheim, G., Taqqu M. (eds) Theory and Applications of Long-Range Dependence. Birkhäuser, Boston, 229–250 (2003)
[Gje94] Gjessing, H.: Wick calculus with applications to anticipating stochastic differential equations. Manuscript, Univ. of Bergen (1994)
[Gol84] Goldman, A.: Estimations analytiques concernant le mouvement brownien fractionnaire à plusieurs paramètres. (Analytic estimations concerning a fractional Brownian motion.) (French) C. R. Acad. Sci. Paris Sér. I Math., **298**, 91–93 (1984)
[Gor77] Gorodeckii, V.V.: On convergence to semi-stable Gaussian processes. Theory Prob. Appl. **22**, 498–508 (1978)
[GRV03] Gradinaru, M., Russo, F., Vallois, P.: Generalized covariations, local time and Stratonovich Itô's formula for fractional Brownian motion with Hurst index $H \geq 1/4$. Ann Prob., **31**, 1772–1820 (2003)
[GR80]   Gradshteyn, I.S.; Ryzhik, I.M.: Table of Integrals, Series, and Products. Academic Press, New York (1980)
[GA98]   Grecksch, W., Anh, V.: Approximation of stochastic differential equations with modified fractional Brownian motion. Z. Anal. Anwend., **17**, 715–727 (1998)
[GA99a] Grecksch, W., Anh, V.: Approximation of stochastic Hammerstein integral equation with fractional Brownian motion input. Monte Carlo Methods Appl., **5**, 311–323 (1999)
[GA99b] Grecksch, W., Anh, V. A parabolic stochastic differential equation with fractional Brownian motion input. Stat. Prob. Lett., **41**, 337–346 (1999)

[GN96] Grippenberg, G., Norros, I.: On the prediction of fractional Brownian motion. J. Appl. Prob., **33**, 400–410 (1996)

[GuNu05] Guerra, J., Nualart, D.: The $1/H$-variation of the divergence integral with respect to fractional Brownian motion for $H > 1/2$ and fractional Bessel processes. Stoch. Proc. Appl., **115**, 91–115 (2005)

[GK05] Gushchin, A.A., Küchler, U.: On recovery of a measure from its symmetrization. Theory Prob. Appl. **49**, 323–333 (2005)

[HM06a] Herbin, E., Merzbach, E.: A characterization of the set-indexed fractional Brownian motion by increasing paths. C. R., Math., Acad. Sci. Paris **343**, 767–772 (2006)

[HM06b] Herbin, E., Merzbach, E.: A set-indexed fractional Brownian motion. J. Theor. Probab. **19**, 337–364 (2006)

[HH76] Hida, T., Hitsuda, M.: Gaussian Processes. American Mathem. Society: Translations of Math. Monographs (1976)

[Hit68] Hitsuda, M.: Representation of Gaussian processes equivalent to Wiener process. Osaka J. Mathematics, **5**, 299–312 (1968)

[Ho96] Ho, H.-C. On central and noncentral limit theorems in density estimation for sequences of long-range dependence. Stoch. Proc. Appl., **63**, 153-174 (1996)

[HH03] Ho, H.-C., Hsing, T.: A decomposition for generalized U-statistics of long memory linear processes. In: Doukhan, P., Oppenheim, G., Taqqu, M. (eds) Theory and Applications of Long-Range Dependence. Birkhäuser, Boston, 143–156 (2003)

[HC93] Hodges, S., Carverhill, A.: Quasimean reversion in an efficient stock market: the characterization of economics equilibria which support Black-Scholes option pricing. The Economic J., **103**, 395-405 (1993)

[HOUZ96] Holden, H., Øksendal, B., Ubøe, J., Zhang, T.: Stochastic Partial Differential Equations. Birkhäuser, Boston (1996)

[Hu05] Hu,Y.: Integral tranformations and anticipative calculus for fractional Brownian motions. Mem. Amer. Math. Soc., **175** (2005)

[HN04] Hu,Y., Nualart, D.: Some processes associated with fractional Bessel processes. Preprint No. 347, Univ. de Barcelona (2004)

[HN06] Hu,Y., Nualart, D.: Rough path analysis via fractional calculus. Manuscript. arXiv:math.PR/0602050 v1 2Feb (2006)

[HO02] Hu, Y., Øksendal, B.: Chaos expansion of local time of fractional Brownian motions. Stoch. Anal. Appl., **20**, 815–837 (2002)

[HO03] Hu, Y., Øksendal, B.: Fractional white noise calculus and applications to finance. Infin. Dimens. Anal. Quantum Prob. Relat. Top., **6**, 1–32 (2003)

[HOS05] Hu, Y., Øksendal, B., Salopek, D., M.: Weighted local time for fractional Brownian motion and applications to finance. Stochastic Anal. Appl., **23**, 15–30 (2005)

[HOS03] Hu, Y., Øksendal, B., Sulem, A.: Optimal portfolio in a fractional Black and Scholes market. Infin. Dimens. Anal. Quantum Probab. Relat. Top. **6**, 519–536 (2003)

[HC78] Huang, S.T., Cambanis, S.: Stochastic and multiple Wiener integrals for Gaussian processes. Ann. Prob. **6**, 585–614 (1978)

[Hur51] Hurst, H.E.: Long-term storage capacity in reservoirs. Trans. Amer Soc. Civil. Eng., **116**, 400–410 (1951)

[HBS65] Hurst, H.E., Black, R.P., Simaika, Y.M.: Long Term Storage in Reservoirs. An Experimental Study. Constable, London (1965)

[HP04a] Hüsler, J., Piterbarg, V.: Limit theorem for maximum of the storage process with fractional Brownian motion input. Stoch. Proc. Appl., **114**, 231–250 (2004)

[HP04b] Hüsler, J., Piterbarg, V.: On the ruin probability for physical fractional Brownian motion. Stoch. Proc. Appl., **113**, 315–332 (2004)

[IK81] Ibragimov, I.A., Khas'minski, R.Z.: Statistical Estimation: Asymptotic Theory. Springer, Berlin (1981)

[IT99] Iglói, E., Terdik, Gy.: Bilinear stochastic systems with fractional Brownian motion input. Ann Appl. Prob., **9**, 46–77 (1999)

[INA06] Inoue, A., Nakano, Y., Anh, V.: Linear filtering of systems with memory and application to finance. Journal of Appl. Math. and Stoch. Anal., **2006**, 1–26 (2006)

[IN07] Inoue, A., Nakano, Y.: Optimal long term investment model with memory. Appl. Math. Optimization, **55**, 93–122 (2007).

[INA07] Inoue, A., Nakano, Y., Anh, V.: Binary market models with memory. Stat. Prob. Lett., **77**, 256–264 (2007).

[Itô42] Itô, K.: Differential equations determining Markov processes. Zenkoku Shijo Sugaku Danwakai, **244**, 1352–1400 (1942)

[Itô44] Itô, K.: Stochastic integral. Proc. Imp. Acad. Tokyo, **20**, 519–524 (1944)

[Itô51] Itô, K.: Stochastic differential equations. Memoirs of the Amer. Math. Soc. **4** (1951)

[Jac79] Jacod, J.: Calcul stochastique et problème de martingales. Lect. Notes. Math., **714** Springer, Berlin (1979)

[JS87] Jacod, J., Shiryaev, A.N.: Limit Theorems for Stochastic Processes. Springer, Berlin (1987)

[JM83] Jain, N.C., Monrad, D.: Gaussian measures in B_p^1. Ann. Prob., **11**, 46–57 (1983)

[Jost06] Jost, C.: Transformation formulas for fractional Brownian motion. Stoch. proc. Appl., **116**, 1341–1357 (2006)

[Jum93] Jumarie, G.: Stochastic differential equations with fractional Brownian motion input. Int. J. Syst. Sci., **24**, 1113–1131 (1993)

[Kam96] Kamont, A.: On the fractional anisotropic Wiener field. Prob. Math. Stat., **16**, 85–98 (1996)

[KS98] Karatzas, I, Shreve, S.E.: Methods of Mathematical Finance. Springer, New York (1998)
[KK97] Kasahara, Y., Kosugi, N.: A limit theorem for occupation times of fractional Brownian motion. Stoch. Proc. Appl., **67**, 161–175 (1997)
[KM96] Kasahara, Y., Matsumoto, Y.: On Kallianpur-Robbins law for fractional Brownian motion. J. Math. Kyoto Univ., **36**, 815–824 (1996)
[KO99] Kasahara, Y., Ogawa, N.: A note of the local time of fractional brownian motion. J. Theor. Prob., **12**, 207–216 (1999)
[KK73] Kawada, T., Kôno, N.: On the variation of Gaussian processes. In: Proc. 2nd Japan–USSR Symp. Prob. Theory, Lect. Notes Math., **330**, 176–192 (1973)
[KKA98a] Kleptsyna, M.L., Kloeden, P.E., Anh, V.V.: Existence and uniqueness theorems for stochastic differential equations with fractional Brownian motion. Probl. Inf. Transm. **34**, 54–56 (1998)
[KKA98b] Kleptsyna, M., Kloeden, P., Anh, V.: Nonlinear filtering with fractional Brownian motion. Probl. Inf. Transm., **34**, 150–160 (1998)
[KLeB99] Kleptsyna, M.L., Le Breton, A.: Extension of the Kalman–Bucy filter to elementary linear systems with fractional Brownian noises. Stat. Infer. Stoch. Proc., **5**, 249–271 (2002)
[KLeB02] Kleptsyna, M.L., Le Breton, A.: Statistical analysis of the fractional Ornstein–Uhlenbeck type process. Stat. Infer. Stoch. Proc., **5**, 229-248 (2002)
[KLeBR99] Kleptsyna, M.L., Le Breton, A., Roubaud, M.C.: An elementary approach to filtering in systems with fractional Brownian observation noise. In: Grigelionis B. (ed) Prob. Theory Math. Stat. (Proc. 7th Vilnius Conf.), TEV, Vilnius, 373–392 (1999)
[KLeBR00] Kleptsyna, M.L., Le Breton, A., Roubaud, M.C.: General approach to filtering with fractional Brownian noises – application to linear systems. Stoch. Stoch. Rep., **71**, 119–140 (2000)
[Kli98] Klingenhöfer, F.: Differential equations with fractal noise. PhD Thesis, University of Jena (1998)
[KZ99] Klingenhöfer, F., Zähle, M.: Ordinary differential equations with fractal noise. Proc. Amer. Math. Soc., **127**, 1021–1028 (1999)
[KP92] Kloeden, P.E., Platen, E.: Numerical Solution of Stochastic Differential Equations. Springer, Berlin (1992)
[KK04] Klüppelberg, C., Kühn, Ch.: Fractional Brownian motion as a weak limit of Poisson shot noise processes – with applications to finance. Stoch. Proc. Appl., **113**, 333–351 (2004)
[KP94] Kohatsu-Higa, A., Protter, P.: The Euler scheme for SDE's driven by semimartingales. In: Stochastic Analysis

on Infinite-Dimensional Spaces, Pitman Res. Notes in Math. Ser. **310**, 141–151 (1994)
[Kol31] Kolmogoroff, A.N.: Über die analytischen Methoden in der Wahrscheinlichkeitsrechnung. Math. Annalen, **104**, 415–458 (1931)
[Kol40] Kolmogorov, A.N.: The Wiener spiral and some other interesting curves in Hilbert space. Dokl. Akad. Nauk SSSR, **26**, 115–118 (1940)
[Kon37] Kondurar, V.: Sur l'intégrale de Stieltjes. Recueil Math., **2**, 361–366 (1937)
[Kono96] Kôno, N.: Kallianpur–Robbins law for fractional Brownian motion. Prob. Theory Math. Stat. Proc. 7th Japan-Russia Symp. World Scientific, Singapore, 229–236 (1996)
[Kop84] Kopp, P.E.: Martingales and Stochastic Integrals. Cambridge, Cambridge Univ. (1984)
[KS03] Koul, H.I., Surgailis,D. : Robust estimation in regressian models with long-memory errors. In: Doukhan, P., Oppenheim, G., Taqqu, M. (eds) Theory and Applications of Long-Range Dependence. Birkhäuser, Boston, 339–355 (2003)
[KK99] Kozachenko, Y.V., Kurchenko, O.O.: An estimate for the multiparameter fractional Brownian motion. Theory Stoch. Proc., **5 (21)**, 113–119 (1999)
[KM06] Kozachenko, Y.V., Mishura Y.S.: Maximal upper bounds of moments of stochastic integrals and the solutions of stochastic differential equations involving fractional Brownian motion with Hurst index $H < 1/2$. I (Ukrain.) Prob. Theory Math. Stat., **75**, 47–59 (2006)
[KM07] Kozachenko, Y.V., Mishura Y.S.: Maximal upper bounds of moments of stochastic integrals and the solutions of stochastic differential equations involving fractional Brownian motion with Hurst index $H < 1/2$. II (Ukrain.) Prob. Theory Math. Stat., **76**, 53–69 (2007)
[KM00] Krvavych,Y., Mishura,Y.: The stochastic Fubini theorem for integrals containing random integrand and fractional Brownian motion as integrator. Theory Stoch. Proc., **6 (22)**, 79–89 (2000)
[KMV05] Kukush, A., Mishura, Y., Valkeila, E. Statistical inference with fractional Brownian motion. Stat. Infer. Stoch. Proc., **8**, 71–93 (2005)
[Ku99] Kuznetsov, Yu.A.: The absence of arbitrage in a model with fractal Brownian motion. Russ. Math. Surv., **54**, 847–848 (1999)
[LN03] Lanjri, Z.N., Nualart, D.: Smoothness of the law of the supremum of the fractional Brownian motion. Electron. Commun. Prob., **8**, 102–111 (2003)

[Leb95] Lebedev, V.A.: Fubini theorems for depending on parameter stochastic integrals with respect to L^0-valued random measures. Theory Prob. Appl., **40**, 313–323 (1995)

[LeB98] Le Breton, A.: Filtering and parameter estimation in a simple linear system driven by a fractional Brownian motion. Stat. Prob. Lett., **38**, 263–274 (1998)

[LT02] León, J.A., Tudor, C.: Semilinear fractional stochastic differential equations. Bol. Soc. Mat. Mexicana, **8**, 205–226 (2002)

[LL00] León, J.R., Ludeña, C.: Estimating the diffusion coefficient for diffusions driven by fBm. Stat. Inference Stoch. Process., **3**, 183–192 (2000).

[LP95] Leu, J.-S., Papamarcou, A.: On estimating the spectral exponent of fractional Brownian motion. IEEE Trans. Inf. Theory, **41**, 233–244 (1995)

[Le48] Lévy, P.: Processus stochastiques et mouvement Brownien. Gauthier–Villars, Paris (1948) Second edition (1965)

[Lif95] Lifshitz, M.A.: Gaussian Random Functions (Russian). TViMS, Kiev (1995)

[Lin95] Lin, S.J.: Stochastic analysis of fractional Brownian motions. Stoch. Stoch. Rep., **55**, 121–140 (1995)

[Lind93] Lindstrøm, T.: Fractional Brownian fields as integrals of white noise. Bull. London Math. Soc., **25**, 83–88 (1993)

[LS86] Liptzer, R.S., Shiryaev, A.N.: Theory of Martingales. (Russian). Nauka, Moscow (1986)

[LS01] Liptser, R.S., Shiryaev, A.N.: Statistics of Random Processes. 1: General Theory. Applications of Mathematics. 5. Springer, Berlin (2001)

[Lo91] Lo, A.W.: Long-term memory in stock market prices. Econometrica **59**, 1279–1313 (1991)

[Mac81a] Maccone, C.: Time rescaling and Gaussian properties of the fractional Brownian motions. Nuovo Cimento B (11), **65**, 259–271 (1981)

[Mac81b] Maccone, C.: Eigenfunction expansion for fractional Brownian motions. Nuovo Cimento B (11), **61**, 229–248 (1981)

[MS93] Maheswaran, S., Sims, C.A.: Empirical implications of arbitrage-free asset markets. In: Philips, P.C.B. (eds) Models, Methods, and Applications of Econometrics. Basil Blackwell (1993)

[Mae03] Maejima, M.: Limit theorems for infinite variance sequences. In: Doukhan, P., Oppenheim, G., Taqqu, M. (eds) Theory and Applications of Long-Range Dependence. Birkhäuser, Boston, 157–164 (2003)

[Mal97] Malliavin, P.: Stochastic Analysis. Springer, Berlin (1997)

[Man97] Mandelbrot, B.B.: Fractals and Scaling in Finance. Discontinuity, Concentration, Risk. Springer, New York (1997)

[MvN68] Mandelbrot, B.B., van Ness J.W.: Fractional Brownian motions, fractional noices and applications. SIAM Review, **10**, 422–437 (1968)

[MN03] Maslowski, B., Nualart, D.: Evolution equations driven by a fractional Brownian motion. J. Funct. Anal., **202**, 277–305 (2003)

[Mas93] Masry, E.: The wavelet transform of stochastic processes with stationary increments and its application to fractional Brownian motion. IEEE Trans. Inf. Theory, **39**, 260–264 (1993)

[Mas96] Masry, E.: Convergence properties of wavelet series expansions of fractional Brownian motion. Appl. Comput. Harmon. Analysis, **3**, 239–253 (1996)

[MS99] Massoulie, L., Simonian, A.: Large buffer asymptotics for the queue with FBM input. J. Appl. Probab. **36**, 894–906 (1999).

[MMV01] Memin, J., Mishura, Yu., Valkeila, E.: Inequalities for the moments of Wiener integrals with respect to fractional Brownian motions. Stat. Prob. Lett., **51**, 197–206 (2001)

[MP80] Métivier, M., Pellaumail, J.: Stochastic integration. Academic Press, New York (1980)

[Me76] Meyer, P.A.: Un cours sur les intégrales stochastiques. Séminaire Prob. X, In: Lect. Notes Math., **511**, 246–400 (1976)

[MN00] Mikosch, T., Norvaiša, R.: Stochastic integral equations without probability. Bernoulli, **6**, 401–434 (2000)

[Mis03] Mishura, Yu.: Quasilinear stochastic differential equations with fractional-Brownian component. Prob. Theory Math. Stat., **68**, 95–106 (2003)

[Mis04] Mishura, Yu.: Fractional stochastic integration and Black-Scholes equation for fractional Brownian motion with stochastic volatility. Stoch. Stoch. Rep., **76**, 363–381 (2004)

[Mis05] Mishura, Yu.: An estimate of ruin probabilities for the long range dependence models. Prob. Theory Math. Stat., **72**, 93–100 (2005)

[MisIl03] Mishura, Yu., Ilchenko S.: Itô formula for fractional Brownian fields. Prob. Theory Math. Stat., **69**, 141–153 (2003)

[MisIl04] Mishura, Yu., Ilchenko S.: Generalized two-parameter Lebesgue–Stieltjes integrals and their applications to fractional Brownian fields. Ukr. Math. J.,**56**, 435–450 (2004)

[MisIl06] Mishura, Yu., Ilchenko S.: Stochastic integrals and stochastic differential equations involving fractional Brownian field. Theory Prob. Math. Stat., **75**, 85–101 (2006)

[MN04] Mishura, Y., Nualart, D.: Weak solutions for stochastic differential equations with additive fractional noise. Stat. Prob. Lett., **70**, 253–261 (2004)

[MP07] Mishura, Yu., Posashkov S.: Existence and uniquenes of solution of mixed stochastic differential equation driven by fractional

Brownian motion and Wiener process. Theory Stoch. Proc., **13(29)**, 152–165 (2007)
[MR01] Mishura, Yu., Rudomino-Dusyatska, N.: Consistency of drift parameter estimates in fractional-Brownian diffusion models. Theory Stoch. Proc., **7(23)**, 103–112 (2001)
[MS07a] Mishura, Yu., Shevchenko G.M.: The rate of convergence for Euler approximations of solutions of stochastic differential equations driven by fBm. http://arxiv.org/abs/0705.1773
[MS07b] Mishura, Yu., Shevchenko G.M.: On differentiability of solution to stochastic differential equation with fractional Brownian motion. Theory Stoch. Proc., **13(29)**, 243–250 (2007)
[MV00] Mishura, Yu., Valkeila, E.: An isometric approach to generalized stochastic integrals. J. Theor. Prob., **13**, 673–693 (2000)
[MV02] Mishura, Yu. S., Valkeila E.: The absence of arbitrage in a mixed Brownian-fractional Brownian model. Proc. of the Steklov Institute of Math., **237**, 215–224 (2002)
[MV06] Mishura, Yu. S., Valkeila E.: An extension of the Lévy characterization to fractional Brownian motion. Preprint, Helsinki Univ. Techn. A514 (2006)
[Mol69] Molchan, G.: Gaussian processes with spectra which are asymptotically equivalent to a power of λ. Theory Prob. Appl. **14**, 530–532 (1969)
[Mol03] Molchan, G.: Historical comments related to fractional Brownian motion. In: Doukhan, P., Oppenheim, G., Taqqu, M. (eds) Theory and Applications of Long-Range Dependence. Birkhäuser, Boston, 39–42 (2003)
[MG69] Molchan, G., Golosov, J.: Gaussian stationary processes with asymptotic power spectrum. Soviet Math. Dokl. **10**, 134–137 (1969)
[MY67] Monin, A., Yaglom, A.M.: Statistical Hydromechanics. Mechanics of Turbulence. **I-II**. (Russian) Nauka, Moscow (1967)
[MS03] Moulines, E., Soulier, P.: Semi-parametric spectral estimation for fractional processes. In: Doukhan, P., Oppenheim, G., Taqqu, M. (eds) Theory and Applications of Long-Range Dependence. Birkhäuser, Boston, 251–301 (2003)
[Mur92] Muraoka, H.: A fractional Brownian motion and the canonical representations of its coefficient processes. Soochow J. Math., **18**, 361–377 (1992)
[Nar98] Narayan, O.: Exact asymptotic queue length distribution for fractional Brownian traffic. Adv. Perform. Analysis, **1**, 39–63 (1998)
[NN06] Neunkirch, A., Nourdin, I.: Exact rate of convergence of some approximation schemes associated to SDEs driven by a fractional Brownian motion. Preprint, available online at http://arxiv.org/abs/math.PR/0601038 (2006)

[Nie04] Nieminen, A.: Fractional Brownian motion and martingale-differences. Stat. Prob. Lett., **70**, 1–10 (2004)

[Nor95] Norros, I.: On the use of the fractional Brownian motion in the theory of connectionless networks. IEEE J. Sel. Areas Commun., **13**, 953–962 (1995)

[Nor97] Norros, I.: Four approaches to the fractional Brownian storage. In: Lévy Véhel, J., Lutton, E., Tricot, C. (eds) Fractals in Engineering. Springer, 154–169 (1997)

[Nor99] Norros, I.: Busy periods of fractional Brownian storage: a large deviations approach. Adv. Perform. Analysis, **2**, 1–19 (1999)

[NVV99] Norros, I., Valkeila, E., Virtamo, J.: An elementary approach to a Girsanov formula and other analytical results on fractional Brownian motions. Bernoulli **5(4)**, 571–587 (1999)

[Nrv99] Norvaiša, R.: p-variation and integration of sample functions of stochastic proceses. In: Grigelionis B. (ed) Prob. Theory Math. Stat. (Proc. 7th Vilnius Conf.), TEV, Vilnius, 521–540 (1999)

[Nou05] Nourdin, I.: Schémas d'approximation associés à une équation différentielle dirigée par une fonction höldérienne; cas du mouvement brownien fractionnaire. C. R. Acad. Sci. Paris Sér. I Math., **340**, 611–614 (2005)

[NV98] Novikov, A., Valkeila, E.: On some maximal inequalities for fractional Brownian motions. Preprint 186, University of Helsinki (1998)

[Nua95] Nualart, D.: The Malliavin Calculus and Related Topics. Springer, Berlin (1995)

[Nua98] Nualart, D.: Analysis on Wiener space and anticipating stochastic calculus. In: Lect. Notes Math., **1690**, 123–227 (1998)

[Nua03] Nualart, D.: Stochastic integration with respect to fractional Brownian motion and applications. In: González-Barrios, J.M. et al. (eds), Stochastic Models. Contemp. Math., **336**, AMS, Providence, RI, 3–39, (2003)

[Nua06] Nualart, D.: Fractional Brownian motion: stochastic calculus and applications. In: Proceed. Intern. Congress Mathem. Madrid, Spain. European Mathem. Society (2006)

[NO02] Nualart, D., Ouknine, Y.: Regularization of differential equations by fractional noise. Stoch. Proc. Appl., **102**, 103–116 (2002)

[NO03a] Nualart, D., Ouknine, Y.: Besov regularity of stochastic integrals with respect to the fractional Brownian motion with parameter $H > 1/2$. J. Theor. Prob., **16**, 451–470 (2003)

[NO03b] Nualart, D., Ouknine, Y.: Stochastic differential equations with additive fractional noise and locally unbounded drift. Progress in Probability, **56**, 353–365 (2003)

[NP95] Nualart, D., Pardoux, E.: Stochastic calculus with anticipating integrands. Prob. Theory Rel. Fields, **78**, 535–581 (1995)

[NR00] Nualart, D. Răşcanu, A.: Differential equations driven by fractional Brownian motion. Collect. Math., **53**, 55–81 (2000)

[NS05] Nualart, D., Saussereau, B.: Malliavin calculus for stochastic differential equations driven by a fractional Brownian motion. Preprint No. 371, Univ. de Barcelona (2005)

[Oks03] Øksendal, B.: Stochastic Differential Equations. An Introduction with Applications. Springer, Berlin (2003)

[Oks07] Øksendal, B.: Fractional Brownian motion in finance. In: Jensen, B.S., Palokangas, T. (eds) Stochastic economic dynamics. CBS Press, Copenhagen (2007)

[OZ01] Øksendal, B., Zhang, T.: Multiparameter fractional Brownian motion and quasi-linear stochastic partial differential equations, Stoch. Stoch. Rep. **71**, 141–163 (2001)

[PT02a] Pérez-Abreu, V., Tudor, C.: Multiple stochastic fractional integrals: a transfer principle for multiple stochastic fractional integrals. Bol. Soc. Mat. Mexicana, **3**, 187–203 (2002)

[PT00a] Pipiras, V., Taqqu, M.: Convergence of the Weierstrass-Mandelbrot process to fractional Brownian motion. Fractals, **8**, 360–384 (2000)

[PT00b] Pipiras, V., Taqqu, M.: Integration questions related to fractional Brownian motion. Prob. Theory Rel. Fields, **118**, 251–291 (2000)

[PT01] Pipiras, V., Taqqu, M.S.: Are classes of deterministic integrands for fractional Brownian motion on an interval complete? Bernoulli, 7, 873–897 (2001)

[PT02b] Pipiras, V., Taqqu, M.S.: Deconvolution of fractional Brownian motion. J. of Time Series Analysis, 23, 487–501 (2002)

[PT03] Pipiras, V., Taqqu M.S.: Fractional calculus and its connection to fractional Brownian motion. In: Doukhan, P., Oppenheim, G., Taqqu, M. (eds) Theory and Applications of Long-Range Dependence. Birkhäuser, Boston, 165–202 (2003)

[Po99] Podlubny, I.: Fractional Differential Equations. An Introduction to Fractional Derivatives, Fractional Differential Equations, Some Methods of Their Solution and Some of Their Applications. Academic Press, San Diego (1999)

[Pos05] Posashkov, S. Optimal filtering in systems with fractional-Brownian noises. Theory Prob. Math. Stat, **72**, 120–128 (2005)

[Pri98] Privault, N.: Skorokhod stochastic integration with respect to a non-adapted process on Wiener space. Stoch. Stoch. Rep., **65**, 13–39 (1998)

[Pro90] Protter, P.: Stochastic Integration and Differential Equations. A New Approach. Applications of Mathematics, **21**. Springer, Berlin (1990)

[RZ91] Ramanathan, J., Zeitouni, O.: On the wavelet transform of fractional Brownian motion. IEEE Trans. Inform. Theory, **37**, 1156–1158 (1991)

[RLT95] Reed, I.S., Lee, P.C., Truong, T.K.: Spectral representation of fractional Brownian motion in n dimensions and its properties. IEEE Trans. Inf. Theory, **41**, 1439–1451 (1995)

[Rog97] Rogers, L.C.G.: Arbitrage with fractional Brownian motion, Mathem. Finance, **7**, 95–105 (1997)

[Ros87] Rosen, J.: The intersection local time of fractional Brownian motion in the plane. J. Multivariate Analysis, **23**, 37–46 (1987)

[RV93] Russo, F., Vallois, P.: Forward, backward and symmetric stochastic integration. Prob. Theory Rel. Fields, **97**, 403–421 (1993)

[RV95a] Russo, F., Vallois, P.: Itô formula for C^1-functions of semimartingales. Prob. Theory Rel. Fields, **104**, 27–41 (1995)

[RV95b] Russo, F., Vallois, P.: The generalized covariation process and Itô formula. Stoch. Proc. Appl., **59**, 81–104 (1995)

[RV98] Russo, F., Vallois, P.: Forward and backward calculus with respect to finite quadratic variation processes. Preprint, Institut Galilée, Mathématiques, Université Paris (1998)

[RV00] Russo, F., Vallois, P.: Stochastic calculus with respect to continuous finite quadratic variation processes. Stoch. Stoch. Rep., **70**, 1–40 (2000)

[Ruz00] Ruzmaikina, A.A.: Stieltjes integrals of Hölder continuous functions with applications to fractional Brownian motion. J. Stat. Phys., **100**, 1049–1069 (2000)

[Sai92] Sainty, Ph.: Construction of a complex-valued fractional Brownian motion of order N. J. Math. Phys., **33**, 3128–3149 (1992)

[Sal98] Salopek, D.M.: Tolerance to arbitrage. Stock. Proc. Appl. **76**, 217–230 (1998)

[SKM93] Samko, S.G., Kilbas, A.A., Marichev, O.I.: Fractional Integrals and Derivatives. Theory and Applications. Gordon and Breach Science Publishers, New York (1993)

[ST94] Samorodnitsky, G., Taqqu, M.S.: Stable Non-Gaussian Random Processes. Chapman & Hall, New York (1994)

[Sch99] Schoenmakers, J.G.M., Kloeden, P.: Robust option replication for a Black–Scholes model extended with nondeterministic trends. J. Appl. Math. Stoch. Anal.,**12**, 113–120 (1999)

[Seb95] Sebastian, K. L.: Path integral representation for fractional Brownian motion. J. Phys. A, Math. Gen., **28**, 4305–4311 (1995)

[Sh96] Shieh, N.-R.: Limit theorems for local times of fractional Brownian motions and some other self-similar processes. J. Math. Kyoto Univ., **36**, 641–652 (1996)

[Shi99] Shiryaev, A.N.: Essentials of Stochastic Finance: Facts, Models and Theory. World Scientific, Singapore (1999)

[Shi01] Shiryaev A.N.: Arbitrage and replication for fractal models. Preprint, MaPhySto, Aarhus (2001)
[Sin97] Sinaj, Ya.G.: Distribution of the maximum of a fractional Brownian motion. Russ. Math. Surv., **52**, 359–378 (1997)
[Sin94] Singer, P.: An integrated fractional Fourier transform. J. Comp. Appl. Math., **54**, 221–237 (1994)
[SL95] Sithi, V.M., Lim, S.C.: On the spectra of Riemann–Liouville fractional Brownian motion. J. Phys. A, Math. Gen., **28**, 2995–3003 (1995)
[Sko75] Skorokhod, A.V.: On a generalization of a stochastic integral. Theory Prob. Appl., **20**, 219–233 (1975)
[SS70] Skorohod, A.V., Slobodenyuk, N.P.: Limit Theorems for Random Walks (Russian). Naukova Dumka, Kiev (1970)
[Sot01] Sottinen, T.: Fractional Brownian motion, random walks and binary market models. Finance Stoch., **5**, 343–355 (2001)
[SV03] Sottinen, T., Valkeila E.: On arbitrage and replication in the fractional Black–Scholes pricing model. Stat. Decisions, **21**, 93–107 (2003)
[Sur03a] Surgailis, D.: CLTs for polynomials of linear sequences: diagram formula with illustrations. In: Doukhan, P., Oppenheim, G., Taqqu, M. (eds) Theory and Applications of Long-Range Dependence. Birkhäuser, Boston, 111–128 (2003)
[Sur03b] Surgailis, D.: Non-CLTs: U-statistics, multinomial formula and approximations of multiple Wiener–Itô integrals. In: Doukhan, P., Oppenheim, G., Taqqu, M. (eds) Theory and Applications of Long-Range Dependence. Birkhäuser, Boston, 129–142 (2003)
[SW03] Surgailis, D., Woyczynski, W.A.: Limit theorems for the Burgers equation initialized by data with long-range dependence. In: Doukhan, P., Oppenheim, G., Taqqu, M. (eds) Theory and Applications of Long-Range Dependence. Birkhäuser, Boston, 507–523 (2003)
[Tal95] Talagrand, M.: Hausdorff measure of trajectories of multiparameter fractional Brownian motion. Ann. Prob., **23**, 767–775 (1995)
[Tal96] Talagrand, M.: Lower classes for fractional Brownian motion. J. Theor. Prob., **9**, 191–213 (1996)
[TX96] Talagrand, M., Xiao, Y.: Fractional Brownian motion and packing dimension. J. Theor. Prob., **9**, 579–593 (1996)
[Taq75] Taqqu, M.: Weak convergence to fractional Brownian motion and to the Rosenblatt process. Z. Wahr. verw. Gebiete, **31**, 287–302 (1975)
[Taq77] Taqqu, M.: Law of the iterated logarithm for sums of non linear functions of Gaussian variables that exhibit long-range dependence. Z. Wahr. verw. Gebiete, **40**, 203–238 (1977)

[Taq03] Taqqu, M.: Fractional Brownian motion and long-range dependence. In: Doukhan, P., Oppenheim, G., Taqqu, M. (eds) Theory and Applications of Long-Range Dependence. Birkhäuser, Boston, 5–38 (2003)

[TTW95] Taqqu, M.S., Teverovsky, V., Willinger, W.: Estimators for long-range dependence: an empirical study. Fractals, **3**, 785–798 (1995)

[TT97] Taqqu, M.S., Teverovsky, V.: On estimating long-range dependence in finite and infinite variance series. In: Adler, R., Feldman, R., Taqqu, M.S. (eds) A Practical Guide to Heavy Tails: Statistical Techniques and Applications. Birkhäuser, Boston, 177–217 (1997)

[Thao03] Thao, T.H.: A note on fractional Brownian motion. Vietnam J. Math., **31**, 255–260 (2003).

[TTV03] Tindel, S., Tudor, C., Viens, F.: Stochastic evolution equations with fractional Brownian motion. Prob. Theory Rel. Fields, **127**, 186–204 (2003)

[Tud03] Tudor, C.A.: Weak convergence to the fractional Brownian sheet in Besov spaces. Bull. Braz. Math. Soc. (N.S.), **34**, 389–400 (2003)

[TT03] Tudor, C., Tudor, M.: On the two-parameter fractional Brownian motion and Stieltjes integrals for Hölder functions. J. Math. Anal. Appl., **286**, 765–781 (2003)

[TV03] Tudor, C.A., Viens, F.G.: Itô formula and local time for the fractional Brownian sheet. Electronic J. Prob., **8**, 1–31 (2003)

[Vage96] Våge, G.: A general existence and uniqueness theorem for Wick-SDEs in $(Y)^n_{-1,k}$. Stoch. Stoch. Reports, **58**, 259–284 (1996)

[Wang03] Wang, W.: Weak convergence to fractional Brownian motion in Brownian scenery. Prob. Theory Rel. Fields, **126**, 203–220 (2003)

[WRL03] Wang, X.-T., Ren, F.-Y., Liang, X.-Q.: A fractional version of the Merton model. Chaos, Solut. Fractals, **15**, 455–463 (2003)

[Wat95] Watson, G.N.: A Treatise on the Theory of Bessel Functions. Cambridge Mathematical Library edition. Cambridge University Press, Cambridge (1995)

[Wie23] Wiener N.: Differential space. J. Math. Phys. **2**, 131–134 (1923)

[WTT99] Willinger, W., Taqqu, M., Teverovsky, V.: Stock market process and long-range dependence. Finance Stoch., **3**, 1–13 (1999)

[Wyss00] Wyss, W.: The fractional Black–Scholes equation. Fraction. Calc. Appl. Anal., **3**, 51–61 (2000)

[Xiao91] Xiao, Y.: Multiple points of fractional Brownian motion and Hausdorff dimension. (Chinese) Chin. Ann. Math., Ser. A, **12**, 612–618 (1991)

[Xiao96] Xiao, Y.: Packing measure of the sample paths of fractional Brownian motion. Trans. Amer. Math. Soc., **348**, 3193–3213 (1996)
[Xiao97a] Xiao, Y.: Packing dimension of the image of fractional Brownian motion. Stat. Prob. Lett., **33**, 379–387 (1997)
[Xiao97b] Xiao, Y.: Hausdorff measure of the graph of fractional Brownian motion. Math. Proc. Camb. Philos. Soc., **122**, 565–576 (1997)
[Yag57] Yaglom, A.M.: Certain types of random fields in n-dimensional space similar to stationary stochastic processes. Theory Prob. Appl., **2**, 292–338 (1957)
[Yag87] Yaglom, A.M.: Correlation Theory of Stationary and Related Random Functions I. Springer, New York (1987)
[Yin96] Yin, Z.-M.: New methods for simulation of fractional Brownian motion. J. Comp. Phys., **127**, 66–72 (1996)
[Yor76] Yor, M.: Sur les intégrales stochastiques optionelles et une suite remarquable de formules exponentielles. (French) Séminaire Prob. X, In: Lect. Notes Math. **511**, Springer, Berlin, 481–500 (1976)
[Yor88] Yor, M.: Remarques sur certaines constructions des mouvements browniens frationnaires. (Remark on certain constructions of fractional Brownian motions.) (French) Séminaire Prob. XXII, In: Lect. Notes Math., **1321**, Springer, Berlin, 217–224 (1988)
[Yos78] Yoshihara, K.-I.: A weak convergence theorem for functionals of sums of martingale differences. Yokohama Math. J., **26**, 101–107 (1978)
[Zah98] Zähle, M.: Integration with respect to fractal functions and stochastic calculus. I. Prob. Theory Rel. Fields, **111**, 333–374 (1998)
[Zah99] Zähle, M.: On the link between fractional and stochastic calculcus. In: Grauel, H., Gundlach, M. (eds) Stochastic Dynamics, Springer, 305–325 (1999)
[Zah01] Zähle, M.: Integraton with respect to fractal functions and stochastic calculus. II. Math. Nachr., **225**, 145–183 (2001)
[Zah02a] Zähle, M.: Long range dependence, no arbitrage and the Black-Scholes formula. Stochastics and Dynamics, **2**, 265–280 (2002)
[Zah02b] Zähle, M.: Forward integrals and stochastic differential equations. In: Dalang, R.C., Dozzi, M. (eds) Sem. Stoch. Anal, Random Fields Appl. III, Progress in Prob., **52**, Birkhäuser, 293–302 (2002)
[Zah05] Zähle, M.: Stochastic differential equations with fractal noise. Math. Nachr., **278**, 1097–1106 (2005)
[Zha96] Zhang, L.: Some liminf results on increments of fractional Brownian motion. Acta Math. Hung., **71**, 215–240 (1996)
[Zha97] Zhang, L.: A note on liminfs for increments of a fractional Brownian motion. Acta Math. Hung., **76**, 145–154 (1997)

Index

(g)-transformable, 61
C-function, 287
Δ^2-condition, 287
b-self-similar, 7

arbitrage, 302
arbitrage opportunity, 306
asymptotic efficiency, 359

backward integral, 161
Besov space, 73, 128
Bessel function of the first kind, 64
Black–Scholes equation, 322
   fractional, 326
bracket, 205
Burkholder–Davis–Gundy inequalities, 47

capital, 305
chain rule for stochastic derivative, 145
complex alternative, 328
complex hypothesis, 328
composition formulas for fractional
   integrals, 2
conditionally Gaussian pair, 296
confidence level, 330
convolution, 59
critical areas, 330
critical values, 330

density process, 66
derivative operator, 160
directional derivative, 145
discounted gain, 305

divergence operator, 161
Dudley integral, 42

entropy maximal estimates, 41
errors of the first and of the second
   kind, 330

field with independent increments, 119
fractional analog of the Burgers
   equation, 321
fractional Brownian motion, 7
   approximation, 71
   backward, 11
   forward, 11
   geometric, 302
   Mandelbrot–van Ness representation, 9, 23
   multi-parameter, 117
fractional derivative
   Marchaud, 3
      two-parameter, 119
   Riemann–Liouville, 2
      two-parameter, 119
   Weyl representation, 3
fractional Doleans exponent, 191
fractional noise, 15
fractional Wick exponent, 256
fundamental martingale, 27

Gaussian subspaces, 59
generalized Lebesgue–Stieltjes integral, 123
generalized quadratic variation, 205
Girsanov theorem, 67

Index

goodness-of-fit test, 335

Hölder continuous functions, 2
Hardy–Littlewood theorem, 1
Hellinger process, 70
Hermite functions, 12
Hermite polynomials, 12
Hurst effect, 301
Hurst law, 301
Hurst phenomenon, 301

integration-by-parts formula
  for fractional derivatives, 4
  for fractional integrals, 2
Itô formula, 182
  for $H \in (0, 1/2)$, 185
  for fractional fields, 186
  for Wick integrands, 184

Lévy theorem, 27, 94
Lenglart inequality, 48
likelihood ratio, 329
local asymptotic normality, 358
local martingale, 66
local solution, 205
long memory, 59
long-memory Gaussian processes, 59
long-range dependence, 8

Markov-type strategy, 306
martingale, 27
maximum likelihood estimate, 345
metric ε-capacity, 41
metric ε-entropy, 42
Mittag–Leffler function, 351
mixed version of the Black–Merton–Scholes model, 305
modulus of continuity, 87
modulus of uniform continuity, 87
Molchan martingale, 27

observation process, 291
observed trajectory, 328
optimal filter, 291
optimal filtering problem, 292
Orlicz space, 280

portfolio, 305

quasi-likelihood ratio, 331

random walk, 80
rate of convergence of Euler approximations, 243
regularly varying, 91
rescaled adjusted range statistic or R/S-statistic, 301
Riemann–Liouville fractional integral
  on (a, b), 1
  on $\mathbb{R}$, 1
  two-parameter, 118

self-financing strategy, 305, 306
semi-metric, 41
semi-norm, 163
semimartingale, 65, 71
signal process, 291
Skorohod integral, 158, 161
Skorohod space, 80
Skorohod topology, 81
slowly varying at ∞, 80
smooth functionals, 146
Sobolev space, 38
Sobolev–Slobodeckij space, 205
spectral density function, 8
spectral representation, 8
stochastic derivative, 145
stochastic differential equation
  Euler approximations, 243
  mixed, 225
  moment estimates for solution, 56
  pathwise, 197
  quasilinear of Skorohod type, 255
  weak solution, 263
  with additive Wiener integral, 262
  with fractional white noise, 241
stochastic Fubini theorem, 57, 174
stochastic integral
  generalized
    of first order, 166
  generalized forward, 205
  generalized of kth order, 168
  Skorohod, 158, 161
  Stratonovich, 146
  symmetric Stratonovich, 161
  w.r.t. fBm
    forward, 161
    Wick, 144
    Wiener, 16
stopping time, 46

Stratonovich integral, 146
  symmetric, 161
strong martingale, 120
strong Molchan martingale, 121

testing procedure, 330
The Fourier transform, 5
two-parameter left Riemann–Stieltjes
  integral, 135
two-sided Wiener process, 9

uniform modulus, 87

weak solution, 263
Weierstrass–Mandelbrot process, 79
white noise, 10
Wick products, 141
Wiener field, 120
Wiener integral
  generalized, 165
  w.r.t. fBm, 16
    moment inequalities, 35

Lecture Notes in Mathematics

For information about earlier volumes
please contact your bookseller or Springer
LNM Online archive: springerlink.com

Vol. 1735: D. Yafaev, Scattering Theory: Some Old and New Problems (2000)
Vol. 1736: B. O. Turesson, Nonlinear Potential Theory and Weighted Sobolev Spaces (2000)
Vol. 1737: S. Wakabayashi, Classical Microlocal Analysis in the Space of Hyperfunctions (2000)
Vol. 1738: M. Émery, A. Nemirovski, D. Voiculescu, Lectures on Probability Theory and Statistics (2000)
Vol. 1739: R. Burkard, P. Deuflhard, A. Jameson, J.-L. Lions, G. Strang, Computational Mathematics Driven by Industrial Problems. Martina Franca, 1999. Editors: V. Capasso, H. Engl, J. Periaux (2000)
Vol. 1740: B. Kawohl, O. Pironneau, L. Tartar, J.-P. Zolesio, Optimal Shape Design. Tróia, Portugal 1999. Editors: A. Cellina, A. Ornelas (2000)
Vol. 1741: E. Lombardi, Oscillatory Integrals and Phenomena Beyond all Algebraic Orders (2000)
Vol. 1742: A. Unterberger, Quantization and Non-holomorphic Modular Forms (2000)
Vol. 1743: L. Habermann, Riemannian Metrics of Constant Mass and Moduli Spaces of Conformal Structures (2000)
Vol. 1744: M. Kunze, Non-Smooth Dynamical Systems (2000)
Vol. 1745: V. D. Milman, G. Schechtman (Eds.), Geometric Aspects of Functional Analysis. Israel Seminar 1999-2000 (2000)
Vol. 1746: A. Degtyarev, I. Itenberg, V. Kharlamov, Real Enriques Surfaces (2000)
Vol. 1747: L. W. Christensen, Gorenstein Dimensions (2000)
Vol. 1748: M. Ruzicka, Electrorheological Fluids: Modeling and Mathematical Theory (2001)
Vol. 1749: M. Fuchs, G. Seregin, Variational Methods for Problems from Plasticity Theory and for Generalized Newtonian Fluids (2001)
Vol. 1750: B. Conrad, Grothendieck Duality and Base Change (2001)
Vol. 1751: N. J. Cutland, Loeb Measures in Practice: Recent Advances (2001)
Vol. 1752: Y. V. Nesterenko, P. Philippon, Introduction to Algebraic Independence Theory (2001)
Vol. 1753: A. I. Bobenko, U. Eitner, Painlevé Equations in the Differential Geometry of Surfaces (2001)
Vol. 1754: W. Bertram, The Geometry of Jordan and Lie Structures (2001)
Vol. 1755: J. Azéma, M. Émery, M. Ledoux, M. Yor (Eds.), Séminaire de Probabilités XXXV (2001)
Vol. 1756: P. E. Zhidkov, Korteweg de Vries and Nonlinear Schrödinger Equations: Qualitative Theory (2001)
Vol. 1757: R. R. Phelps, Lectures on Choquet's Theorem (2001)
Vol. 1758: N. Monod, Continuous Bounded Cohomology of Locally Compact Groups (2001)

Vol. 1759: Y. Abe, K. Kopfermann, Toroidal Groups (2001)
Vol. 1760: D. Filipović, Consistency Problems for Heath-Jarrow-Morton Interest Rate Models (2001)
Vol. 1761: C. Adelmann, The Decomposition of Primes in Torsion Point Fields (2001)
Vol. 1762: S. Cerrai, Second Order PDE's in Finite and Infinite Dimension (2001)
Vol. 1763: J.-L. Loday, A. Frabetti, F. Chapoton, F. Goichot, Dialgebras and Related Operads (2001)
Vol. 1764: A. Cannas da Silva, Lectures on Symplectic Geometry (2001)
Vol. 1765: T. Kerler, V. V. Lyubashenko, Non-Semisimple Topological Quantum Field Theories for 3-Manifolds with Corners (2001)
Vol. 1766: H. Hennion, L. Hervé, Limit Theorems for Markov Chains and Stochastic Properties of Dynamical Systems by Quasi-Compactness (2001)
Vol. 1767: J. Xiao, Holomorphic Q Classes (2001)
Vol. 1768: M. J. Pflaum, Analytic and Geometric Study of Stratified Spaces (2001)
Vol. 1769: M. Alberich-Carramiñana, Geometry of the Plane Cremona Maps (2002)
Vol. 1770: H. Gluesing-Luerssen, Linear Delay-Differential Systems with Commensurate Delays: An Algebraic Approach (2002)
Vol. 1771: M. Émery, M. Yor (Eds.), Séminaire de Probabilités 1967-1980. A Selection in Martingale Theory (2002)
Vol. 1772: F. Burstall, D. Ferus, K. Leschke, F. Pedit, U. Pinkall, Conformal Geometry of Surfaces in S^4 (2002)
Vol. 1773: Z. Arad, M. Muzychuk, Standard Integral Table Algebras Generated by a Non-real Element of Small Degree (2002)
Vol. 1774: V. Runde, Lectures on Amenability (2002)
Vol. 1775: W. H. Meeks, A. Ros, H. Rosenberg, The Global Theory of Minimal Surfaces in Flat Spaces. Martina Franca 1999. Editor: G. P. Pirola (2002)
Vol. 1776: K. Behrend, C. Gomez, V. Tarasov, G. Tian, Quantum Comohology. Cetraro 1997. Editors: P. de Bartolomeis, B. Dubrovin, C. Reina (2002)
Vol. 1777: E. García-Río, D. N. Kupeli, R. Vázquez-Lorenzo, Osserman Manifolds in Semi-Riemannian Geometry (2002)
Vol. 1778: H. Kiechle, Theory of K-Loops (2002)
Vol. 1779: I. Chueshov, Monotone Random Systems (2002)
Vol. 1780: J. H. Bruinier, Borcherds Products on $O(2,1)$ and Chern Classes of Heegner Divisors (2002)
Vol. 1781: E. Bolthausen, E. Perkins, A. van der Vaart, Lectures on Probability Theory and Statistics. Ecole d' Eté de Probabilités de Saint-Flour XXIX-1999. Editor: P. Bernard (2002)
Vol. 1782: C.-H. Chu, A. T.-M. Lau, Harmonic Functions on Groups and Fourier Algebras (2002)

Vol. 1783: L. Grüne, Asymptotic Behavior of Dynamical and Control Systems under Perturbation and Discretization (2002)
Vol. 1784: L. H. Eliasson, S. B. Kuksin, S. Marmi, J.-C. Yoccoz, Dynamical Systems and Small Divisors. Cetraro, Italy 1998. Editors: S. Marmi, J.-C. Yoccoz (2002)
Vol. 1785: J. Arias de Reyna, Pointwise Convergence of Fourier Series (2002)
Vol. 1786: S. D. Cutkosky, Monomialization of Morphisms from 3-Folds to Surfaces (2002)
Vol. 1787: S. Caenepeel, G. Militaru, S. Zhu, Frobenius and Separable Functors for Generalized Module Categories and Nonlinear Equations (2002)
Vol. 1788: A. Vasil'ev, Moduli of Families of Curves for Conformal and Quasiconformal Mappings (2002)
Vol. 1789: Y. Sommerhäuser, Yetter-Drinfel'd Hopf algebras over groups of prime order (2002)
Vol. 1790: X. Zhan, Matrix Inequalities (2002)
Vol. 1791: M. Knebusch, D. Zhang, Manis Valuations and Prüfer Extensions I: A new Chapter in Commutative Algebra (2002)
Vol. 1792: D. D. Ang, R. Gorenflo, V. K. Le, D. D. Trong, Moment Theory and Some Inverse Problems in Potential Theory and Heat Conduction (2002)
Vol. 1793: J. Cortés Monforte, Geometric, Control and Numerical Aspects of Nonholonomic Systems (2002)
Vol. 1794: N. Pytheas Fogg, Substitution in Dynamics, Arithmetics and Combinatorics. Editors: V. Berthé, S. Ferenczi, C. Mauduit, A. Siegel (2002)
Vol. 1795: H. Li, Filtered-Graded Transfer in Using Noncommutative Gröbner Bases (2002)
Vol. 1796: J.M. Melenk, hp-Finite Element Methods for Singular Perturbations (2002)
Vol. 1797: B. Schmidt, Characters and Cyclotomic Fields in Finite Geometry (2002)
Vol. 1798: W.M. Oliva, Geometric Mechanics (2002)
Vol. 1799: H. Pajot, Analytic Capacity, Rectifiability, Menger Curvature and the Cauchy Integral (2002)
Vol. 1800: O. Gabber, L. Ramero, Almost Ring Theory (2003)
Vol. 1801: J. Azéma, M. Émery, M. Ledoux, M. Yor (Eds.), Séminaire de Probabilités XXXVI (2003)
Vol. 1802: V. Capasso, E. Merzbach, B. G. Ivanoff, M. Dozzi, R. Dalang, T. Mountford, Topics in Spatial Stochastic Processes. Martina Franca, Italy 2001. Editor: E. Merzbach (2003)
Vol. 1803: G. Dolzmann, Variational Methods for Crystalline Microstructure – Analysis and Computation (2003)
Vol. 1804: I. Cherednik, Ya. Markov, R. Howe, G. Lusztig, Iwahori-Hecke Algebras and their Representation Theory. Martina Franca, Italy 1999. Editors: V. Baldoni, D. Barbasch (2003)
Vol. 1805: F. Cao, Geometric Curve Evolution and Image Processing (2003)
Vol. 1806: H. Broer, I. Hoveijn. G. Lunther, G. Vegter, Bifurcations in Hamiltonian Systems. Computing Singularities by Gröbner Bases (2003)
Vol. 1807: V. D. Milman, G. Schechtman (Eds.), Geometric Aspects of Functional Analysis. Israel Seminar 2000-2002 (2003)
Vol. 1808: W. Schindler, Measures with Symmetry Properties (2003)
Vol. 1809: O. Steinbach, Stability Estimates for Hybrid Coupled Domain Decomposition Methods (2003)
Vol. 1810: J. Wengenroth, Derived Functors in Functional Analysis (2003)

Vol. 1811: J. Stevens, Deformations of Singularities (2003)
Vol. 1812: L. Ambrosio, K. Deckelnick, G. Dziuk, M. Mimura, V. A. Solonnikov, H. M. Soner, Mathematical Aspects of Evolving Interfaces. Madeira, Funchal, Portugal 2000. Editors: P. Colli, J. F. Rodrigues (2003)
Vol. 1813: L. Ambrosio, L. A. Caffarelli, Y. Brenier, G. Buttazzo, C. Villani, Optimal Transportation and its Applications. Martina Franca, Italy 2001. Editors: L. A. Caffarelli, S. Salsa (2003)
Vol. 1814: P. Bank, F. Baudoin, H. Föllmer, L.C.G. Rogers, M. Soner, N. Touzi, Paris-Princeton Lectures on Mathematical Finance 2002 (2003)
Vol. 1815: A. M. Vershik (Ed.), Asymptotic Combinatorics with Applications to Mathematical Physics. St. Petersburg, Russia 2001 (2003)
Vol. 1816: S. Albeverio, W. Schachermayer, M. Talagrand, Lectures on Probability Theory and Statistics. Ecole d'Eté de Probabilités de Saint-Flour XXX-2000. Editor: P. Bernard (2003)
Vol. 1817: E. Koelink, W. Van Assche (Eds.), Orthogonal Polynomials and Special Functions. Leuven 2002 (2003)
Vol. 1818: M. Bildhauer, Convex Variational Problems with Linear, nearly Linear and/or Anisotropic Growth Conditions (2003)
Vol. 1819: D. Masser, Yu. V. Nesterenko, H. P. Schlickewei, W. M. Schmidt, M. Waldschmidt, Diophantine Approximation. Cetraro, Italy 2000. Editors: F. Amoroso, U. Zannier (2003)
Vol. 1820: F. Hiai, H. Kosaki, Means of Hilbert Space Operators (2003)
Vol. 1821: S. Teufel, Adiabatic Perturbation Theory in Quantum Dynamics (2003)
Vol. 1822: S.-N. Chow, R. Conti, R. Johnson, J. Mallet-Paret, R. Nussbaum, Dynamical Systems. Cetraro, Italy 2000. Editors: J. W. Macki, P. Zecca (2003)
Vol. 1823: A. M. Anile, W. Allegretto, C. Ringhofer, Mathematical Problems in Semiconductor Physics. Cetraro, Italy 1998. Editor: A. M. Anile (2003)
Vol. 1824: J. A. Navarro González, J. B. Sancho de Salas, $\mathscr{C}^\infty$ – Differentiable Spaces (2003)
Vol. 1825: J. H. Bramble, A. Cohen, W. Dahmen, Multiscale Problems and Methods in Numerical Simulations, Martina Franca, Italy 2001. Editor: C. Canuto (2003)
Vol. 1826: K. Dohmen, Improved Bonferroni Inequalities via Abstract Tubes. Inequalities and Identities of Inclusion-Exclusion Type. VIII, 113 p, 2003.
Vol. 1827: K. M. Pilgrim, Combinations of Complex Dynamical Systems. IX, 118 p, 2003.
Vol. 1828: D. J. Green, Gröbner Bases and the Computation of Group Cohomology. XII, 138 p, 2003.
Vol. 1829: E. Altman, B. Gaujal, A. Hordijk, Discrete-Event Control of Stochastic Networks: Multimodularity and Regularity. XIV, 313 p, 2003.
Vol. 1830: M. I. Gil', Operator Functions and Localization of Spectra. XIV, 256 p, 2003.
Vol. 1831: A. Connes, J. Cuntz, E. Guentner, N. Higson, J. E. Kaminker, Noncommutative Geometry, Martina Franca, Italy 2002. Editors: S. Doplicher, L. Longo (2004)
Vol. 1832: J. Azéma, M. Émery, M. Ledoux, M. Yor (Eds.), Séminaire de Probabilités XXXVII (2003)
Vol. 1833: D.-Q. Jiang, M. Qian, M.-P. Qian, Mathematical Theory of Nonequilibrium Steady States. On the Frontier of Probability and Dynamical Systems. IX, 280 p, 2004.
Vol. 1834: Yo. Yomdin, G. Comte, Tame Geometry with Application in Smooth Analysis. VIII, 186 p, 2004.

Vol. 1835: O.T. Izhboldin, B. Kahn, N.A. Karpenko, A. Vishik, Geometric Methods in the Algebraic Theory of Quadratic Forms. Summer School, Lens, 2000. Editor: J.-P. Tignol (2004)
Vol. 1836: C. Năstăsescu, F. Van Oystaeyen, Methods of Graded Rings. XIII, 304 p, 2004.
Vol. 1837: S. Tavaré, O. Zeitouni, Lectures on Probability Theory and Statistics. Ecole d'Eté de Probabilités de Saint-Flour XXXI-2001. Editor: J. Picard (2004)
Vol. 1838: A.J. Ganesh, N.W. O'Connell, D.J. Wischik, Big Queues. XII, 254 p, 2004.
Vol. 1839: R. Gohm, Noncommutative Stationary Processes. VIII, 170 p, 2004.
Vol. 1840: B. Tsirelson, W. Werner, Lectures on Probability Theory and Statistics. Ecole d'Eté de Probabilités de Saint-Flour XXXII-2002. Editor: J. Picard (2004)
Vol. 1841: W. Reichel, Uniqueness Theorems for Variational Problems by the Method of Transformation Groups (2004)
Vol. 1842: T. Johnsen, A. L. Knutsen, K_3 Projective Models in Scrolls (2004)
Vol. 1843: B. Jefferies, Spectral Properties of Noncommuting Operators (2004)
Vol. 1844: K.F. Siburg, The Principle of Least Action in Geometry and Dynamics (2004)
Vol. 1845: Min Ho Lee, Mixed Automorphic Forms, Torus Bundles, and Jacobi Forms (2004)
Vol. 1846: H. Ammari, H. Kang, Reconstruction of Small Inhomogeneities from Boundary Measurements (2004)
Vol. 1847: T.R. Bielecki, T. Björk, M. Jeanblanc, M. Rutkowski, J.A. Scheinkman, W. Xiong, Paris-Princeton Lectures on Mathematical Finance 2003 (2004)
Vol. 1848: M. Abate, J. E. Fornaess, X. Huang, J. P. Rosay, A. Tumanov, Real Methods in Complex and CR Geometry, Martina Franca, Italy 2002. Editors: D. Zaitsev, G. Zampieri (2004)
Vol. 1849: Martin L. Brown, Heegner Modules and Elliptic Curves (2004)
Vol. 1850: V. D. Milman, G. Schechtman (Eds.), Geometric Aspects of Functional Analysis. Israel Seminar 2002-2003 (2004)
Vol. 1851: O. Catoni, Statistical Learning Theory and Stochastic Optimization (2004)
Vol. 1852: A.S. Kechris, B.D. Miller, Topics in Orbit Equivalence (2004)
Vol. 1853: Ch. Favre, M. Jonsson, The Valuative Tree (2004)
Vol. 1854: O. Saeki, Topology of Singular Fibers of Differential Maps (2004)
Vol. 1855: G. Da Prato, P.C. Kunstmann, I. Lasiecka, A. Lunardi, R. Schnaubelt, L. Weis, Functional Analytic Methods for Evolution Equations. Editors: M. Iannelli, R. Nagel, S. Piazzera (2004)
Vol. 1856: K. Back, T.R. Bielecki, C. Hipp, S. Peng, W. Schachermayer, Stochastic Methods in Finance, Bressanone/Brixen, Italy, 2003. Editors: M. Fritelli, W. Runggaldier (2004)
Vol. 1857: M. Émery, M. Ledoux, M. Yor (Eds.), Séminaire de Probabilités XXXVIII (2005)
Vol. 1858: A.S. Cherny, H.-J. Engelbert, Singular Stochastic Differential Equations (2005)
Vol. 1859: E. Letellier, Fourier Transforms of Invariant Functions on Finite Reductive Lie Algebras (2005)
Vol. 1860: A. Borisyuk, G.B. Ermentrout, A. Friedman, D. Terman, Tutorials in Mathematical Biosciences I. Mathematical Neurosciences (2005)

Vol. 1861: G. Benettin, J. Henrard, S. Kuksin, Hamiltonian Dynamics – Theory and Applications, Cetraro, Italy, 1999. Editor: A. Giorgilli (2005)
Vol. 1862: B. Helffer, F. Nier, Hypoelliptic Estimates and Spectral Theory for Fokker-Planck Operators and Witten Laplacians (2005)
Vol. 1863: H. Führ, Abstract Harmonic Analysis of Continuous Wavelet Transforms (2005)
Vol. 1864: K. Efstathiou, Metamorphoses of Hamiltonian Systems with Symmetries (2005)
Vol. 1865: D. Applebaum, B.V. R. Bhat, J. Kustermans, J. M. Lindsay, Quantum Independent Increment Processes I. From Classical Probability to Quantum Stochastic Calculus. Editors: M. Schürmann, U. Franz (2005)
Vol. 1866: O.E. Barndorff-Nielsen, U. Franz, R. Gohm, B. Kümmerer, S. Thorbjønsen, Quantum Independent Increment Processes II. Structure of Quantum Lévy Processes, Classical Probability, and Physics. Editors: M. Schürmann, U. Franz, (2005)
Vol. 1867: J. Sneyd (Ed.), Tutorials in Mathematical Biosciences II. Mathematical Modeling of Calcium Dynamics and Signal Transduction. (2005)
Vol. 1868: J. Jorgenson, S. Lang, $Pos_n(R)$ and Eisenstein Series. (2005)
Vol. 1869: A. Dembo, T. Funaki, Lectures on Probability Theory and Statistics. Ecole d'Eté de Probabilités de Saint-Flour XXXIII-2003. Editor: J. Picard (2005)
Vol. 1870: V.I. Gurariy, W. Lusky, Geometry of Müntz Spaces and Related Questions. (2005)
Vol. 1871: P. Constantin, G. Gallavotti, A.V. Kazhikhov, Y. Meyer, S. Ukai, Mathematical Foundation of Turbulent Viscous Flows, Martina Franca, Italy, 2003. Editors: M. Cannone, T. Miyakawa (2006)
Vol. 1872: A. Friedman (Ed.), Tutorials in Mathematical Biosciences III. Cell Cycle, Proliferation, and Cancer (2006)
Vol. 1873: R. Mansuy, M. Yor, Random Times and Enlargements of Filtrations in a Brownian Setting (2006)
Vol. 1874: M. Yor, M. Émery (Eds.), In Memoriam Paul-André Meyer - Séminaire de Probabilités XXXIX (2006)
Vol. 1875: J. Pitman, Combinatorial Stochastic Processes. Ecole d'Eté de Probabilités de Saint-Flour XXXII-2002. Editor: J. Picard (2006)
Vol. 1876: H. Herrlich, Axiom of Choice (2006)
Vol. 1877: J. Steuding, Value Distributions of L-Functions (2007)
Vol. 1878: R. Cerf, The Wulff Crystal in Ising and Percolation Models, Ecole d'Eté de Probabilités de Saint-Flour XXXIV-2004. Editor: Jean Picard (2006)
Vol. 1879: G. Slade, The Lace Expansion and its Applications, Ecole d'Eté de Probabilités de Saint-Flour XXXIV-2004. Editor: Jean Picard (2006)
Vol. 1880: S. Attal, A. Joye, C.-A. Pillet, Open Quantum Systems I, The Hamiltonian Approach (2006)
Vol. 1881: S. Attal, A. Joye, C.-A. Pillet, Open Quantum Systems II, The Markovian Approach (2006)
Vol. 1882: S. Attal, A. Joye, C.-A. Pillet, Open Quantum Systems III, Recent Developments (2006)
Vol. 1883: W. Van Assche, F. Marcellàn (Eds.), Orthogonal Polynomials and Special Functions, Computation and Application (2006)
Vol. 1884: N. Hayashi, E.I. Kaikina, P.I. Naumkin, I.A. Shishmarev, Asymptotics for Dissipative Nonlinear Equations (2006)
Vol. 1885: A. Telcs, The Art of Random Walks (2006)
Vol. 1886: S. Takamura, Splitting Deformations of Degenerations of Complex Curves (2006)

Vol. 1887: K. Habermann, L. Habermann, Introduction to Symplectic Dirac Operators (2006)
Vol. 1888: J. van der Hoeven, Transseries and Real Differential Algebra (2006)
Vol. 1889: G. Osipenko, Dynamical Systems, Graphs, and Algorithms (2006)
Vol. 1890: M. Bunge, J. Funk, Singular Coverings of Toposes (2006)
Vol. 1891: J.B. Friedlander, D.R. Heath-Brown, H. Iwaniec, J. Kaczorowski, Analytic Number Theory, Cetraro, Italy, 2002. Editors: A. Perelli, C. Viola (2006)
Vol. 1892: A. Baddeley, I. Bárány, R. Schneider, W. Weil, Stochastic Geometry, Martina Franca, Italy, 2004. Editor: W. Weil (2007)
Vol. 1893: H. Hanßmann, Local and Semi-Local Bifurcations in Hamiltonian Dynamical Systems, Results and Examples (2007)
Vol. 1894: C.W. Groetsch, Stable Approximate Evaluation of Unbounded Operators (2007)
Vol. 1895: L. Molnár, Selected Preserver Problems on Algebraic Structures of Linear Operators and on Function Spaces (2007)
Vol. 1896: P. Massart, Concentration Inequalities and Model Selection, Ecole d'Été de Probabilités de Saint-Flour XXXIII-2003. Editor: J. Picard (2007)
Vol. 1897: R. Doney, Fluctuation Theory for Lévy Processes, Ecole d'Été de Probabilités de Saint-Flour XXXV-2005. Editor: J. Picard (2007)
Vol. 1898: H.R. Beyer, Beyond Partial Differential Equations, On linear and Quasi-Linear Abstract Hyperbolic Evolution Equations (2007)
Vol. 1899: Séminaire de Probabilités XL. Editors: C. Donati-Martin, M. Émery, A. Rouault, C. Stricker (2007)
Vol. 1900: E. Bolthausen, A. Bovier (Eds.), Spin Glasses (2007)
Vol. 1901: O. Wittenberg, Intersections de deux quadriques et pinceaux de courbes de genre 1, Intersections of Two Quadrics and Pencils of Curves of Genus 1 (2007)
Vol. 1902: A. Isaev, Lectures on the Automorphism Groups of Kobayashi-Hyperbolic Manifolds (2007)
Vol. 1903: G. Kresin, V. Maz'ya, Sharp Real-Part Theorems (2007)
Vol. 1904: P. Giesl, Construction of Global Lyapunov Functions Using Radial Basis Functions (2007)
Vol. 1905: C. Prévôt, M. Röckner, A Concise Course on Stochastic Partial Differential Equations (2007)
Vol. 1906: T. Schuster, The Method of Approximate Inverse: Theory and Applications (2007)
Vol. 1907: M. Rasmussen, Attractivity and Bifurcation for Nonautonomous Dynamical Systems (2007)
Vol. 1908: T.J. Lyons, M. Caruana, T. Lévy, Differential Equations Driven by Rough Paths, Ecole d'Été de Probabilités de Saint-Flour XXXIV-2004 (2007)
Vol. 1909: H. Akiyoshi, M. Sakuma, M. Wada, Y. Yamashita, Punctured Torus Groups and 2-Bridge Knot Groups (I) (2007)
Vol. 1910: V.D. Milman, G. Schechtman (Eds.), Geometric Aspects of Functional Analysis. Israel Seminar 2004-2005 (2007)
Vol. 1911: A. Bressan, D. Serre, M. Williams, K. Zumbrun, Hyperbolic Systems of Balance Laws. Lectures given at the C.I.M.E. Summer School held in Cetraro, Italy, July 14–21, 2003. Editor: P. Marcati (2007)
Vol. 1912: V. Berinde, Iterative Approximation of Fixed Points (2007)

Vol. 1913: J.E. Marsden, G. Misiołek, J.-P. Ortega, M. Perlmutter, T.S. Ratiu, Hamiltonian Reduction by Stages (2007)
Vol. 1914: G. Kutyniok, Affine Density in Wavelet Analysis (2007)
Vol. 1915: T. Bıyıkoğlu, J. Leydold, P.F. Stadler, Laplacian Eigenvectors of Graphs. Perron-Frobenius and Faber-Krahn Type Theorems (2007)
Vol. 1916: C. Villani, F. Rezakhanlou, Entropy Methods for the Boltzmann Equation. Editors: F. Golse, S. Olla (2008)
Vol. 1917: I. Veselić, Existence and Regularity Properties of the Integrated Density of States of Random Schrödinger (2008)
Vol. 1918: B. Roberts, R. Schmidt, Local Newforms for $GSp(4)$ (2007)
Vol. 1919: R.A. Carmona, I. Ekeland, A. Kohatsu-Higa, J.-M. Lasry, P.-L. Lions, H. Pham, E. Taflin, Paris-Princeton Lectures on Mathematical Finance 2004. Editors: R.A. Carmona, E. Çinlar, I. Ekeland, E. Jouini, J.A. Scheinkman, N. Touzi (2007)
Vol. 1920: S.N. Evans, Probability and Real Trees. Ecole d'Été de Probabilités de Saint-Flour XXXV-2005 (2008)
Vol. 1921: J.P. Tian, Evolution Algebras and their Applications (2008)
Vol. 1922: A. Friedman (Ed.), Tutorials in Mathematical BioSciences IV. Evolution and Ecology (2008)
Vol. 1923: J.P.N. Bishwal, Parameter Estimation in Stochastic Differential Equations (2008)
Vol. 1924: M. Wilson, Littlewood-Paley Theory and Exponential-Square Integrability (2008)
Vol. 1925: M. du Sautoy, L. Woodward, Zeta Functions of Groups and Rings (2008)
Vol. 1926: L. Barreira, V. Claudia, Stability of Nonautonomous Differential Equations (2008)
Vol. 1927: L. Ambrosio, L. Caffarelli, M.G. Crandall, L.C. Evans, N. Fusco, Calculus of Variations and Non-Linear Partial Differential Equations. Lectures given at the C.I.M.E. Summer School held in Cetraro, Italy, June 27–July 2, 2005. Editors: B. Dacorogna, P. Marcellini (2008)
Vol. 1928: J. Jonsson, Simplicial Complexes of Graphs (2008)
Vol. 1929: Y. Mishura, Stochastic Calculus for Fractional Brownian Motion and Related Processes (2008)

Recent Reprints and New Editions

Vol. 1618: G. Pisier, Similarity Problems and Completely Bounded Maps. 1995 – 2nd exp. edition (2001)
Vol. 1629: J.D. Moore, Lectures on Seiberg-Witten Invariants. 1997 – 2nd edition (2001)
Vol. 1638: P. Vanhaecke, Integrable Systems in the realm of Algebraic Geometry. 1996 – 2nd edition (2001)
Vol. 1702: J. Ma, J. Yong, Forward-Backward Stochastic Differential Equations and their Applications. 1999 – Corr. 3rd printing (2007)
Vol. 830: J.A. Green, Polynomial Representations of GL_n, with an Appendix on Schensted Correspondence and Littelmann Paths by K. Erdmann, J.A. Green and M. Schocker 1980 – 2nd corr. and augmented edition (2007)